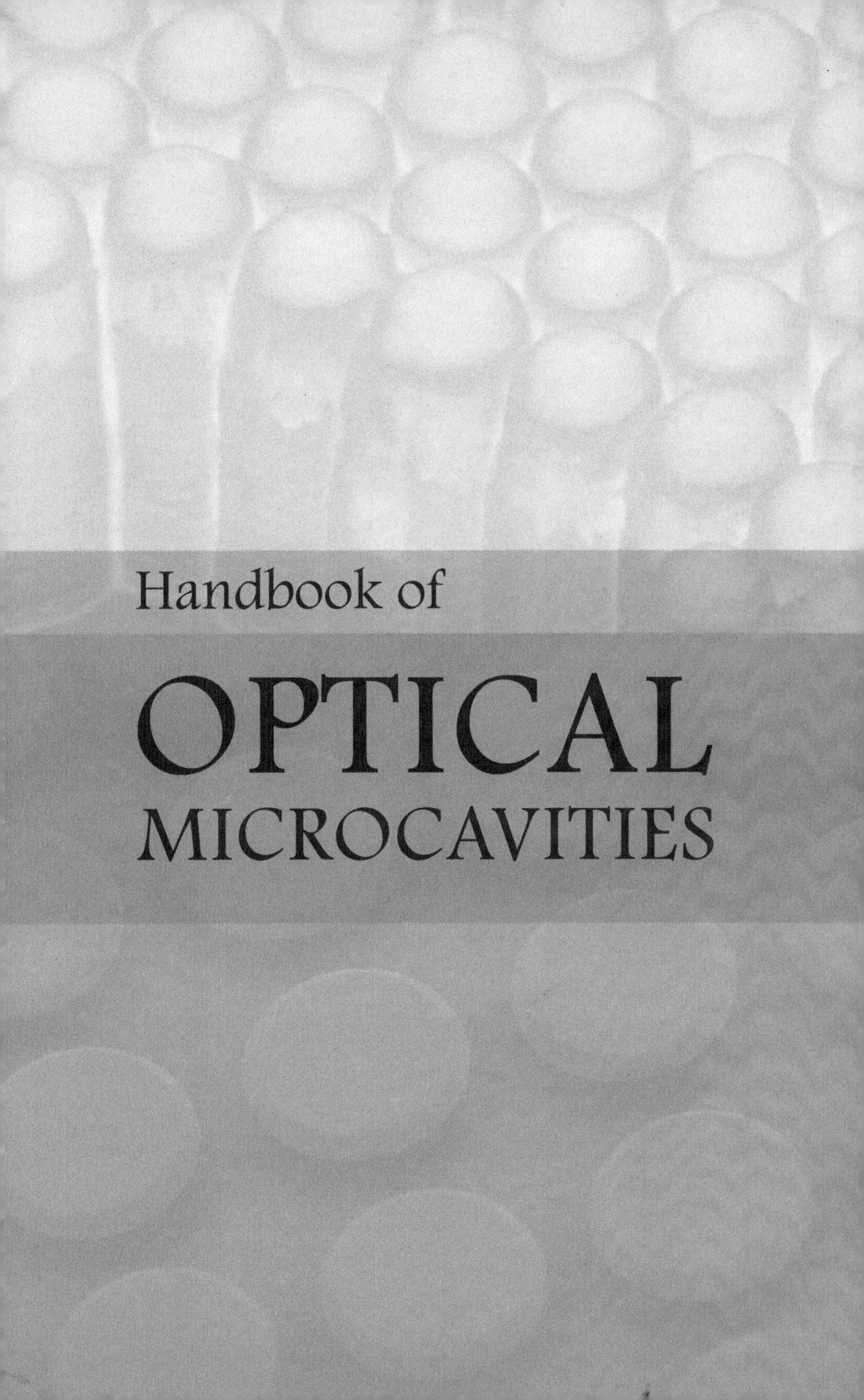

Handbook of OPTICAL MICROCAVITIES

Handbook of OPTICAL MICROCAVITIES

edited by
Anthony H. W. Choi

PAN STANFORD PUBLISHING

Published by

Pan Stanford Publishing Pte. Ltd.
Penthouse Level, Suntec Tower 3
8 Temasek Boulevard
Singapore 038988

Email: editorial@panstanford.com
Web: www.panstanford.com

British Library Cataloguing-in-Publication Data
A catalogue record for this book is available from the British Library.

Handbook of Optical Microcavities

ISBN 978-981-4463-24-9 (Hardcover)
ISBN 978-981-4463-25-6 (eBook)

Printed in the USA

Contents

Preface

An optical microcavity constitutes an integral part of a laser diode, confining light within an optical structure to establish optical resonances. At present, edge-emitting and vertical-cavity surface-emitting lasers are the most common types of laser diodes in the market, both of which are based on Fabry–Pérot cavities. However, novel forms of laser structures, based on alternative modes of optical confinement such as whispering-gallery modes and photonic bandgaps have been demonstrated in recent years, allowing light to be confined within smaller dimensions with high finesse.

Apart from different mechanisms of optical confinement, laser diodes based on different material systems give rise to devices operating at different wavelengths for different applications. For instance, the short-wavelength violet laser diodes based on the III-nitrides have been widely used for high-density optical storage in blu-ray players, while long-wavelength As-based infrared (IR) laser diodes are deployed in optical communication networks.

The chapters in this book cover a comprehensive range of topics pertaining to resonance in optical cavities by leading researchers in the field. Subjects include theory, design, simulation, fabrication, and characterization of micrometer and nanometer scale structures and devices that support cavity resonance via various mechanisms such as Fabry–Pérot, whispering gallery, photonic bandgap, and plasmonic modes. Optical cavities resonating from UV to IR wavelengths based on prominent III–V material systems, including nitrides of Al, In, and Ga; ZnO; and GaAs, are covered in this book.

The book begins with a comprehensive review of photonic crystal microcavities and microlasers, followed by a discussion on the simulation of planar photonic resonators. Chapters 3 to 10 pertain to optical cavities based on the III-nitride material system, including that of UV lasers; nitride microdisk cavities grown on Si substrate; MOCVD-grown nitride distributed Bragg reflectors; photonic crystal light-emitting diodes (LEDs); nanostructures in GaN-Si materials; light-coupling to surface plasmons; nanosphere-patterned microcavities; and epitaxial lift-off of GaN thin films. Chapters 11 and 12 deal with optical microvacities based on GaAs and ZnO, respectively, while Chapter 13 analyzes nanoscale ring lasers.

It was a privilege to take responsibility for putting together this book, with the honor to invite distinguished researchers in field to contribute their expertise and insights in this intriguing technology. I would like to express my immense gratitude to all the contributors of the book. Without their kind and strong support, the project would not have been accomplished.

Anthony H. W. Choi
Summer 2014

Chapter 1

Photonic Crystal Microcavities and Microlasers

Tomoyuki Yoshie

Nano Precision Medical, Inc., Emeryville, Suite 393, 5858 Horton Street, California, USA

yoshie@alumni.caltech.edu

This chapter presents different forms of photonic crystal microcavities, photonic crystal microlasers, dynamic properties of photonic crystal microlasers, current-injection photonic crystal microlasers, and photonic crystal microcavities for sensing and cavity quantum electrodynamics (QED). The chapter also focuses on the progress in photonic crystal microcavities and lasers in which light confinement is of wavelength scale.

1.1 Introduction

Photonic crystals are optical analogs to atomic crystals in which electronic waves are represented in terms of Bloch waves. Typical photonic crystal systems have translation properties in the permittivity, that is, $\varepsilon(\mathbf{r}) = \varepsilon(\mathbf{r} + \mathbf{a}_n)$, where $\mathbf{r}$ is a position vector and $\mathbf{a}_n$ is any primitive translation vector. Quasi photonic crystals, which are geometries with rotational symmetry, are not discussed in this

Handbook of Optical Microcavities
Edited by Anthony H. W. Choi

ISBN 978-981-4463-24-9 (Hardcover), 978-981-4463-25-6 (eBook)
www.panstanford.com

chapter. Photonic bands are formed either due to multiple periodic scattering or simply due to interference. Because of this nature, sub-wavelength accuracy in the geometry is often desired. In 1987, Yablonovitch [1] and John [2] proposed photonic crystal geometries to study the suppression of spontaneous emission and the photonic Anderson localization in a randomly fluctuating periodic geometry.

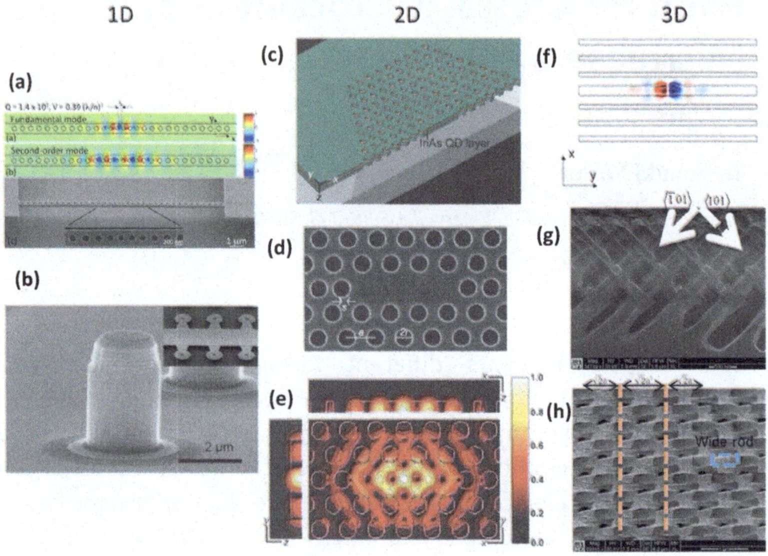

Figure 1.1 Photonic crystal microcavities: 1D beam/pillar, 2D slab, and 3D photonic crystal microcavities. (a) 1D photonic crystal nanobeam microcavity. Reprinted with permission from Ref. [3], © American Institute of Physics. (b) 1D photonic crystal pillar microcavity. Reprinted with permission from Ref. [4], © American Institute of Physics. (c–e) 2D photonic crystal slab microcavity. Adapted from Ref. [5]. (f–h) 3D photonic crystal microcavity. Adapted from Ref. [6].

Microcavity modes are formed when a perturbation, $\Delta\varepsilon(\mathbf{r})$, in the geometry induces certain allowed states in a photonic bandgap via disturbed interference. Localized lightwaves in photonic crystals, or photonic crystal microcavities, can also be expressed as a superposition of Bloch or Wannier waves. While whispering gallery modes (WGMs) are traveling waves, photonic crystal microcavity modes are standing waves that require forward and backward propagation waves, for example. All-direction optical confinement can be realized in a 3D photonic crystal, that is, light is trapped due to omni-directional optical confinement by a complete photonic bandgap. In order to create 3D optical confinement in a simple manner, total internal reflection is often combined with photonic bandgap effect. Popular 1D and 2D photonic crystals have 1D and 2D translation symmetries in a high-index slab, and some perturbations are introduced in the lattices to form microcavity states. Figure 1.1 shows examples of 1D pillar, 1D beam, 2D slab, and 3D photonic crystal microcavities, which will be discussed in the following section [3–6].

1.2 Photonic Crystal Microcavity Modes

Microcavity modes in photonic crystals are induced by a perturbation in the lattice. In this section, examples of photonic crystal microcavity modes are shown, starting from simple yet useful microcavities, that is, 1D and 2D slab photonic crystal microcavities, and then 3D photonic crystal microcavities and surface Bloch mode microcavities are discussed.

1.2.1 1D Photonic Crystal Microcavity

An extensive amount of study in 1D photonic crystal microcavity has been done through the study of 1D photonic crystal micropillars—small-diameter, vertical-cavity surface-emitting lasers (VCSELs), which are typically a half-wavelength ($\lambda/2n$) or one-wavelength (λ/n) space sandwiched between two distributed Bragg mirrors, that is, multiple stacks of a pair of two dissimilar $\lambda/4n$ long transparent media; see Figs. 1.1(b) and 1.2(a). Light is confined around the gain media spacing of $\lambda/2n$ or λ/n length for lasing. Due to small gain volume, high reflectivity is required. Multiple stacks of two alternating transparent materials of $\lambda/4n$ length correspond to a 1D photonic crystal, and the spacer becomes a defect. During the periodic index modulation, light gets trapped because the frequency of the defect state moves into a 1D photonic bandgap, or the so-called stop band. One-dimensional photonic crystal pillars are able to support 3D optical confinement due to 2D total internal reflection at the sidewalls of the pillar as well as due to the 1D photonic bandgap effect. VCSELs were invented by Iga *et al.* in the late 1970s [7]. One-dimensional photonic crystal micropillar lasers were developed between the late 1980s and early 1990s [8]. For example, a micropillar laser of 3 μm diameter showed ~0.4 nm linewidth above threshold, corresponding to loaded Q ~ 2500 [9]. Careful modeling and fabrication led to the demonstration of strong coupling between a single quantum dot and a single-mode photonic field based on a high-Q micropillar microcavity [10] and to intriguing quantum optics experiments such as indistinguishable photons from a single-photon source embedded in a micropillar microcavity [11]. It has been found that although the mode volume is larger than 3D photonic crystal microcavities, the quality factor can exceed 10^5 for a diameter less than 3 μm [12] as seen in Fig. 1.2(a) [12,13].

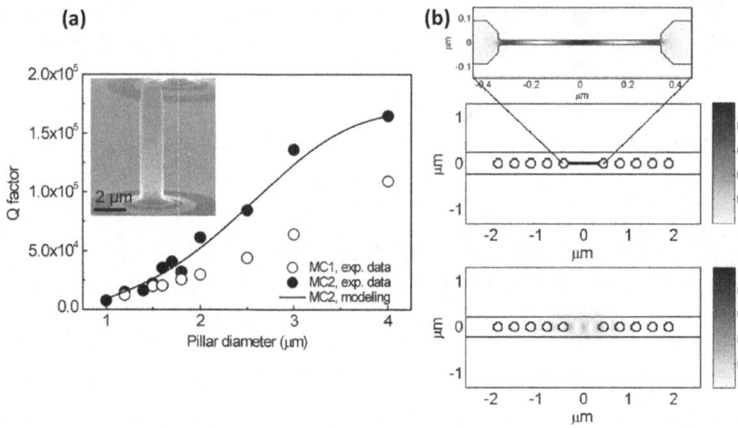

Figure 1.2 1D photonic crystal microcavities. (a) 1D photonic crystal pillar microcavity, and its Q factor dependence on pillar diameter. Reprinted with permission from Ref. [12], © American Institute of Physics. (b) 1D photonic crystal slot microcavity, and 1D photonic crystal defect microcavity. Reprinted with permission from Ref. [13], © American Physics Society.

Another popular form of 1D photonic crystal is a 1D nanobeam photonic crystal, whose lattice is defined in a planar single-mode wire waveguide. Fan *et al.* [14] found that a defect in a 1D array of holes in a narrow beam supported high-Q microcavity modes. Robinson *et al.* [13] showed an interesting way to squeeze the mode volume by using a boundary of dissimilar permittivities at which the component of displacement vector along the permittivity discontinuity is continuous, that is, $\boldsymbol{D}_{1|v} = \boldsymbol{D}_{2|v}$, or $\boldsymbol{E}_{1|v} = (\varepsilon_2/\varepsilon_1)\,\boldsymbol{E}_{2|v}$, so that the electric field in a slot is enhanced, resulting in reduction of mode volume; see Fig. 1.2(b). This is also a new approach in building high-sensitivity microcavity optical sensors.

Enhancement of quality factors was proposed by Notomi *et al.* who used small digital shifts of lattice of 1D ladder geometry and obtained $Q = 200 \times 10^6$ and $V = 1.4(\lambda/n)^3$ [15]. Deotare *et al.* [3] demonstrated optically pumped lasing from high-Q 1D nanobeam microcavities as shown in Fig. 1.1(a), in which simulated $Q = 1.4 \times 10^7$ and measured $Q = 7.5 \times 10^5$ for mode volume $V = 0.39(\lambda/n)^3$ [3]. At present, the use of this type of 1D photonic crystal microcavity is most popular as it can be easily fabricated and connected with planar single-mode wire waveguides. Simple and excellent mechanical properties of 1D photonic crystal nanobeam have attracted a lot of researchers in the field of cavity optomechanics, and 1D photonic crystal beam geometries have provided a popular platform for cavity optomechanics; see Ref. [16] as an example.

1.2.2 2D Photonic Crystal Slab Microcavity

Two-dimensional photonic crystal slab microcavities are the most extensively studied photonic crystal microcavities because their fabrication is relatively simple and they provide excellent confinement of light in a small volume. Typically, three-layer wafer is used to build 2D photonic crystal slab microcavities; a thin slab layer for single-mode in-slab optical guiding, a sacrificial layer for securing index contrast, and a substrate for chip handling. A silicon-on-insulator (SOI) wafer is a perfect choice for silicon-based photonic crystal slab, in which integration of photonic crystal devices is realized by high-resolution lithography, anisotropic silicon etching, and sacrificial oxide-selective etching. Painter *et al.* demonstrated the first 2D photonic crystal slab microlasers on InGaAsP multi-quantum well (MQW) slab wafer [17], and this example is regarded as an important milestone in 2D photon crystal slab systems [5]. This work provided great opportunities in advancing nanophotonic technology and applications. Yoshie *et al.* have shown strong coupling between a single quantum dot and a single-mode localized field on the 2D photonic crystal slab system [5]. This work provided great opportunities in handling quantum states in a solid-state form and advancing quantum information sciences.

Geometric symmetry determines the microcavity mode structure. As an example, for triangular lattice, monopole, dipole, quadrupole, and hexapole modes were identified (see Refs. [18,19]). As understood from the 2D tiling problem, the following three types of lattices are allowed: (i) triangular lattice, (ii) square lattice, and (iii) hexagonal lattice (or honeycomb lattice). As a triangular lattice has the most round first Brillouin zone surface, this lattice has the largest photonic bandgap among the three. In order to confine light tightly in a small space, triangular lattice is often the first choice. A slab waveguide with mirror symmetry about the middle plane (i.e., a symmetric slab waveguide) supports transverse electric (TE) and transverse magnetic (TM) modes. Perforated symmetric photonic crystal slab supports TE-like and TM-like modes, which are expressed as photonic Bloch bands in photonic crystal dispersion (i.e, ω–k space).

Slab waveguides confine light vertically, and sufficiently deep 2D photonic crystal lattices can be defined for lateral confinement. However, the structure "slab on substrate" requires deep etching holes and large index contrast. Otherwise, light is diffracted down toward the substrate [20]. Due to lack of mirror symmetry about the middle plane, the quality factor tends to be reduced. Simple-defect and modified-defect 2D photonic crystal slab microcavities showed promising performance (high-Q and small-mode volume) in 3D finite-difference time-domain (FDTD) modeling [21], leading to demonstration of lasing action in InGaAsP MQW 2D photonic crystal defect microcavities [17].

Momentum space analysis [22,23] has enabled one to significantly improve the quality factor of 2D photonic crystal slab microcavities by suppressing spatial Fourier component of field inside escape light cone (indicated by a dashed circle in Fig. 1.3a), resulting in lesser loss of radiation power. Figure 1.3(a) shows spatial Fourier distribution of filed component for a single-defect photonic crystal microcavity [23]. The peaks are outside the light cone, therefore, the confinement of light is more or less good. However, small amplitude inside the escape light cone contributes to radiation loss as total internal reflection does not take place. Simple-defect microcavities can be modified to enhance the Q factor. Vuckovic *et al.* successfully improved the Q factor to 10,000 by eliminating radiation loss within the escape light cone [22]. In the same year, Srinivasan *et al.* showed an important method to design high-Q microcavity by gradually confining microcavity modes along a certain direction with the help of a shallow light trap induced by gradual lattice change [23]. In 2003, Akahane *et al.* [25] used the still-popular high-Q L3 photonic crystal microcavities to demonstrate experimental $Q = 45{,}000$; see Fig. 1.3(b) [24,28]. In 2005, Song *et al.* used double-hetero 2D photonic crystal slab structures to obtain experimental $Q = 600{,}000$ and simulated $Q = 2 \times 10^6$ [25]. In 2006, two groups used mode gap methods to produce experimental $Q \sim 10^6$ [26,27]; see Figs. 1.3(c,d). Mode gap

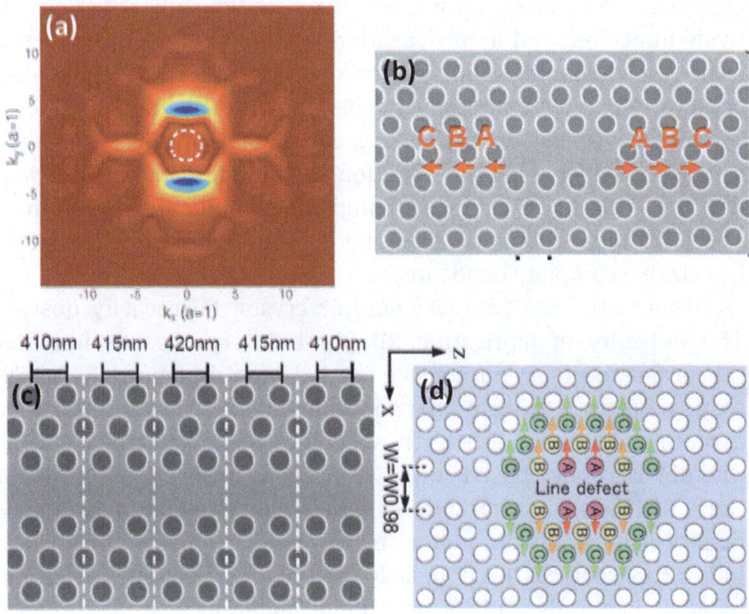

Figure 1.3 2D photonic crystal slab microcavities. (a) Fourier analysis of x dipole mode in a single defect of triangular lattice on a slab. Reprinted with permission from Ref. [24], © Optical Society of America. (b) L3 2D photonic crystal slab microcavity. Reprinted with permission from Ref. [29], © Optical Society of America. (c) Ultrahigh-Q heterojunction microcavities induced by small digital changes of lattice constant (or by mode gap). Reprinted with permission from Ref. [28], © Optical Society of America. (d) Ultrahigh-Q microcavity induced by mode gap. Reprinted with permission from Ref. [27], © American Institute of Physics.

method is used to introduce small local lattice modification (such as lattice constant and lattice shift) and induce a gradual trap to form ultrahigh-Q microcavity modes whose mode volumes are typically $1 \sim 2(\lambda/n)^3$. Although simulated Q factors can exceed 10^7, experimental Q factors are limited to $\sim 1 \times 10^6$, mainly due to absorption.

1.2.3 3D Photonic Crystal Microcavity

Three-dimensional photonic crystals appear to be the most-complicated structures by nature among 1D, 2D, and 3D photonic crystals. A complete photonic bandgap—in which light is destructively interfered and its propagation is inhibited—requires certain lattice type and index contrast, and so occurs in certain 3D photonic crystals. A complete photonic bandgap is fundamentally unique, as 1D and 2D systems do not possess complete bandgaps—frequency ranges in which light propagation is inhibited in all directions due to 3D Bragg reflection. A complete photonic bandgap can be open in simple cubic and diamond lattices, and a diamond lattice has sizable complete bandgaps.

There are very few 3D photonic crystal microcavity designs. The difficulty of fabricating 3D photonic crystals has hindered the progress of 3D photonic crystal device designing. Relatively easy ways to fabricate 3D photonic crystals include 3D self-assembly of spherical dielectrics, such as opals and silica, and 3D photolithography based on two-photon absorption. However, these methods fail to produce complete bandgap materials in high-quality semiconductor materials. Inversed geometries from low-index 3D structures are popularly used to create complete photonic bandgap materials made of amorphous materials.

High-Q microcavities are obtained from layer-by-layer approach, that is, the method to place one layer by one layer. Figure 1.4 shows experimental results. Kyoto group [29] used wafer bonding with alignment, and Tokyo group [30] inserted and stacked one layer by one layer in a space inside four alignment pillars. Woodpile photonic crystals are fabricated by layer-by-layer technique in these experiments. An advantage of using this technique is the ability of employing high-quality semiconductors, whereas other deposition-based approaches are usually difficult to include high-quality semiconductors. In 2004, Ogawa *et al.* [29]

used layer-by-layer method to produce woodpile photonic crystals consisting of 17 layers and demonstrate $Q \sim 3000$ at 1.5 μm (see Fig. 1.4). In the same year, Qi *et al.* showed a defect mode in silicon-based 3D photonic crystal [31]. In 2011, Tandaechanurat *et al.* demonstrated experimental $Q \sim 38{,}500$ and lasing from their woodpile photonic crystals with quantum dots by optical pumping at 7 K [30].

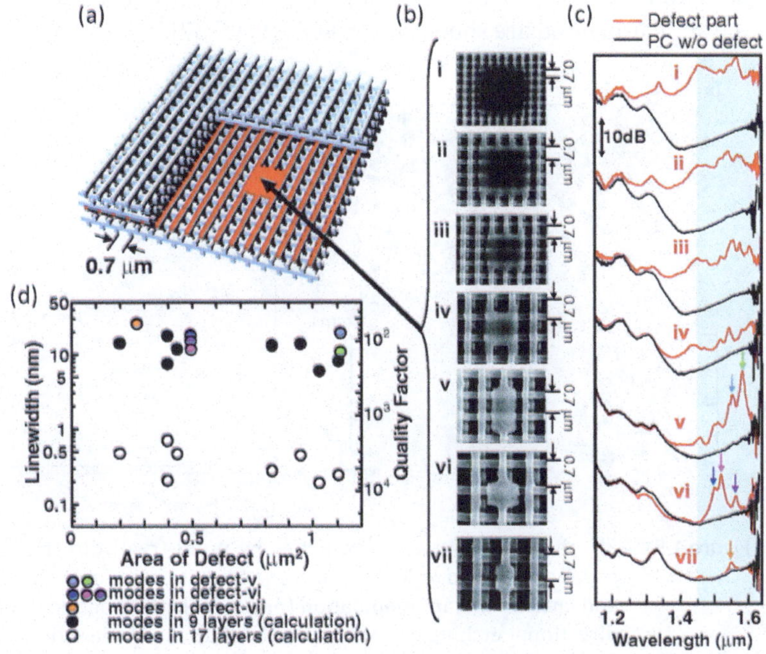

Figure 1.4 3D photonic crystal microcavities. (a) Defect in a woodpile. (b) SEM images of defects. (c) PL spectra dependence on defect size. (d) Linewidth versus area of defect. (a–d) Reprinted with permission from Ref. [30], © American Association for the Advancement of Science.

Multidirectional etching method is a simple way to fabricate 3D photonic crystals. The technique was used to create Yablonovite 3D photonic crystals [32,33]. However, it is still a challenge to build microcavities inside 3D photonic crystals using this fabrication method. Tang *et al.* showed how to resolve the challenge by pulling a waveguide mode in a mode gap induced by lattice constant modulation and proving the possibility of microcavities in a complete bandgap from 2D etching. In Fig. 1.5(a), a wide rod (blue) is found from ⟨101⟩′ direction, and this waveguide supports waveguide modes as shown in Fig. 1.5(c). A change of lattice constant in the middle portion (orange) along ⟨–101⟩′ direction induces the mode pulling effect into mode gaps, resulting in microcavity formation seen in Fig. 1.1(f). The structure can be constructed by anisotropic etching from two directions, that is, etching from ⟨101⟩′ and ⟨–101⟩′ directions for the case of Fig. 1.5(a). 3D photonic crystals based on the similar method are shown in the Refs. [6,33–37].

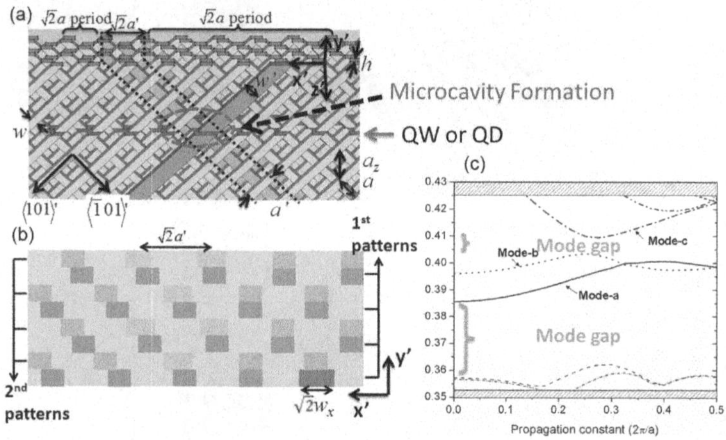

Figure 1.5 3D photonic crystal microcavities by multi-(two) directional etching. (a) Woodpile photonic crystal with wide rod (blue) and lattice constant modulation (orange). (b) Two patterns for directional etching. (c) Mode profile of dipole mode. Adapted from Ref. [39].

1.2.4 Surface Bloch Mode Microcavity

Optical surface Bloch modes are optical analog to electronic surface Bloch modes appearing within an electronic bandgap due to a termination of periodic atomic potential. Tamm [38] and Shockley [39] are two definite pioneers in electronic surface states. The investigation of optical surface Bloch modes was pioneered by Yariv *et al.* [40] who studied 1D surface modes. Meade *et al.* analyzed optical surface modes in a complete photonic bandgap material made of a 3D Yablonovite crystal [41]. Ishizaki and

Noda experimentally studied optical surface modes on a woodpile 3D photonic crystal by measuring the wave dispersion relation via a prism coupler [42]. Yang *et al.* demonstrated optically pumped lasing action from finite-length surface in 2D triangular lattice [43]. On the other hand, surface Bloch modes are expected to show high sensitivity S in microcavity optical sensing, as they have large field overlap with low-index media (e.g., air and water) [44]. Lu *et al.* constructed surface mode microcavity lasers on an edge of 2D photonic crystal slab for optical sensing by trapping light of edge surface propagation mode via a mode gap induced by lattice modulation (see Fig. 1.6) [45].

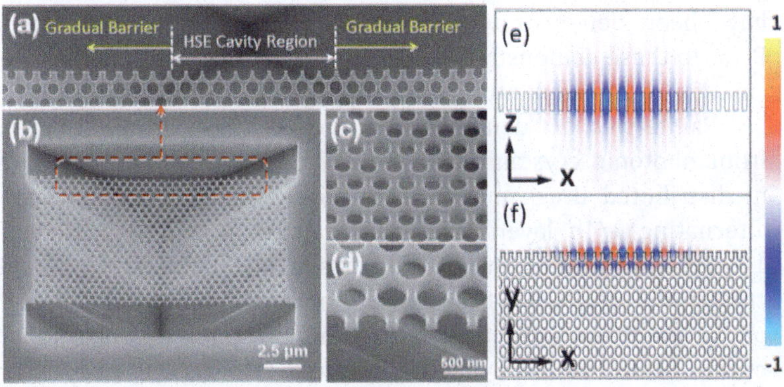

Figure 1.6 Surface Bloch mode microcavities. (a–d) Surface mode microcavity defined in a 2D triangular lattice photonic crystal slab. (e, f) mode profiles from side and from top. Reprinted with permission from Ref. [47], © American Institute of Physics.

Other important type of surface mode microcavities is based on surface plasmon that requires opposite signs of real part of permittivity and is realized by the use of dielectric–metal interface. However, surface plasmon modes are generally lossy due to large optical absorption. Although surface plasmon microcavities have small mode volume, they support low Q modes only. Therefore, it would be intriguing to see the development of high-Q dielectric-based surface mode microcavities—which seem to be possible in photonic crystal surface Bloch mode microcavities.

1.3 Optically Pumped Photonic Crystal Microlasers

Miniaturization of lasers is an important research focal point in nanophotonic, as it provides abundant amount of opportunities in handling light in a small volume. Lasers can be used for a variety of purposes, such as optical source, optical sensors, and optomechanical actuators. Optically pumped photonic crystals have been demonstrated in 1D beam, 1D pillar, 2D slab, and 3D photonic crystal systems. Although reliable electrically pumped photonic crystal microlaser is still a challenge, electrically pumped 2D slab photonic crystal microcavities in telecommunication wavelength have been demonstrated by KAIST, Stanford, and NTT groups. Prior to these demonstrations, Colombelli *et al.* have demonstrated current-injection quantum cascade photonic crystal microlasers.

Painter *et al.* demonstrated the first defect mode lasing from photonic crystals in 1999 [17,46]. Although a λ/n spacer in distributed Bragg mirror stack consisting of index-contrasted alternating $\lambda/4n$ layers may be regarded as a defect mode in 1D photonic crystal, the demonstration of the 2D slab photonic crystal defect lasers has opened an opportunity of realizing planar integration of ultra-compact-mode-volume microlasers and micron-scale optoelectronic circuits and of designing microlasers from engineering defect modes. The first photonic crystal lasers are realized from the mode volume of $0.4(\lambda/2n)^3$ and Q factor of 250.

In 2002, quantum dot photonic crystal defect microlasers [47] were demonstrated from coupled modes of 2D photonic crystal defect cavities with InAs quantum dots and showed smaller threshold than the first photonic crystal microlasers with quantum well active media; input pump power at threshold was 120 μW, which would be encouraging for developing thresholdless lasing; see Figs. 1.7(a,b). Due to rapid improvement of designs in 2D photonic crystal slab microcavities, laser threshold has been reduced significantly. One of notable low threshold lasing from photonic crystal microcavities was demonstrated by Strauf *et al.* who used the L3 cavity design with InAs quantum dots; see Figs. 1.7(c,d) [48]. The threshold pump power was about 120 nW, due to large Q factor and tuning. The absorbed threshold was estimated to be as low as 4 nW.

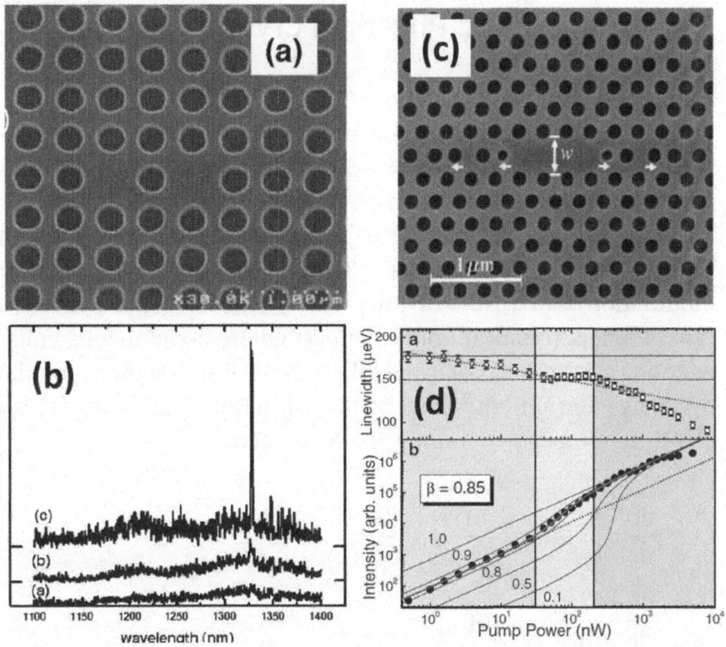

Figure 1.7 Photonic crystal quantum dot microlasers. (a, b) First photonic crystal microlasers using quantum dot gain. Adapted from Yoshie *et al.* [49] (author use). (c, d) Ultralow-threshold photonic crystal defect microlasers. Reproduced with permission from Ref. [50], © American Physics Society. (a, c) Scanning electron micrographs. (b) Spectrum dependence on pump power of the device c. (d) Linewidth and output power dependence on pump power.

Technology of 1D photonic crystal microcavities, which is one of the important planar nanophotonic systems, has also been advanced. Although micropillars and wire-like beams possess similar 1D periodic structure, 1D photonic crystal beams can be well-packaged for integrated nanophotonics.

Three-dimensional photonic crystal microcavities also became lasers in 2011 [30]. The quality factor was in a range of 10^4, which was sufficient to achieve lasing action. Although, in its lasing characteristics, 3D photonic crystal microlasers do not show evident differences from 1D and 2D photonic crystal microlasers, it would be curious when lasing happens in perfectly omnidirectional optical confinement.

1.4 High-Speed Photonic Crystal Microlasers

A unique feature of laser diodes is direct-current modulation, and we can modulate laser light by modulated current. Then NTT group led by Yamamoto showed [49–51] that microlasers could directly modulate light output at extremely high frequency. This is not intuitive as the speed of laser is proportional to the photon number that is small for microlasers. The high-frequency direct-current modulation is a result of microcavity effects—high β mode and Purcell effect (quantum optics effect) where spontaneous emission coupling factor is β. We can explain the high-frequency modulation by using the rate equations of photon number per volume, p, and electron number per volume, n, are written as

$$\frac{\mathrm{d}p}{\mathrm{d}t} = \left(\Gamma G(n,p) - \frac{1}{\tau_p}\right)p + \frac{\beta}{\tau_{\mathrm{sp}}}n$$
$$\frac{\mathrm{d}n}{\mathrm{d}t} = \frac{I}{eV} - \frac{n}{\tau_{\mathrm{sp}}} - G(n,p)p. \tag{1.1}$$

where Γ, $G(n, p)$, τ_{p}, τ_{sp}, and I are overlap factor, optical gain rate per photon, photon lifetime, and spontaneous emission rate in cavity. In textbook, the term $\beta n/\tau_{\mathrm{sp}}$ is often neglected, but it is important to include it in microcavity. For simplicity, we use linear gain model, that is, $G(n, p) = A(n - n_{\mathrm{th}})$. The Einstein's relation requests that the spontaneous emission rate coupled to lasing mode $(An\Gamma)$ equals the stimulated emission rate $(\beta n/\tau_{\mathrm{sp}})$, yielding

$$A = \frac{\beta}{\Gamma\tau_{\mathrm{sp}}} = \frac{F_{\mathrm{p}}\beta}{\Gamma\tau_{\mathrm{sp,0}}} \tag{1.2}$$

where F_{p} and $\tau_{\mathrm{sp,0}}$ are Purcell factor and spontaneous emission rate in free space. The differential gain coefficient, A, is enhanced by a factor of $F_{\mathrm{p}}\beta$, which can be significant in microlasers. We can show [52] that the relaxation frequency is approximated to

$$f_{\mathrm{RO}} \cong \frac{1}{2\pi}\sqrt{\frac{Ap_0}{\tau_{\mathrm{p}}}} = \frac{1}{2\pi}\sqrt{\frac{(\beta F_{\mathrm{p}})p_0}{\Gamma\tau_{\mathrm{sp,0}}\tau_{\mathrm{p}}}} \tag{1.3}$$

where p_0 is the photon density at laser operation point. The relaxation oscillation frequency is obtained by finding a pole of spectrum function $S(\omega)=|p_1(\omega)/(i_1(\omega)/eV)|^2$ where small modulated functions of photon density, carrier density, and pump rate are $p = p_0 + p_1\exp(i\omega t)$, $n = n_0 + n_1\exp(i\omega t)$ and $I = i_0 + i_1\exp(i\omega t)$.

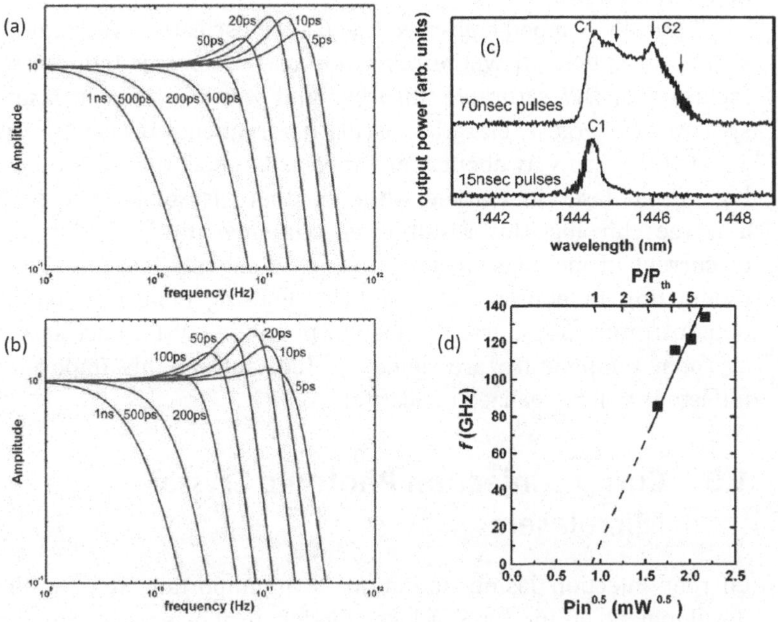

Figure 1.8 High speed photonic crystal microlasers. Adapted from Refs. [54,55]. (a) Frequency response of amplitude modulation as a function of modified spontaneous emission lifetime for Q = 500, $P/P_{\mathrm{th}} = 5$, $\lambda = 1.5$ μm. (b) Frequency response as a function of modified spontaneous emission lifetime for Q = 1000. (c) PL spectra for pulse pumping tightly focused on the center of cavity for 15 ns and 70 ns pulses. (d) Relaxation oscillation frequency as a function of a square-root of input power (or $P_{\mathrm{in}}/P_{\mathrm{th}}$).

Conducting precise small signal modulation analysis based on the equation (1.1) and linear gain model, we can plot frequency responses, as seen in Fig. 1.8 (a) and (b). Large spontaneous emission coupling (β) factor and large Purcell factor (F_{p}) are important to raise the relaxation oscillation frequency in microcavity lasers. Large

Purcell factor is realized by increasing the ratio Q/V so we would like to achieve large Q factor from a mode with a small mode volume. However, as you see in Fig. 1.7(a) and (b), photon decay rate needs be sufficiently large so that it would not become a bottleneck. It is possible to achieve 200 GHz-level modulation by appropriate choice of Q and V values.

Optically pumped high-speed relaxation oscillation was demonstrated in 2004 [53] where relaxation oscillation was induced by inclusion of QW saturable absorber and was observed in lasing spectra. The largest relaxation oscillation frequency measured was 135 GHz, which was about four times as large as that of the best high-speed laser in market. Although there is some price such as large chirping, this result is encouraging. Direct modulation of current in microlasers has not been demonstrated, but direct modulation in photonic crystal light emitting diodes (LEDs) is demonstrated [54]. Current-injection photonic crystal microlasers at room temperature are demonstrated, and so this important milestone will be realized in near future.

1.5 Current-Injection Photonic Crystal Microlasers

Current injection lasing operation is an important step in the development of any type of laser diodes. It allows us to simplify the implementation of generating laser light, modulate laser light from modulation of current flowing through a laser diode, and provide a plausible way of manipulating coherent light sources.

Electrically pumped photonic crystal microlasers were first demonstrated in a sophisticated 2D photonic crystal microcavity with quantum cascade active media by Colombelli *et al.* in 2003 [55]. The work used a way to localize light based on vertical plasmonic confinement and lateral photonic crystal defect-based optical confinement from no-photonic-bandgap materials, in which the lateral confinement of light is made by pulling a mode in a dispersion point that is located in an unsupported propagation range in the surrounding photonic crystal.

In 2004, Park *et al.* [56] demonstrated the first telecommunication-wavelength electrically driven 2D photonic crystal slab microlasers emitting at 1520 nm enabled by InGaAsP MQW active materials and $0.69(\lambda/n)^3$ mode volume single defect mode by careful geometric design (with center post) achieving high Q factor and efficient current flow through the tiny mode volume and the post.

Using quantum dot active media potentially allows us to obtain low threshold lasers, and electrically driven photonic crystal microcavities provide novel quantum emitters, such as single photon sources. Ellis *et al.* demonstrated the first electrically pumped quantum dot photonic crystal microlasers in 2011 [57]. The devices are unique, in that current injection is defined laterally by Si and Be implantation followed by annealing. The threshold currents are 181 nA and 287 nA at 50 K and 150 K. The spontaneous emission coupling factors are measured to be between 0.29 and 0.41. Room temperature operation could not be realized on this device, possibly due to insufficient electronic confinement.

A challenge in efficiently providing current to an active volume for population inversion in photonic crystal microcavity modes is that metallic electrode placement near the modes deteriorates the quality factor. Because of this constraint, it is not a simple task to raise the quantum efficiency of light emission and maintain large population inversion, as leak current is typically large and heat sink is insufficient.

After the work of Park *et al.*, it took eight years to achieve room-temperature, continuous-wave operation of telecommunication-wavelength photonic crystal microlasers. Matsuo *et al.* demonstrated the first room-temperature, continuous-wave operation of photonic crystal microlasers in 2012 due to excellent optical and electronic confinement (see Fig. 1.9) [58]. An embedded box-shaped active volume allows electronic confinement in an optical mode volume and suppresses optical absorption outside the mode volume. The quality factor was large: $Q \sim 44{,}600$ at threshold. The threshold is 390 μA, and the effective threshold is 9.4 μA. Due to the suppressed leakage current, the temperature raise of the microlaser is suppressed to below 15 K. Room-temperature, continuous-wave lasing operation is an important milestone, as it offers opportunities to reliably create innovative microcavity light emitters as well as microlasers.

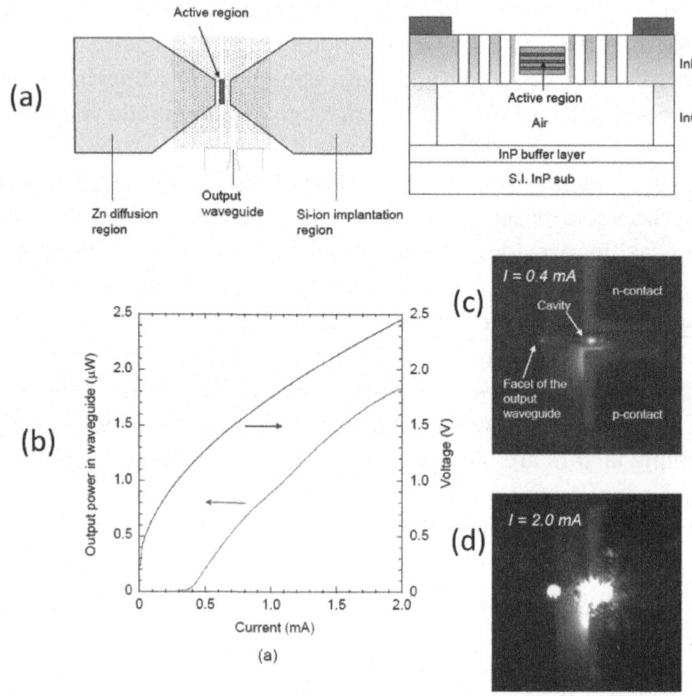

Figure 1.9 Room-temperature, continuous-wave, current-injection photonic crystal microlasers. (a) Profiles of doping, active region in the photonic crystal microlaser. (b) Output power and voltage as a function of driving current. (c) Image below threshold. (d) Image above threshold. Reprinted with permission from Ref. [60], © Optical Society of America.

1.6 Optical Sensing in Photonic Crystal Microcavity

Microcavity optical sensing is to detect changes in optical resonances due to a perturbation induced by analytes (see Fig. 1.10). This can be analyzed by using a perturbation theory. Unperturbed microcavity geometry, $\varepsilon(\mathbf{r})$, supports an optical mode $E_j(r)$ at resonance frequency ω_j and a perturbation to the microcavity geometry, $\Delta\varepsilon(r)$, induces the frequency shift by

$$\Delta\omega_j \cong -\frac{\omega_j}{2}\frac{\int \Delta\varepsilon(r)\left|E_j(r)\right| d^3r}{\int \varepsilon(r)\left|E_j(r)\right| d^3r}. \tag{1.4}$$

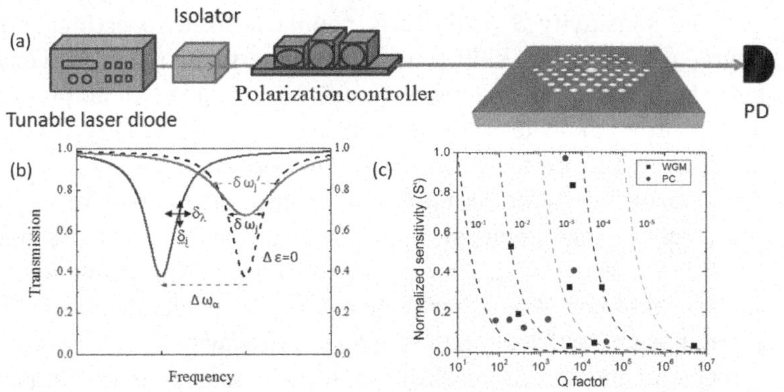

Figure 1.10 Optical microcavity sensing basics. Adapted from Ref. [61] (author use). (a) Basic optical microcavity sensing setup using tunable laser, isolator, polarization controller, tapered fiber coupled microcavity, and photodiode. (b) Modification of microcavity response due to a perturbation by analytes. (c) Minimum detectable index change $\Delta n_{\min} = (S'Q)^{-1}$ as a function of Q factor and normalized sensitivity S'.

A large resonance wavelength shift is possible by making large overlap of optical field with analytes. Reduction of quality factor is seen in the first-order perturbation when absorption is increased. As shown in Refs. [46,59], the quality factor is reduced from Q to

$$Q' \cong \frac{Q}{1+QR} \tag{1.5}$$

where

$$R = \frac{\int \mathrm{Im}[\Delta\varepsilon(r)]|E_j(r)|\mathrm{d}^3 r}{\int \varepsilon(r)|E_j(r)|\mathrm{d}^3 r}. \tag{1.6}$$

A minimum detectable index change $\Delta n_{\min}$ can be defined as index change which makes a resonance shift equal to resonance linewidth, yielding

$$\Delta n_{\min} = \frac{1}{S'Q} \tag{1.7}$$

where the normalized sensitivity is defined as

$$S' = \frac{S}{\lambda} = \frac{1}{\lambda}\frac{\mathrm{d}\lambda}{\mathrm{d}n}. \tag{1.8}$$

The sensitivity $S = d\lambda/dn$ is popularly used to estimate the degree of wavelength shift caused by the perturbation. However, in fact, the normalized sensitivity S', not S, is a fundamental quantity, and in order to achieve ultrasensitive optical microcavity sensors, maximizing the product $S'Q$ is indeed important.

Whispering gallery modes are currently the only microcavities that realize single-molecule detection. Although the WGMs have small S' value, the quality factor can exceed 10^8. As a result, optical sensing based on WGMs has the capability of detecting adsorption event of each particle or molecule on a microcavity.

One of pioneering work in photonic crystal chemical sensors was done by Loncar *et al.* [60] who immersed 2D photonic crystal slab microcavity lasers in water, isopropyl alcohol, and methanol and observed lasing wavelength shifts. When immersed in water, lasers were red-shifted—understood by Eq. (1.4). The mode used for the sensing is a dipole mode that is modified from a single defect by introduction of dislocation [23]. Although the quality factor drops as the refractive index of the surrounding media increases, the microcavity mode has large overlap of the field with the surrounding media containing analytes.

1.7 Microcavity-QED in Photonic Crystal

A small mode volume microcavity provides a single-mode photon field at frequency ω that is represented by a superposition of Fock states and is a great tool to manipulate photonic quantum states. On the other hand, a single quantum dot provides electronic quantum states. When these two quantum systems are coupled via photon–electron interaction, they form a joint quantum space. Therefore, a photonic crystal microcavity coupled with a single quantum dot provides wonderful opportunities in investigating quantum optics and quantum information sciences in solid-state form.

The coupling is often categorized, based on the strength of coupling relative to other energy dissipation processes from the coupled system, that is, strong and weak couplings. Purcell effect is a quantum optics phenomenon in the weak coupling regime. In high Q/V mode, the spontaneous emission rate is increased by a factor of the Purcell factor $F_p = 3Q(\lambda/n)^3/(4\pi^2 V)$. The Purcell effect allows us to drive photonic crystal lasers at high frequency.

A strongly coupled joint system is a great test bed in studying quantum optics. In 2004, strong coupling from a photonic crystal microcavity with a single quantum dot was demonstrated [5] as well as from a 1D photonic crystal micropillar with a single quantum dot [10]. Although single quantum dot transition energy can be tuned by temperature change, resonance energy can be tuned by removal or addition of resonator material. Mosor *et al.* developed the red-shift tuning method by condensing xenon on surface of photonic crystal microcavity without hurting the quality factor and showed strong coupling [61]. Hennessy *et al.* [62] showed digital etching method to blue-shift resonance and demonstrated strong coupling by using the digital etching method and deterministic quantum dot placement method [63]. Reflectivity response from the coupled system is tuned by pump power and probed, which is an important technique to detect the state of the strong coupling system [64]. A quantum phase gate up to $\pi/4$, which is an essential component in quantum logic operations, is also demonstrated based on a photonic crystal microcavity with a single quantum dot [65].

1.8 Conclusions

Photonic crystal microcavities are unique in that the mode volume can be as small as $(\lambda/n)^3$, while the quality factor is maintained to be as high as one million. Photonic crystal microlasers and microcavities provide great opportunities in optical sensing, quantum information sciences, and on-chip optical communication/interconnect, to name a few.

Acknowledgment

I would like to thank all researchers who have advanced technology in photonic crystals. Without their great efforts, it is impossible to see this great advance in research in photonic crystal microcavity and microlasers. I also would like to deeply apologize for resorting to omitting, from this writing, many important works in photonic crystal microcavities and microlasers due to page limits. Some of our works have been supported by the National Science Foundation.

References

1. Yablonovitch, E. (1987). Inhibited spontaneous emission in solid-state physics and electronics, *Phys. Rev. Lett.*, **58**, pp. 2059–2062.
2. John, S. (1987). Strong localization of photons in certain disordered dielectric superlattices, *Phys. Rev. Lett.*, **58**, pp. 2486–2489.
3. Deotare, P. B., McCutcheon, M. W., Frank, I. W., Khan, M., and Loncar, M. (2009). High quality factor photonic crystal nanobeam cavities, *Appl. Phys. Lett.*, **94**, p. 121106.
4. Schneider, C., Heindel, T., Huggenberger, A., Niederstrasser, T. A., Reitzenstein, S., Forchel, A., and Kamp, M. (2012). Microcavity enhanced single photon emission from an electrically driven site-controlled quantum dot, *Appl. Phys. Lett.*, **100**, p. 091108.
5. Yoshie, T., Scherer, A., Hendrickson, J., Khitrova, G., Gibbs, H. M., Rupper, G., Ell, C., Shchekin, O. B., and Deppe, D. G. (2004). Vacuum Rabi splitting with a single quantum dot in a photonic crystal nanocavity, *Nature*, **432**, pp. 200–203.
6. Tang., L., and Yoshie, T. (2011). Light localization in woodpile photonic crystal built via two-directional etching method, *IEEE J. Quantum Electron.*, **47**, pp. 1028–1035.
7. Soda, H., Iga, K., Kitahara, C., and Suematsu, Y. (1979). GaInAsP/InP surface emitting injection lasers, *Jpn. J. Appl. Phys.*, **18**, pp. 2329–2330.
8. Jewell., J. L., Harbison, J. P., Scherer, A., Lee, Y. H., and Florez, L. T. (1991). Vertical-cavity surface-emitting lasers: Design, growth, fabrication, characterization, *IEEE J. Quantum Electron.*, **27**, pp. 1332–1346.
9. Scherer, A., Jewell, J. L., Lee, Y. H., Harbison, J. P., and Florez, L. T. (1989). Fabrication of microlasers and microresonator optical switches, *Appl. Phys. Lett.*, **55**, pp. 2724–2726.
10. Reithmaier, J. P., Sek, G., Loffler, A., Hofmann, C., Kuhn, S., Reitzenstein, S., Keldysh, L. V., Kulakovskii, V. D., Reinecke, T. L., and Forchel, A. (2004). Strong coupling in a single quantum dot-semiconductor microcavity system, *Nature*, **432**, pp. 197–200.
11. Santori, C., Fattal, D., Vuckovic, J., Solomon, G. S., and Yamamoto, Y. (2002). Indistinguishable photons from a single-photon device, *Nature*, **419**, pp. 594–597.

12. Reitzenstein, S., Hofmann, C., Gorbunov, A., Strauß, M., Kwon, S. H., Schneider, C., Loffler, A., Hofling, S., Kamp, M., and Forchel, A. (2007). AlAs/GaAs micropillar cavities with quality factors exceeding 150,000, *Appl. Phys. Lett.*, **90**, p. 251109.

13. Robinson, J. T., Manolatou, C., Chen, L., and Lipson, M. (2005). Ultrasmall mode volumes in dielectric optical microcavities, *Phys. Rev. Lett.*, **95**, p. 143901.

14. Fan, S., Winn, J. N., Devenyi, A., Chen, J. C., Meade, R. D., and Joannopoulos, J. D. (1995). Guided and defect modes in periodic dielectric waveguides, *J. Opt. Soc. Am. B*, **12**, pp. 1267–1272.

15. Notomi, M., Kuramochi, E., and Taniyama, H. (2008). Ultrahigh-Q nanocavity with 1D photonic gap, *Opt. Express*, **16**, pp. 11095–11102.

16. Eichenfield, M., Chan, J., Camacho, R. M., Vahala, K. J., and Painter, O. (2009). Optomechanical crystals, *Nature*, **462**, pp. 78–82.

17. Painter O. J., Lee, R. K., Scherer, A., Yariv, A., O'Brien, J. D., Dapkus, P. D., and Kim, I. (1999). Two-dimensional photonic band-gap defect mode laser, *Science*, **284**, pp. 1819–1821.

18. Painter, O., Srinivasan, K., O'Brien, J. D., Scherer, A., and Dapkus, P. D. (2001). Tailoring of the resonant mode properties of optical nanocavities in two-dimensional photonic crystal slab waveguides, *J. Opt. A*, **3**, pp. S161–S170.

19. Kim, S. H., and Lee, Y. H. (2003). Symmetry relations of two-dimensional photonic crystal cavity modes, *IEEE J. Quantum Electron.*, **39**, pp. 1081–1085.

20. D'Urso, B., Painter, O., O'Brien, J., Tombrello, T., Yariv, A., and Scherer, A. (1998). Modal reflectivity in finite-depth two-dimensional photonic-crystal microcavities, *J. Opt. Soc. Am. B*, **15**, pp. 1155–1159.

21. Painter, O., Vuckovic, J., and Scherer, A. (1999). Defect modes of two-dimensional photonic crystal in an optically thin dielectric slab, *J. Opt. Soc. Am. B*, **16**, pp. 275–285.

22. Vuckovic, J., Loncar, M., Mabushi, H., and Scherer, A. (2002). Optimization of the Q factor in photonic crystal microcavities, *IEEE J. Quantum Electron.*, **38**, pp. 850–856.

23. Srinivasan, K., and Painter, O. (2002). Momentum space design of high-Q photonic crystal optical cavities, *Opt. Express*, **10**, pp. 670–684.

24. Akahane, Y., Asano, T., Song, B. S., and Noda, S. (2003). High-Q photonic nanocavity in a two-dimensional photonic crystal, *Nature*, **425**, pp. 944–947.

25. Song, B. S., Noda, S., Asano, T., and Akahane, Y. (2005). Ultra-high-Q photonic double-heterostructure nanocavity, *Nat. Mater.*, **4**, pp. 207–210.

26. Kuramochi, E., Notomi, M., Mitsugi, S., Shinya, A., and Tanabe, T. (2006). Ultrahigh-Q photonic crystal nanocavities realized by the local width modulation of a line defect, *Appl. Phys. Lett.*, **88**, p. 041112.

27. Asano, T., Song, B. S., and Noda, S. (2006). Analysis of the experimental Q factors (~1 million) of photonic crystal nanocavities, *Opt. Express*, **14**, pp. 1996–2002.

28. Akahane, Y., Asano, T., Song, B. S., and Noda, S. (2005). Fine-tuned high-Q photonic-crystal nanocavity, *Opt. Express*, **13**, pp. 1202–1214.

29. Ogawa, S., Imada, M., Yoshimoto, S., Okano, M., and Noda, S. (2004). Control of light emission by 3D photonic crystals, *Science*, **305**, pp. 227–229.

30. Tandaechanurat, A., Ishida, S., Guimard, D., Nomura, M., Iwamoto, S., and Arakawa, Y. (2011). Lasing oscillation in a three-dimensional photonic crystal nanocavity with a complete bandgap, *Nat. Photonics*, **5**, pp. 91–94.

31. Qi, M., Lidorikis, E., Rakich, P. T., Johnson, S. G., Joannopoulos, J. D., Ippen, E. P., and Smith, H. I. (2004). A three-dimensional optical photonic crystal with designed point defects, *Nature*, **429**, pp. 538–542.

32. Yablonovitch, E., Gmitter, T. J., and Leung, K. M. (1991). Photonic band structure—the face centered-cubic case employing nonspherical atoms, *Phys. Rev. Lett.*, **67**, pp. 2295–2298.

33. Cheng, C. C., Arbet-Engels, Scherer, A., and Yabronovitch, E. (1996). Nanofabricated 3D photonic crystals operating at optical wavelengths, *Phys. Scripta*, **T68**, pp. 17–20.

34. Takahashi, S., Okano, M., Imada, M., and Noda, S. (2006). Three-dimensional photonic crystals based on double-angled etching and wafer-fusion techniques, *Appl. Phys. Lett.*, **89**, p. 123106.

35. Takahashi, S., Suzuki, K., Okano, M., Imada, M., Nakamori, T., Ota, Y., Ishizaki, K., and Noda, S. (2009). Direct creation of three-dimensional photonic crystals by a top-down approach, *Nat. Mater.*, **8**, pp. 721–725.

36. Tang, L., and Yoshie, T. (2007). Ultra-high-Q three-dimensional photonic crystal nano-resonators, *Opt. Express*, **15**, pp. 17254–17263.

37. Tang, L., and Yoshie, T. (2010). Woodpile photonic crystal fabricated in GaAs by two-directional etching method, *J. Vac. Sci. Tech. B*, **28**, pp. 301–303.

38. Tamm, V. I. (1932). Uber eine mogliche art der elektronenbindung an kristalloberflachen, *Zeitschrift fur Physik*, **76**, pp. 849–850.
39. Shockley, W. (1939). On the surface states associated with a periodic potential, *Phys. Rev.*, **56**, pp. 317–323.
40. Yariv, A., Yeh, P., and Hong, C. S. (1977). Electromagnetic propagation in periodic stratified media. I. General theory, *J. Opt. Soc. Am.*, **67**, pp. 423–438.
41. Meade, R. D., Brommer, K. D., Rappe, A. M., and Joannopoulos, J. D. (1991). Electromagnetic Bloch waves at the surface of photonic crystal, *Phys. Rev. B.*, **44**, pp. 10961–10964.
42. Ishizaki, K., and Noda, S. (2009). Manipulation of photons at the surface of three-dimensional photonic crystals, *Nature*, **460**, pp. 367–370.
43. Yang, J. K., Kim, S. H., Kim, G. H., Park, H. G., Lee, Y. H., and Kim, S. B. (2004). Slab-edge modes in two-dimensional photonic crystals, *Appl. Phys. Lett.*, **84**, pp. 3016–3018.
44. Yoshie, T., Tang, L., and Su, S. Y. (2011). Optical microcavities: Sensing down to single molecules and atoms, *Sensors*, **11**, pp. 1972–1991.

Chapter 2

Simulation of Planar Photonic Resonators

Stefan Declair and Jens Förstner

Department of Physics, Computational Nanophotonics, University of Paderborn, Warburger Street 100, 33098 Paderborn, Germany

sdeclair@mail.upb.de

In the last years, optoelectronics and photonics became a new and important field in physics. The fundamental basis is the design of properly designed optical resonators with sizes from nanometers as for instance plasmonic nanostructures up to micrometers like a microdisk (MD) resonator or a photonic crystal cavity (PhCC). The range of applications is broad: those systems are already used for probing, investigating, and manipulating the fundamental physics of light–matter interaction. Devices for single photon generation, high finesse filters, slow light generation, and so on are also realized. Still, this is a current research topic in modern physics in the area of nanooptics, nanophotonics, and optoelectronics. But there is still a problem: if one wants to build devices which are operating in a specific regime, for instance coupling regime with embedded semiconductor heterostructures such as quantum dots (QDs), then this device has to be tested. To overcome trial-and-error procedures, numerical analysis became a crucial tool in designing optoelectronic

Handbook of Optical Microcavities
Edited by Anthony H. W. Choi

ISBN 978-981-4463-24-9 (Hardcover), 978-981-4463-25-6 (eBook)
www.panstanford.com

devices. Nowadays, large-scale computers and computer clusters allow simulations of various problems of optoelectronics and photonics and treat the electromagnetic together with dynamic material equations for large simulation volumes with nanoscale-level spatial discretization for reliable results in acceptable times.

As a method for near-field resolved numerical analysis of optoelectronic devices, this chapter gives an overview over the applying of finite-difference time-domain (FDTD) method [1–3] to photonics. As the name of the method already reveals, this method works in the time domain. Thus, information in a broad range of frequencies can be obtained with one single simulation. Stability issues and limitations are discussed, like the termination of the computational domain and the order of accuracy. For specific optical resonators, a dielectric MD and a PhCC, numerical results obtained with the FDTD method are shown. Since the inclusion of active material, which provides an additional optical polarization field, is an important point in designing optoelectronic devices, a quantum mechanical resonator treated in the second quantization scheme is embedded in a PhCC. In the strong coupling regime, this coupled system shows vacuum Rabi splitting even in this semiclassical approach. For the case of high excitation intensity, nonlinearities of the active material are treated with the FDTD approach as well.

For the sake of completeness it should also be mentioned that there is also frequency domain, which can be utilized for numerical investigation, such as finite element method (FEM) [4,5] operating in frequency domain. Hence, one needs to run various simulations to obtain a complete spectrum. Both methods have their own advantages and disadvantages. However, the method of choice is highly problem dependent.

2.1 Maxwell's Equations

In 1865, James Clark Maxwell published the basic set of equations for electric and magnetic fields $\boldsymbol{E}$ and $\boldsymbol{H}$, respectively, today famously called Maxwell's equations [6]. These equations connect the electric and magnetic properties of the fields and show that the sources of the electric and magnetic fields are the scalar function ρ, the charge density, and the vector function $\boldsymbol{j}$, the current density. Since Maxwell's equations are form-invariant under unit transformation, such transformations change definitions and constants [7,8]. Throughout this chapter, SI units are used.

The equations in their differential form are written as [9]:

$$\nabla \times \boldsymbol{E}(r,t) = -\dot{\boldsymbol{B}}(r,t) \tag{2.1}$$

$$\nabla \times \boldsymbol{B}(\boldsymbol{r},t) = \mu_0 \boldsymbol{j}(\boldsymbol{r},t) + \mu_0 \varepsilon_0 \dot{\boldsymbol{E}}(\boldsymbol{r},t) \tag{2.2}$$

$$\nabla \cdot \boldsymbol{B}(\boldsymbol{r},t) = 0 \tag{2.3}$$

$$\nabla \cdot \boldsymbol{E}(\boldsymbol{r},t) = \frac{\rho(\boldsymbol{r},t)}{\varepsilon_0}, \tag{2.4}$$

where $\boldsymbol{E}(\boldsymbol{r}, t)$ is the electric field strength, $\boldsymbol{B}(\boldsymbol{r}, t)$ the magnetic flux density, $\rho(\boldsymbol{r}, t)$ the charge density, $\boldsymbol{j}(\boldsymbol{r}, t)$ the current density at position $\boldsymbol{r}$ and time t, ε_0, and μ_0 are the free space permittivity (vacuum permittivity) and free space permeability (vacuum permeability), respectively.

Maxwell's equations can be rewritten in integral form [9]:

$$\oint_{\partial S} \boldsymbol{E}(\boldsymbol{r},t) \cdot dl = -\partial_t \Phi_{B,S} \tag{2.5}$$

$$\oint_{\partial S} \boldsymbol{B}(\boldsymbol{r},t) \cdot d\boldsymbol{l} = \mu_0 I_S + \mu_0 \varepsilon_0 \partial_t \Phi_{E,S} \tag{2.6}$$

$$\oiint_{\partial V} \boldsymbol{B}(\boldsymbol{r},t) \cdot d\boldsymbol{A} = 0 \tag{2.7}$$

$$\oiint_{\partial V} \boldsymbol{E}(\boldsymbol{r},t) \cdot d\boldsymbol{A} = \frac{1}{\varepsilon_0} \rho_{\partial V}. \tag{2.8}$$

Here, $\Phi_{B,S}$ and $\Phi_{E,S}$ are the magnetic and electric flux through a surface S, I_S is the electric current through surface S, and $\rho_{\partial V}$ is the charge density inside the boundary of volume V. The vectorial length and area elements are dl and dA, respectively.

The differential forms of Maxwell's equations (2.1) to (2.4) can be transformed into the integral forms (2.5) to (2.8) with vector-analytical methods using Kelvin–Stokes' and Gauss' theorem, see Refs. [8,9].

Equations (2.1)–(2.4) have to be extended to obey optical and magnetic properties of material on a nonatomic level. Calculations on the atomic level are utterly impossible to perform. Therefore, an averaged quantity of the electric and magnetic response of the considered material is introduced, which flattens possible fluctuations in the fields due to atomic-sized objects. Thus, the electric field is modified with the dielectric constant ε_r and the magnetic field is modified with magnetic permeability μ_r, which typically depend on position $\boldsymbol{r}$. Actually, these two quantities are tensors of second rank and frequency-dependent, but in this framework, these characteristics are not taken into account. Since the interest lies on nonmagnetic materials, μ_r is always equal to 1. The quantity $\mu_r(\boldsymbol{r})$ only differs weakly from unity, so that approximation to 1 is justified. For the sake of completeness, $\mu_r(\boldsymbol{r})$ will still appear in the following equations. Therefore, Eqs. (2.1)–(2.4) extend to the macroscopic Maxwell's equations via the constitutive relationships [8]:

$$\boldsymbol{B}(\boldsymbol{r},t) = \mu_0\mu_r(\boldsymbol{r})\boldsymbol{H}(\boldsymbol{r},t) + \boldsymbol{M}(\boldsymbol{r},t),\ \boldsymbol{D}(\boldsymbol{r},t) = \varepsilon_0\varepsilon_r(\boldsymbol{r})\boldsymbol{E}(\boldsymbol{r},t) + \boldsymbol{P}(\boldsymbol{r},t), \tag{2.9}$$

where $\boldsymbol{H}$ and $\boldsymbol{D}$ are the magnetic field strength and dielectric displacement field at position $\boldsymbol{r}$ at time t. Magnetization $\boldsymbol{M}(\boldsymbol{r}, t)$ and polarization $\boldsymbol{P}(\boldsymbol{r}, t)$ are not taken into account here. With this, Eqs. (2.1)–(2.4) transform to the macroscopic Maxwell equations:

$$\nabla \times \boldsymbol{E}(\boldsymbol{r},t) = -\dot{\boldsymbol{B}}(\boldsymbol{r},t) \tag{2.10}$$

$$\nabla \times \boldsymbol{H}(\boldsymbol{r},t) = \boldsymbol{j}(\boldsymbol{r},t) + \dot{\boldsymbol{D}}(\boldsymbol{r},t) \tag{2.11}$$

$$\nabla \cdot \boldsymbol{B}(\boldsymbol{r},t) = 0 \tag{2.12}$$

$$\nabla \cdot \boldsymbol{D}(\boldsymbol{r},t) = \rho(\boldsymbol{r},t), \tag{2.13}$$

where $\rho(\boldsymbol{r}, t)$ and $\boldsymbol{j}(\boldsymbol{r}, t)$ are now macroscopic charge and current densities, respectively. The time-dependent Maxwell's curl equations (2.1) and (2.2) together with the constitutive relationships (2.9), and assuming that the fields are known at an initial position in time and space, form a hyperbolic partial differential equations

(PDE). This set of hyperbolic PDEs has a unique solution, uniquely depending on the initial values. For numerical evaluation, one uses the curl equations with $\boldsymbol{E}(\boldsymbol{r}, t)$ and $\boldsymbol{H}(\boldsymbol{r}, t)$ for reasons which are discussed in Section 2.2:

$$\nabla \times \boldsymbol{E}(\boldsymbol{r},t) = -\frac{1}{\mu_0 \mu_r(\boldsymbol{r})} \dot{\boldsymbol{B}}(\boldsymbol{r},t) \tag{2.14}$$

$$\nabla \times \boldsymbol{H}(\boldsymbol{r},t) = \boldsymbol{j}(\boldsymbol{r},t) + \varepsilon_0 \varepsilon_r(\boldsymbol{r}) \dot{\boldsymbol{D}}(\boldsymbol{r},t). \tag{2.15}$$

Discretization of Eqs. (2.14) and (2.15) is fundamentally shown in Section 2.2.1 for numerical evaluation. For further information, see Refs. [2,3].

2.2 Numerics with Maxwell's Equations: The FDTD

To numerically evaluate Eqs. (2.10)–(2.13), one needs to discretize time and space, since continuous problems cannot be solved on computers due to limited computational resources like memory. The discretization scheme in space is based on the idea of Yee [10], where the different field components of different field quantities are placed at different positions on a staggered grid, the so-called Yee-cube (see Fig. 2.1, left). Electric field components are centered on the edges of the electric grid, the primary grid, whereas magnetic field components are centered on the edges of the magnetic grid, the secondary grid. From Eqs. (2.5)–(2.8), the magnetic field has to be on a grid and the corner needs to be in the center of the grid for the electric field. For example, in equation (2.5) one has to calculate the closed line integral of the electric field along a closed curve to obtain the magnetic flux through area bounded by this closed curve (see also Fig. 2.1, right). The proposal from Yee exactly fulfills the conditions, given by the integral forms of Maxwell's equations. Namely, the secondary grid is shifted half a grid edge length in each direction, so that the magnetic field components centered on the surfaces of the primary grid and vice versa. This section shows how to discretize Maxwell's equations and follows the work given in Refs.

[2,3]. For further information and reading, the interested reader is referred to these references.

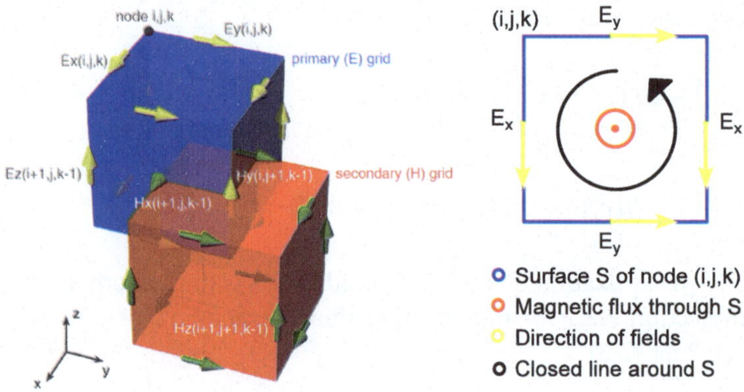

Figure 2.1 Left: Primary and secondary grid of the Yee-cube at node (i, j, k) [11]. Electric field components live on the edges of the primary grid (blue), while magnetic fields live on the edges of the secondary grid (red). The arrangement of both grids, staggered half a cell length in all directions, let electric field be localized face-centered on the secondary grid, and magnetic field be located face-centered on the primary grid. Right: Scheme of the closed line integral at node (i, j, k). Due to the staggered arrangement, the closed line integral, exemplarily for equation (2.5) can be evaluated easily.

2.2.1 Update Equations of the FDTD

Spatially the computational volume is separated into voxels, the 3D analog to the 2D pixel, which are indexed with natural number triples (i, j, k) for the x, y, z-direction. Neighboring cells have incremented or decremented indices depending on the direction. An edge length of a grid cell is $\Delta\mu$, $\mu = x, y, z$, so that neighboring field quantities are separated by one discretization length. Now, to address the field components in a numeric scheme, the following nomenclature is used:

$$\begin{aligned}(\boldsymbol{E}(\boldsymbol{r}))_\mu &= E_{\mu\, ijk}\\ &= E_\mu\left(\left(i+\frac{1}{2}\delta_{\mu x}\right)\Delta x,\ \left(j+\frac{1}{2}\delta_{\mu y}\right)\Delta y,\ \left(k+\frac{1}{2}\delta_{\mu z}\right)\Delta z\right) \quad (2.16)\end{aligned}$$

$$(\boldsymbol{H}(\boldsymbol{r}))_\mu = H_{\mu\, ijk}$$

$$= H_\mu\left(\left(i+\frac{1}{2}-\frac{1}{2}\delta_{\mu x}\right)\Delta x, \left(j+\frac{1}{2}-\frac{1}{2}\delta_{\mu y}\right)\Delta y, \left(j+\frac{1}{2}-\frac{1}{2}\delta_{\mu z}\right)\Delta z)\right),$$

(2.17)

where $\delta_{\mu\nu}$ is the Kronecker symbol, which is equal to 1 if $\mu = \nu$ and 0 otherwise.

With this spatial discretization scheme one can now discretize the spatial differential operator ∂_μ. Since the derivative is a quantity obtained by a limes, discretization is needed to let the differential operator fit on the spatial grid. So the derivative is now written like in its derivation, but with central differences. Exemplarily, $\partial_x(\boldsymbol{H}(\boldsymbol{r}, t))_y$ then yield

$$\partial_x H_{y\, i+\frac{1}{2} jk+\frac{1}{2}} = \frac{1}{\Delta x}\left(H_{y\, i+1\, jk+\frac{1}{2}} - H_{y\, ijk+\frac{1}{2}}\right). \tag{2.18}$$

The macroscopic Maxwell's equations (2.10)–(2.13) now have to be discretized in time with central differences, too, due to the partial derivatives in time. Temporal integration is then done with the so-called leap-frog algorithm, where electric field quantities at time t are calculated from neighboring field quantities from the magnetic quantities at $t + \frac{1}{2}\Delta t$ and $t - \frac{1}{2}\Delta t$, where Δt is the discretization in time. See Fig. 2.2 for a coarse scheme. Therefore, the numerical scheme for all the field components is extended with another natural numbered index, n, for the time, i.e.,

$$(\boldsymbol{E}(r,t))_x = E^n_{x\, i+\frac{1}{2} jk} = E_x\left(\left(i+\frac{1}{2}\right)\Delta x, j\Delta y, k\Delta z, n\Delta t\right) \tag{2.19}$$

$$(\boldsymbol{H}(r,t))_x = H^{n+\frac{1}{2}}_{x\, ijk} = H_x\left(i\Delta x, \left(j+\frac{1}{2}\right)\Delta y, \left(k+\frac{1}{2}\right)\Delta z, \left(n+\frac{1}{2}\right)\Delta t\right).$$

(2.20)

A generic discretization of the partial temporal derivative with central differences of $(\boldsymbol{E}(\boldsymbol{r}, t))_x$ and $(\boldsymbol{H}(\boldsymbol{r}, t))_z$ then yields:

$$\partial_t E^n_{x\,i+\frac{1}{2}jk} = \frac{1}{\Delta t}\left(E^{n+\frac{1}{2}}_{x\,i+\frac{1}{2}jk} - E^{n-\frac{1}{2}}_{x\,i+\frac{1}{2}jk}\right) \tag{2.21}$$

$$\partial_t H^{n+\frac{1}{2}}_{z\,i+\frac{1}{2}j+\frac{1}{2}k} = \frac{1}{\Delta t}\left(H^{n+1}_{z\,i+\frac{1}{2}j+\frac{1}{2}k} - H^{n}_{z\,i+\frac{1}{2}j+\frac{1}{2}k}\right). \tag{2.22}$$

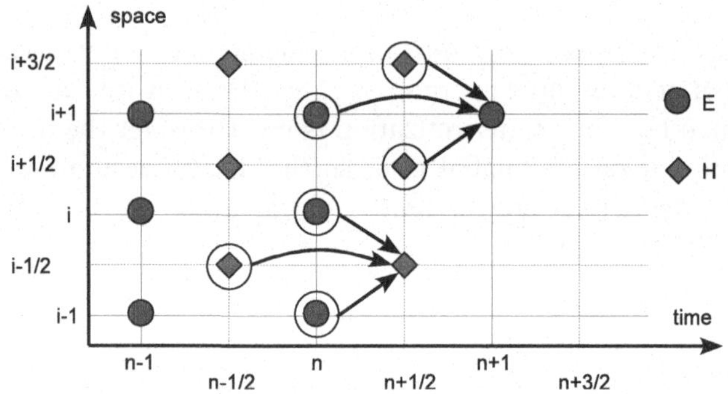

Figure 2.2 Update scheme for the fields (leap-frog scheme). Time stepping is in horizontal direction, while spatial stepping is in vertical direction (direction according to arrows). To update the magnetic field $H^{n+\frac{1}{2}}_{i-\frac{1}{2}}$, the field one full time step Δt earlier, $H^{n-\frac{1}{2}}_{i+\frac{1}{2}}$, and the electric fields half a time step before and half a space step back and forth each, E^n_{i-1} and E^n_{i+1}, are needed. Update of the electric field is similar, but fields are exchanged and half a step has to be added to time and space. (Here, two half steps in time and space are shown to have the update scheme separated.)

With these tools one can now derive update equations to calculate the components of electric and magnetic field on all grid points at the next unknown time step. To do so, one uses the curl equations (2.12)–(2.13) and replaces the derivatives in time and space in equations (2.18) and (2.21) and (2.22). This yields the following update equations exemplarily for the x-component of the electric and magnetic field:

$$E^{n+1}_{x\,i+\frac{1}{2}jk} = E^{n}_{x\,i+\frac{1}{2}jk} - \frac{\Delta t}{\varepsilon_0 \varepsilon_{i+\frac{1}{2}jk}}$$

$$\times\left(\frac{1}{\Delta y}\left(H^{n+\frac{1}{2}}_{z\,i+\frac{1}{2}j+\frac{1}{2}k}-H^{n+\frac{1}{2}}_{z\,i+\frac{1}{2}j-\frac{1}{2}k}\right)-\frac{1}{\Delta z}\left(H^{n+\frac{1}{2}}_{y\,i+\frac{1}{2}jk+\frac{1}{2}}-H^{n+\frac{1}{2}}_{y\,i+\frac{1}{2}jk-\frac{1}{2}}\right)-j^{n+\frac{1}{2}}_{x\,i+\frac{1}{2}jk}\right) \tag{2.23}$$

$$H^{n+\frac{1}{2}}_{x\,ij+\frac{1}{2}k+\frac{1}{2}}=H^{n-\frac{1}{2}}_{x\,ij+\frac{1}{2}k+\frac{1}{2}}-\frac{\Delta t}{\mu_0}$$

$$\times\left(\frac{1}{\Delta y}\left(E^{n}_{z\,ij+1k+\frac{1}{2}}-E^{n}_{z\,ijk+\frac{1}{2}}\right)-\frac{1}{\Delta z}\left(E^{n}_{y\,ij+\frac{1}{2}k+1}-E^{n}_{y\,ij+\frac{1}{2}k}\right)\right), \tag{2.24}$$

from which one can calculate the unknown $(\boldsymbol{E})_\mu$ at time $t + \Delta t$ using $(\boldsymbol{E})_\mu$ at time t at the same position in the Yee-cube and H on the surrounding positions.

These update equations are now based on a centered difference scheme in time as well as in space. As one may have already noticed, the spatial derivatives $\partial_\mu(\boldsymbol{E}(\boldsymbol{r}, t))_\mu$ did not emerge. The reason is that this update scheme based on the Yee-cube naturally fulfills the divergence equations (2.12) and (2.13). Hence, the full set of electrodynamic equations (2.10)–(2.13) is complete with temporally and spatially update equations (2.23) and (2.24) of the FDTD.

2.2.2 Accuracy and Stability

Discretization of the Maxwell's equations (2.14) and (2.15) induces errors in the temporal and spatial derivatives, since they are continuous, but the FDTD grid in time and space is discrete. A Taylor expansion shows the order of the error in terms of the spatial discretization. Assume a function $f(x)$ probed by a grid with discretization Δx and use central differences to express a derivative in x. For the left-aligned (right-aligned) function values, $f\left(x-\frac{\Delta x}{2}\right) f\left(x+\frac{\Delta x}{2}\right)$, the Taylor expansion equals

$$f\left(x+\frac{\Delta x}{2}\right)=f(x)+\frac{\Delta x}{2}f^{(1)}(x)+\frac{1}{2}\left(\frac{\Delta x}{2}\right)^2 f^{(2)}(x)+\frac{1}{6}\left(\frac{\Delta x}{2}\right)^3 f^{(3)}(\chi_+) \tag{2.25}$$

$$f\left(x-\frac{\Delta x}{2}\right)=f(x)-\frac{\Delta x}{2}f^{(1)}(x)+\frac{1}{2}\left(\frac{\Delta x}{2}\right)^2 f^{(2)}(x)-\frac{1}{6}\left(\frac{\Delta x}{2}\right)^3 f^{(3)}(\chi_-), \tag{2.26}$$

where $f^{(n)}$ is the nth derivative of f and $\chi_{\pm} \in \left[x, x \pm \frac{\Delta x}{2}\right]$. Now, subtracting equation (2.26) from equation (2.25) and dividing by Δx, one gets

$$f^{(1)}(x) = \frac{f\left(x + \frac{\Delta x}{2}\right) - f\left(x - \frac{\Delta x}{2}\right)}{\Delta x} - \frac{1}{3}\left(\frac{\Delta x}{2}\right)^3 f^{(3)}(\chi), \tag{2.27}$$

with $\chi \in \left[x - \frac{\Delta x}{2}, x + \frac{\Delta x}{2}\right]$. So the centered first derivative approximation equation (2.27) is of $\mathcal{O}(\Delta x^2)$.

The time integration to get $\boldsymbol{E}(\boldsymbol{r}, t)$ and $\boldsymbol{H}(\boldsymbol{r}, t)$ for the new time steps in equations (2.23) and (2.24) is done according to Fig. 2.2. Here, $\boldsymbol{E}(\boldsymbol{r}, t)$ ($\boldsymbol{H}(\boldsymbol{r}, t)$) at time $t = (n + 1)\Delta t$ (for ($\boldsymbol{H}(\boldsymbol{r}, t)$: $t = \left(n + \frac{1}{2}\right)\Delta t$) is calculated with the previous value at $t = (n - 1)\Delta t$ (for ($\boldsymbol{H}(\boldsymbol{r}, t)$: $t = \left(n - \frac{1}{2}\right)\Delta t$) and the surrounding $\boldsymbol{H}(\boldsymbol{r}, t)$ ($\boldsymbol{E}(\boldsymbol{r}, t)$) at time $t = \left(n + \frac{1}{2}\right)\Delta t$ (for $\boldsymbol{E}(\boldsymbol{r}, t)$: $t = n\Delta t$). Expressing this procedure with the functions $f(t)$ and $g(t)$, which are defined at temporal positions $t = n\Delta t$ and $t = \left(n + \frac{1}{2}\right)\Delta t$, respectively, and assuming that $\mathcal{F}(\cdot)$ is some functional, the update scheme looks like

$$f((n+1)\Delta t) = f(n\Delta t) + \mathcal{F}\left(g\left(\left(n + \frac{1}{2}\right)\Delta t\right)\right) \tag{2.28}$$

$$g\left(\left(n + \frac{1}{2}\right)\Delta t\right) = g\left(\left(n - \frac{1}{2}\right)\Delta t\right) + \mathcal{F}(f(n\Delta t)). \tag{2.29}$$

Since the time integration is of second order in the discretization, too, the stability criterion for the FDTD implementation, the so-called Courant-criterion, can be exploited to the maximum. The Courant-criterion says that

$$\Delta t \leq s \cdot \frac{1}{c_0\sqrt{\Delta x^{-2} + \Delta y^{-2} + \Delta z^{-2}}} \tag{2.30}$$

ensures stability of the implementation. In the case of different orders of accuracy for the time and space derivative approximations, the factor s has to be $s \ll 1$, resulting in a small temporal discretization. In the FDTD case, s is mostly chosen to be ≈ 1.

If s equals 1, equation (2.30) is called the magic time step. In a 1D FDTD implementation, the simulation equals the analytic solution exactly. The reason is the numerical dispersion, which is induced by the spatial grid. Numerical dispersion in general means that a wave propagating in the spatial grid will be slowed down compared to its natural speed, c_0. In a 1D case, one can choose the time step according to the magic time step, $\Delta t = \frac{\Delta x}{c}$. Hence, the time, which is needed for a propagating wave to go $c_0 \Delta t$ equals exactly equation (2.30) with $s = 1$, and the speed of the wave is exactly c_0, so no numerical dispersion occurs. In higher-dimensional FDTD simulations, the magic time step would only ensures no numerical dispersion along the principle axes of the simulation volume (if a uniform grid is provided). The other directions then carry a numerical dispersion, which are calculated via

$$\left[\frac{1}{c_0 \Delta t}\sin\left(\frac{\omega \Delta t}{2}\right)\right]^2 = \sum_{\mu = x,y,z}\left[\frac{1}{\Delta \mu}\sin\left(\frac{\tilde{k}_\mu \Delta \mu}{2}\right)\right]^2, \tag{2.31}$$

where $\tilde{k}_\mu$ are the components of the numerical wave vector (see Ref. [3]).

In other words, the temporal discretization is determined by the spatial discretization and s. Since equation (2.31) is of the same order of accuracy as equation (2.29), the numerical dispersion can be reduced drastically when a smaller spatial discretization is used. The problem is that when reducing the spatial discretization by $\Delta\mu_{\text{coarse}} = m \cdot \Delta\mu_{\text{fine}}$, memory raises with $\mathcal{O}(m^3)$ and computational time raises with $\mathcal{O}(m^4)$. Accordingly, a blind reduction of $\Delta\mu$ has no use, in fact one has to think about the problem one wants to solve. A good value to start with is $\Delta\mu = 0.1\lambda_0$, with λ_0 being the smallest wavelength in the material with the highest dielectric permittivity and the highest conductivity one wants to detect. However, this is of course no rule, but a rule of thumb. Another point to think of is, what is the smallest geometrical shape one wants to resolute. Usually, curvatures need a higher resolution to let the waves in the computational domain interpret the structure as a curved structure. In Cartesian grids, curvatures always result in staircasing (Fig. 2.3 left), which affects resonant frequencies drastically [12,13]. Non-Cartesian grids can be applied to solve the problem. For cylindrical structures, the best choice is a cylindrical coordinate system (if one has point symmetry to the origin). Therefore, transform equations (2.12) and (2.13) into cylindrical coordinates and then perform the discretization in the new coordinate system, like in the free available software package MEEP [1]. Another method is to use (non-)conformal meshes [3], a polygonal mesh without orthogonal principal axis like in the frequency-domain method FEM [4,14], the discontinuous Galerkin time-domain (DGTD) method [15–19], or the finite integration technique (FIT) [20]. To overcome a lack in accuracy due to staircasing effects, one can use an averaging method for the permittivity within the cells (Section 2.2.3).

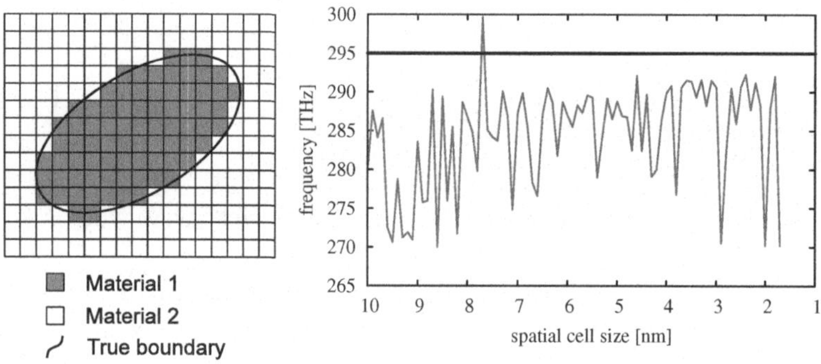

Figure 2.3 Left: 2D scheme of the staircasing. The black solid line shows the interface between two media with different dielectric permittivity (white and gray). Spatial cells, which contain the interface, are filled with only one of the two materials, if no averaging routine is provided. The shape of the originally curved object gets staircased. Right: Staircasing effect on the resonance of an eigenmode in a circular resonator with metallic boundary (2D calculation in a MATLAB code). A Cartesian (uniform) grid is used to map the perfectly circular structure. The spatial cell size is decreased from 10 nm down to 1.7 nm in 0.1 nm steps. The vertical axes show the resonant frequency for a certain eigenmode (red: numerically calculated resonance, black: analytical value).

Figure 2.3, right, shows the dependency of the numerical frequency of a perfectly circular resonator structure in a metallic environment on the size of the spatial grid cells, used in a 2D FDTD. The radius of the resonator structure is set to $R = 361$ nm and the permittivity is set to the Gallium Arsenide (GaAs) value of $\varepsilon = 11.56$. The usage of a Cartesian grid yields the mapping of the dielectric structure to be staircased at the material interface. Therefore, an increase or decrease in the cell size changes the effective resonator shape, hence the frequency of the resonance changes. The analytically calculated value of the fourth resonance is $f = 295$ THz (black). As can be seen, the numerically calculated values for the resonances (red) approach asymptotically the analytical value, in fact it stays below the exact value, even for cell sizes much smaller than the expected wavelength. Thus, convergence for a curved structure in a Cartesian grid is not expected along this chapter, but sufficiently small spatial probing will be chosen to ensure experimentally immeasurable frequency changes.

2.2.3 Subpixel Averaging for Isotropic Dielectric Media

In the setup of the spatial grid and the implantation of dielectric material in arbitrary shape, it happens that Yee-cells are not filled totally with one and the same dielectric material, hence one cell has two or even more different values for the dielectric permittivity. The question arising is what value is assigned to this cell. Obviously, one can, in the case of two dielectric materials in one cell, calculate the amount of each material filling this cell, and then fill the whole cell with the most frequently present. This results, especially for circular structures, in a staircasing (see Fig. 2.3).

Also, this reduces the accuracy of the algorithm to first order. To overcome this problem, one uses an averaging method for the permittivity, based on the filling factor of the different permittivities in the corresponding cell. In general, such averaging schemes can drop the accuracy down to first order in the spatial discretization. For two reasons, the method of choice is the subpixel averaging scheme developed in Ref. [21]. First, the averaging scheme is chosen to be in a way that, treating the problem in a perturbative approach with perturbative parameter s, for $s \to 0$ the first error vanishes, providing second-order accuracy. In fact, the result is then an enhanced accuracy, up to second order for special cases and specific fields, but in general a higher accuracy than without averaging. Second, this averaging scheme can be extended to full anisotropic materials, too [22].

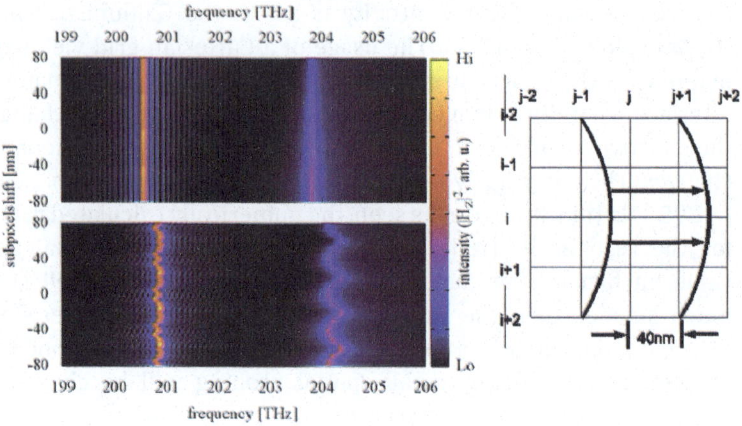

Figure 2.4 Left: Comparison of a simulation of a cylindrical object with a radius of $r = 1$ µm. Top: No ε-averaging. Bottom: With ε-averaging. Chasing the resonance occurring at $f = 200.58$ THz, one can clearly see blue shift of the resonance up to $\Delta f \approx 100$ GHz with a periodicity of 40 nm, equal to the spatial discretization, resulting from an efficiently smaller object due to staircasing. The decreasing intensity of the resonance originates from a shifted source, but not shifted monitor to record the time-dependent field. Right: Scheme of the subpixel shift of the dielectric cylindrical object. As the true boundary passes a cell, the cell itself is either fully filled or half filled.

The performance of the averaging process is shown in Fig. 2.4, left. Here, a 2D FDTD calculation with a cylindrical dielectric object with a radius of $r = 1$ µm, embedded in vacuum, is done. The behavior of a specific mode around $f = 200.58$ THz is chosen to be tracked, while the cylindrical object is shifted in subpixel steps over 4 cells (Fig. 2.4, right, shown schematically). The spatial discretization corresponds to $\Delta x = \Delta y = 40$ nm. In the bottom case, no ε-averaging routine is used in the simulation. A periodic variation up to $\Delta f \approx 100$ GHz in the resonance frequency of the system with a periodicity of the cell length, as well as a modification in the intensity of the resonance, is observed. In fact, shifting the dielectric object in subpixel steps changes the geometry, since cells, which were filled before, are no longer filled (or vice versa) after the shift. In general, the staircasing reduces the effective size of the cylindrical object, thus increases the resonance frequency. The upper case, where the ε-averaging procedure is applied, shows no shift in the resonance. Still, the intensity in the resonance changes with a periodicity as before without the averaging process, but the deviation is of lower order. Generally, the decreased intensity of the resonances originates from the larger distance between the excitation source and the recording monitor for the time-depending fields. Observing the second resonance at f = 204.11 THz in the frequency window of interest in Fig. 2.4, the amplitude of the oscillation of the resonance with the subpixel shift attributes to a spatiospectral response of other symmetry compared to the mode at 200.58 THz. Applying the averaging routine flattens the oscillatory behavior here as well.

2.2.4 CPML: Truncate the Computational Volume

Since the computational resources are limited, one has to, next to temporal and spatial discretization, truncate the computational volume in space. Setting all the field values out of the simulation volume hard to zero would end up as a perfect electric (magnetic) conductor, PEC (PMC). Hence, all fields inside the computational volume are then reflected from the boundary, since no fields would be able to penetrate the outer regions. The result then is that the implanted energy will stay inside the simulation volume forever (despite usage of lossy materials). Since this is unphysical, one at best needs absorbing boundary conditions (ABC) which totally absorb the energy of incident waves, independent of the polarization, frequency, and incident angle, in fact are 100% transmittive. Furthermore, the absorbing boundary should be effective when scattering objects or sources, especially point sources, are close. Additionally, since the ABC area also consists of cells which need to update $\boldsymbol{E}(\boldsymbol{r}, t)$ and $\boldsymbol{H}(\boldsymbol{r}, t)$ in every time step with additional effort, the absorbing area should not be too large to not unnecessarily enlarge the computational domain and hence increase the computation time.

The first widely used numerically stable, second-order accurate ABC was formulated by Mur [33], but the disadvantage is that it only absorbs well-specific polarizations. Berenger then introduced a highly efficient ABC [24], the so-called perfectly matched layer (PML), based on formulation of split fields, which are orthogonal to each other, resulting in 12 coupled first PDEs with loss parameters. The big advantage of the PML is the universal applicability to cut the simulation volume consisting not only of isotropic lossless dielectric material, but also anisotropic, nonlinear, and dispersive materials, just to mention some. Later, a formulation with stretched coordinates was developed. This gave the possibility to also apply the PML concept to other orthogonal coordinate systems than the Cartesian one. The split-field PML proposed by Berenger and its stretched analog is based on a mathematical model. The uniaxial PML (UPML) based on an anisotropic PML yields a physical formulation [25]. The performance of Berenger's split-field PML and the UPML is the same, but they differ in Gauss' law. Since the use of a classical tensor coefficient can lead to large reflection errors in the low frequency range, the complex frequency shifted (CFS) PML overcomes this problem [26,27]. However, Roden and Gedney implemented a more efficient version using stretched coordinates from Berenger based on a recursive convolution formulation, the convolutional PML (CPML). The result is more efficient, more accurate than the UPML, and provides better applicability to terminate regions with generalized media.

The stretched coordinates are built with the CFS tensor coefficient s_μ. Here, $s_\mu = \kappa_\mu + \frac{\sigma_\mu}{\alpha_\mu + i\omega\varepsilon_0}$, where the parameter ranges are $\kappa_\mu \geq 1$, $\sigma_\mu \geq 0$, and $\mu = x, y, z$. Berenger's stretched coordinates metrics is obtained for $\kappa_\mu = 1$ and $\alpha_\mu = 0$. The update equation within the CPML material exemplarily for the x-component of the electric field yields

$$E^{n+1}_{x\,i+\frac{1}{2}jk} = E^{n}_{x\,i+\frac{1}{2}jk} - \frac{\Delta t}{\varepsilon_0 \varepsilon_{i+\frac{1}{2}jk}\mu_0}\left(\frac{1}{\kappa_{yj}\Delta y}\left(H^{n+\frac{1}{2}}_{z\,i+\frac{1}{2}j+\frac{1}{2}k} - H^{n+\frac{1}{2}}_{z\,i+\frac{1}{2}j-\frac{1}{2}k}\right) - \frac{1}{\kappa_{zk}\Delta z}\left(H^{n+\frac{1}{2}}_{y\,i+\frac{1}{2}jk+\frac{1}{2}} - H^{n+\frac{1}{2}}_{y\,i+\frac{1}{2}jk-\frac{1}{2}}\right)\right) - \frac{\Delta t}{\varepsilon_0 \varepsilon_{i+\frac{1}{2}jk}\mu_0}\left(\Psi^{n+\frac{1}{2}}_{E_{x,y}\,xi+\frac{1}{2}jk} - \Psi^{n+\frac{1}{2}}_{E_{x,z}\,xi+\frac{1}{2}jk}\right) \tag{2.32}$$

with

$$\Psi^{n+\frac{1}{2}}_{E_{x,y}\,xi+\frac{1}{2}jk} = b_{yj}\Psi^{n-\frac{1}{2}}_{E_{x,y}\,xi+\frac{1}{2}jk} + \frac{c_{yj}}{\Delta y}\left(H^{n+\frac{1}{2}}_{z\,i+\frac{1}{2}j+\frac{1}{2}k} - H^{n+\frac{1}{2}}_{z\,i+\frac{1}{2}j-\frac{1}{2}k}\right) \tag{2.33}$$

and

$$\alpha_\mu = \alpha_{\mu,\max}\left(1-\frac{x}{d}\right)^{m_\alpha}$$

$$b_\mu = e^{-\left(\frac{\sigma_\mu}{\kappa_\mu}+\alpha_\mu\right)\frac{\Delta t}{\varepsilon_0}}$$

$$c_\mu = \frac{\sigma_\mu}{\kappa_\mu\left(\sigma_\mu+\kappa_\mu\alpha_\mu\right)}\left(b_\mu-1\right), \tag{2.34}$$

where $0 \leq x \leq d$, the boundary to the CPML is located at 0, d is the thickness of the CPML, and m_α determines the scaling order. A detailed derivation can be found in Ref. [27].

Within this chapter, open boundary conditions are used, utilizing a CPML with parameters as given in Table 2.1, if not mentioned otherwise.

Table 2.1 List of CPML parameters

CPML parameter	Value
$\alpha_{\max}$	0.05
$\sigma_{\max}$	0.8
$\kappa_{\max}$	5
m	3
m_α	1

Note: For more information, refer Ref. [27].

2.3 Nanoscale Optical Resonators

There are various kinds of optical resonators. This chapter concentrates on microcavities, optical resonators in the (sub-) micrometer range. Optical resonators can be classified into two large groups due to distinct confinement methods: first, material without a photonic bandgap (PBG) like a microtoroid, an MD, or a slab waveguide, where the confinement of the light is purely relying on the principle total internal reflection (TIR), and second, a PBG material with a defect to confine the light, like a PhCC in a 2D dielectric slab, pierced with air holes. Here, in the slab plane confinement is due to Bragg reflection while TIR causes the confinement perpendicular to the slab.

In the following, an MD resonator and a planar PhCC, consisting of a dielectric slab with air holes in a hexagonal lattice and omitting three air holes inline in ΓK direction, representing the cavity, are discussed. For the first case, an analytic solution holds and is discussed shortly. For the latter case, only numerical analysis is done.

2.3.1 The MD Resonator

The MD resonator is a good example for a resonant structure without a PBG, because periodicity is not provided. As will be seen in this chapter, the electromagnetic problem of a (perfectly) cylindrical dielectric object can be solved analytically, if one omits the post on which the MD is standing to separate the dielectric layer forming the disk from the substrate material. As will be seen later, the post will not affect the field distribution in the MD. The ratio between thickness of the membrane and radius of MDs can thereby vary over some orders of magnitude [28,29].

The spatiospectral response is now derived for monochromatic excitation. Regarding Maxwell's equations in their time-harmonic form, utilizing the curl equations and assume, that no free charges or currents are present ($\boldsymbol{j}(\boldsymbol{r}, t) \equiv 0, \rho(\boldsymbol{r}, t) \equiv 0$) in a nonmagnetic material with $\mu_r(\boldsymbol{r}) = 1$, one can rewrite equations (2.1) and (2.2) using equation (2.9) to

$$\nabla \times \boldsymbol{E}(\boldsymbol{r},\omega) = -\frac{i\omega}{\varepsilon_0 c_0^2} \boldsymbol{H}(\boldsymbol{r}) \cdot e^{-i\omega t} \tag{2.35}$$

$$\nabla \times \boldsymbol{H}(\boldsymbol{r},\omega) = -i\omega\varepsilon_0\varepsilon_r(\boldsymbol{r})\boldsymbol{E}(\boldsymbol{r}) \cdot e^{-i\omega t}. \tag{2.36}$$

Decoupling equations (2.35) and (2.36) results in wave equations for $\boldsymbol{E}$ and $\boldsymbol{H}$

$$\nabla \times \nabla \times \boldsymbol{E}(\boldsymbol{r}) = \varepsilon_r(\boldsymbol{r}) \left(\frac{\omega}{c_0}\right)^2 \boldsymbol{E}(\boldsymbol{r}), \tag{2.37}$$

$$\nabla \times \varepsilon_r(\boldsymbol{r})^{-1} \nabla \times \boldsymbol{H}(\boldsymbol{r}) = \left(\frac{\omega}{c_0}\right)^2 \boldsymbol{H}(\boldsymbol{r}), \tag{2.38}$$

which can be used to derive the Helmholtz equation, exemplarily for the electric field $\boldsymbol{E}$ (in areas of constant $\varepsilon_r(\boldsymbol{r})$) using standard vector calculus:

$$(\Delta + n^2 \boldsymbol{k}^2)\boldsymbol{E}(\boldsymbol{r}) = 0. \tag{2.39}$$

The dispersion relation $\boldsymbol{k}^2 = \left(\frac{\omega}{c}\right)^2$, is used here, $\boldsymbol{k}$ is the wave vector, $\Delta = \Sigma_\mu \partial_\mu^2$, $\mu = x, y, z$, denotes the Laplacian operator and it is assumed, that $\varepsilon_r(\boldsymbol{r}) = n^2$, since the focus lies on investigating pure dielectric materials without resonances or losses.

Without loss of generality, one first considers an infinitely extended dielectric slab in the x–y plane, whose permittivity is piecewise constant at the interfaces, a finite thickness t in z-direction and propagating fields in the x-direction. The thickness is an important quantity if one is interested in single-mode operating devices, whose basis is a planar waveguiding system. Light is kept within the thin dielectric slab due to TIR. Using the fields propagating in x-direction (with wave vector k_x) and equation (2.39), one gets the TE-case ($E_y = E_y(z) \cdot e^{ik_x x}$)

$$\Bigg(\partial_z^2 + \underbrace{\left(\varepsilon_r \left(\frac{\omega}{c_0}\right)^2 - k_x^2\right)}_{(*)}\Bigg)\bar{E}_y = 0. \tag{2.40}$$

For the permittivity, $\varepsilon_r = \varepsilon$ for $|z| < \frac{t}{2}$ and $\varepsilon_r = 1$ for $|z| > \frac{t}{2}$ applies. So for both spatial domains, one can rewrite the (*)-marked part in equation (2.40) into

$$(*) = \kappa^2 = \varepsilon\left(\frac{\omega}{c_0}\right)^2 - k_x^2, \text{ for } |z| < \frac{t}{2} \tag{2.41}$$

$$(*) = (i\gamma)^2 = i^2\left(k_x^2 - \left(\frac{\omega}{c_0}\right)^2\right), \text{ for } |z| > \frac{t}{2}. \tag{2.42}$$

Since this problem has spatial symmetry in z-direction, symmetric as well as antisymmetric solution can be derived. To obtain exponentially decaying solutions outside and standing waves inside the slab, κ and γ need to be larger than 0. Using the boundary conditions at the interface of the two different dielectric media (the field and its derivative with respect to space have to be continuous), one derives symmetry-dependent transcendental equations in order to calculate the wave vector

$$\tan\xi = \frac{\gamma d}{2\xi}, \text{ for even modes and} \tag{2.43}$$

$$\cot\xi = -\frac{\gamma d}{2\xi}, \text{ for odd modes.} \tag{2.44}$$

where $\xi = \frac{\kappa d}{2}$. Rewriting $\gamma d = d^2\left(\left(\frac{\omega}{c_0}\right)^2(\varepsilon-1)-\kappa^2\right)$ using some algebraic transformations, the graphical solutions for equations (2.43) and (2.44) result in Fig. 2.5, where every intersection is a valid value for κ and γ for the given system (the used parameters are given in the caption). A discrete set of valid eigenvalues emerges, depending on the slab thickness t. More thick devices support more modes.

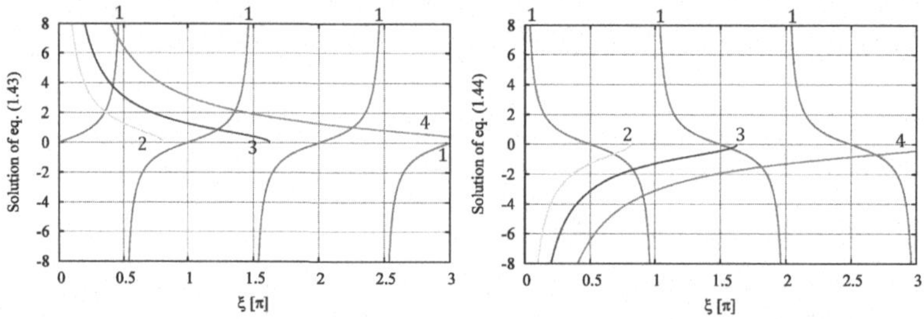

Figure 2.5 Graphical solution of equations (2.43) and (2.44). The intersections show the valid values for κ, hence for γ. The RHS is shown by curves marked 1. Used parameters are ε = 11.56, λ_0 = 1000 nm, t = 250 nm (2), 500 nm (3), and 1000 nm (4). Single mode operation is only possible for t = 250 nm. For increasing thickness, more modes (higher harmonics) are occupied.

For the in-plane solution one considers a dielectric cylinder with permittivity $\varepsilon_r = \varepsilon$ and radius R, infinitively extended in z-direction, to calculate the field distribution, hence to understand the in-plane confinement of electromagnetic radiation. Since the slab is assumed to be in the x–y plane, one now uses the Helmholtz equation, equation (2.39), and transforms the problem into cylindrical coordinates $(x, y) \rightarrow (r, \phi)$ yielding

$$\left(\frac{1}{r}\partial_r(r\partial_r)+\frac{1}{r^2}\partial_\phi^2+\varepsilon\left(\frac{\omega}{c_0}\right)^2\right)E_y(r,\phi)=0. \tag{2.45}$$

Here, one uses a product ansatz for $(E)_y\ (r,\ \phi) = R(r)\cdot\Phi(\phi) = R(r)\cdot e^{(\pm iM\phi)}$ (ϕ-harmonic with azimuthal mode order M; the ± occurs because of double degeneracy for propagation clockwise or counterclockwise) to separate the variables r and ϕ. Utilizing the product ansatz and the common substitution to a unitless, effective radius ρ, $r \rightarrow \frac{c_0}{n\omega}\rho$, one finally finds the differential equation for the radial component of $(E)_y$ to be a Bessel differential equation

$$\left(\partial_\rho^2+\frac{1}{\rho}\partial_\rho+\left(1-\left(\frac{M}{\rho}\right)^2\right)\right)R(\rho)=0, \tag{2.46}$$

whose solutions are the Mth order Bessel function of first kind $J_M(\rho)$ inside the cylinder ($\rho \leq R$) and a (complex) superposition of the Mth order Bessel functions of first kind $J_M(\rho)$ and second kind $Y_M(\rho)$, the Mth order Hankel function $H_M(\rho) = J_M(\rho) + iY_M(\rho)$, outside the cylinder ($\rho > R$). The radial mode order is coded in the effective radius ρ, where the wave vector enters as $k_N = \frac{n\omega N}{c_0}$. The order N now depends on which solution one chooses from the equation, resulting from the continuity at the interface between the disk and the environment. Figure 2.6 shows the analytical field distributions for a cylinder with metallic environment ($(E)_y \equiv 0$ outside) and different combinations of M and N. As can be seen, the field is mostly concentrated in the rim of the cylinder (except for large radial mode orders and/or small azimuthal mode orders). For a dielectric environment, the field is expected to be localized even closer to the interface, because the field leaks out of the resonator and decays exponentially in the environmental region. That motivates the assumption that the post, on which the MD is standing freely, has no influence on the field distribution.

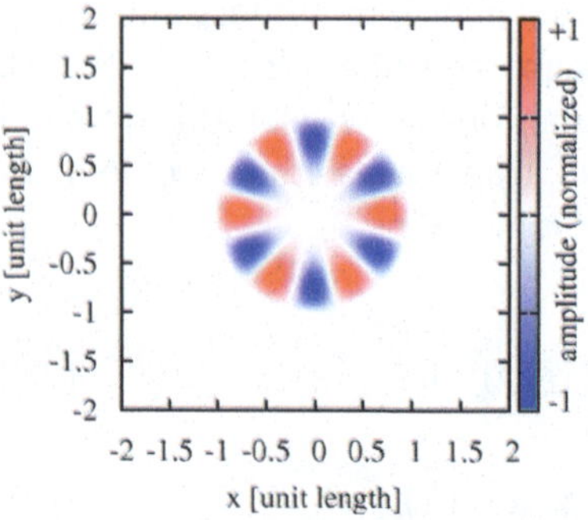

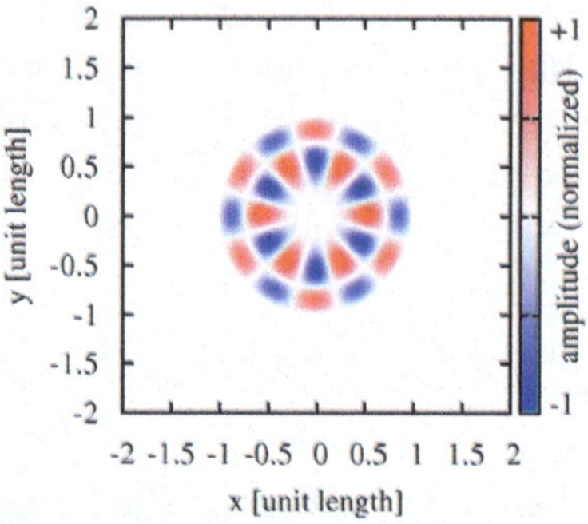

Figure 2.6 Analytic solution of equation (2.46) for the case of a metallic environment for different cases of N and M for $TE_{N,M}$ (here z-component of the magnetic field). A unit radius is assumed for the disk and the amplitude is normalized to 1. Left: $TE_{1,6}$. Right: $TE_{2,6}$. The center of the disk has zero field, even for higher mode orders in radial direction.

2.3.1.1 Simulation of an MD resonator

In the previous part it was shown that the spatial profile of the field confined in an MD resonator is concentrated near the rim of the disk. The advantage is that the simulation of a device can now be performed without the standing post, since no field is concentrated in the center of the MD. To perform a simulation, it is therefore convenient to place an exciting light source near the MD-environment interface inside the MD. Multiple TIRs at the material-vacuum interface cause the standing wave inside the resonator. Since an MD resonator provides a whole bunch of different modes with different quantum numbers M and N, referred to the azimuthal and radial mode order, respectively, a broad band excitation is used to spectrally excite a broad range of M and N.

As an example, an MD resonator in the x–y plane is numerically analyzed. Here, a 2D simulation is performed, where the z-direction is treated periodically and the simulation domain in the x–y plane is truncated with a CPML. The cylindrical dielectric resonator consists of material with a dielectric constant of $\varepsilon = 11.56$ (for instance for GaAs bulk material) with a radius of $R = 1\ \mu m$. As in the previous part, the focus lies in TE-like modes. Hence, a z-polarized magnetic point source is positioned at $(0.9R, 0, 0)$. Monitoring the TE-component of the magnetic field at $(0.9R, 0, 0)$ results in Fig. 2.7, left. The H_z-component in arbitrary units is shown over the time in picoseconds. The excitation pulse can be clearly identified. Also, the reflections from the MD-vacuum interface can be seen in the first 600 fs. Letting the system evolve some more time, the resulting broadband spectrum is illustrated in Fig. 2.7, right. The absolute value of the H_z-component in arbitrary units is plotted over the frequency in THz. Here, equidistant peaks with shortest distance can be identified as the whispering gallery modes with $N = 1$ and different M, increasing from low to high frequency. The equidistant peaks with larger distance between each other are the whispering gallery modes (WGMs) with higher azimuthal mode order. The discrete Fourier spectrum is not converged, yet, due to the short simulation time and the bunch of different modes with different quality factors (Q-factor). However, the spectral peak positions can be estimated very well. As an alternative way to get spectral information, the filter diagonalization method (FDM) harmonic inversion [30,31] can be used. This method decomposes a time signal into a finite sum of exponentially decaying oscillations and can efficiently extract the amplitude, frequency, decay rate, and phase of all involved modes and gives very good results after simulation times shorter than the resonator decay time (see Section A.1).

Figure 2.7 Left: Temporal response of the H_z-component of a microdisk resonator surrounded by air (normalized to 1, monitored close to the rim of the microdisk). After a short incident pulse, multiple reflections arrive at the monitoring position due to the TIR, propagating along the circumference of the microdisk before the waves start to superimpose to a strongly disordered signal due to all the different confined modes within the system. Right: Spectral response (segment) of a microdisk resonator. Equidistant peaks of different amplitude and distance belong to confined modes with different radial and azimuthal quantum numbers N and M, respectively.

For instance, in an experiment, one can read out the frequency difference between the $N = 1$ peaks and approximate the MD radius. Assuming that a WGM with a high azimuthal mode order M has its field maxima almost at the rim of the MD, then the optical pathway for this mode along the rim equals c_0M/f and the radius can be determined via $R = \frac{c_0}{2\pi n_{\mathrm{MD}}\Delta f}$, where Δf is the frequency difference between neighboring modes with a radial quantum number $N = 1$. For the used example here, $\Delta f \approx 15$ THz which results in a radius $R \approx 936$ nm, which is quite a good result for the performed approximations. Nevertheless, the higher the azimuthal quantum number, the better the approximation.

To get information about the spatial field distribution of the device, a narrowband excitation is performed. Aiming the modes at 186, 200, 203, and 221 THz, one simulation per frequency is performed with a quasi-CW excitation. After the pulse has vanished, the field at every grid point is recorded over one period. As a result, Fig. 2.8 shows the corresponding field distributions from left to right. As expected, the equidistant peaks with the shortest distance belong to the $N = 1$ quantum number and with increasing frequency, the quantum number M increased by 1 as well from $M = 9$ to $M = 10$. The other two modes have quantum numbers $N = 2$ and two subsequent azimuthal mode orders $M = 7$ and $M = 8$.

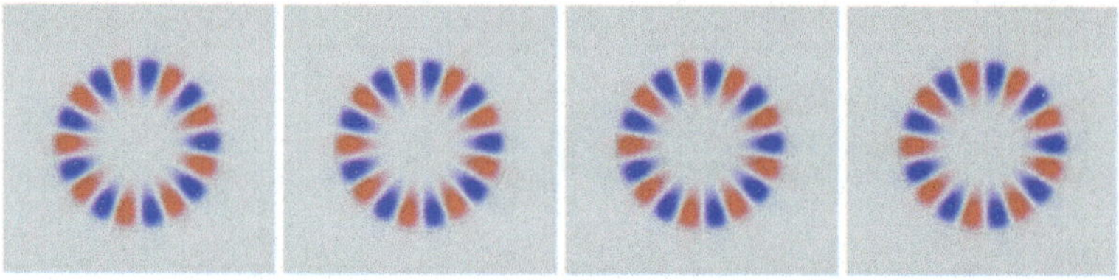

Figure 2.8 Normalized TE spatiospectral response (H_z) of a microdisk resonator (2D simulation). The amplitude of the mode is color-coded (red 1, blue −1) and the material-air interface is illustrated with thin gray lines. From left to right: $\mathrm{TE}_{1,9}$ at 186 THz, $\mathrm{TE}_{1,10}$ at 200 THz, $\mathrm{TE}_{2,7}$ at 203 THz, and $\mathrm{TE}_{2,8}$ at 221 THz.

2.3.2 The Photonic Crystal Cavity

In nature, the concept of photonic crystals is not a new feature. One of the most famous representatives is the butterfly *Morpheus rhodopteron* with its shiny blue wings or the opal, which can please every woman's temper. The effect is no biological or chemical pigmentation, rather a physical one. Multiple reflections of incident light on successive layers lead to constructive and destructive interference, depending on wavelength and angle of incidence.

Within the last decades, PhCs have inspired great interest due to their applications not only in research, but also in various commercial areas accessible in the future such as optical information processing. In this case, electronic manipulation of data is then substituted by optical or optoelectronic manipulation [32]. Scientists all over the world deal with the experimental realization of optical and optoelectronic devices, miniaturization, integrability, and usability. Therefore, PhC is a promising candidate, when once a device is characterized, the characteristics can be scaled up and down in size, because Maxwell's equations (2.14) and (1.15) scale with length [33].

In a PhC, the optical analog to a conductive crystal for charge carriers, photons behave in the same manner as charge carriers do in the conductive crystal [34–36]. The atomic (molecular) lattice is substituted by different dielectric materials, which form the crystal, hence the potential for the photons. The compound material can then form a PBG for the electromagnetic wave for a specific frequency range. The prohibition of propagation can be manipulated to be in all directions, for all polarizations, for smaller or larger frequency ranges, depending on the designing rules applied to the compound material [33,37,38]. Thus, manipulation of light can be performed on the nanoscale, which is the basic key for small-sized photonic devices for future quantum information processing computers.

To derive an equation to access the fields in arbitrarily shaped and arbitrarily composited dielectric structures, one explicitly needs spatial variation of the dielectric constant even on small length scales (in the range of the material wavelength, or even smaller), therefore $\varepsilon_r \equiv \varepsilon_r(r)$. Attention needs to be paid in equation (2.13), since the divergence is taken from the dielectric displacement field $\boldsymbol{D}(\boldsymbol{r}, t)$ to get the charge density. Now, inserting equation (2.9) results in

$$\nabla \cdot \boldsymbol{D}(\boldsymbol{r},t) = \nabla \cdot (\varepsilon_r(r)\boldsymbol{E}(\boldsymbol{r},t)) = \rho(\boldsymbol{r},t), \tag{2.47}$$

while equations (2.12), (2.14), and (2.15) remain unchanged. Derivation of the wave equation for the time-harmonic magnetic field

$\boldsymbol{H}(\boldsymbol{r}, t) = \boldsymbol{H}(\boldsymbol{r}) \cdot e^{-i\omega t}$ and electric field $\boldsymbol{E}(\boldsymbol{r}, t) = \boldsymbol{E}(\boldsymbol{r}) \cdot e^{-i\omega t}$, assuming no sources of electromagnetic radiation in the standard way [8,39,40], yields [33]

$$\nabla\times(\varepsilon_r(\boldsymbol{r})^{-1}\nabla\times\boldsymbol{H}(\boldsymbol{r}))=\hat{\Theta}_H\boldsymbol{H}(\boldsymbol{r})=\left(\frac{\omega}{c_0}\right)^2\boldsymbol{H}(\boldsymbol{r}), \tag{2.48}$$

$$\varepsilon_r(\boldsymbol{r})^{-1}\nabla\times\nabla\times\varepsilon_r(\boldsymbol{r})^{-1}\boldsymbol{D}(\boldsymbol{r})=\hat{\Theta}_D\boldsymbol{D}(\boldsymbol{r})=\left(\frac{\omega}{c_0}\right)^2\varepsilon_r(\boldsymbol{r})^{-1}\boldsymbol{D}(\boldsymbol{r}). \tag{2.49}$$

Maxwell's equations are now rewritten as an eigenvalue problem with the linear operator $\hat{\Theta}_H(\hat{\Theta}_D)$, corresponding eigenvalue $\left(\frac{\omega}{c_0}\right)^2$ and eigenvector $\boldsymbol{H}(\boldsymbol{r})$ ($\boldsymbol{D}(\boldsymbol{r})$). Equation (2.49) is divided by an extra $\varepsilon_r(\boldsymbol{r})$ to keep $\hat{\Theta}_D$ Hermitian. This is an important point because the eigenvalues are then real and positive semi-definite. Additionally, the fact that $\hat{\Theta}_X$ is a linear operator cannot be underrated, since (linear) superpositions of found solutions are also solutions. For mathematical convenience, equation (2.48) is normally used with $\hat{\Theta}_H = \hat{\Theta}$. Access to $\boldsymbol{D}(\boldsymbol{r}, t)$ is gained by solving equation (2.48), and subsequently using equation (2.11) in its time-harmonic form. The solutions of equation (2.48) have some more interesting and useful properties. For example, the analogy to quantum mechanics, Schrödinger's equation, which allows to transfer some properties from quantum mechanics directly to classical electrodynamics (find a comparison in Table 2.2), like symmetries and orthogonality of the solutions of equation (2.48). The latter one is the consequence of the Hermitian property of $\hat{\Theta}_H$.

Table 2.2 Quantum mechanics vs. electrodynamics: a comparison

	Quantum mechanics	Electrodynamics
Field	$\Psi(r)\cdot e^{-\frac{iEt}{\hbar}}$	$H(r)\cdot e^{-i\omega t}$
Eigenvalue problem	$\hat{H}\Psi = E\Psi$	$\hat{\Theta}H = \left(\frac{\omega}{c_0}\right)^2 H$
Hermitian operator	$-\frac{\hbar^2}{2m}\nabla^2 + V(r)$	$\nabla\times\varepsilon_r(r)^{-1}\nabla\times$
Potential	$V(r)$	$\varepsilon(r)$

Source: Adapted from Ref. [33].

Due to the similarities between the electronic and electromagnetic eigenvalue problem, methods known from electronic system can be used for the electromagnetic problem, too, like a tight-binding formalism [41,42]. This answers directly the question for the origin of the PBG.

Last but not least one has to mention the photonic density of states (PhDoS), which gives the number of allowed photonic modes per frequency interval $\delta\omega$. Inside a PBG (yellow region in Fig. 2.9), the density of states is zero, because no propagating states are allowed within this frequency range. Photonic states that would be located spectrally in the PBG region are pushed out of this gap and are localized above and below the PBG (blue areas in Fig. 2.9). But modes with a complex wave vector are allowed in the PBG region, since these modes are not propagating modes, but evanescent ones due to the space-harmonic fields $\boldsymbol{H}(\boldsymbol{r}, t) = \boldsymbol{H}(t) \cdot e^{i\boldsymbol{k}\cdot\boldsymbol{r}}$ ($\boldsymbol{E}(\boldsymbol{r}, t) = \boldsymbol{E}(t) \cdot e^{i\boldsymbol{k}\cdot\boldsymbol{r}}$). Evanescent modes in the PBG are localized with a defect in the periodicity of the crystal, as is depicted as the red line in Fig. 2.9. The figure shows how one can make use of the PBG and the enlarged PhDoS to store light in a small volume (comparable to the cubic material wavelength of the confined photons) over a long time, when introducing a defect in the PhC—a PhCC. In a PhCC, one makes use of the increased density of states around the PBG. Evanescent modes are allowed to penetrate the PBG region. In a defect, like a missing hole in a periodic arrangement of air holes in a 2D dielectric slab, this evanescent mode can manifest itself and become a highly confined mode, a cavity mode. Cavity modes reflect the symmetry of the PhC, for example a rotational symmetry of the lattice leads to degenerate modes. These modes have the same frequencies, but are not necessarily orthogonal. This degeneracy can be lifted by destroying a symmetry property, for example the air holes are not perfectly circular shaped, but more of elliptical shape [43]. Then the lifted degeneracy manifests itself in polarization properties of the harmonic modes and the degenerate spectral line splits into two polarization-dependent eigenmodes.

Nowadays, this concept has already been applied to various kinds of PhCs. Few examples are the 2D dielectric slabs with periodic arrangements of air holes exhibiting a hexagonal symmetry, the 3D Yablonovite, a woodpile stack of layers of parallel oriented dielectric columns, alternating in height in a 90° angle or the inverse opal structure.

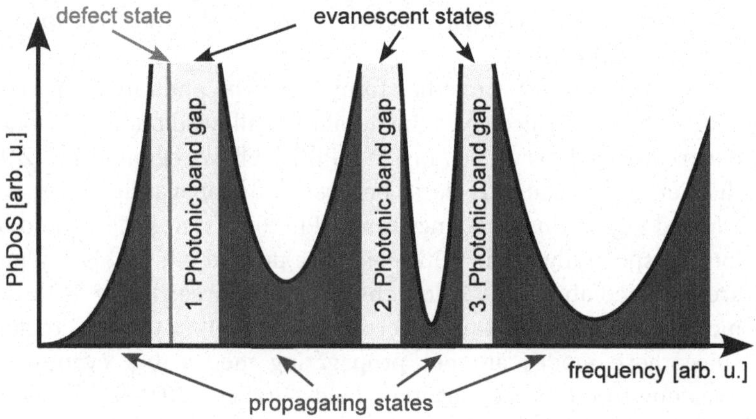

Figure 2.9 Photonic density of states (PhDoS) vs. frequency in arbitrary units. The blue region shows propagating modes with a real wave vector $\boldsymbol{k}$. Near the PBGs (1, 2, and 3 shown in figure), the photonic density of states raises, since the states from the band are pushed outside, hence increasing the PhDoS. In these photonic gaps, only modes with an imaginary wave vector are allowed, since they represent evanescent modes, which only decay and show no propagation. In these regions, one can confine the light introducing a defect, like a point or line defect, thus increasing the density of states for a certain frequency matching the resonance of the defect (vertical line shown in PBG 1). Adapted from Ref. [33].

In the following, the focus lies on a planar PhC, consisting of a dielectric slab, which is pierced with air holes in a hexagonal lattice. In such a system, confinement is based on fulfilling the Bragg condition of reflection, like in a distributed Bragg reflector. As a result of periodicity, energetically forbidden ranges for the electromagnetic waves emerge in the dispersion relation $\omega(\boldsymbol{k})$, called the PBG. Figure 2.10, left, shows a typical band structure, calculated with MIT Photonic Bands (MPB) [44], a frequency-domain method based on plane wave expansion. The PhC has a lattice constant of a = 300 nm, a hole radius r = 0.3 × a, and the dielectric permittivity of the slab is ε = 10.595, consistent with a GaAs bulk material at T = 4 K [45]. A spatial (reciprocal) discretization of 64 points per lattice constant (between high-symmetry points) is used together with a numerical error of $|\Delta f - f_0|$ $= 10^{-5}$ per iteration to ensure accurate results. The calculation is done for a 2D system, thus infinite extension perpendicular to the air hole plane. In this case, the solution of equation (2.48) can be split into a pure TE- and TM-component. Normalized frequencies are plotted against the wave vector along the Brillouin zone (see Fig. 2.10, right, for an illustration of the real space and reciprocal space lattice together with the irreducible Brillouin zone, marked with a blue shade). Frequencies normalized to the lattice constant a are plotted over the wave vector $|\boldsymbol{k}|$ according to the irreducible Brillouin zone. Using normalized frequencies is a good measure, since equation (2.48) scales with length. The light cone is shown with a black and the TE (TM) bands are shown in red (blue). Note the gaps between the first two bands for TE and TM, and especially the overlapping gap, marked with a gray shade. In this region, neither TE nor TM modes are allowed, therefore this is called a complete band gap. In this region, propagation of modes is forbidden and the wave vector is purely complex.

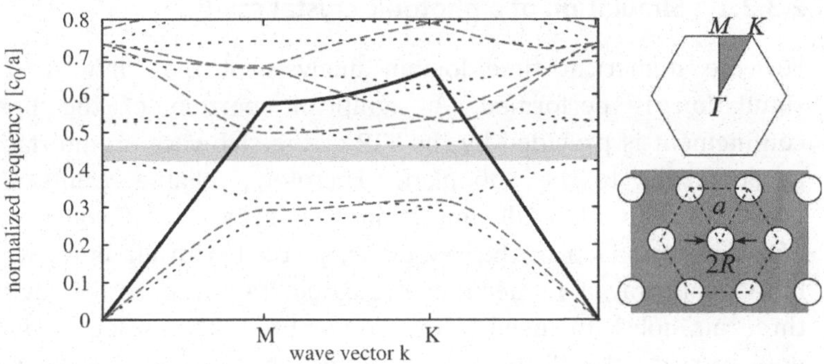

Figure 2.10 Left: Band structures for a 2D PhCC with a hexagonal lattice of air holes in a dielectric slab. Points of high symmetry are marked with common labels. In the band structure (normalized frequencies in c_0/a, over the wave vector $|\boldsymbol{k}|$) is illustrated, as calculated with MPB [44]. Dashed curves: TE-modes. Dotted curves: TM-modes. Solid curve: Light cone. A complete band gap (TE- and TM-band have a PBG at the same frequencies). Right: Real space lattice (bottom) with air holes (white) and dielectric slab (shaded) and the irreducible Brillouin zone (shaded) in the reciprocal space lattice (top).

In reality, one always deals with a 3D device. The confinement in the third dimension is provided by TIR. The thickness t of the device is normally chosen to be in the range of $t \approx \frac{\lambda}{2}$, λ is the vacuum wavelength, divided by the refractive index of the desired material, to have only the lowest optical mode in this direction occupied (single mode operation). Thus, all components mix and one cannot uncouple the different polarizations. Hence, only TE- and TM-like band gaps occur. The introduction of a defect in the periodic lattice concentrates electromagnetic energy due to the increased PhDoS at certain frequencies. Spatially, the light is confined in the defect region, where it can be concentrated in the material part, referring to modes located energetically near the material band (band below gap) or in the air part, referring to modes located energetically close to the air band (band above gap) [33]. In the following, a line defect is introduced into the hexagonal lattice of air holes and the arising PhCC is numerically analyzed.

2.3.2.1 Simulation of a photonic crystal cavity

For the numerical time-domain analysis of the PhCC, a 3D simulation is performed. A significant portion of the light confinement is provided by the TIR at the slab-vacuum interface, perpendicular to the slab plane. Therefore, reliable results are only obtained with a full simulation. Exemplarily, a PhC consisting of a dielectric slab in the x–y plane, pierced with air holes in a hexagonal array, is numerically evaluated. To introduce the cavity, three air holes arranged inline are omitted. This defect should now increase the PhDoS and confine light efficiently. Since the light is expected to be confined in the defect region, the system is excited with a broadband light source placed inside the defect. Since the air hole array has hexagonal symmetry, the distance between next-neighboring air holes is always exactly one lattice constant a. A discretization on a Cartesian coordinate system using a spatially uniform discretization has the disadvantage when the distance in x-direction is a, the distance in y-direction is $\sqrt{3a}$. Hence, centers of air holes and also the discretized circular shape of the holes are different from hole to hole and therefore the hexagonal and circular symmetry is not provided. To overcome crucial point of the device itself, a nonuniform grid is used. For the discretization, it then follows $dy = \frac{\sqrt{3}}{2}dx$ to ensure to have exactly the same number of grid points in each direction between neighboring air holes.

To investigate TE-like modes, a z-polarized magnetic point dipole is placed in a position without any symmetry to ensure that the spatial overlap of the source and the modes of interest is non-zero. To estimate the spectral range of the source, a band structure calculation can be performed, using the plane-wave-based method of the MIT Photonic Bands package [44]. Monitoring the field inside the defect also at a nonsymmetric position ensures measuring the contributions of all resonant modes of the defect. For the slab, a dielectric constant of $\varepsilon = 10.595$ (similar to GaAs bulk material at $T = 4$ K [45]) is used. The PhC parameters are chosen to be similar to the ones used in Ref. [46] to have a TE-like PBG in the desired frequency range. Therefore, the lattice constant is $a = 300$ nm, the slab thickness is $t = 180$ nm, and the air hole radius is $r = 0.6 \times a$. ABC (CPML) are applied in all three directions.

The temporal response looks like in Fig. 2.11, left. As can be seen, straight after the excitation pulse vanishes, multiple reflected waves arrive at the monitor position due to the different layers of air holes around the defect region, where the waves are Bragg reflected. The frequencies the band gap propagate out and are absorbed by the CPML, thus the signal decays fast in the beginning and only the line defect are left. A spectral analysis of the time signal shows several modes in the band gap. It is shown that for a cavity of this kind, up to 6 modes are supported [47], depending on the ratios of the parameters, which characterize the PhCC. Here, the spectral response, see Fig. 2.11 right, shows 5 distinct peaks, plus a shoulder on the right side of the central peak which is identified to be the sixth mode between 275 and 340 THz. The wiggles on the left and right side of the gray shade, the PBG, are typical artifacts from regions outside the photonic band. To investigate the mode structure of the mode at the edge of the low-frequency band edge (which intrinsically has the highest Q-factor), the system is excited with a narrowband pulse around the resonance at $f = 279$ THz. Figure 2.12, left, shows the spatio-spectral response of the TE-like component (here: E_x, E_y, and H_z from left to right). As can be seen, the field is concentrated in the dielectric slab, therefore are called material modes. In contrast, the mode at the high-frequency band edge at f = 337 THz is concentrated in the air region (see Fig. 2.12, right) and thus such modes are called air modes.

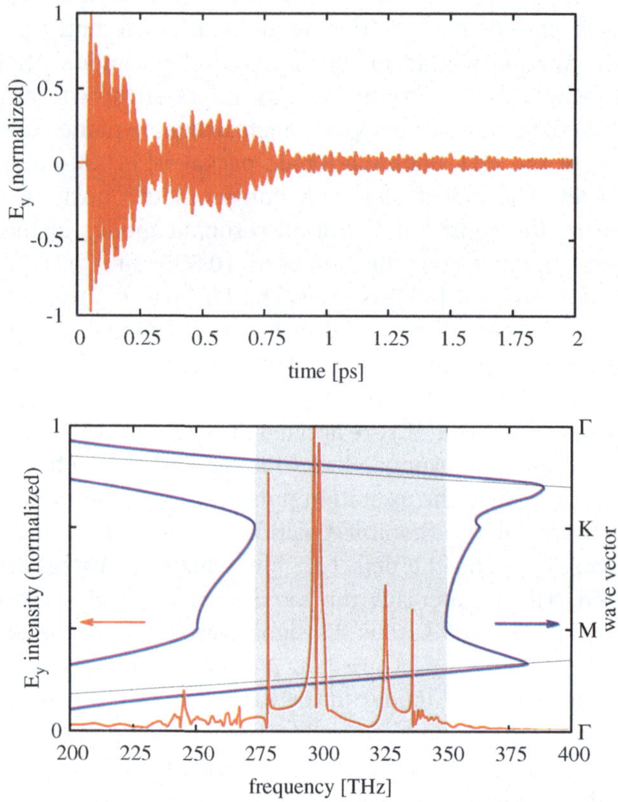

Figure 2.11 Left: Temporal response of the L3-type defect cavity in a PhC with a hexagonal lattice of air holes in a dielectric slab (normalized to 1, monitored within the defect area). After the incident pulse, multiple waves from the different layers of air holes around the defect arrive in the defect area due to Bragg reflection. After a short time (≈800 fs) die leaky modes propagated out of the simulation domain and only the defect modes are left. Right: Spectral response (red), covering the first band gap (gray-shaded region). Six modes are identified, as expected [47]. The wiggles on the left and right side of the band gap are artifacts due to the bands. Blue: First two photonic band. The black line illustrates the light cone.

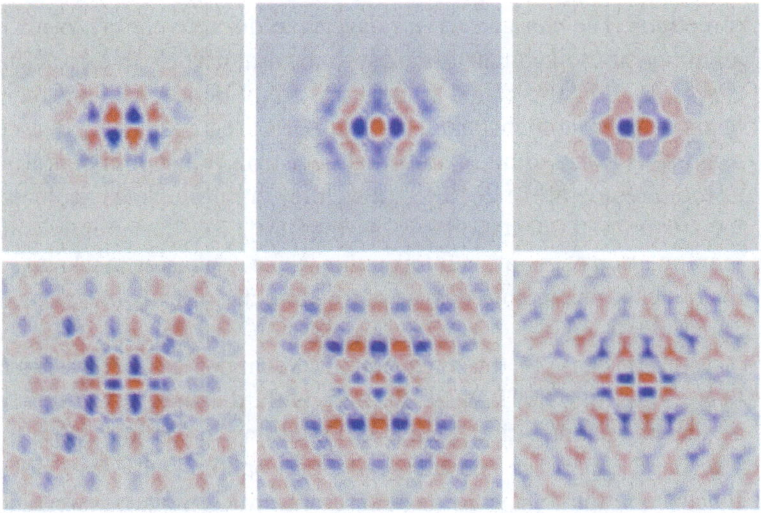

Figure 2.12 Normalized TE-like spatiospectral response of the of the L3-type defect cavity in a PhC with a hexagonal lattice of air holes in a dielectric slab. The amplitude of the mode is color-coded (red 1, blue −1) and the material-air interface is illustrated with thin gray lines. Top: Mode at 279 THz (fundamental mode). Bottom: Mode at 337 THz (higher-order mode). Due to the vicinity of the two modes to the dielectric band (lower) and air band (upper), the defect modes concentrate their energy in the dielectric and air region, respectively. From left to right: E_x, E_y, and H_z.

2.4 Active Material

Today with the available technology, semiconductor structures can be made smaller in size than the wavelength of the light, which is used to transfer the energy of the light to the electronic system or manipulate the electronic system in any way. This gives the possibility for compound structures consisting of many different semiconductor materials to use all the different properties at once and use the combined electronic band structure properties. An important representative of semiconductor heterostructures is the QD, for instance, self-assembled QDs with measures from tens to some hundreds nanometers scaling in each direction with a density of up to 10^4 atoms or molecules. Those QDs look like flat pyramidal structures, which confine the charge carriers in all three spatial directions. The charge carriers have then discrete energy levels in all dimensions like in an atomic systems or a particle in a box. That is why a QD is also called an artificial atom. Hence, this effectively leads to a zero-dimensional structure. QDs are already widely used, e.g., in QD lasers, diodes, and single electron transistors operating at room temperature [5,48,49]. Also, they are promising candidates for quantum information processing [50], e.g., for storage and manipulation of quantum information, since dephasing mechanisms for the charge and spin degree of freedom are already investigated and understood well, see for example [51].

A first step is the inclusion of a simple two-level system (TLS) into the FDTD grid, treated with dynamic equations in a second quantization scheme. Equations of motion for the interband coherence and the conduction band density, the optical Bloch equations, provide an additional optical polarization field. Therefore, a self-consistent inclusion is needed, where the optical polarization field, generated by the active material due to excitation of the electromagnetic field, is fed back into the electromagnetic field, calculated by the FDTD method.

In this section, the numerical inclusion of this self-consistency issue due to the integration of the optical Bloch equations is discussed. A nonlocal model, using carrier wave functions distributed on the spatial FDTD grid, is used. The effect of such an atomic resonator in an electromagnetic resonator shows the formation of cavity-polaritons and beating between the cavity and the atomic system.

2.4.1 The Quantum Dot Model

In a first approximation, a QD at low temperatures can be considered as a TLS like an atom with only two energetically allowed levels: a ground state and an excited state. The excited state normally is due to an exciton resonance. An exciton is a bound electron-hole pair, which can be excited optically. The electron from the valence band is excited into the conduction band via the incident optical field and a hole is left. Now, since the two particles have opposite charge (and momentum), the Coulomb interaction bounds them together. Like in an atom, an exciton Bohr radius a_{0_X} is defined as $a_{0_X} = \frac{4\pi\varepsilon_0\varepsilon_r\hbar^2}{\mu\varepsilon^2}$,

e is the elementary charge (defined positive). The excited state will be denoted with $|1\rangle$ while the ground state will be denoted by $|0\rangle$. The main reason for this notation is the possible usage of QD exciton states for quantum information, where the logical bits 0 and 1 are mapped to the quantum-logical qubits $|0\rangle$ and $|1\rangle$ on the Bloch sphere [32] (the quantum-logical states $|0\rangle$ and $|1\rangle$ are referred to the south and north poles of the sphere, respectively). For the derivation of the equations of motions for the microscopic polarization and the conduction band density of the QD, a second quantization scheme is used [52,53]. Therefore, fermionic ladder operators (creation operator $a_\mu^\dagger$, annihilation operator a_μ) are defined. The subscript μ labels the two different bands with V and C for the valence and conduction band, respectively. Since no dispersion of the bands and spin for the electrons is assumed, labels for the wave vector and spin are omitted. An interband coherence, or microscopic polarization, between the ground and excited state is then given by $a_{vb}^\dagger a_{cb}$, where an electron in the valence band is destroyed and electron is created in the conduction band at the same time. The two bands are coupled via a dipole matrix element $\boldsymbol{d}_{\mu\mu'}$. Generally, for a n-level resonator with normalized wave functions Ψ_μ for the charge carriers, $d_{\mu\mu'}$ is defined as

$$\boldsymbol{d}_{\mu\mu'} = -e\int_{\mathbb{R}^3} dr\, \Psi_\mu^*(\boldsymbol{r})\boldsymbol{r}\Psi_{\mu'}(\boldsymbol{r}). \tag{2.50}$$

where r is the spatial coordinate and * means complex conjugation.

To calculate the time dynamics of such a quantum mechanical system, one needs the Hamilton operator $\hat{H}_{\text{tot}}$, which lets the system evolve in time according to the Heisenberg equation of motion of a time-dependent operator $\hat{A}$

$$\frac{d}{dt}\hat{A} = -\frac{1}{i\hbar}\left[\hat{H}_{\text{tot}}, \hat{A}\right]_{-} \partial_t \hat{A}, \tag{2.51}$$

where $[\hat{A},\hat{B}]_- = \hat{A}\hat{B} - \hat{B}\hat{A}$ is the commutator. The last term vanishes for not explicitly time-depending operators. For the QD system, the Hamilton operator with only the free part and light–matter interaction is given by

$$\hat{H}_{\text{tot}} = \hat{H}_{\text{free}} + \hat{H}_{\text{light-matter}}$$

$$= \overbrace{\hbar \Sigma_{\mu} \omega_{\mu} \hat{a}^{\dagger}_{\mu} \hat{a}_{\mu}} + \overbrace{\frac{i}{\hbar} \boldsymbol{E}(t) \cdot \Sigma_{\mu\mu'} \boldsymbol{d}_{\mu\mu'} \hat{a}^{\dagger}_{\mu} \hat{a}_{\mu'}}. \tag{2.52}$$

$\boldsymbol{E}(t)$ is the external light field. For the interband coherence p and conduction band density n, defined as the expectation value of the number operator $\hat{a}^{\dagger}_{c}\hat{a}_{c}$, one gets with equation (2.52)

$$\dot{p} = i\omega_{\text{gap}} p + i\Omega_{\text{R}}(1-2n) - \gamma_{\text{p}} p \tag{2.53}$$

$$\dot{n} = -2\Im\left(\Omega_{\text{R}} p^{*}\right) - \gamma_{\text{n}} n \tag{2.54}$$

The quantity $\Omega_{\text{R}} = \hbar^{-1} d_{\mu\mu'} \cdot E$ is the *Rabi* frequency, ω_{gap} is the gap frequency of the QD material, and γ_p and γ_n are the phenomenologically introduced dephasing rates for the coherence p and the conduction band density $n = n_c$, respectively. The equation of motion for the density is limited to the conduction band density. Using charge carrier conservation, the valence band density n_v is eliminated via $n_c + n_v = 1$. In the calculations, $\gamma_n = 2\gamma_p$ (coherent limit).

2.4.2 Quantum Dots in the FDTD: Back-Coupling to the Electromagnetic Field and Limits

A QD provides a microscopic polarization after optical or electronic excitation. This microscopic polarization couples back to the surrounding light field via the macroscopic polarization, which enters Maxwell's equations in equation (2.11) via an additional term in the displacement field $\boldsymbol{D}(\boldsymbol{r}, t) = \varepsilon_0 \varepsilon_r(r)\, \boldsymbol{E}(\boldsymbol{r}, t) + \boldsymbol{P}(\boldsymbol{r}, t)$:

$$\nabla \times \boldsymbol{H}(\boldsymbol{r},t) = \mu_0 \mu_r(\boldsymbol{r}) \boldsymbol{j}(\boldsymbol{r},t) + \frac{\mu_r(\boldsymbol{r})}{c_0^2} \dot{\boldsymbol{D}}(\boldsymbol{r},t) \tag{2.55}$$

$$= \mu_0 \mu_r(\boldsymbol{r}) j(r,t) + \frac{\mu_r(\boldsymbol{r})}{c_0^2} \left(\varepsilon_0 \varepsilon_r(\boldsymbol{r}) \dot{E}(r,t) + \dot{\boldsymbol{P}}(\boldsymbol{r},t) \right) \tag{2.56}$$

where the macroscopic polarization $\boldsymbol{P}(\boldsymbol{r}, t)$ is calculated by the sum of all microscopic polarizations, the coherences, of the QD, weighted with the corresponding dipole matrix element:

$$P(r,t)=\sum\nolimits_{\mu\mu'} d_{\mu\mu'}\langle a_\mu^\dagger a_{\mu'}\rangle \Psi_\mu^*(r)\Psi_{\mu'}(r). \tag{2.57}$$

This macroscopic polarization feeds the light field which itself again excites the QD.

Since the QD is embedded in the spatial and temporal FDTD grid, the stepping for the update equation needs to be matched. The spatial stepping is done straight forward. For the temporal stepping, an extrapolation method needs to be used. Exemplarily, for a 1D case, the modified update equation for $\boldsymbol{E}(\boldsymbol{r}, t)|_i^{n+1}$ at Yee-cell i and time $(n + 1)\Delta t$, taking the macroscopic polarization $\boldsymbol{P}(\boldsymbol{r}, t)|_i$ from the QD into account, is

$$E_i^{n+1} = E_i^n - \frac{\Delta t}{\varepsilon_0 \varepsilon_i \Delta x}\left(H_{i+\frac{1}{2}}^{n+\frac{1}{2}} - H_{i-\frac{1}{2}}^{n+\frac{1}{2}}\right) - \left(P_i^{n+1} - P_i^n\right) \tag{2.58}$$

As can be seen from equation (2.58), the values for $\boldsymbol{P}(\boldsymbol{r}, t)|_i$ at $t = (n + 1)\Delta t$ are necessary to update the electric field, because both quantities are defined on the same temporal grid. Using a standard fourth-order Runge–Kutta integrator (see Section A.2) [54] to integrate the equations of motion for the QD, the electric field at times $t = \left(n + \frac{1}{2}\right)\Delta t$ and $t = (n + 1)\Delta t$ are needed, but not provided by the FDTD method. The missing values can be obtained by standard extrapolation methods, like linear extrapolation or n^{th} polynomial Lagrange extrapolation, shortly depicted in Section A.3 [54]. Alternatively, a field update utilizing the FDTD update scheme equations (2.23) and (2.24) can be used to calculate the electric field at the unknown time steps. A comparison of the performance of the different extrapolation schemes is shown in Fig. 2.13. The plot shows the coherence of the QD on a very short time scale only covering some time steps, but nicely shows the differences between the schemes. No extrapolation (red), meaning a constant electric field is used, and linear extrapolation (green) show comparably strong deviations from the more accurately working third-order Lagrange extrapolation scheme (blue). The extrapolation scheme taken from the FDTD update (magenta) is consistent with the Lagrange scheme and therefore the method of choice.

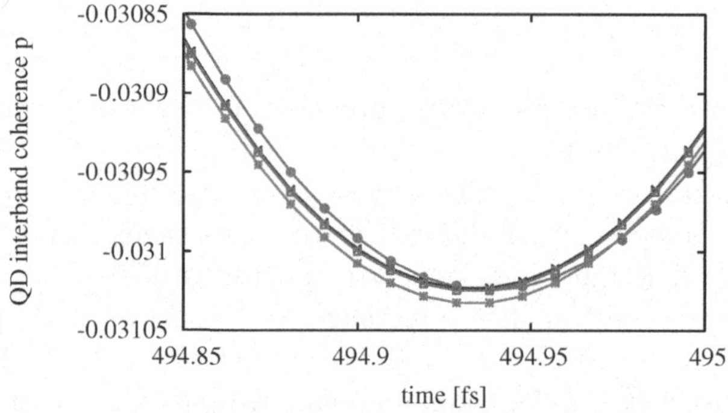

Figure 2.13 Comparison of different extrapolation schemes for calculating (unknown) future values for the electric field, used in the fourth Runge–Kutta integrator (Section A.2 [54]). QD interband coherence vs. time in fs on a very short time window. (-●-): No extrapolation, $E^{n+1} = E^{n+\frac{1}{2}} = E^n$. (-✱-): Linear extrapolation. (-✶-): Lagrange extrapolation of third order (Section A.3 [54]). (-□-): Extrapolation via FDTD update with Maxwell's equations. The Lagrange and Maxwell extrapolation scheme agree nicely, while no and only linear extrapolation show comparably large deviations.

To describe a physically meaningful picture, the interband coherence and conduction band density have to be damped. In the presence of the fields, occupying the spatial FDTD grid, the QD starts to decay radiatively once excited, because the optical polarization field of the QD is feedback into the FDTD lattice. Also the update equations transport the energy away, which leads in an exponential decay for the microscopic polarization. This radiative decay constant can be calculated analytically [55], assuming an infinitely thin quantum well (QW) structure and linear excitation, yielding

$$\Gamma_{\text{rad}}^{\text{QW}} = \frac{\omega d_{\text{VC}}^2}{2\varepsilon_0 c_0 \hbar a_{0_x}^2}, \tag{2.59}$$

where ω is the angular frequency. This result is valid for a dense grid of QDs in two dimensions, simulating a QW-like structure. Numerically, a simulation domain containing one single QD in its center which apply a CPML in propagation direction and periodic boundary conditions in the perpendicular plane to simulate a QW structure is set up. The length between the QD and its periodic neighbor then is used as the exciton Bohr radius a_{0x} in equation 2.59. If a delta-like QD is assumed, the radiative decay constant can also be derived analytically [56] for a single QD. Therefore, Maxwell's equations for the full electric field are solved in reciprocal space with a Green's tensor approach and using the Lippmann–Schwinger equation to determine the frequency-dependent electric field in real space. Calculating the averaged electric field at the QD position and inserting it into the macroscopic polarization with a renormalized susceptibility (renormalized to the exciting electric field at the QD position), the radiative decay constant of the QD yields:

$$\Gamma_{\text{rad}}^{\text{QD}} = \frac{d_{\text{VC}}^2 \omega^3}{6\pi\varepsilon_0 \hbar c_0^3}. \tag{2.60}$$

Figure 2.14 depicts the numerical performance of a QD array in the 3D FDTD grid, depending on the distance between the QDs. In x-direction, the computational domain is set to 800 nm with CMPL boundary (10 cells).

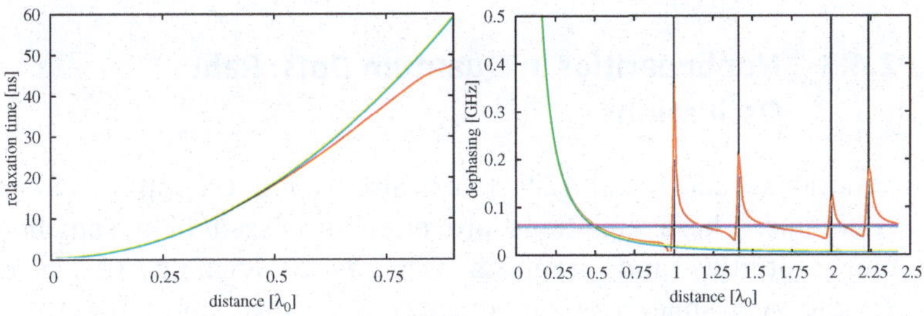

Figure 2.14 Left: Relaxation time in ns vs. QD distance in units of the resonant wavelength λ_0 of the QD. Green: QW limit, calculated via equation (2.59). Red: Numerical data from FDTD. Clear deviations from the QW limit can be seen with the bare eye to emerge at QD distances larger than $0.3\lambda_0$. Right: Radiative dephasing in GHz of the QD array vs. QD distance in λ_0 (red). Clear resonance features (black vertical lines) are visible due to the radiative coupling between the QDs in the square lattice array. Noninteger resonances emerge from coupling in diagonal directions ($\sqrt{3}$, $\sqrt{5}$, etc.). Green: Analytical QW limit, equation (2.59) [55]. Blue: Analytical value for an isolated QD, equation (2.60) [56].

Periodic boundary conditions are applied in y- and z-direction and the edge length of the computational domain was gradually increased in the same manner (horizontal axis, in units of the (vacuum) emission wave length λ_0 of the QD) with the QD placed in the center to get an infinitively extended square lattice of QDs, mimicking the infinitively thin QW structure. On the left side, the relaxation time of the QD array in ns is depicted over the resonant wavelength λ_0. Green shows the QW limit equation (2.59), while red shows the numerical data. One can clearly see the deviation from equation (2.59), starting above $0.3\lambda_0$, where the performed approximations break down. On the right side, the dephasing in GHz is plotted against the distance in λ_0. The blue line shows the analytical value for an isolated QD equation (2.60), in green the QW limit is shown and red shows the numerical data for the QD. At all integer distances, clear dips, and resonant features occur. The radiative coupling (constructive interference of the emitted spherical waves) between the QDs suppress the emission, hence the dephasing drops. The resonant features emerging at noninteger distances arise from the radiative coupling between QDs which are not next neighbored [55].

2.4.3 Nonlinearities in Quantum Dots: Rabi Oscillations

Equations (2.53) and (2.54) are also called the optical Bloch equations. These equations are often discussed in a somehow more intuitive and graphical way: as a dynamic 3D vector $\mathbf{s}$ in the Bloch sphere. As components of $\mathbf{s}$, the inversion of the QD – $w = 1 - 2n$ and absorptive part u (dispersive part v) of the coherence $p = u - iv$ appear:

$$s = \begin{pmatrix} u \\ v \\ w \end{pmatrix} = \begin{pmatrix} u \\ v \\ 2n-1 \end{pmatrix} \tag{2.61}$$

As can be seen, $|\mathbf{s}| = 1$ for all possible u, v, and w. So the possible orbit for the Bloch vector is a unit sphere, on which it rotates with angular velocity ω around the w-axis (can be eliminated with the rotating wave approximation, RWA).

Under certain assumptions, the coupled equations (2.53) and (2.54) can be solved analytically. First assumption is that one can write equations (2.53) and (2.54) in RWA, eliminating the fast oscillations of the excitation field and the coherence and just leaving the slowly varying envelopes (denoted with a bar). This yields a new set of differential equations:

$$\dot{\bar{p}} = i\bar{\Omega}_R(1-2n) \tag{2.62}$$

$$(1-2n) = -2\Im(\bar{\Omega}_R\rho^*). \tag{2.63}$$

The Rabi frequency $\bar{\Omega}_R$ defines the pulse area $\Theta(t)$ at a time t as

$$\Theta(t) = \int_{-\infty}^{t} dt'\,\bar{\Omega}_R(t') \quad \text{with} \quad \bar{\Omega}_R(t') = \frac{d_{\mu\mu'}\cdot\bar{E}}{\hbar}. \tag{2.64}$$

The barred quantity $\bar{\boldsymbol{E}}$ denotes the envelope function of the external light field ($\boldsymbol{E} = \bar{\boldsymbol{E}}\cdot\cos(\omega t+\phi)$, ϕ is the phase). As a second condition for solvability, the phenomenologically introduced dephasing rates are set to zero and $\Omega_R \in \mathbb{R}$ (for resonant excitation). This set of equations is now solvable analytically with

$$\bar{p} = \frac{i}{2}\sin\frac{\theta(t)}{2},\ 1-2n = \cos\frac{\theta(t)}{2} \tag{2.65}$$

resulting in

$$n = \sin^2\frac{\theta(t)}{2}. \tag{2.66}$$

If the pulse area $\Theta(t\to\delta)$ = $2m\pi$, the system performs m full flops of the conduction band density, means inverting the QD and destroying occupancy again m times. In the picture of the Bloch sphere this means that the Bloch vector s rotates m times on the edge of the sphere and returns to his originating position where it started.

Introducing a phenomenological dephasing rate γ affects the Rabi oscillations. The inversion of the QD is not complete and after the excitation pulse vanishes, the n decays to zero according to γ.

Dephasing always occurs in such systems because the QDs are never isolated, but coupled to a phonon bath of the substrate or other QDs via the long range Coulomb interaction, the Förster coupling. Figure 2.15 depicts the dynamics of equations (2.53) and (2.54), calculated in a 3D FDTD simulation. Red (green) shows the CB density (VB density), blue the interband coherence between VB and CB, and the gray shade shows the temporal presence of the excitation pulse. The left row is calculated without a phenomenological damping for the QD, while the right row has a damping rate of γ = 10 THz, which is chosen to be quite large to show the dephasing effect on a short time scale.

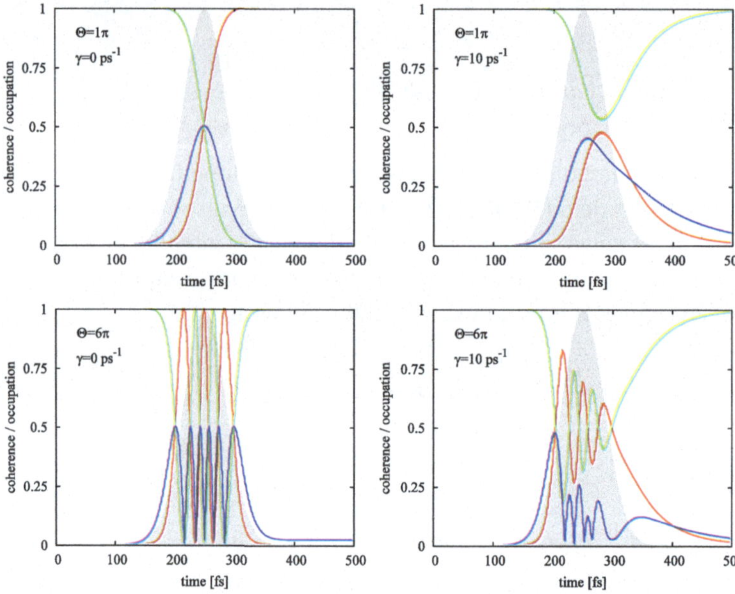

Figure 2.15 Rabi oscillations in a TLS, calculated with the FDTD method and self-consistent backcoupling. From top to bottom: Densities of conduction band (red), valence band (green), coherence between conduction and valence band (blue), and the excitation pulse (normalized, gray shade). Parameters for the TLS are: Dipole matrix element d_{VC} = 3 eÅ ($\boldsymbol{d}_{VC} || \boldsymbol{E}$ is assumed), gap energy ε_{gap} = 1.5 eV, phenomenological dephasing γ = 0 (left), and γ = 10 THz (right). Radius of Gaussian wave function for the carriers r = 30 nm. Parameters for the excitation pulse are center time t_0 = 250 fs, pulse width τ = 50 fs, and energy ε = 1.5 eV. The slight oscillations, covering the whole dynamics of p and n, are a no-RWA effect.

In the top line, a pulse area of $\Theta = \pi$ is chosen and according to equation (2.66), the system performs the expected inversion of the QD in the undamped case. In the damped case, the density flopping is suppressed and the occupation of the conduction band (valence band) decays to zero (one). Choosing a pulse area of $\Theta = 6\pi$ causes the CB density to flop three times in a row during the presence of the light field in the undamped case. Like in the previous case, the damping constant prevents full inversion of the QD and the conduction band density (valence band density) decays to zero (one) after the excitation pulse vanished.

As discussed in the last Section 2.4.2, a QD will decay radiatively in the FDTD without a phenomenologically introduced dephasing. Hence, a physical picture is obtained automatically. However, a phenomenological damping to obey dephasing due to many-body effects like electron–phonon coupling should be introduced.

2.4.4 A Single Quantum Dot in a Photonic Crystal Cavity

As in the case of two coupled classical oscillators, like pendulums with coupling constant g, the QD and the field start to oscillate with two different frequencies, depending on the single frequencies and the coupling constant. The coupled system can be decoupled into two orthogonal states: Considering a system of two oscillators in the states $|\Psi_1\rangle$ and $|\Psi_2\rangle$, coupled with the coupling constant g and eigenfrequencies f_1 and f_2, the result is a beating between those two oscillators when the eigenfrequencies are equal. The problem can be formulated as

$$\begin{pmatrix} f_1 & g \\ g & f_2 \end{pmatrix}\begin{pmatrix} \Psi_1 \\ \Psi_2 \end{pmatrix} = \begin{pmatrix} f+g & 0 \\ 0 & f-g \end{pmatrix}\begin{pmatrix} \Psi_0 \\ \Psi_- \end{pmatrix} \tag{2.67}$$

for $f_1 = f_2 = f_3$. Diagonalizing the matrix, thus uncoupling the oscillations into two orthogonal normal oscillations Ψ_+ and Ψ_-, yields an line splitting $\Delta = 2g$, depending on the coupling strength g. Applying this to a system of a QD strongly coupled to a PhCC mode and small intensities of the field (linear response of the QD), normal mode splitting (vacuum Rabi splitting) is observed. The energy in the system is pumped from the field to the QD and back and so forth. This beating, visible in the time regime, can be observed in frequency domain as well, where the beating frequency is expressed by a symmetric line splitting around the isolated resonance. Both eigenmodes of the coupled system refer now to normal modes of a decoupled system in the basis of the even and odd modes Ψ_1 and Ψ_2. The amount of line splitting is determined by the light–matter interaction, $g = \frac{\boldsymbol{d}_{\text{VC}}\boldsymbol{E}_{\text{vac}}}{\hbar}$ ($\boldsymbol{d}_{\text{VC}} || \boldsymbol{E}$ is assumed), where the vacuum field amplitude is defined as $|\boldsymbol{E}_{\text{vac}}|^2 = \frac{\pi\hbar f}{\varepsilon_0\varepsilon_r V_{\text{eff}}}$, ε_r is the dielectric constant of the QD and V_{eff} is the effective modal volume of the cavity mode, defined via the following equation:

$$V_{\text{eff}} = \int_{\mathbb{R}} \mathrm{d}r \frac{\boldsymbol{D}(\boldsymbol{r}) \cdot \boldsymbol{E}(\boldsymbol{r})}{\max\limits_{r \in \mathbb{R}} (\boldsymbol{D}(\boldsymbol{r}) \cdot \boldsymbol{E}(\boldsymbol{r}))}. \tag{2.68}$$

In general, the line splitting in such system also depends strongly on the spatial location and also on the direction of the dipole vector of the QD. Taking this into account, the line splitting can be written as

$$\Delta_1 = 2g \frac{E(r) \cdot d_{\text{VC}}}{|E_{\text{max}}||d_{\text{max}}|} \tag{2.69}$$

where the subindex indicates the coupling to one single QD.

A crucial condition for the visibility of the strong coupling regime is the ratio between the line width of the cavity and the line splitting. Thus, the line widths of the uncoupled oscillators, namely the cavity mode and the QD, have to be smaller than the line splitting (strong coupling condition). Since the life time of the interband polarization of the QD is fairly long, in the nanosecond regime, the QD line width is negligible. The minimum Q-factor of the cavity can be estimated by $Q_{\text{min}} \gtrsim \frac{f}{2\pi g}$. Thus, for d_{VC} = 3 eÅ, $V_{\text{eff}} \approx (\lambda/n)^3$, f_{cav} = 279.2 THz, and ε_{QD} = 10.595 one gets $Q_{\text{min}} \gtrsim 3500$. To ensure to be in the strong coupling regime, the method of gentle confinement [46,57] is applied to the PhCC from Section 2.3.2. Modifying the next-neighbored air holes of the defect region in the ΓK-direction, the envelope becomes similar to a Gaussian function. This implies less nonvanishing contributions inside the light cone and therefore an increased Q-factor.

Less wave vectors fulfill the refraction condition at the slab-air interface perpendicular to the slab and are confined in the defect region. As a result of gentle confinement, the first next-neighbored air holes along the ΓK-direction are shifted by

$s = 0.185 \times a$ away from the cavity center and their radii are shrunk to $\dot{r} = 0.2 \times a$, which increases the cavity Q-factor from <1000 to ≈9000. Also, the frequency of the fundamental mode decreases slightly to $f = 277.47$ THz due to an effectively increased defect area. Together with $Q = 9000$, the cavity decay rate corresponds to $\gamma < 10$ GHz, which is sufficient to observe strong coupling between the fundamental cavity mode and the single QD.

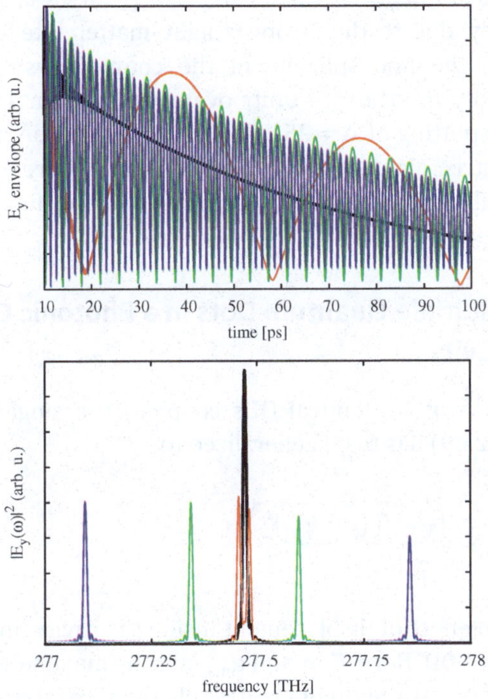

Figure 2.16 Strong coupling in a single QD-cavity system. Left: Beating between the QD and the single cavity mode (only the envelope is plotted). The beating frequency scales with the dipole matrix element in a linear manner and manifests itself in the spectral response in a line splitting equal to the inverse beating frequency (right). Analytical values according to equation (2.69) are given in parenthesis. Black: Uncoupled system. Red: $d_{VC} = 3$ eÅ. The transfer time between the cavity and the QD is ≈39.422 ps which equals a line splitting of $\Delta = 25.367$ (24.324 GHz). Green: $d_{VC} = 30$ eÅ. From the transfer time of ≈3.934 ps, resulting in a numerical line splitting of 254.194 GHz (243.24 GHz). Blue: $d_{VC} = 90$ eÅ. A transfer time of 1.3131 ps yields a line splitting of $\Delta = 761.557$ GHz (729.72 GHz).

Figure 2.16, left, shows the beating of the temporal response of the coupled system with intensity in arbitrary units over the time in ps for different values of the dipole matrix element d_{VC} = 3 eÅ, 30 eÅ, and 90 eÅ of the QD (red, green, and blue, respectively, the black illustrates the uncoupled system). As can be seen, the transfer time of the energy between the QD and the electromagnetic field decreases linearly with increasing dipole moment from $t_{\text{transfer}} \approx 39.422$ ps to $t_{\text{transfer}} \approx 1.3131$ ps for d_{VC} = 3 eÅ and d_{VC} = 90 eÅ, respectively, due to the stronger light–matter interaction. On the right side, the line splitting of the coupled system is plotted. The intensity in arbitrary units over frequency in THz shows the clear line splitting of $\Delta = 25.269$ GHz for d_{VC} = 3 eÅ. For the larger dipole matrix elements, the line splittings increase linearly to 761.557 GHz. Thus, the transfer time and the line splitting are inverse proportional.

2.4.5 Multiple Quantum Dots in a Photonic Crystal Cavity

For the case of N identical QDs coupled to a single PhCC mode, equation (2.69) has to be generalized to

$$\Delta_N = \Delta_1 \sqrt{\sum_{i=1}^{N} \frac{|\boldsymbol{E}_i(\boldsymbol{r}) \cdot \boldsymbol{d}_{i,\text{VC}}|^2}{|E_{\text{max}}|^2 \, |\boldsymbol{d}_{\text{max}}|^2}} \tag{2.70}$$

to obey the effect of displacement inside the cavity and a tilt dipole vector of the QD. Here, $\boldsymbol{E}_{\text{max}}$ and $\boldsymbol{d}_{\text{max}}$ are the maximum values of the electric field and the dipole matrix element, respectively, of the ith QD at position $\boldsymbol{r}$.

Numerically, this behavior can be observed in a PhCC with embedded QDs as well. For this, the previously modified L3-type defect is used (see Section 2.4.4) to enhance the Q-factor of the cavity and thus enhance the light–matter interaction with the method of gentle confinement [46,57]. A cut along the x-axis shows a Gaussian-like envelope function of the confined electric field, see the red graph in Fig. 2.17, modulated with a sine function, which exhibits several maxima (minima) inside the defect region. The green line shows the (scaled) dielectric profile of the PhC slab. Placing a single QD at labels A and B results in the spectral response shown in

Fig. 2.18 (left). Black illustrates the uncoupled resonance while green and red show position-depending line splittings for the position of the QD at A and B, respectively. Since the field at A is approximately 87.5% of the amplitude at B, the line splitting is reduced to ≈87.5% as well. If one now subsequently embeds more QDs to the system, say start with one QD at B, then place two identical QDs positions A and C and then place three identical QDs at A, B, and C, one can numerically analyze the line splitting and compare the analytically expected values. For a single QD positioned at B the line splitting, obtained with equation (2.69), yields $\Delta_{1,\text{ana}}$ = 24.324 GHz. In the simulation, the line splitting results in $\Delta_{1,\text{num}}$ = 25.269 GHz. The deviation of approximately 1 GHz can be explained due to a coarse spatial sampling. Hence, the overlap between the carrier wave function of the QD and the light field confined in the defect region of the PhCC contains only very few points. In the case, where two identical QDs are located at positions A and C, equation (2.70) applies.

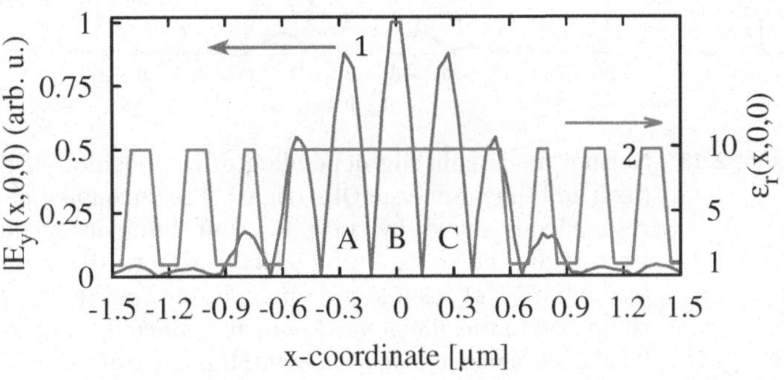

Figure 2.17 Absolute value of the field profile [$|\boldsymbol{E}_y|$, normalized, (1)] of the fundamental mode of a L3-type defect cavity in a hexagonal array of air holes in a dielectric slab and its dielectric profile $\varepsilon_r(x)$ (2) versus the x-coordinate in μm. At the three positions A, B, and C, where the absolute value of the amplitude is maximal inside the defect region, QDs can be strongly coupled to this photonic mode.

Analytically, one gets $\Delta_{2,\text{ana}}$ = 30.143 GHz compared to $\Delta_{2,\text{num}}$ = 30.099 GHz for the numerical value of the line splitting. Placing a third identical QD at the field maximum at position B, the line splitting gets enhanced to $\Delta_{3,\text{num}}$ = 37.927 GHz. Analytically, equation (2.70) yields $\Delta_{3,\text{ana}}$ = 38.733 GHz.

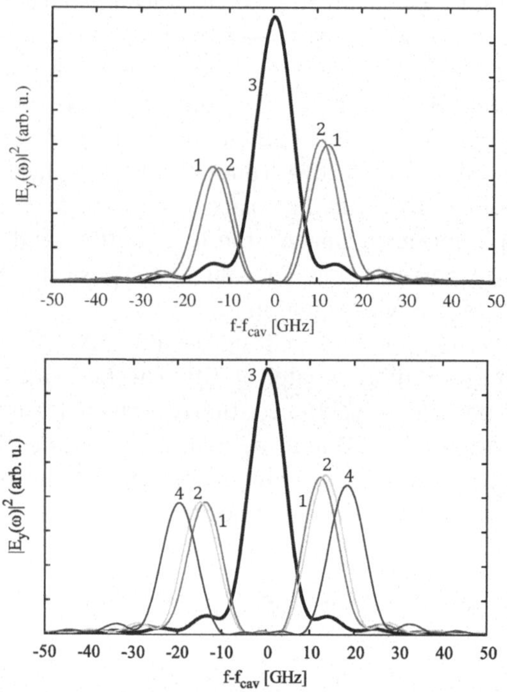

Figure 2.18 Normal mode splitting depending of the position of the QD (left) and the number of QDs (right). The uncoupled photonic resonance is shown by curve (3). Left: Influence of the field strength on the line splitting. Placing a single QD in the local maximum B (2) results in a line splitting reduced by 12.5%, compared to the global maximum at A, since the amplitude at B is 12.5% smaller than at A. Right: Influence of the number of identical QDs placed at different positions in the field maxima A, B, and C. Analytical values $\Delta_{i,\mathrm{num}}$ according to equation (2.70) are given in parenthesis. (1): One single QD at B, yielding $\Delta_{1,\mathrm{num}}$ = 25.269 GHz (24.324 GHz). (2): Two identical QDs at A and C results in $\Delta_{1,\mathrm{num}}$ = 30.099 GHz (30.143 GHz). (4): Three identical QDs at A, B, and C. The line splitting equals $\Delta_{3,\mathrm{num}}$ = 37.927 GHz (38.733 GHz).

Epilog

Using the FDTD method to simulate high-quality modes in electromagnetic resonators is a demanding task. The stair casing effect, introduced due to a Cartesian coordinate system, on which the photonic device is discretized spatially, induced errors in the spectral response. This reduces the accuracy of the numerical method down to $\mathcal{O}(\Delta x)$. However, applying a proper averaging procedure for grid cells filled with more than one dielectric material can reduce the error dramatically and the intrinsic order of accuracy of $\mathcal{O}(\Delta x^2)$ can be restored completely in special cases. The inclusion of optically active material such as semiconductor QDs, which provide additional optical polarization fields which need to be feedback self-consistently, results in required values of the fields at unknown times, but can be provided by means of interpolation. Normal mode splitting of a single photonic mode in a high-Q resonator with one or more quantum mechanical oscillators can be simulated and the results fit very well with the analytical values.

The investigation of the energy transfer between different, well-separated QDs in a PhCC or in a system of coupled PhCCs can be numerically analyzed. This helps to understand the purely light-mediated coupling between the photonic modes and quantum mechanical oscillators. Many-body effects such as long-range Coulomb interaction, the Förster coupling, and phonons coupled via a deformation potential for acoustic phonons and the Fröhlich term for the longitudinal optical phonons to the electronic system of the QD can be implemented and analyzed. Also, optical forces, high-harmonic generation in metals, and field enhancements for enhanced light–matter interaction can be calculated very accurately. Using time-domain simulations to develop, test, and improve such systems for future communication and quantum computation devices, and the flexibility of the implementation makes the FDTD a powerful tool to simulate complex optoelectronic devices.

Appendix A: Analysis Tools and Numerical Methods

For spectral analysis of the data obtained by the time-domain simulation, an effective alternative to the discrete Fourier Transform (DFT) is introduced with the FDM harmonic inversion [30]. The technique to numerically integrate systems of inhomogeneous, ordinary differential equations (ODE) is also shown, additionally to the extrapolation scheme, which is based on Lagrange interpolation. The latter techniques are mentioned in Section 2.4.

A.1 FDM Harmonic Inversion

Next to the standard DFT, spectral analysis is also possible with the Fast Fourier Transform (FFT) [58] or the FDM harmonic inversion [30,31,59,60]. The latter method relies on the fact, that one can transform the spectral analysis of a time-evolving dataset $f(t)$ of finite, exponentially decaying sum of oscillations into a matrix equation, which can be diagonalized to efficiently extract spectral properties from time data like resonant frequencies f, phases ϕ, and decay rates γ:

$$f(t) = \sum_{k<\infty} A_k e^{-i(2\pi f_k - i\gamma_k)t + \phi_k}, \quad (A.1)$$

where A_k is the (complex) amplitude of the oscillation with index k. In general, the computational performance of the harmonic inversion is essentially better than that of the DFT, but one has to be careful with large sets of oscillations. The great advantage of the FDM is that the time signal does not need to decay to zero to get a fully converged spectrum. The complexity of the FDM, for N time points and J frequencies, is $\mathcal{O}(NJ + J^3)$. Compared, the DFT uses orthogonality relations of oscillations of different frequency. The complexity here is $\mathcal{O}(N^2)$.

A.2 Runge–Kutta Algorithm

For the integration of the system of inhomogeneous, ODE of first order for the interband coherence and conduction band density of the nonlinear treated QD, the Runge–Kutta algorithm of fourth order is used, since it is well established and of great stability. The integration steps will only be depicted shortly, a more detailed description can be found in Ref. [54].

Assume an inhomogeneous ODE in the form

$$\dot{y}(x) = f(y(x), x). \quad (A.2)$$

Basically, the algorithm transforms the equation (A.2) into an integral equation:

$$y(x+\Delta x) = \int_x^{x+\Delta x} f(y(\xi), \xi) d\xi \approx y(x) + \Delta x \sum_{i=1}^{j} y_i f(y_i(x_i), x_i). \quad (A.3)$$

where Δx is the discretization. In an explicit Runge–Kutta algorithm, $f(y_i(x_i), x_i) \approx k_i(y, x)$ is set. In fourth order those functions are given by the weightings γ_i:

$$\gamma_1 k_1(y,x) = \frac{1}{3} f(y,x),$$

$$\gamma_2 k_2(y,x) = \frac{1}{3} f\left(y + \frac{k_1}{2}, x + \frac{\Delta x}{2}\right),$$

$$\gamma_3 k_3(y,x) = \frac{1}{3} f\left(y + \frac{k_2}{2}, x + \frac{\Delta x}{2}\right),$$

$$\gamma_4 k_4(y,x) = \frac{1}{6} f\left(y + \frac{k_3}{2}, x + \Delta x\right). \tag{A.4}$$

A.3 Lagrange Interpolation

Since the Runge–Kutta integration algorithm, Section A.2, for the time-depending interband coherence and conduction band density equations of the QD, equations (2.53) and (2.54), needs the future time steps at $t + \frac{1}{2}\Delta t$ and $t + \Delta t$, which are not provided by the FDTD algorithm, one needs to extrapolate these values from earlier time steps. One method for the extrapolation, actually not only used for interpolation, but also applicable for this purpose, is the Lagrange interpolation scheme, which is a polynomial interpolation scheme. In general, one creates a polynomial $p(x)$ of order $(n - 1)$, whose i^{th} element equals the function, which wants to be interpolated, at the ith element:

$$p(x_i) = \sum_j^n l_{jn}(x_i) f(x_j). \tag{A.5}$$

With $l_{jn}(x_i) = \delta_{ij}$, this condition is fulfilled. Hence, the coefficients $l_{jn}(x_i)$ take the form

$$l_{jn}(x) = \prod_{i=1, i\neq j}^{n} \frac{x - x_i}{x_j - x_i}. \tag{A.6}$$

For the extrapolation of the needed future time steps, it is enough to use a third-order scheme, resulting in (for reasons of readability, the spatial indices are omitted)

$$E^{n+\frac{1}{2}} = \frac{3}{8}E^{n-2} - \frac{5}{4}E^{n-1} + \frac{15}{8}E^{n}$$

$$E^{n+1} = E^{n-2} - 3E^{n-1} + 3E^{n}. \tag{A.7}$$

References

1. Oskooi, A. F., Roundy, D., Ibanescu, M., Bermel, P., Joannopoulos, J. D. and Johnson, S. G. (2010). MEEP: A flexible free-software package for electromagnetic simulations by the FDTD method. *Comp. Phys. Commun.*, **181**, pp. 687–702.
2. Schneider, J. B. (2010). *Understanding the finite-difference time-domain method.* http://www.eecs.wsu.edu~schneidj/ufdtd/.
3. Taflove, A. and Hagness, S. C. (2005). *Computational Electrodynamics—The Finite-Difference Time-Domain Method,* 3rd Ed. (Artech House, Boston).
4. Kakihara, K., Kono, N., Saitoh, K. and Koshiba, M. (2006). Full-vectorial finite element method in a cylindrical coordinate system for loss analysis of photonic wire bends, *Opt. Exp.*, **14**(23), pp. 11128–11141.
5. Zhuang, L., Guo, L. and Chou, S. Y. (1998). Silicon single-electron quantum-dot transistor switch operating at room temperature. *Appl. Phys. Lett.*, **72**(10), pp. 1205–1207.
6. Maxwell, J. C. (1865). A dynamical theory of the electromagnetic field. *Phil. Trans. R. Soc. Lond.*, **155**, pp. 459–512.
7. Born, M. and Wolf, E. (2006). *Principles of Optics*, 7th Ed., 4th Printing (Cambridge).
8. Jackson, J. D. (2006). *Klassische Elektrodynamik*, 4th Ed. (Gruyter Berlin).
9. Griffiths, D. J. (2008). *Introduction to Electrodynamics*, 3rd Ed. (Pearson Benjamin Cummings).
10. Yee, K. S. (1966). Numerical solution of initial boundary value problems involving Maxwell's equations in isotropic media. *IEEE Trans. Ant. Propag.*, **14**, pp. 302–307.
11. http://www.speag.com/products/semcad/components/what-is-fdtd/ (2010).
12. Okada, N. and Cole, J. B. (2010). Simulation of whispering gallery modes in the Mie regime using the nonstandard finite-difference time domain algorithm. *J. Opt. Soc. Am. B*, **27**(4), pp. 631–639.

13. Qui, S.-L. and Li, Y.-P. (2006). Q-factor instability and its explanation in the staircased FDTD simulation of high-Q circular cavity. *JOSA B*, **26**(9), pp. 1664–1674.

14. Zschiedrich, L., Burger, S., Kettner, B. and Schmidt, F. (2006). Advanced finite element method for nano-resonators. *Proc. SPIE*, **6115**, doi: 10.1117/12.646252.

15. Hesthaven, J. S. and Warburton, T. (2002). Nodal high-order methods on unstructured grids: 1. Time-domain solution of Maxwell's equations. *J. Comput. Phys.*, **181**, pp. 186–221.

16. Hesthaven, J. S. and Warburton, T. (2007). *Nodal Discontinuous Galerkin Methods: Algorithms, Analysis, and Applications*, 1st Ed. (Springer).

17. Ji, X., Lu, T., Cai, W. and Zhang, P. (2010). Discontinuous Galerkin time domain (DGTD) methods for the study of 2-D waveguide-coupled microring resonators. *J. Lightwave Technol.*, **23**, p. 3864.

18. Stannigel, K., König, M., Niegemann, J. and Busch, K. (2008). Discontinuous Galerkin time-domain computations of metallic nanostructures, *Opt. Exp.*, **17**, pp. 14934–14947.

19. Tang, G., Panetta, R. L. and Yang, P. (2010). Application of a discontinuous Galerkin time domain method to simulation of optical properties of dielectric particles. *Appl. Opt.*, **49**, pp. 2827–2840.

20. Weiland, T. (1977). A discretization method for the solution of Maxwell's equations for six-component fields. *Electron. Commun. AEUE*, **31**(3), pp. 116–120.

21. Farjadpour, A., Roundy, D., Rodriguez, A., Ibanescu, M., Bermel, P., Joannopoulos, J. D., Johnson, S. G. and Burr, G. (2006). Improving accuracy by subpixel smoothing in FDTD. *Opt. Lett.*, **31**, pp. 2972–2974.

22. Oskooi, A. F., Kottke, C. and Johnson S. G. (2009). Accurate finite-difference time-domain simulation of anisotropic media by subpixel smoothing. *Opt. Lett.*, **34**, pp. 2778–2780.

23. Mur, G. (1981). Absorbing boundary conditions for the finite-difference approximation of the time-domain method electromagnetic field equations. *IEEE Trans. Electromag. Compat. EMC*, **23**(4), pp. 377–382.

24. Berenger, J. P. (1994). A perfectly matched layer for the absorption of electromagnetic waves. *J. Comput. Phys.*, **114**(2), pp. 195–200.

25. Sacks, Z. S., Kingsland, D. M., Lee, R. and Lee, J. F. (1995). A perfectly matched anisotropic absorber for use as an absorbing boundary condition. *IEEE Trans. Ant. Propag.*, **43**, pp. 1460–1463.

26. Kuzuoglu, M. and Mittra, R. (1996). Frequency dependence of the constitutive parameters of causal perfectly matched anisotropic absorbers. *IEEE Microw. Guided W.*, **6**, pp. 447–449.

27. Roden, J. A. and Gedney, S. D. (2000). Convolutional PML (CPML): An efficient FDTD implementation of the CFS-PML for arbitrary media. *Microwave Opt. Tech. Lett.*, **27**(6), pp. 334–339.

28. Piegdon, K. A., Declair, S., Förstner, J., Meier, T., Matthias, H., Urbanski, M., Kitzerow, H.-S., Reuter, D., Wieck, A. D., Lorke, A. and Meier, C. (2010). Tuning quantum-dot based photonic devices with liquid crystals. *Opt. Exp.*, **18**(8), pp. 7946–7954.

29. Song, Q., Cao, H., Ho, S. T. and Solomon, G. S. (2009). Near-IR subwavelength microdisk lasers. *Appl. Phys. Lett.*, **94**, p. 061109.

30. Mandelshtam, V. A. and Taylor, H. S. (1998). Harmonic inversion of time signals. *J. Chem. Phys.*, **107** (17), pp. 6756–6769. See also erratum (1998), ibid. **109**(10), p. 4128.

31. Mandelshtam, V. A. (2003). On harmonic inversion of cross-correlation functions by the filter diagonalization method. *J. Theor. Comput. Chem.*, **2**(4), pp. 497–505.

32. Nielson, M. M. and Chuang, I. L. (2004). *Quantum Computation and Quantum Information* (Cambridge).

33. Joannopoulus, J. D., Johnson, S. G., Winn, J. N. and Meade, R. D. (2008). *Photonic Crystals: Molding the Flow of Light*, 2nd Ed. (Princeton University Press).

34. Joannopoulos, J. D., Villeneuve, P. R. and Fan, S. (1997). Photonic crystals. *Solid State Commun.*, **102**(2/3), pp. 165–173.

35. Sajeev, J. (2987). Strong localization of photons in certain disordered dielectric superlattices. *Phys. Rev. Lett.*, **58**(23), pp. 2486–2489.

36. Yablonovitch, E. (1987). Inhibited spontaneous emission in solid-state physics and electronics. *Phys. Rev. Lett.*, **58**, pp. 2059–2062.

37. Heebner, J., Grover, R. and Ibrahim T. A. (2008). *Optical Microresonators: Theory, Fabrication and Applications* (Springer).

38. Lourtioz, J.-M., Benisty, H., Berger, V., Gerard, J.-M., Maystre, D. and Tchelnokov, A. (2008). *Photonic Crystals: Towards Nanoscale Photonic Devices*, 2nd Ed., Translated by de Fornel, M. P. (Springer).

39. Blume, S. (1991). *Theorie Elektromagnetischer Felder*, 7th Ed., edited by Karl-Heinrich Wittich (Hüthig).

40. Brandt, S. and Dahmen, H. D. (2005). *Elektrodynamik: Eine Einführung in Experiment und Theorie*, 4th Ed. (Springer).

41. Leuenberger, D., Ferrini, R. and Houdré, R. (2004). *Ab initio* tight-binding approach to photonic-crystal based coupled cavity waveguides. *J. Appl. Phys.*, **95**(3), pp. 806–809.

42. Bayindir, M., Temelkuran, B. and Özbay, E. (2000). Tight-binding description of the coupled defect modes in three-dimensional photonic crystals. *Phys. Rev. Lett.*, **84**(10), pp. 2140–2143.
43. Tang, Y., Minatairov, A. M., Merz, J. L., Tokranov, V. and Oktyabrsky, S. (2005). Characterization of 2D-photonic crystal nanocavities by polarisation-dependent photoluminescence. *Proc. 5th IEEE Conf. Nanotechnology,* **1**, pp. 35–38.
44. Johnson, S. G. and Joannopoulos, J. D. (2001). Block-iterative frequency-domain methods for Maxwell's equations in a planewave basis. *Opt. Exp.*, **8**(3), pp. 173–190.
45. NSM. NSM Archive—*Optical properties of semiconductors.* http://www.ioffe.rssi.ru/SVA/NSM/Semicond/GaAs/optic.html, 2001.
46. Akahane, Y., Asano, T., Song, B.-S. and Noda S. (2003). High-Q nanocavity in a two-dimensional photonic crystal. *Nature,* **425**, pp. 944–947.
47. Chalcraft, A. R. A., Lam, S., O'Brien, D., Kraus, T. F., Sahin, M., Szymanski, D., Sanvitto, D., Oulton, R., Skolnick, M. S., Fox, A. M., Whittaker, D. M., Liu, H.-Y. and Hopkinson, M. (2007). Mode structure of the L3 photonic crystal cavity. *Appl. Phys. Lett.,* **90**, p. 241117.
48. Bimberg, D., Grundmann, M. and Ledentsov, N. N. (1998). *Quantum Dot Heterostructures,* 1. Aufl. pp. 177, 279 (Wiley).
49. Ustinov, V. M., Zhukov, A. E. and Egorov, A. Y. (2003). *Quantum Dot Lasers* (Oxford University Press).
50. Ladd, T. D., Jelezko, F., Laflamme, R., Nakamura, Y., Monroe, C. and O'Brian, J. L. (2010). Quantum computers. *Nature,* **464**, pp. 45–53.
51. Grodecka, A., Weber, C., Knorr, A. and Machnikowski, P. (2006). Interplay and optimization of decoherence mechanisms in the optical control of spin quantum bits implemented on a semiconductor quantum dot. *Phys. Rev. B,* **73**(12), p. 125306.
52. Haug, H. and Koch, S. W. (2001). *Quantum Theory of the Optical and Electronic Properties of Semiconductors* (World Scientific Publishing).
53. Kuhn, T. (1998). *Density Matrix Theory of Coherent Ultrafast Optics and Kinetics of Semiconductors,* eds. Schöll, E., pp. 173–187 (Chapman Hall).
54. Press, W. H., Teukolsky, S. A., Vetterling, W. T. and Flannery, B. O. (2007). *Numerical Recipes–The Art of Scientific Computing,* 3rd Ed. (Cambridge).
55. Zimmermann, K. (2003). *Zur Theorie der elektromagnetischen Kopplung in periodischen Quantenpunkt-Strukturen* (Diploma thesis, Technische Universität Berlin).

56. Eichmann, R. (2002). *Lineare optische Eigenschaften dielektrisch strukturierter Halbleiter* (PhD thesis, Phillipps-Universität Marburg).

57. Akahane, Y., Asano, T., Song, B.-S. and Noda S. (2004). Fine-tuned high-Q photonic crystal nanocavity based on a gentle confinement of light. *Quantum Electronics Conference, 2004. (IQEC). International*, pp. 882–883.

58. Johnson, S. G. and Frigo, M. (2008). *Fast Fourier Transforms*, eds. Burrus, C. S., Chapter 11 "Implementing FFTs in Practice" (Connexions).

59. Chen, R. and Guo, H. (1999). Efficient calculation of matrix elements in low storage filter diagonalization. *J. Chem. Phys.*, **111**(2), pp. 464–471.

60. Wall, M. R. and Neuhauser, D. (1995). Extraction, through filter-diagonalization, of general quantum eigenvalues or classical normal mode frequencies from a small number of residues or a short-time segment of a signal. I. Theory and application to a quantum-dynamics model. *J. Chem. Phys.*, **102**(20), pp. 8011–8022.

Chapter 3

Ultraviolet GaN-Based Lasers with Micro-/Nanostructures

Hao-Chung Kuo, Min-Hsiung Shih, Chien-Chung Lin, Tien-Chang Lu, Shing-Chung Wang, Chun-Yen Chang, Cheng-Chang Chen, and Yow-Gwo Wang

Department of Photonics, National Chiao Tung University, Hsinchu 30010, Taiwan, R.O.C.

hckuo@faculty.nctu.edu.tw

3.1 Introduction to GaN-Based Lasers

3.1.1 Nitride-Based Semiconductor Lasers

The research of blue light-emitting devices began from the development of gallium nitride light-emitting diodes in the early 1970s [1–3]. Two decades later, with several efforts devoted to the growth of nitride materials, doping concentration, efficiency of quantum wells, and so on, Nakamura et al. [4] developed the first prototype high-brightness (>100 times greater than previous alternatives, about 1 candela) GaN blue light-emitting diodes in 1993. Then, in 1995, they also developed the first successful GaN electron-injection laser [5]. This makes nitride laser diodes begin to step on the road toward the commercial applications of high-information-density storage, laser printer, compact projector, and so

Handbook of Optical Microcavities
Edited by Anthony H. W. Choi

ISBN 978-981-4463-24-9 (Hardcover), 978-981-4463-25-6 (eBook)
www.panstanford.com

on. In these years, the performance of nitride laser diodes has been improved to emit power as high as 100 mW at room temperature with a low threshold only 3 kA/cm^2 [6], and the devices have been already successfully commercialized to be applied to digital versatile disc (DVD). This next-generation DVD would enlarge the storage capacity from 4.7 gigabytes which current DVD owns to 27 gigabytes.

3.1.2 Edge-Emitting Lasers

The major type of commercial nitride-based laser diodes is now edge-emitting laser. This kind of structure is currently the most mature and producible one among all nitride laser devices. It is featured by a stripe-type long cavity and a pair of cleaved facet as the mirrors as shown in Fig. 3.1. Typically, the cavity length of the structure is about several hundred micrometers. The light in the cavity is resonant in the horizontal direction and emits from the cleaved mirrors. The standard epilayer structure of edge-emitting lasers (EELs) is usually grown on sapphire and composes of *p-n* junction, multiple quantum wells (active region), and some cladding layer for photons and electrons. Over this decade, several improvements and researches on the growth and optimization of its structure were reported [6–9]. One is to improve the great amounts of defects and threading dislocations existing in the grown nitride materials on sapphire substrate due to the large lattice-mismatch between the nitride material and sapphire. The epitaxial lateral overgrowth (ELOG) is a very efficient method to reduce threading dislocation and defect density [6,7]. It is a method using the patterned SiO_x to make GaN laterally coalesce and bend threading dislocations to reduce the dislocation density. Generally, ELOG could decrease dislocation density down to the order of 108 cm^{-2}. Other one is to raise the low confinement factor of the optical field in the structure. Because the index difference between $Al_xGa_{1-x}N$ and GaN is small, the guiding of optical field in active region is usually weak. Nakamura et al. provided a solution that using modulation doped strain superlattice layer improves the optical confinement [6,10]. The other is to reduce spontaneous polarization and piezoelectric field in multiple quantum wells. Recently, Schmidt et al. [11] demonstrated nonpolar laser diode on m-plane GaN substrate. The laser diode emits violet light ~410 nm and has the threshold current density of 3.7–8.2 KA/cm^2. The spontaneous and piezo-electric polarization effect was eliminated and the performance of laser diode could be further improved using this lattice matched substrate. The fabrication of flat and high-reflectivity mirrors is also a key point for achieving an excellent lasing performance. In order to obtain a flat facet, the etching and cutting techniques were rapidly developed.

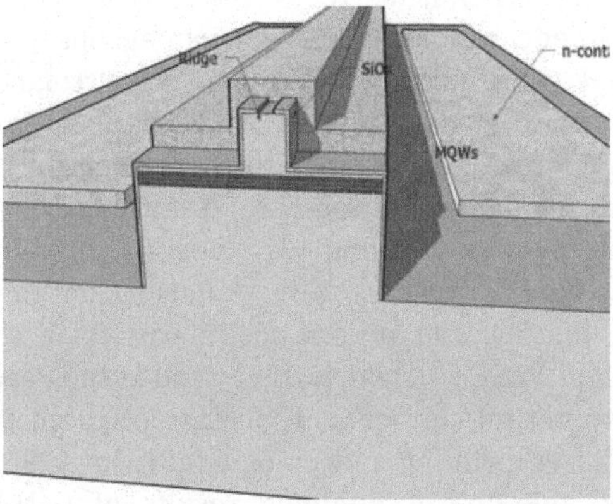

Figure 3.1 The schematic diagram of an edge-emitting laser diode.

3.2 Surface-Emitting Lasers

3.2.1 Vertical Cavity Surface-Emitting Lasers

Although the optimization of the edge-emitting laser keeps going, some properties of this kind of laser are unfavorable. One of these properties is its elliptic beam shape. On one hand, the coupling efficiency would be low as the elliptic beam is coupled into optical fiber (typically in the form of circular core). On the other hand, for the application of storage, the elliptic beam not only makes each writing pixel larger but also raises expenses for correcting light shape. Usually, this kind of laser shows slightly large divergence angle to be over ten degree. This is also disadvantageous to the projection. Furthermore, the side-emitting laser devices also make the testing of devices a tough task. The wafer should be cut into several stripes (several laser devices on one strip) before the testing. For a commercial product, the complicated testing would result in a poor producing efficiency and be disadvantageous. Therefore, in order to have a superior laser device, Iga [12] demonstrated a new kind of laser diodes, vertical cavity surface emitting laser (VCSEL), in 1977. VCSEL is a vertical-emitting-type laser. It is formed by sandwiching a few-lambda cavity in a pair of reflectors, usually in the form of distributed Bragg reflector (DBR), with a very high reflectivity (>99.9%) as shown in Fig. 3.1. In contrast to EELs, photons in the cavity of VCSEL are vertically in resonance and emit outside perpendicularly to the surface of the structure. This laser diode can have many advantageous properties than conventional edge-emitting laser, such as circular beam shape, lower divergence angle, 2D laser array possible, efficient testing, low threshold, and so on. Owing to these superior performances, VCSELs had become very attractive and started to be applied to the commercial products at long wavelength range. In fact, short-wavelength VCSELs are also very promising for the applications of storage, display, and projection. In particular, the use of 2D arrays of blue VCSELs could further reduce the read-out time in high density optical storage and increase the scan speed in high-resolution laser printing technology [13]. In 1996, Redwing et al. reported the optical pumping results of GaN-based VCSEL structure at room temperature for the first time [14]. Their VCSEL structure consisted of all epitaxially grown layers, including a 10 μm GaN active region and 30-period $Al_{0.40}Ga_{0.60}N/Al_{0.12}Ga_{0.88}N$-based top and bottom mirror. The threshold pumping energy was as high as 2.0 MW/cm^2 due to the low mirror reflectivity of about 84%~93% and the thick GaN active layer. Subsequently, the $In_{0.1}Ga_{0.9}N$ VCSEL was fabricated and observed lasing action at 77 K by Arakawa et al. in 1998 [15]. The structure was constructed by a hybrid-type cavity consisting of an $In_{0.1}Ga_{0.9}N$ active layer grown on a 35-pair $Al_{0.34}Ga_{0.66}N$/GaN-based bottom DBR with the reflectivity of 97%, and a 6-pair TiO_2/SiO_2 dielectric top DBR with the reflectivity of 98%. In addition, Song et al. using laser lift-off (LLO) technology demonstrated the dielectric-type VCSEL structure consisting of 10-pair SiO_2/HfO_2 top and bottom dielectric DBRs and InGaN MQWs [16]. The cavity quality factor was as large as 600 due to the high reflectivity of the top and bottom DBRs exceeding 99.5% and 99.9%, respectively. Moreover, the room temperature (RT) lasing at blue wavelength in the hybrid-type GaN-based VCSEL under optical pumping has been reported by Someya et al. in 1999 [17]. The structure was formed by an InGaN MQW sandwiched between nitride-based and oxide-based DBRs. Thereafter, Carlin et al. reported the crack-free fully epitaxial nitride microcavity in 2005 [18]. The cavity was formed by lattice-matched AlInN/GaN-based DBR with reflectivity up to 99%. In the following years, several groups report the optically pumped GaN-based VCSELs [19–22]. We also reported optically pumped GaN-based VCSEL structures in dielectric-type cavity [23,24] and hybrid-type cavity structures [25–27]. Figure 3.2 shows the structure of a vertical-cavity surface-emitting laser diode.

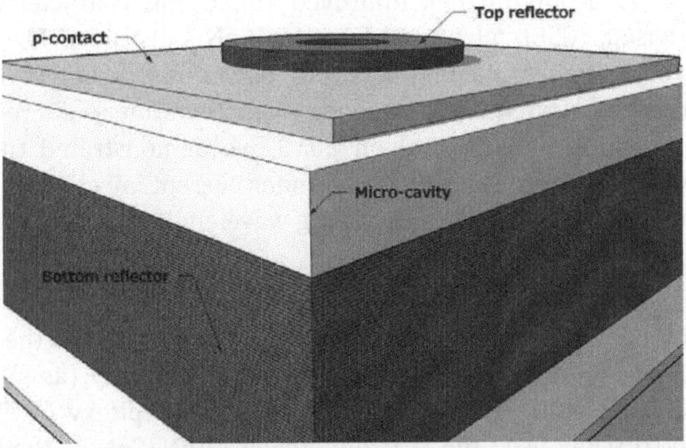

Figure 3.2 The schematic diagram of a vertical-cavity surface-emitting laser diode. Although there are many challenges in fabrication of nitride-based VCSELs, the corresponding structural designs for nitride-based VCSELs can be classified into three major types. Reprinted from Ref. [28].

The electrically pumped GaN-based VCSEL has not been reported until 2008. Lu et al. first demonstrated electrically pumped GaN-based VCSEL with hybrid DBRs at 77 K [29]. The GaN-based VCSEL structure consisted of a 10-pair InGaN/GaN MQW active layer embedded in a GaN hybrid microcavity of 5λ optical thickness and sandwiched between an epitaxial AlN/GaN DBR and a Ta_2O_5/SiO_2 dielectric DBR with reflectivity 99.4% and 99%, respectively. A 240 nm Indium tin oxide (ITO) was deposited on top of the aperture to serve as the transparent contact layer. Continuous wave (CW) laser action was achieved at a threshold injection current of 1.4 mA in a 10 μm aperture at 77 K. The laser emitted a blue wavelength at 462 nm with a narrow linewidth of about 0.15 nm. In the same year, Higuchi et al. demonstrated CW lasing GaN-based VCSEL at room temperature [30]. The structure of the VCSEL consisted of top and bottom dielectric SiO_2/Nb_2O_5 DBRs and a two-pair InGaN/GaN quantum well active layer. The optical cavity consisted of a 7λ-thick GaN semiconductor layer. A 50-nm ITO layer was deposited as a *p*-type ohmic contact and current spreading layer. The LLO technique was used to remove the sapphire substrate, and GaN-based VCSEL was mounted on a Si substrate by wafer bonding. The threshold current was 7.0 mA for an 8 μm aperture device and the emission wavelength was approximately 414 nm. In 2009, the same group further improved the lasing characteristics of GaN-based VCSEL fabricated by using GaN substrates. The results showed a higher maximum output power and a longer lifetime than that fabricated using a sapphire substrate due to the reduction of dislocation density [31]. In 2010, we demonstrated the GaN-based VCSEL with a hybrid cavity under current injection operating at room temperature with the lasing wavelength of 412 nm [32].

3.2.1.1 Fully epitaxial grown VCSEL structure

In 2005, J. F. Carlin [18] and E. Feltin [33] demonstrated the wholly epitaxial and high-quality nitride-based microcavity (as shown in Fig. 3.3(a)) using metal organic vapor phase epitaxy (MOVPE or MOCVD). They used the lattice-matched AlInN/GaN as the bottom and top reflectors to avoid cracks happened due to the accumulation of the strain after stacking large pairs of layers. The reflectivity of AlInN/GaN could be achieved as high as 99.4%. They showed that the 3/2-lambda cavity emitted a very narrow emission with a linewidth of 0.52 nm, corresponding to a quality factor of ~800.

3.2.1.2 VCSEL structure with two dielectric mirrors

Compared to epitaxial grown reflectors, the fabrication of dielectric mirrors is relatively simple. Furthermore, the large index difference of dielectric mirrors makes them easily have wide stop band (>50 nm) and high reflectivity (>99%) by coating just few stacks of 1/4-lambda-thick layers. Therefore, using dielectric mirrors to accomplish nitride-based VCSELs had begun attractive. Song et al. [16], Tawara et al. [21], and J. T. Chu et al. [23,24] successively reported the structure (as shown in Fig. 3.3(b)) after 2000. They employed some process techniques such as wafer bonding and LLO to make dielectric mirrors be coated onto both sides of nitride-based cavity. They showed a microcavity could have a very high-quality factor to be greater than 400 and achieve lasing action using optical pumping.

3.2.1.3 VCSEL structure with hybrid mirrors

The so-called hybrid mirrors are a combination of two different kinds of reflectors, for example, a dielectric mirror and an epitaxial reflector. Typically, the fabrication of this structure is to grow bottom reflector and cavity using MOCVD and then coat dielectric mirror to complete VCSEL structure (as shown in Fig. 3.3(c)). The hybrid-cavity nitride-based VCSEL formed by the dielectric mirror and the epitaxial grown high-reflectivity GaN/$Al_xGa_{1-x}N$ DBR was reported earlier. In 1999, Someya et al. [17] used 43 pairs of $Al_{0.34}Ga_{0.66}N$/GaN as the bottom DBR and reported the lasing action at ~400 nm. Then, Zhou et al. [19] also employed a bottom DBR of 60 pairs $Al_{0.25}Ga_{0.75}N$/GaN and observed the lasing action at 383.2 nm. Both these AlGaN/GaN DBR structures required large numbers of pairs due to the relatively low refractive index contrast between $Al_xGa_{1-x}N$ and GaN. Therefore, recently some groups began to study the AlN/GaN for application in nitride VCSEL. The DBR structure using AlN/GaN has higher refractive index contrast ($\Delta n/n = 0.16$) that can achieve high reflectivity with relatively less numbers of pairs. It has wide stop band that can easily align with the active layer emission peak to achieve lasing action. However, the AlN/GaN combination has also relatively large lattice mismatch (~2.4%) and the difference in thermal expansion coefficients between GaN (5.59×10^{-6}/K) and AlN (4.2×10^{-6}/K) that tends to cause cracks in the epitaxial film during the growth of the AlN/GaN DBR structure and could result in the reduction of reflectivity and increase in scattering loss. With the mature of epitaxy techniques, the high-reflectivity AlN/GaN DBR structure with relatively smooth surface morphology has become possible with just 20 or 30 pairs [34].

In comparison of these three VCSELs, it does not require complicated process such as LLO technique to complete a hybrid VCSEL device. This means the fabrication of such structure is stable and reliable comparing to other structures. Thus, the hybrid structure is more advantageous in the aspects of fabrication and commercialization.

In fact, the investigation of the characteristics of the GaN-based VCSELs has gradually attracted more attentions. Kako et al. [35]

investigated the coupling efficiency of spontaneous emission (β) and the polarization property of the nitride VCSEL and obtained a high β value of 1.6×10^{-2} and a strong linear polarization of 98%. Tawara et al. [21] also found a high β value of 10^{-2} in the nitride VCSEL with two dielectric mirrors. Honda et al. reported the estimation of high characteristics temperature of GaN-based VCSEL [36]. These all mean the development of nitride-based VCSEL and the understanding of the laser performance has become more and more important.

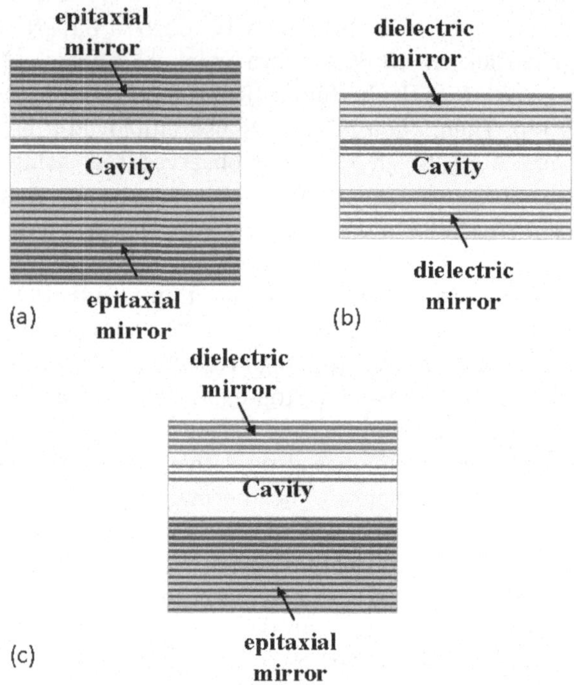

Figure 3.3 Schematic diagrams of three nitride-based VCSEL structures. Reprinted from Ref. [28].

3.2.2 Design and Fabrication of the the Hybrid-Type GaN-Based VCSELs

The design of a typical VCSEL structure should consider three dominant components: the microcavity length, the location of active region, and how many pairs of DBR layers we should coat. To achieve lasing action of VCSEL, a careful evaluation and design for the active region and the DBRs are quite important. In this section, we discuss the design of nitride-based VCSEL with hybrid cavity. The hybrid cavity is a cavity sandwiched by an epitaxial grown AlN/GaN DBR and a Ta_2O_5/SiO_2 dielectric DBR. The design for each component in our VCSEL structure is determined by considering the analysis of the reflectance of DBRs and the electric field distribution inside our structure. The cavity length was designed to be number of lambda in optical length ($\lambda \sim 410$ nm). In addition to the consideration of high-reflectivity mirror, the location of active region, usually in the form of MQWs, is also an important issue to the fabrication of GaN-based VCSEL. The active region plays a role of the gain medium. How to effectively make photons to oscillate with active medium is an important consideration. If we put the gain medium at the node of optical field, the interaction between light and electrons would be very weak, and, therefore, the gain medium could not efficiently provide gain for lasing action. On the contrary, the lasing threshold would be significantly lowered if the active region was put at anti-node of optical field. The same mechanism can be applied in other component of the cavity.

3.2.3 The Optically Pumped Hybrid-Type GaN VCSELs at Room Temperature

The structure of the GaN-based VCSEL was grown in a vertical-type metal organic chemical vapor deposition system(EMCORE D-75) with a fast rotating disk, which can hold one 2-in. wafer. The polished optical-grade 2-in. C-face (0001) sapphire was used as substrate for the epitaxial growth of the VCSEL structure. Trimethylindium, trimethylgallium, trimethylaluminum, and ammonia were used as the In, Ga, Al, and N sources, respectively. Initially, a thermal cleaning process was carried out at 1080°C for 10 min in a stream of hydrogen ambient before the growth of epitaxial layers. After depositing a 30 nm–thick GaN nucleation layer at 530°C, the temperature was raised up to 1045°C for the growth of a 1 μm–thick GaN buffer layer. Then a 25 pairs AlN/GaN DBR structure was grown at 1040°C under the fixed chamber pressure of 100 Torr similar to the previous reported growth condition [34,37]. Then a 380 nm thick *n*-type GaN, followed by a 10-pair $In_{0.2}Ga_{0.8}N$/GaN(2.5 nm/7.5 nm) multiple quantum well and a 100 nm thick *p*-type GaN were grown to form a 3λ cavity. Finally, an eight-pair Ta_2O_5/SiO_2 dielectric mirror was deposited by the e-gun as the top DBR reflector. The schematic diagram and the scanning electron microscopy (SEM) image of the overall VCSEL structure are shown in Figs. 3.4(a) and 3.4(b).

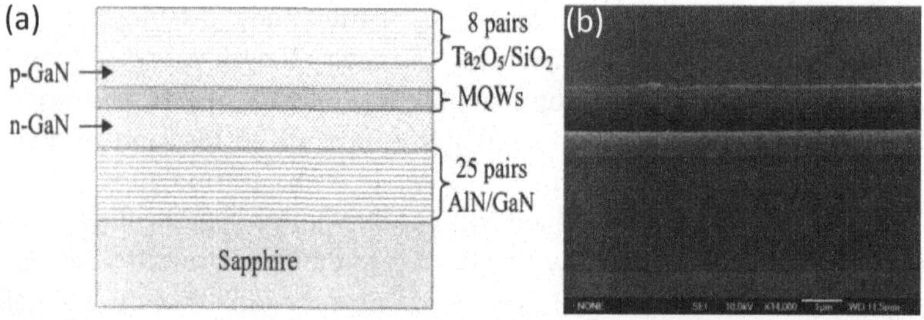

Figure 3.4 (a) The schematic diagram of the overall vertical-cavity surface-emitting laser structure. (b) The SEM image of the full structure.

During these processes, the reflectivity spectrum of the AlN/GaN DBR structure and the Ta_2O_5/SiO_2 dielectric mirror was measured by the n & k ultraviolet-visible spectrometer with normal incidence at room temperature. Figure 3.5 shows the reflectivity spectrum of AlN/GaN DBR and Ta_2O_5/SiO_2 DBR, respectively. The peak reflectance of the top and bottom DBR was 97.5% and 94% at 450 nm, respectively.

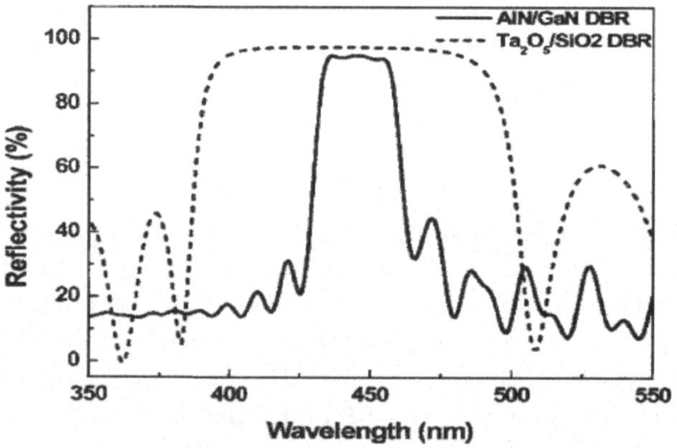

Figure 3.5 The reflectivity spectrum of the top DBR (Ta_2O_5/SiO_2) and the bottom DBR (AlN/GaN); the peak reflectivity of the top DBR is about 97.5% and bottom DBR is about 94% at 450 nm, respectively.

The emission spectrum of the GaN VCSEL structure was all measured using a microscopy system (WITec, alpha snom) at room temperature. The photoluminescence (PL) emission was excited by a 325 nm He–Cd laser with a spot size of about 10 μm diameter. The optical pumping of the sample was performed using a frequency-tripled Nd:YVO_4 355 nm pulsed laser with a pulse width of ~0.5 ns at a repetition rate of 1 kHz. The pumping laser beam with a spot size of 50 μm was incident normal to the VCSEL sample surface. The light emission from the VCSEL sample was collected using an imaging optic into a spectrometer charge couple device (Jobin-Yvon Triax 320 Spectrometer) with a spectral resolution of ~0.1 nm for spectral output measurement.

Figure 3.6 shows the PL spectrum from the GaN-based VCSEL at room temperature. A narrow emission peak with full width at half maximum of 1.4 nm corresponding to the cavity resonant mode at 448 nm was observed. It indicates the emission peak was well aligned with vertical cavity formed by the high reflectance of AlN/GaN DBR and the Ta_2O_5/SiO_2 dielectric mirror. The cavity quality factor estimated from the emission linewidth of 1.4 nm is about 320. This value agrees with the VCSEL cavity formed by AlN/GaN DBR and Ta_2O_5/SiO_2 DBR.

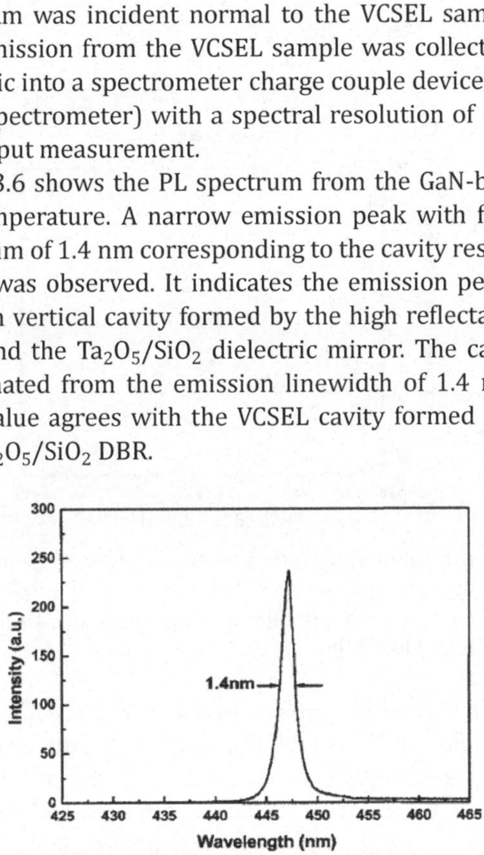

Figure 3.6 The PL spectrum of the GaN-based VCSEL at room temperature.

The light emission intensity from the VCSEL as a function of the pumping energy is shown in Fig. 3.7(a) and replotted it in a semilogarithm scale, as shown in Fig. 3.7(b). A distinct threshold characteristic was observed at the threshold pumping energy (E_{th}) of about 1.5 (μJ) corresponding to an energy density of 53 mJ/cm^2.

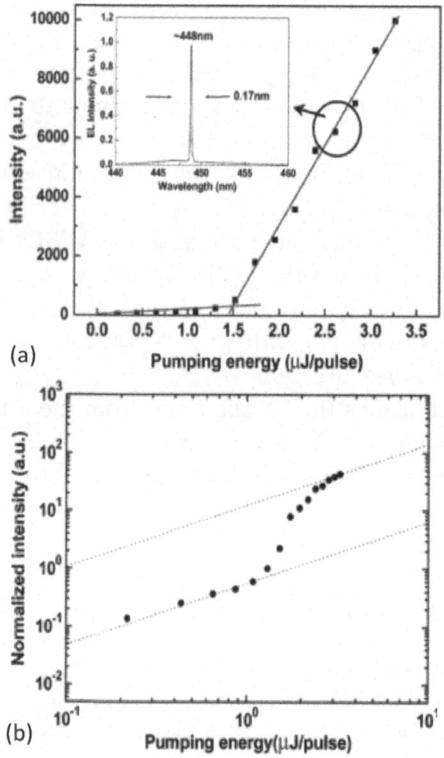

Figure 3.7 (a) Light output intensity as a function of the pumping energy at room temperature. The inset is the spectrum of laser emission at the pumping energy of $1.71E_{th}$. (b) Replotted (a) in a semilogarithm scale.

Then the laser output increased linearly with the pumping energy beyond the threshold. The carrier density at the threshold is estimated to be about 3 × 1020 cm^{-3}, assuming the reflectivity of the top mirror at pumping wavelength of 355 nm was 40%, the absorption coefficient of the GaN was about 10^5 cm^{-1} at 355 nm [38] and the quantum efficiency was 10% [17]. We estimated the threshold gain(gth) of our VCSEL cavity using the equation [17,21]

$$g_{th} \cong (1/2N_w L_w)\ln(1/R_1 R_2)$$

where N_w is the number of quantum wells, L_w is the width of each quantum well, and R_1 and R_2 are the reflectivity of the top and bottom mirrors, respectively. We obtained the required threshold gain which is about 1.45 × 10^4 cm^{-1}. The parameters used in the estimations of carrier density at threshold and threshold gain are listed in the Table 3.1. The threshold gain value is roughly in agreement with the gain value estimated based on Nakamura's [3] report of the gain coefficient of InGaN at our threshold carrier density and slightly higher than the gain value of Park's report [38].

Table 3.1 The parameters used in the estimation of carrier density at thershold and threshold gain

The reflectivity of the top mirror at 355 nm	40%
The absorption coefficient of the GaN at 355 nm	10^5 cm^{-1}*
The quantum efficiency	10%†
The number of quantum wells (N_w)	10
The width of each quantum well (L_w)	3 nm
The reflectivity of the top mirrors (R_1)	97.5%
The reflectivity of the bottom mirrors (R_2)	94.5%

*Ref. [13]
†Ref. [5]

The β value of our VCSEL estimated from Fig. 3.4(b) is about 5 × 10^{-2}. This value is three orders of magnitude higher than that of the typical edge-emitting semiconductor lasers (normally about 10^{-5}) indicating the enhancement of the spontaneous emission into a lasing mode by the high-quality factor microcavity effect in the VCSEL structure.

Figure 3.8 shows the variation of emission spectrum with the increasing pumping energy. A dominant laser emission line at 448 nm appears above the threshold pumping energy. The laser emission spectral linewidth reduces with the pumping energy above the threshold energy and approaches 0.25 nm at the pumping energy of $2.52E_{th}$.

Figure 3.9 shows the laser emission intensity as a function of the angle of the polarizer at the pumping energy of $1.71E_{th}$. The variation of the laser emission intensity with the angle of the polarizer shows nearly a cosine square variation. The degree of polarization (P) is defined as $P = (I_{max} - I_{min})/(I_{max} + I_{min})$, where I_{max} and I_{min} are the maximum and minimum intensity of the nearly cosine square variation, respectively. The result shows that the laser beam has a degree of polarization of about 84%, suggesting a near linear polarization property of the laser emission.

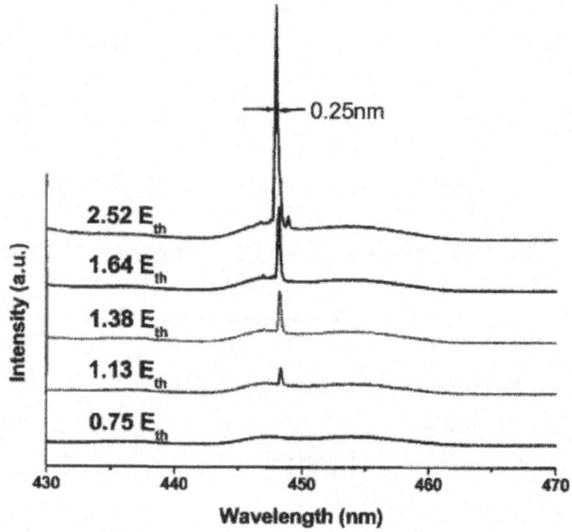

Figure 3.8 The variation of laser emission spectrum with the increasing pumping energy. The laser emission wavelength is 448 nm with a linewidth of about 0.25 nm.

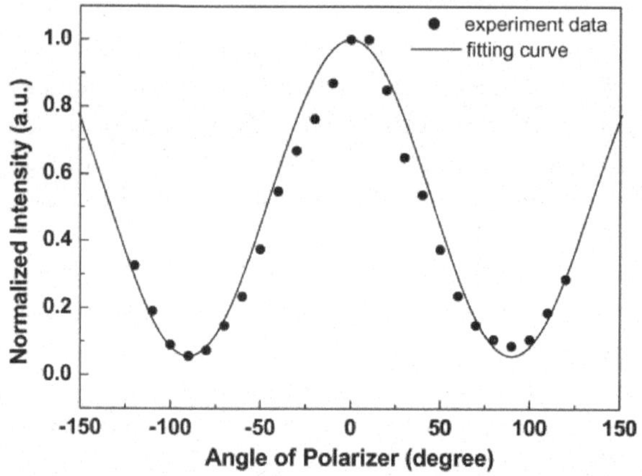

Figure 3.9 Polarization characteristic of the laser emission at the pumping energy of $1.71E_{th}$.

Figure 3.10 shows the seminatural-logarithm plots of the dependence of the threshold pumping energy (ln E_{th}) on the operation temperature (T). The threshold energy gradually increased as the operation temperature rose from 120 K to 300 K. The relationship between the threshold energy and the operation temperature could be characterized by the equation $E_{th} = E_0 \times e^{T/T_0}$, where T_0 is the characteristic temperature and E_0 is a constant. Therefore, we obtain a high characteristic temperature of about 244 K by linear fitting the experiment data. The high T_0 of our sample within the temperature range could be due to various factors. We measured the lasing wavelength variation with the temperature. The result showed a slight red shift of about 1.6 nm as the temperature rose from 120 K to 300 K. The PL emission of the MQW was also measured and the result showed a red shift of about 2.9 nm over the same temperature range. These results suggest that the gain peak almost align with the cavity mode within this temperature variation range. Therefore, the high characteristic temperature obtained from our GaN-based VCSEL sample could be partly attributed to the good alignment of the gain peak and cavity mode. Furthermore, the 10-pair $In_{0.2}Ga_{0.8}N/GaN$ MQW structure which could suppress the carrier leakage from the MQW active layers to the cladding layers and the thick GaN cavity (480 nm in thickness) provides a good heat dissipation path during the high carrier injection and high temperature conditions. We have fabricated a GaN-based VCSEL with hybrid DBR mirrors, consisting of AlN/GaN DBR and Ta_2O_5/SiO_2. We also investigated its performance with emission wavelength at 448 nm and threshold pumping energy density of about 53 mJ/cm^2 under the optical pumping at room temperature. The laser beam has a nearly linear polarization property with a degree of polarization

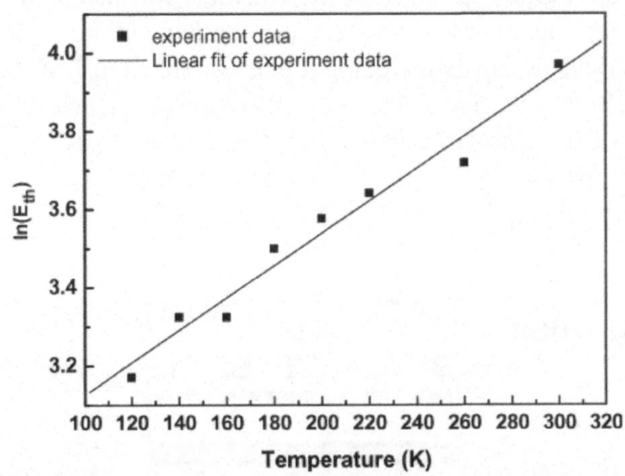

Figure 3.10 Seminatural-logarithm threshold energy as a function of the operation temperature. The solid dot shows the experiment data and the solid line is the linear fit of the experiment data.

a high characteristic temperature of 244 K suggesting potential of about 84%. The laser has a high value of about indicating the coupling coefficient enhancement due to the laser microcavity, and for high temperature applications. The laser also shows a strong spontaneous emission coupling with an estimated coupling efficiency of about 5 × 10^{-2}. The solid dot shows the experiment data and the solid line is the fitting curve.

3.2.4 The Optically Pumped All Dielectric-Type GaN VCSELs at Room Temperature

The VCSEL layer structure grown on a 0001-oriented sapphire substrate by MOCVD included a 30 nm GaN nucleation layer, a 4 μm GaN bulk layer, MQWs consisting of 10 periods of 5 nm GaN barriers and 3 nm $In_{0.1}Ga_{0.9}N$ wells, and a 200 nm GaN cap layer. Six pairs of SiO_2 and TiO_2 were evaporated on the top of the grown VCSEL structure to form the first dielectric DBR. The reflectivity of the SiO_2/TiO_2 DBR at 416 nm was 99.5%. Next, an array of circular SiO_2/TiO_2 DBR mesas with diameters of 60m was formed by standard lithography and chemical wet etching process. To deposit the second dielectric DBR, the structure was subjected to an LLO process using a KrF excimer laser to remove the sapphire substrate. The GaN surface after LLO was further polished and obtained a surface roughness of about 1 nm for deposition of the second dielectric DBR consisting of eight pairs of SiO_2 and Ta_2O_5 as shown in Fig. 3.11. The final thickness of the epitaxial structure after LLO and polishing process was about 4 μm. The Ta_2O_5 was used to reduce the absorption of the pumping beam at a wavelength of 355 nm. The reflectivity of the SiO_2/Ta_2O_5 DBR at 416 nm was 97%.

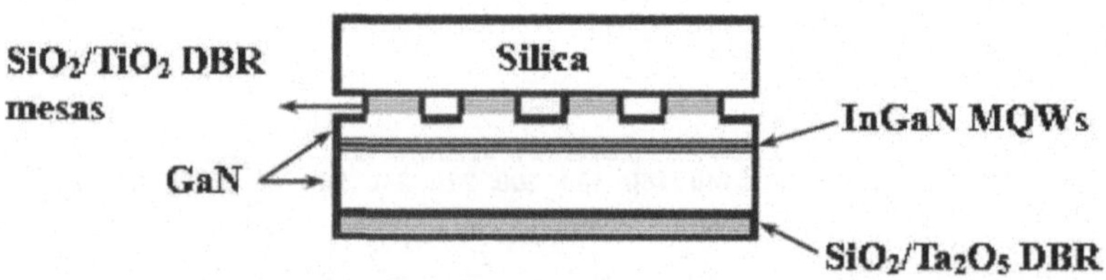

Figure 3.11 The scheme of all dielectric-type VCSEL structure.

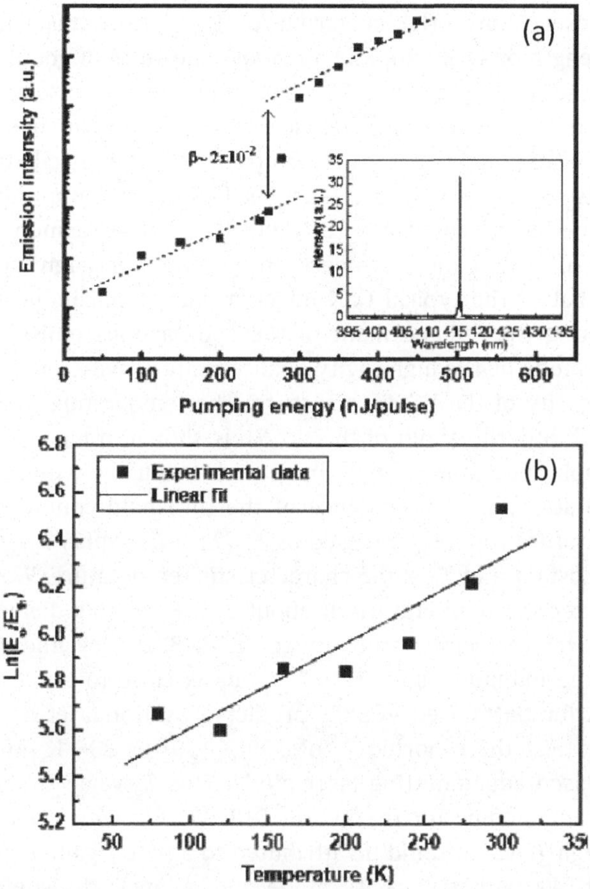

Figure 3.12 (a) Laser emission intensity vs. pumping energy in semilogarithmic scale. The β value estimated from the difference between the two dashed lines is about 2×10^{-2}. The inset shows the spectrum of the laser emission with a wavelength of 415.9 nm and a linewidth of about 0.25 nm. (b) Temperature dependence of the lasing threshold pumping energy (E_{th}) and the estimated T_0 of 278 K. The electrically pumped GaN VCSELs at low temperature (77 K).

The laser emission intensity as the function of pumping energy is shown in Fig. 3.12(a) in a semilogarithmic scale. A clear evidence of threshold behavior was observed at a pumping energy of E_{th} = 270 nJ corresponding to an energy density of 21.5 mJ/cm^2.

The inset shows the spectrum of the laser emission with a wavelength of 415.9 nm and a narrow linewidth of about 0.25 nm. From the semilogarithmic data, the spontaneous emission factor β, which indicated the coupling efficiency of the spontaneous emission to the lasing mode, was estimated from the difference between the heights of the emission intensities before and after lasing was estimated to be about 2×10^{-2}, which was in agreement with the reported results and was nearly three orders of magnitude higher than that of the typical GaN edge-emitting semiconductor lasers, indicating the enhancement of the spontaneous emission into a lasing mode by the high cavity quality factor (cavity Q factor) of the microcavity of the VCSEL structure. The temperature dependence of the lasing threshold of the VCSEL is shown in Fig. 3.12(b). The threshold pumping energy increased gradually with increasing temperature. The dependence of the threshold condition on the temperature can be expressed as $E_{th}(T) = E_0 \exp(T/T_0)$, where E_0 is a constant and T_0 is the characteristic temperature. We obtain a characteristic temperature of about 278 K for this dielectric-type VCSEL for the temperature range of 58–322 K by linearly fitting the experimental result. This T_0 value is close to the T_0 value of 300 K for GaN-based VCSELs predicted by Honda et al. [36] and higher than the reported T_0 of 170 K [39] or 235 K [40] for the GaN-based edge-emitting laser diode. This T_0 value is also higher than the T_0 value for the hybrid DBR VCSEL we reported earlier. The high T_0 value could be attributed to a better gain alignment of the MQWs with the cavity mode and a lower threshold carrier density due to the higher quality factor provided by both dielectric DBRs.

For the electrically pumped VCSEL structure, the ITO layer as a current spreading layer should be considered as one important part of the whole cavity because it would increase the absorption when light resonates in the cavity region. We observe the CW laser action of current injection GaN-based VCSEL at 77 K. The laser has an InGaN MQW active layer embedded in a high-quality hybrid microcavity structure with two high reflectivity mirrors which consist of an epitaxially grown crack-free AlN/GaN DBR and a Ta_2O_5/SiO_2 dielectric DBR. This hybrid microcavity structure allows the incorporation of an intracavity current injection structure that has a highly transparent and conductive ITO current injection layer as the *p* contact and a coplanar *n* contact. The laser action was achieved under CW current injection at 77 K with single mode emission at 462.8 nm at a low threshold current of 1.4 mA and a high degree of polarization.

Figure 3.13(a) shows the schematic of the hybrid microcavity GaN VCSEL structure. The device fabrication steps are as follows: first, the MOCVD reactor was used to grow the device epitaxial structure. The epitaxial structure mainly consists of a 29-pair AlN/GaN DBR and a 5λ (our target wavelength is λ = 460 nm) optical thickness microcavity composed of a 790 nm thick Si-doped *n*-type GaN, 10 pairs of $In_{0.2}Ga_{0.8}N$ (2.5 nm)/GaN (7.5 nm) MQWs, and a 120 nm thick Mg-doped *p*-type GaN layer. The MQWs were located at the antinode of light field in the microcavity for enhancing the coupling of photons and the cavity mode. To obtain a crack-free and high reflectivity AlN/GaN DBR, we insert the AlN/GaN superlattices into the AlN/GaN DBR structure during the epitaxial growth to reduce the biaxial tensile strain in AlN/GaN DBR [41]. The superlattice has 5.5 pairs of AlN/GaN corresponding to a half-wavelength stack and was inserted in every four pairs of AlN/GaN quarter wavelength stacks to form a 29-pair AlN/GaN bottom DBR. The grown DBR showed a high peak reflectivity of R = 99.4% with a spectral bandwidth of 25 nm. The epitaxially grown structure was then processed to form the intracavity coplanar *p* and *n* contacts for current injection, as shown in Fig. 3.13(a). A 200 nm thick SiN_x layer was used to form a current injection and a light-emitting aperture of 10 μm in diameter. An ITO was deposited on top of the aperture to serve as the transparent contact layer. Since the ITO locates within the VCSEL microcavity, the thickness of ITO is set to 240 nm, corresponding to 1λ (optical length is λ = 460 nm) in order to match the resonance phase condition of the microcavity. We annealed the ITO layer at 525°C under the nitrogen ambient to reduce the contact resistance as well as to increase transparency, thus reducing the internal cavity loss. The deposited ITO has a measured transmittance of about 98.6% at λ = 460 nm. Finally, an eight-pair Ta_2O_5/SiO_2 dielectric DBR (measured reflectivity of about 99% at λ = 460 nm) was deposited as the top DBR mirror to complete the full hybrid microcavity VCSEL device.

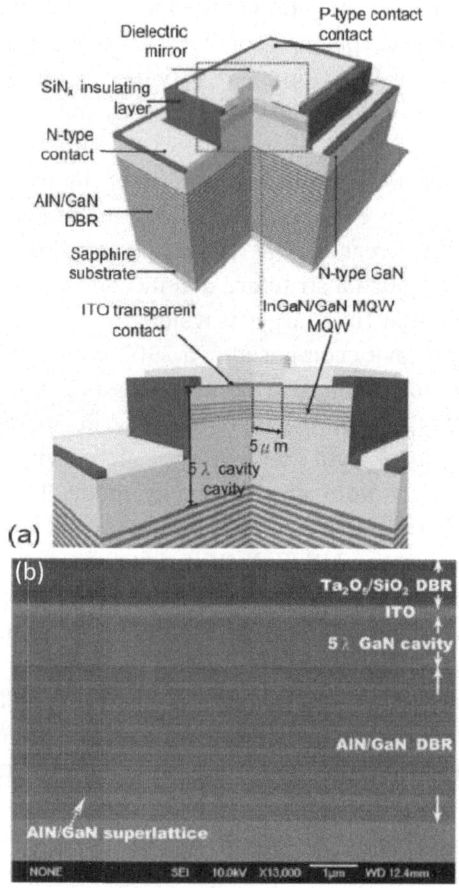

Figure 3.13 (a) The schematic diagram of the GaN VCSEL. (b) The cross-section image of the VCSEL microcavity with hybrid DBRs observed by SEM.

3.2.5 Result and Discussion

Figure 3.13(b) shows the SEM image of the completed VCSEL devices. For VCSEL performance characterization, we diced the fabricated VCSEL devices into an individual device size of 120 × 150 μm^2 and packaged them into the TO can. The packaged VCSEL device was mounted inside a cryogenic chamber for testing under CW current injection condition using a CW current source at 77 K. The emission light was collected by a 25 μm diameter multimode fiber using a microscope with a 40× objective (numerical aperture = 0.6) and fed into the spectrometer. The system has a focal distance of 320 mm and a grating of 1800 g/mm with a spectral resolution of 0.15 nm. The output from the spectrometer was detected by a charge-coupled device (CCD) to record the emission spectrum. The spatial resolution of the imaging system was about 1 μm as estimated by the diffraction limit of the objective lens. The near and far field patterns were measured using a confocal microscopy system with two symmetric lenses having a magnification of 40× and a fiber with a 25 μm diameter located at the focal plane for collection of the light emission from the VCSEL sample.

Figure 3.14 shows the light output power versus CW injection current and current-voltage characteristics of the VCSEL sample at 77 K. The turn on voltage is about 4.1 V, indicating the good electrical contact of the ITO transparent contact layer and the intracavity current injection scheme. The laser light output power showed a distinct threshold characteristic at the threshold current (I_{th}) of about 1.4 mA and then was linearly increased with the injection current beyond the threshold. The threshold current density is estimated to be about 1.8 KA/cm^2 for a current injection aperture of 10 μm in diameter, assuming the current is uniformly injected within the aperture.

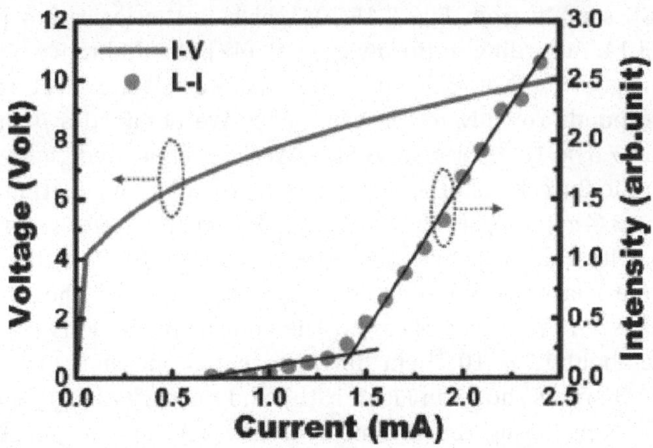

Figure 3.14 The light output intensity vs. injection current and current–voltage characteristics of GaN VCSEL measured under the CW condition at 77 K. The solid line in light output intensity vs. current characteristic is a guide for the eye.

Figure 3.15(a) shows the laser emission spectrum at various injection current levels. A dominant single laser emission line at 462.8 nm appears above the threshold current. Figure 3.15(b) shows the light emission linewidth at various injection current levels. The laser emission spectral linewidth drastically reduces with the injection current above the threshold current and approaches the spectral resolution limit of 0.15 nm at the injection current of $1.7I_{th}$. The inset of Fig. 3.15(a) shows the CCD image of the spatial laser emission pattern across the 10 μm emission aperture at an injection current of 1 mA slightly below the threshold. We observed a nonuniform emission intensity across the emission aperture with several bright emission spots. We believe the nonuniformity in the emission intensity across the aperture could be due to the In nonuniformity that creates a nonuniform spatial gain distribution in the emitting aperture as reported earlier. In fact, the lasing action we observed was mainly from those spots with the brightest intensities as indicated in the inset of Fig. 3.15(a). The spatial dimension of these bright spot clusters is only about a few microns in diameter. As shown in Fig. 3.15(b), the diameter of the near field intensity of the laser spot at the threshold is around 3 μm. A similar result was also observed for the optically pumped GaN VCSELs. We estimated the spontaneous emission coupling factor β of our VCSEL sample from Fig. 3.15(b), which is the logarithm plot of Fig. 3.14. According to Horowicz et al. [42], the difference between the heights of the emission intensities before and after the threshold corresponds roughly to the value of β. We obtained a β value of about 7.5×10^{-2} for our VCSEL. We also estimated the β value from the Purcell factor Fp using the approximation equations [35]: $Fp = (3/4\pi^2) \times \{Q/[V_c/(\lambda/n)^3]\}$ and $\beta = Fp/(1 + Fp)$, where Q is the cavity quality factor, λ is the laser wavelength, V_c is the optical volume of laser emission, and n is the refractive index. The refractive index is 2.45 for the GaN cavity. The optical volume V_c is estimated to be about 1.2×10^{-11} cm^3 for a measured emission spot size of about 3 μm, as shown in Fig. 3.15(b), and the cavity length of about 10.5λ, considering the thickness of the ITO and the penetration depth of the DBRs. The cavity Q is about 1.8×10^3 based on the emission linewidth of 0.25 nm near the threshold. By using these parameters, we obtained an estimated Purcell factor of about

7.9×10^{-2} and an estimated β value of about 7.4×10^{-2}, which is close to the value we obtained above from Fig. 3.15(b). This β value is nearly three orders of magnitude higher than that of the typical edge-emitting laser which is generally in the order of 10^{-4}–10^{-5}.

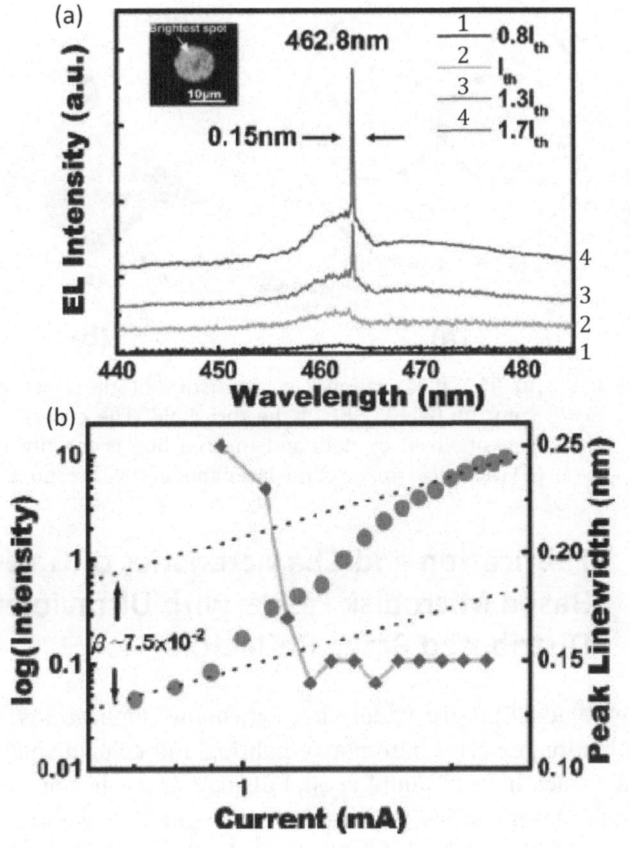

Figure 3.15 (a) The laser emission spectrum at different injection current levels measured at 77 K. (Inset: CCD image of the emission from the aperture). (b) The logarithm light output intensity and the laser emission linewidth vs. the injection current. The dash lines are guides for the eye.

Figure 3.16 shows the polarization characteristics and near field beam patterns of the laser. The laser emission has a degree of polarization of about 80%. A near diffraction-limited far-field pattern with a small divergence angle of about 11.7° in both horizontal and vertical directions was measured.

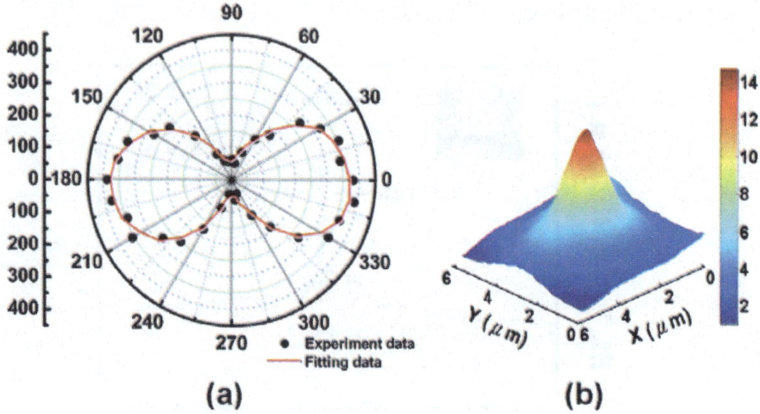

Figure 3.16 (a) The polarization characteristics of the laser emission from the GaN VCSEL at the threshold. The experiment data is symbolized by dots and the red line is the fitting result. (b) Near field image of the laser spot at the threshold.

3.3 Fabrication and Characteristics of GaN-Based Microdisk Lasers with Ultraviolet Distributed Bragg Reflectors

Today, blue LEDs are widely used in many applications such as illumination, exterior automotive lighting, full color display, traffic signals, back light of liquid crystal display, and other applications. The blue laser diodes (LD) can serve as the light source of high density data storage (about 25 GB/disk, 5 times larger than now) in high definition digital versatile disk (HD-DVD) which is the main stream of next generation data storage. The big market of HD-DVD has shown up the importance of blue LDs. Besides, the blue and violet LDs may have many other possible markets, such as high- brightness projector, high-speed printer, medical field, and others.

To extend the applications of GaN-based lasers/LEDs into UV region, there are many studies related to the UV GaN-based DBR [15–20]. Blue light microdisk lasers have also been reported in

GaN-based suspended membranes formed by photoelectrochemical etching [4,21].

3.3.1 Ultraviolet AlN/AlGaN DBR

In general, ultraviolet nitride-based DBRs consist of AlInN/AlGaN or AlGaN/AlGaN material system. Especially, for devices containing pure GaN as an active medium, the ultraviolet DBRs cannot employ GaN as the layer materials because of the strong increase of optical absorption below 360 nm. In this section, we report the growth on 2-in. c-sapphire substrates of high-reflectivity AlN/AlGaN DBRs designed for the ultraviolet spectral region. The structures are grown by MOCVD and consist of AlN/$Al_{0.2}Ga_{0.8}N$ system to increase the difference of refractive index in the DBR structures. The measured peak reflectivity value is 85% for 25 pairs of AlN/AlGaN DBRs. Furthermore, the experimental reflectivity spectra are modeled by transfer matrix theory in order to compare the experimental and theoretical results as shown in Fig. 3.18.

3.3.2 Design and Growth of Ultraviolet AlN/AlGaN DBR

The AlN/AlGaN DBRs were grown in a low-pressure high-speed rotating-disk MOCVD system. Two-inch diameter (0001)-oriented sapphire substrates were used for the growth of DBR samples. During the growth, TMGa and TMAl were used as group III source materials and NH_3 as the group V source material. After thermal cleaning of the substrate in hydrogen ambient for 5 min at 1100°C, a 30 nm thick GaN nucleation layer was grown at 500°C. The growth temperature was raised up to 900°C for the growth of 3.4 μm GaN buffer layer. Then, the AlN/AlGaN DBRs were grown under the fixed chamber pressure of 1.33×10^4 Pa. The DBR consists of 25 pairs of AlN/$Al_{0.2}Ga_{0.8}N$ quarter wavelength layers. By employing in situ optical reflectivity system, the growth rate of the AlN and AlGaN layer can be estimated to be about 0.7 and 2.2 Å/s, respectively. The thicknesses of the total DBR structures were investigated by SEM as shown in Figs. 3.17(a) and 3.17(b). The reflectivity spectrum of the AlN/AlGaN DBRs was measured by the n & k ultraviolet-visible spectrometer with normal incidence at RT [43–45].

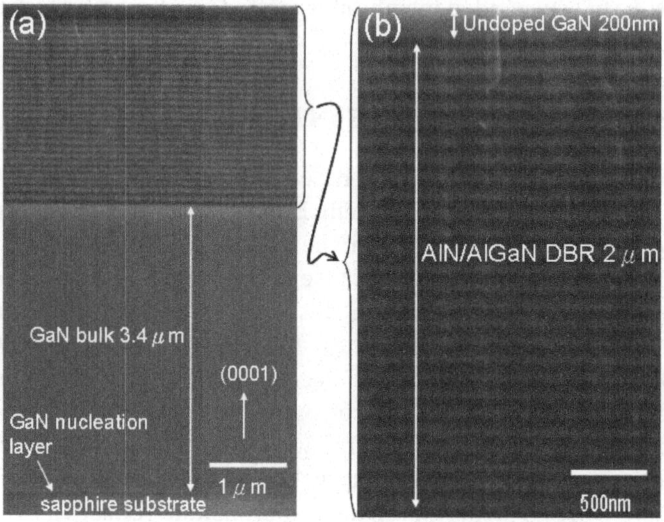

Figure 3.17 (a) An SEM image of UV DBR from cross-section view. The total thickness of 25 pairs of AlN/AlGaN DBR is about 2 μm. (b) The SEM image of the epitaxial gain layer with DBR structure from cross-section view.

From Fig. 3.17, the AlN/AlGaN interfaces are well defined and flat. According to the SEM images, the total thickness of 25-pair $AlN/Al_{0.2}Ga_{0.8}N$ DBRs was measured to be about 2 μm. Therefore, the average bilayer thickness of 87 nm DBR was estimated based on the SEM measurements. Since the bilayer thickness is about 80~90 nm, the cross-sectional SEM images can provide sufficient resolution for the estimation of bilayer thicknesses [33,46]. Furthermore, by a combination of double crystal X-ray diffraction (XRD), growth rate calibrations, and SEM measurements, we can estimate the layer thicknesses of AlN and AlGaN, respectively. The thickness of AlN and AlGaN layers are 45 nm and 42 nm decided by the formula $d_{AlN} = \lambda/4n_{AlN}$ and $d_{AlGaN} = \lambda/4n_{AlGaN}$. Here n_{AlN} and n_{AlGaN} are the refractive indices of AlN and AlGaN, which are 2.033 and 2.197, respectively. Moreover, the GaN nucleation layer and GaN buffer layer can be clearly observed in Fig. 3.17(a).

3.3.3 Analysis of Reflectance Spectra

In Fig. 3.18, experimental measurement shows that the peak reflectivity at 375 nm is about 85% and the stop-band width is

15 nm. To compare the experimental and calculated reflectivity spectra, theoretical simulation based on transfer matrix theory was performed using above layer thicknesses. As for the parameter of refractive index, Adachi model is employed to calculate the refractive index values of GaN, AlN, and $Al_{0.2}Ga_{0.8}N$ layers. The Adachi model can be expressed by the following equation [47]:

$$n(h\nu) = \sqrt{a(x)\left(\frac{h\nu}{E_g}\right)^{-2}\left[2-\sqrt{1+\left(\frac{h\nu}{E_g}\right)}-\sqrt{1-\left(\frac{h\nu}{E_g}\right)}\right]+b(x)} \quad (3.1)$$

where E_g is the band gap of the material, ν is the frequency of the incident light, h is Planck's constant, and $a(x)$ and $b(x)$ are composition-dependent fitting parameters. This equation showed good agreement with measurements on GaN, AlN, and InN for the transparency region [48,49]. In our calculation, we employed the fitting parameters for AlN layer from Brunner et al. [50]. As for the $Al_{0.2}Ga_{0.8}N$ layer, the fitting parameters are more reliable from Laws et al. when the aluminum composition is smaller than 0.38 [51]. Figure 3.18 also shows the calculated reflectivity spectrum of DBR. By comparing the measured and calculated results, the characteristics of simulated reflectivity spectrum of DBR including stop-band width and the phase of the long wavelength oscillations

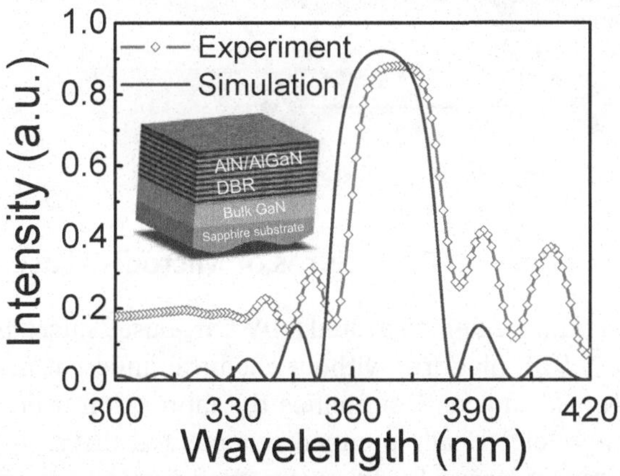

Figure 3.18 Calculated (solid) and measured (dash) reflectivity spectra of ultraviolet AlN/AlGaN DBR measured at room temperature with a stop-band width of about 15 nm and the center wavelength is 375 nm.

are in good agreement with the measured spectrum. Nevertheless, the calculated peak reflectivity is about 85%, which is larger than the measured result. This difference may come from dislocations in the samples and the structural imperfections such as deviations from the designed layer thickness and interface roughness between each epitaxial layer.

3.3.4 The Set-Up of Microphotoluminescence

In Fig. 3.19, the PL emission was excited by a 325 nm He-Cd laser with a spot size of about 30 μm diameter and with an incident power 40 mW. By using the microscopy system (WITec, alpha snom), the emission was collected into a spectrometer/CCD (Jobin-Yvon Triax 320 Spectrometer) with a spectral resolution of ~0.15 nm for spectral output measurement.

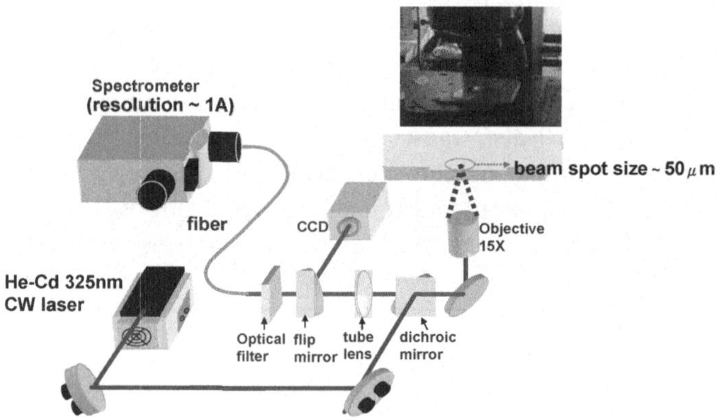

Figure 3.19 Optical pumping system.

3.3.5 Fabrication Procedures of Microdisk Lasers

In this section, we demonstrated a UV GaN-based microdisk laser in AlN/AlGaN DBR platform without chemical undercut process. This type GaN laser not only simplifies the fabrication procedures, but also benefits electrically pumped scheme and thermal conduction of the compact lasers. To achieve a good vertical confinement, an AlN/AlGaN UV DBR structure was designed on the bottom of the microdisk cavity. It can be acted as a mirror to reflect light from the bottom area and played the role as a lower refracted index layer to control the guided modes. Under room temperature optical pumping condition, the lasing action was observed and its lasing peaks were at wavelength 377 nm and 379 nm in the UV range.

GaN microdisk cavities were implanted in an ultraviolet GaN-based AlN/AlGaN DBR structure. This AlN/AlGaN DBR structure was grown by a low pressure MOCVD system. A 3.4 μm thick undoped GaN was first grown on a C-plane (0001) sapphire substrate. Then 25-pair $AlN/Al_{0.2}Ga_{0.8}N$ structure was grown at 900°C, followed by a 200 nm undoped GaN gain layer on the top of the epitaxial structure. The cross-section SEM image of UVDBR was shown in Fig. 3.17. The structural quality of the UVDBR layers was controlled by compensating the compressive and tensile stress in each $\lambda/4$ pair. The growth details were reported in our previous works [51,52]. Figure 3.20 shows the fabrication processes of microdisk structure. A 300 nm silicon nitride layer and a 300 nm polymethyl methacrylate (PMMA) layer were deposited on the top of the GaN wafer as the masks during the processes. Microdisk patterns were defined on the PMMA layer by the electron beam lithography. The patterns were then transferred into the DBR layer by reactive ion etching (RIE) with CHF_3/O_2 mixture and inductive coupled plasma (ICP) etching with Cl_2/Ar mixture. The illustration of a microdisk on the AlN/AlGaN DBR structure is shown in Fig. 3.21.

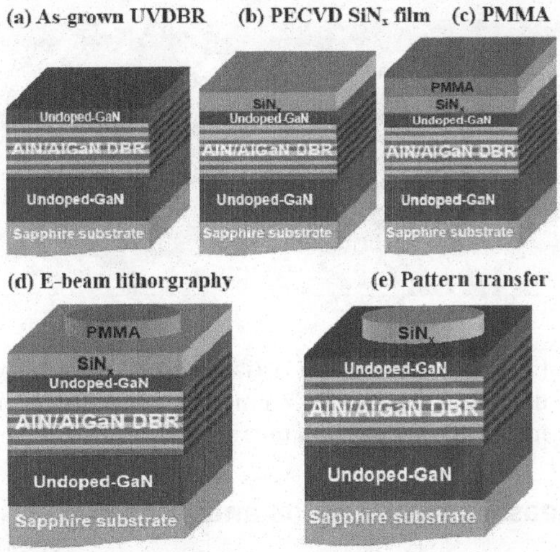

Figure 3.20 Fabrication flowcharts: (a) As-grown sample structure. (b) Deposit SiN_x film by PECVD. (c) Spin on PMMA. (d) E-beam lithography. (e) Microdisk pattern transfer to SiN_x layer.

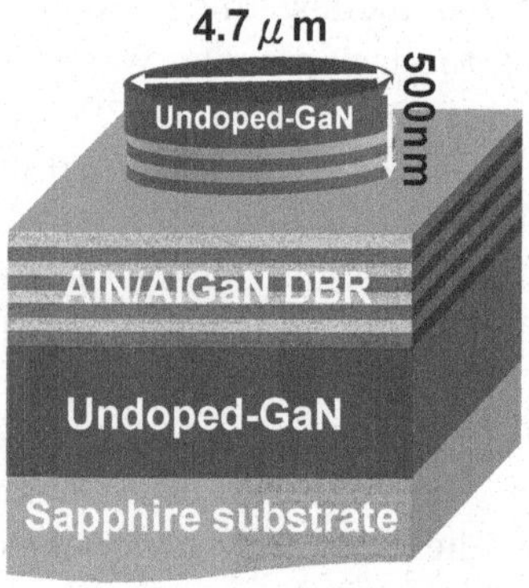

Figure 3.21 Schematic structure of a GaN-based microdisk with AlN/AlGaN DBR.

Figure 3.22(a) is a top-view SEM image of a microdisk array, and Fig. 3.22(b) is the magnified image of a microdisk from angle view. The size of the fabricated microdisk is approximately 4.7 μm in diameter and the etch depth was about 500 nm. It is worth to notice that the chemical undercut or deeply etching steps are unnecessary for this GaN microdisk cavity because of the good vertical confinement from the DBR structure, compare to GaN microdisk cavity without the DBR [11,53,54].

Figure 3.22 (a) An SEM image of a GaN microdisk array with different diameters of 7 μm, 4.7 μm, and 3 μm. (b) A magnified SEM image of a 4.7 μm GaN microdisk cavity from the angle view.

3.3.6 Measurement Results and Analysis

The fabricated GaN-based microdisk cavities were then optically pumped by using a frequency-tripled Nd:YVO_4 355 nm pulsed laser with a pulse width of 0.5 ns and a repetition rate of 1 kHz. The normal incident beam has a spot size of 50 μm which can cover the whole microdisk. The light emission from the device was collected by a 15× objective lens through a multimode fiber and coupled into a spectrometer with the CCD detectors as shown in Fig. 3.19.

Figure 3.23 shows the measured PL spectrum from non-pattern area of the UV DBR sample. The gain peak of the undoped GaN is located around 360 nm wavelength. The lasing action of the GaN microdisk cavity on the DBR structure was observed during the characterization. Figure 3.24 shows the measured spectra from a 4.7 μm microdisk cavity below (red) and above (black) threshold. There are four resonant modes in the spectrum which are labeled with modes A, B, C, and D. Two lasing modes (mode A and mode B) were observed at 377 nm and 379 nm wavelength. Figure 3.25 shows the light-in light-out (L–L) curves of mode A (black) and mode B (red). Their threshold power densities are 0.03 kW/cm^2 and 0.043 kW/cm^2, respectively.

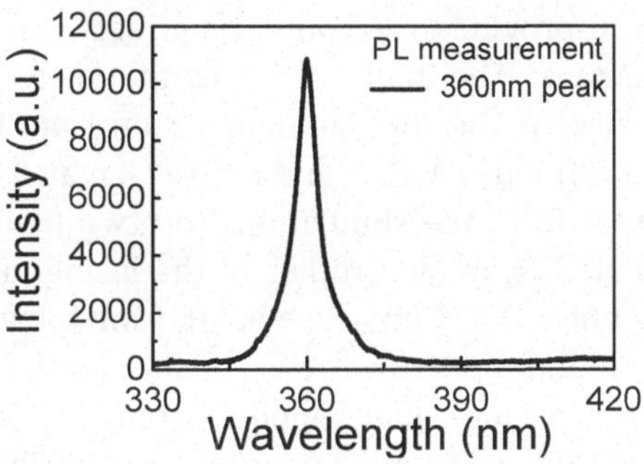

Figure 3.23 The PL spectrum of undoped GaN layer.

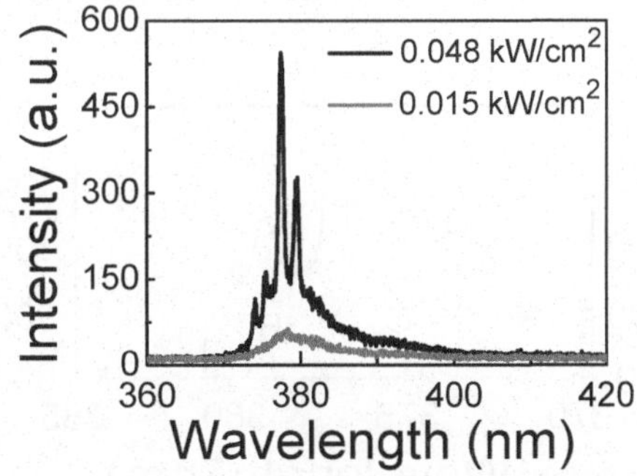

Figure 3.24 The measured spectra from a microdisk laser below and above threshold power. The lasing wavelengths of the microdisk are 377 nm and 379 nm.

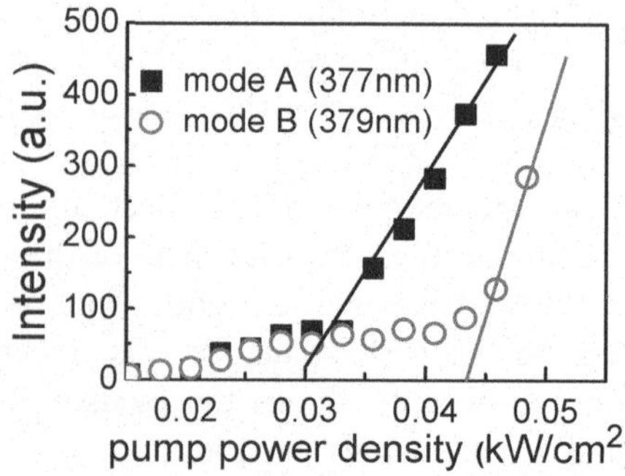

Figure 3.25 The light-in–light-out curve (L–L curve) from the microdisk laser.

To understand more details of the lasing mode of the GaN microdisk, the finite-difference time-domain method (FDTD) with the effective index was used to perform the simulation for this 4.7 μm microdisk cavity.

Figure 3.26 provides the comparison between the simulation and measurement. The blue curve is the measured spectrum, and the red curve is the simulated spectrum from FDTD. The four resonant modes (modes A, B, C, and D) are all match well to the high response modes from the simulation. Top two figures of Fig. 3.27 are the calculated H_z mode profiles of the lasing modes (modes A and B) at 377 nm and 379 nm wavelength from top view. Two lasing modes, mode A and mode B were verified to be the first-order and third-order whispering-gallery modes from the FDTD calculated mode profiles. Since the higher-order whispering-gallery mode usually has a lower Q value, the mode B has a higher threshold power density which is observed from the L–L curves in Fig. 3.25.

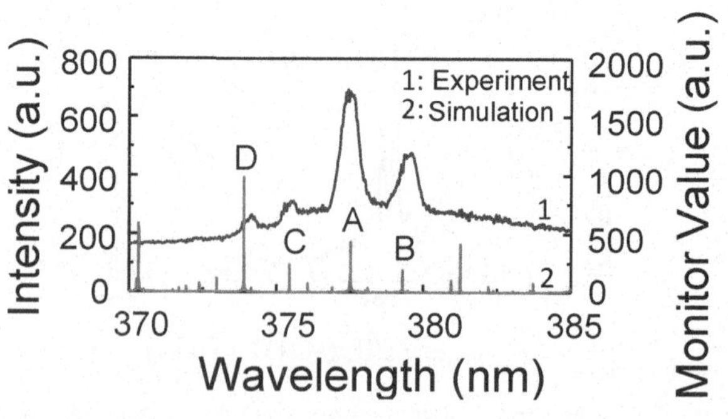

Figure 3.26 Comparison of FDTD simulation (2) and measurement (1).

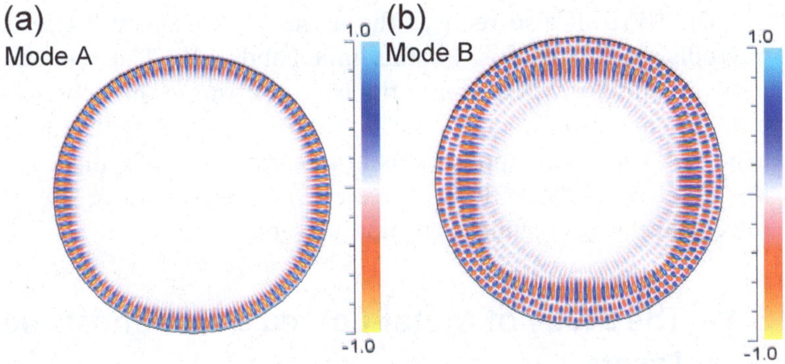

Figure 3.27 (a) The calculated H_z mode profiles for mode A and (b) mode B from FDTD.

In the GaN-based microdisk cavity, the UV DBR plays an important role in selecting the lasing wavelength regime. The gain peak of the GaN without the DBR is around 360 nm wavelength; however, the lasing and resonant modes are around 377 nm wavelength. This 17 nm difference in wavelength is attributed to the reflection of the UV DBR mirror on the bottom of the microdisk cavity as shown in Fig. 3.28. The region of the resonant modes from the cavity also agrees to the bandwidth of the reflectivity spectrum of the DBR in Fig. 3.18. This DBR effect in the GaN epitaxial structure had been observed in our previous works [22].

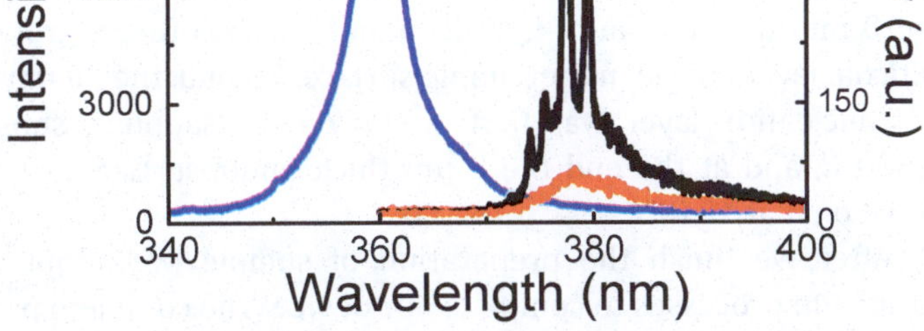

Figure 3.28 17 nm difference in wavelength due to the reflection of the UV DBR.

The UV DBR also reduces the number of resonant modes in microdisk cavity because of the limit bandwidth. The reduction of mode number will decrease the energy waste in the non-lasing modes. Therefore, a high efficient GaN microdisk laser can be expected. For future applications in integrated circuits, this cavity integrated with the DBR structure opens a possibility to tune or select wavelength regime for compact lasers.

3.4 The Study of Metal-Coated GaN Nanostripe Lasers

In this section, we demonstrated lasing in metal-coated GaN nanostripes at room temperature under optical pumping conditions. To make lasing action in this device possible, we employed SiO_2 as an insertion layer between the metal and GaN to improve the quality factor, in conjunction with aluminum providing high reflectivity in the UV region and high thermal conductivity. This is the first time that lasing action has been observed in GaN-based metal-coated nanocavities around ultraviolet wavelengths region at room temperature.

3.4.1 Fabrication of Metal-Coated GaN Nanostripe

We used the undoped GaN layer grown on C-plane (0001) sapphire substrate as gain medium. The thickness of the undoped GaN layer is about 2 μm. MOCVD system (EMCORE D-75) is used to grow undoped GaN layer on the polished optical-grade C-plane (0001), 2-in. sapphire substrate. Trimethylgallium (TMGa) and ammonia (NH_3) were used as the Ga and N sources, respectively.

First, a thermal cleaning process was carried out at 1080°C for 10 min in a stream of hydrogen ambient before the growth of epitaxial layers to clean the sample surface. Second, the 30 nm thick GaN nucleation layer was first grown on the sapphire substrate at 530°C, and at the end the 2 μm thick undoped GaN layer was grown on it at 1040°C.

After we finish the preparation of sample, we adopt some fabrication processes to complete our device. The complete process flow chart for our device is shown in Fig. 3.29. First, we deposit 300 nm thick Si_3N_4 layer on the undoped GaN layer as an etching mask for the following E-beam lithography process. Second, for E-beam lithography process, we use spin coater to put a thin PMMA on the sample, then the nanostripe pattern was defined using E-beam lithography system. After this, we use ICP-RIE dry etching system to transfer the pattern on the PMMA layer onto the Si_3N_4. And the transfer the pattern again onto the undoped GaN layer to form the nanostripe structure on it.

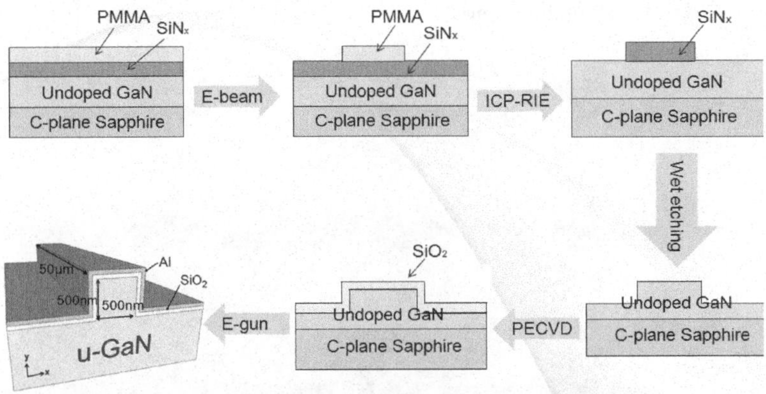

Figure 3.29 Process flow chart for metal-coated GaN nanostripe.

Figure 3.30 shows the SEM image of nanostripe from top view and angle view after these processes. Third, before we start to do the next step, we use wet etching method to clean our sample, washing away the particles created in the previous dry etching processes to promote the performance of our device. After this, we use plasma-enhanced chemical vapor deposition (PECVD) to deposit an SiO_2 layer of 20 nm [55,56] on the GaN nanostripe. At the end, we use E-gun evaporation to deposit a 60 nm thick aluminum layer [57] to complete our device. The SEM image of our device after the deposition of dielectric layer and metal is shown in Fig. 3.31 in angle view. The detail recipe of each process is shown in the following paragraphs, and the schematic diagram of metal-coated GaN nanostripe is shown in Fig. 3.32.

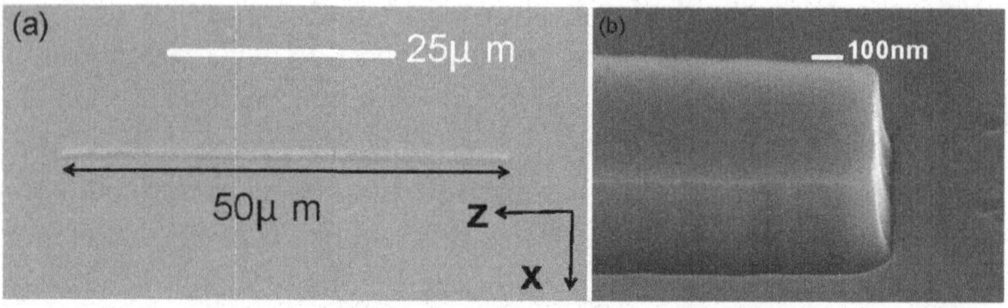

Figure 3.30 SEM image of metal-coated GaN nanostripe. (a) Top view of the GaN nanostripe before deposition of shielding layers. (b) Angle view of one side of the GaN nanostripe before deposition of shielding layers.

Figure 3.31 Angle view of the GaN nanostripe after the deposition of SiO_2 and aluminum layers.

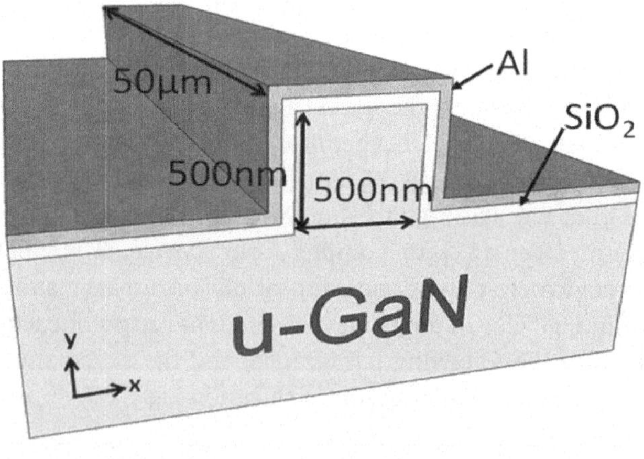

Figure 3.32 Schematic diagram of metal-coated GaN nanostripe.

3.4.2 Characteristics of Optically Pumped Nanostripe Lasers

To explore experimentally about our device, we use the microphotoluminescence system mentioned in Fig. 3.19 to measure its characteristics. We pump our device directly from the metal side to avoid the absorption of undoped GaN layer beneath our device, if we choose to pump the device from the back of the wafer.

To ensure that the lasing action is truly originated from our nanostructure, we use a He-Cd 325 nm continuous-wave laser to pump the undoped GaN layer, before and after the deposition of dielectric layer and metal to see the PL spectrum of them. As shown in Fig. 3.33, the peak wavelength of undoped GaN layer is about 362 nm. However, the spectrum after the deposition of SiO_2 and aluminum layer is totally different, the peak around 362 nm no longer exists because the energy from 350 nm to 380 nm had been totally absorbed by shielding layer. Therefore, we could use Fig. 3.33 to prove that the lasing action is truly from our nanostructure. Moreover, the threshold power density has been presented using the PL spectrum of the metal-coated nanostripe above (gray) and below (blue) as shown in Fig. 3.34. This device has a single lasing mode at 370 nm and the difference between two spectrums further ensure that we observe a lasing action in our structure. From the light-in light-out curves of this mode shown in Fig. 3.35 and the linewidth of the lasing mode, we could observe linear behavior after soft turn-on, which indicates the lasing action, and also the narrowing linewidth shows the transition from spontaneous emission to stimulated emission of the optical mode. These evidences prove the lasing action in our metal-coated GaN nanostripe. The threshold pump power density is about 0.055 kW/cm^2 and the quality factor estimated by the wavelength to linewidth around the transparency is 150. Compared to the sample without metal or SiO_2 shielding layer, the high thermal conductivity and high reflectivity of aluminum make measurement easier and increase the possibility of lasing at room temperature. The nanostripe without shielding layer exhibits high optical loss and lower Q value. It was for these reasons that we cannot observe lasing action in uncoated GaN nanostripe. The reason why we use aluminum instead of other metal such as silver to form our metal-coated device is that aluminum has a higher reflectivity which could show a better optical confinement at UV wavelength region compared with silver [58], even aluminum is also quite lossy at this wavelength region.

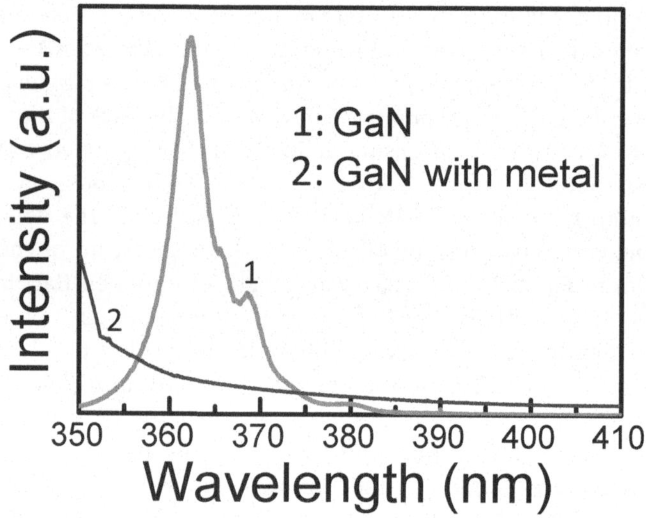

Figure 3.33 PL spectrum of undoped GaN layer with and without metal and dielectric shielding layers.

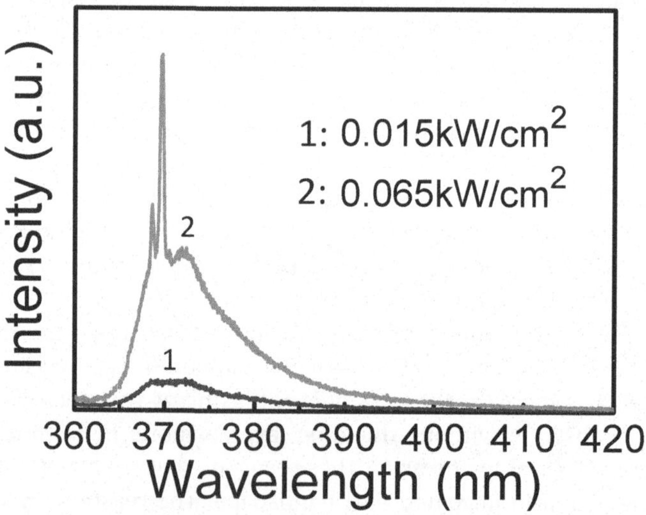

Figure 3.34 PL spectrum of metal-coated GaN nanostripe above (2) and below (1) threshold.

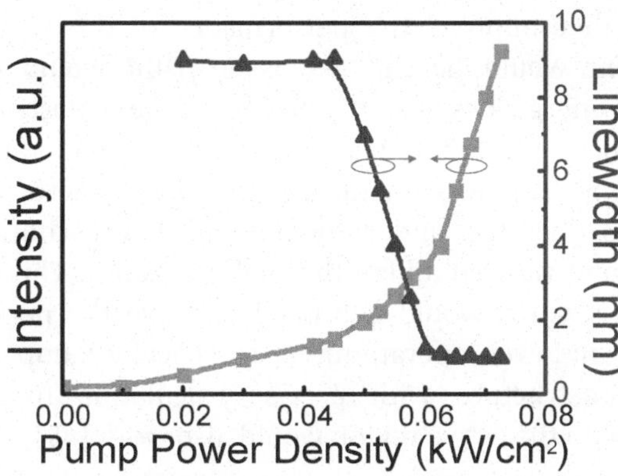

Figure 3.35 Light-in and light-out curve and the linewidth of lasing peak.

To get a better understanding of our experiment results, we use finite element method (FEM) to simulate the optical mode in our metal-coated nanocavity. The simulation model consists of a sapphire layer, an undoped GaN layer, a thin SiO_2 layer, and an aluminum shielding layer. Figures 3.36(a) and (b) show the simulated optical mode profiles for a GaN nanostripe cavity with or without the shielding layers. In Fig. 3.36(a), the model incorporating metal and SiO_2 layers had an optical mode well confined within the nanostripe, demonstrating a clear standing wave pattern with 3.5 nodes. However, for the nanostripe without shielding layers in Fig. 3.36(b), the optical mode leaks into the region of air with a standing wave pattern lacked uniformity and compares with the metal-coated nanostripe. The wavelength of the optical mode shown in Fig. 3.36(a) is 367 nm, and the qualify factor is about 150, which is ten times larger than the case in Fig. 3.36(b). This illustrates the difficulty of achieving GaN nanostripe lasing without dielectric and metal layer coated on it. The small differences between the experimental and simulation results might come from the imprecise fabrication processes and the imprecision of material indices used in the simulation model. From the discussion above, we confirmed that the metal shielding layer in this structure plays an important role in lasing at room temperature. And also, from the thermal conductivity of aluminum (237 W/m·k) and air (0.025 W/m·k), it indicates that under room temperature pulsed condition, metal-coated nanostripe has a better opportunity to observe lasing action before it breaks down than the nanostripe without shielding layers.

To further improve the performance of this device, vertical confinement would be the key issue. DBR would significantly reduce the energy loss from the bottom of the nanostripe as shown in Fig. 3.36(a). For example, C. Y. Lu et al. demonstrated this idea by an *n*-type DBR with metal coated on the microrod [59]. The device was operated under room temperature continuous-wave condition and showed a high thermal resistance. This would form a 3D cavity which would definitely improve the performance of the device and even operate under electrically pumped condition. Also, heterostructure with precise calculation to confine the optical field in the active region would also work [60].

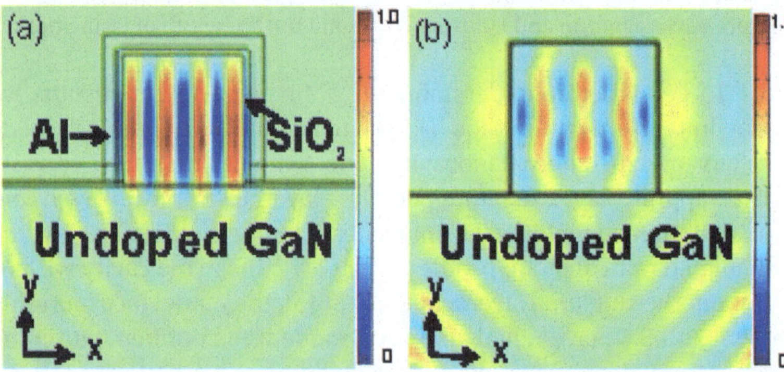

Figure 3.36 (a) The E_z mode profile of the nanostripe without shielding layers. (b) The E_z mode profile of the nanostripe with shielding layers.

Moreover, we adopt effective index method to calculate the band diagram of this structure. The ridge waveguide structure has been split into different part first and calculates the effective indexes of each part, then combined these parts together again and calculate the true effective index of this structure. This way put the complex 2D calculations into one dimensional and still keeps a high accuracy with the experimental result. In our case, we split our nanostripe structure into three parts and then calculate the band diagram. The TE mode band diagram is shown in Fig. 3.37, as you could see, there is only one waveguide band in this structure. Moreover, around 370 nm there is waveguide mode quite fit to our experimental result. We believe that the single mode lasing we observe in the experiment is a combination of TE waveguide mode with surface plasmon mode from the Al/SiO_2/GaN interfaces.

The calculation helps us to clarify the observation result and further prove that there would be only one optical mode with a clear standing wave pattern confined in the nanostripe, which makes lasing action possible.

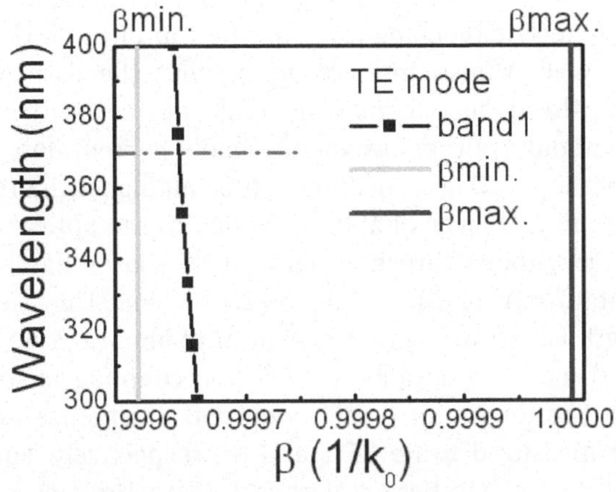

Figure 3.37 TE mode band diagram of nanostripe.

3.5 Conclusions

We have investigated VCSEL of the blue light-emitting devices. To extend the applications of GaN-based lasers/LEDs into UV region, we have demonstrated ultraviolet GaN-based microdisk lasers and the optically pumped GaN-based 2D square lattice of photonic crystal lasers with AlN/AlGaN DBRs. We also fabricated and characterized the large-area ultraviolet GaN-based photonic quasicrystal laser with high efficiency green color emission of semipolar [10,12] $In_{0.3}Ga_{0.7}N/GaN$ MQW by using nanoimprint lithography technology and the regrowth method, and the metal-coated GaN nanostripe laser was investigated and operated at room temperature. Therefore, we will summarize these works contributed from our research.

3.5.1 Vertical Cavity Surface-Emitting Laser

We have demonstrated the CW RT operation of GaN-based VCSELs with hybrid mirrors. The laser structure is composed of a 29-pair high-reflectivity AlN/GaN bottom DBR (99.3%), a 7λ cavity region, a transparent conduct layer (ITO), and a 10-pair SiO_2/Ta_2O_5 dielectric DBR (99%). Among the structure, the laser has a thin ITO layer of 30 nm as the transparent conducting layer combining with a thin heavily doped *p*-type InGaN contact layer to reduce the optical loss while maintaining good current spreading capability. An AlGaN electric blocking layer on the top of the InGaN multiple quantum well is also inserted to prevent the carrier overflow. At 300 K, the quality factors of VCSEL devices with 30 nm ITO layer are about 1600 below the threshold current, lower than the devices without ITO layer of about 1900, but higher than the devices with 210 nm ITO layer of about 700 due to the optical absorption in ITO layer. Above threshold current, the laser has a threshold current at 9.7 mA corresponding to 12.4 kA/cm^2. The laser emission wavelength is 412 nm with a linewidth of about 0.5 nm. The laser has an estimated spontaneous emission coupling factor of about 5×10^{-3}. The degree of polarization and divergence angle of the laser are measured to be 55% and 8°, respectively. Furthermore, the laser device has better electrical characteristics such as the series resistance of about 180 Ω, roll over current density of about 13 kA/cm^2, and damage current density of about 43 kA/cm^2.

3.5.2 GaN-Based Microdisk Laser

We demonstrated a compact GaN-based microdisk laser with the ultraviolet AlN/AlGaN DBR structure in the ultraviolet wavelength region. Two lasing modes were observed at 377 nm and 379 nm wavelength with the low threshold power densities of 0.03 kW/cm^2 and 0.043 kW/cm^2. This DBR structure has strong advantages in simplifying fabrication procedures and tuning lasing wavelength. In the GaN-based microdisk cavity, the UV DBR plays an important role to select the lasing wavelength regime.

The gain peak of the GaN without the DBR is around 360 nm wavelength; however, the lasing and resonant modes are around 377 nm wavelength. This 17 nm difference in wavelength is attributed to the reflection of the UV DBR mirror on the bottom of the microdisk cavity.

3.5.3 Metal-Coated GaN Nanostripe Lasers

This study demonstrated lasing from a metal-coated GaN nanostripe at room temperature, representing the first demonstration of ultraviolet lasing from a metal-coated nanocavity. Using an E-beam lithography technique, we fabricated a GaN nanostripe 500 nm in width, 500 nm in height, and 50 μm in length. We then deposited a thin SiO_2 and aluminum layer on the nanostripe to form a metal-coated nanocavity. We observed lasing action at 370 nm wavelength with a low threshold power density of 0.042 kW/cm^2.

FEM simulation results show that a metal-coated nanostripe cavity could sustain a standing wave pattern, which would not occur in a nanostripe cavity without a metal shielding layer. The wavelengths of the mode were quite close to those of the experimental results. The band diagram of this structure also supports the notion that the nanostripe structure would form an optical mode. This is a clear indication that the metal coating used on the nanocavity played an important role in lasing results at room temperature. The results presented here represent a breakthrough in the integration of metal-coated nanocavities with GaN materials. It could be the highly potential optoelectronic device in the future applications.

References

1. S. Nakamura, M. Senoh, N. Iwasa, and S. Nagahama, "High-bright InGaN blue, green and yellow light-emitting diodes with quantum well structures," *Jpn. J. Appl. Phys.*, vol. **34**, p. L797, 1995.
2. S. Nakamura, T. Mukai, and M. Senoh, "Candela-class high-brightness InGaN/AIGaN double-heterostructure blue-light-emitting diode," *Appl. Phys. Lett.*, vol. **64**, p. 1687, 1994.
3. S. Nakamura, S. Pearton, and G. Fasol, *The Blue Laser Diode: The Complete Story*, Springer-Verlag, Berlin, 2000.
4. E. D. Haberer, R. Sharma, C. Meier, A. R. Stonas, S. Nakamura, S. P. DenBaars, and E. L. Hu, "Free-standing, optically pumped, GaN/InGaN microdisk lasers fabricated by photoelectrochemical etching," *Appl. Phys. Lett.*, vol. **85**, p. 5179, 2004.
5. H. Morkoc, *Nitride Semiconductors and Devices*, Spring Verlag, Heidelberg, 1999.
6. J. I. Pankove, E. A. Miller, and J. E. Berkeyheier, "GaN blue light-emitting diodes," *J. Luminescence*, vol. **5**, p. 84, 1972.
7. H. P. Maraska, D. A. Stevenson, and J. I. Pankove, "Violet luminescence of Mg-doped GaN," *Appl. Phys. Lett.*, vol. **22**, p. 303, 1973.

8. J. I. Pankove, "Blue anti-stokes electroluminescence in GaN," *Phys. Rev. Lett.*, vol. **34**, p. 809, 1975.
9. S. Nakamura, M. Senoh, S. Nagahama, N. Iwasa, T. Yamada, T. Matsushita, Y. Sugimoto, and H. Kiyoku, "Room-temperature continuous-wave operation of InGaN multi-quantum-well-structure laser diodes with a long lifetime," *Appl. Phys. Lett.*, vol. **70**, p. 868, 1997.
10. S. Nakamura, M. Senoh, S. Nagahama, N. Iwasa, T. Yamada, T. Matsushita, H. Kiyoku, Y. Sugimoto, T. Kozaki, H. Umemoto, M. Sano, and K. Chocho, "InGaN/GaN/AlGaN-based laser diodes with modulation-doped strained-layer superlattices grown on an epitaxially laterally overgrown GaN substrate," *Appl. Phys. Lett.*, vol. **72**, p. 211, 1998.
11. A. C. Tamboli, M. C. Schmidt, A. Hirai, S. P. DenBaars, and E. L. Hu, "Photoelectrochemical undercut etching of m-plane GaN for microdisk applications," *J. Electro. Soc.*, vol. **156**, p. 767, 2009.
12. H. E. Li and K. Iga, *Vertical-Cavity Surface–Emitting Laser Devices*, Springer-Verlag, Berlin 2003.
13. T. C. Lu, S. W. Chen, L. F. Lin, T. T. Kao, C. C. Kao, P. Yu, H. C. Kuo, S. C. Wang, and S. Fan, "GaN-based two-dimensional surface-emitting photonic crystal lasers with AlN/GaN distributed Bragg reflectors," *Appl. Phys. Lett.*, vol. **92**, p. 011129, 2008.
14. J. M. Redwing, D. A. S. Loeber, N. G. Anderson, M. A. Tischler, and J. S. Flynn, *Appl. Phys. Lett.*, vol. **69**, p. 1, 1996.
15. T. Someya, K. Tachibana, J. Lee, T. Kamiya, and Y. Arakawa, "Lasing Emission from an $In_{0.1}Ga_{0.9}N$ Vertical Cavity Surface Emitting Laser," *Jpn. J. Appl. Phys.*, vol. **37**, p. L1424, 1998.
16. Y. K. Song, H. Zhou, M. Diagne, I. Ozden, A. Vertikov, A. V. Nurmikko, C. Carter-Coman, R. S. Kern, F. A. Kish, and M. R. Krames, "A vertical cavity light emitting InGaN quantum well heterostructure," *Appl. Phys. Lett.*, vol. **74**, p. 3441, 1999.
17. T. Someya, R. Werner, A. Forchel, M. Catalano, R. Cingolani, and Y. Arakawa, "Room temperature lasing at blue wavelengths in gallium nitride microcavities," *Science*, vol. **285**, p. 1905, 1999.
18. E. Feltin, J. F. Carlin, J. Dorsaz, G. Christmann, R. Butte, M. Laugt, M. llegems, and N. Grandjean, "Crack-free highly reflective AlInN/AlGaN Bragg mirrors for UV applications," *Appl. Phys. Lett.*, vol. **88**, p. 051108, 2006.
19. H. Zhou, M. Diagne, E. Makarona, A. V. Nurmikko, J. Han, K. E. Waldrip, and J. J. Figiel, "Near ultraviolet optically pumped vertical cavity laser," *Electron. Lett.*, vol. **36**, p. 1777, 2000.

20. A. Bhattacharya, S. Iyer, E. Iliopoulos, A. V. Sampath, J. Cabalu, T. D. Moustakas, and I. Friel, "High reflectivity and crack-free AlGaNÕAlN ultraviolet distributed Bragg reflectors," *J. Vac. Sci. Technol. B*, vol. **20**, p. 1229, 2002.
21. T. Tawara, H. Gotoh, T. Akasaka, N. Kobayashi, and T. Saitoh,"GaN-based vertical cavity surface-emitting lasers," *Appl. Phys. Lett.*, vol. **83**, p. 830, 2003.
22. T. Fujii, Y. Gao, R. Sharma, E. L. Hu, S. P. DenBaars, and S. Nakamura, "Increase in the extraction efficiency of GaN-based light-emitting diodes via surface roughening," *Appl. Phys. Lett.*, vol. **84**, p. 855, 2004.
23. J. T. Chu, T. C. Lu, M. You, B. J. Su, C. C. Kao, H. C. Kuo, and S. C. Wang, "Emission characteristics of optically pumped GaN-based vertical-cavity surface-emitting lasers," *Appl. Phys. Lett.*, vol. **89**, p. 121112, 2006.
24. J. T. Chu, T. C. Lu, H. H. Yao, C. C. Kao, W. D. Liang, J. Y. Tsai, H. C Kuo, and S. C. Wang, "Room temperature operation of optically pumped blue-violet GaN-based vertical-cavity surface-emitting lasers fabricated by laser lift-off," *Jpn. J. Appl. Phys.*, vol. **45**, p. 2556, 2006.
25. H. De Neve, J. Blondelle, P. Van Daele, P. Demeester, R. Baets, and G. Borghs, "Recycling of guided mode light emission in planar microcavity light emitting diodes," *Appl. Phys. Lett.*, vol. **70**, p. 799, 1997.
26. M. Boroditsky, T. F. Krauss, R. Coccioli, R. Vrijen, R. Bhat, and E. Yablonovitch, "Light extraction from optically pumped light-emitting diode by thin-slab photonic crystals," *Appl. Phys. Lett.*, vol. **75**, p. 1036, 1999.
27. M. Rattier, H. Benisty, E. Schwoob, C. Weisbuch, T. F. Krauss, C. J. M. Smith, R. Houdre, and U. Oesterle, "Omnidirectional and compact guided light extraction from Archimedean photonic lattices," *Appl. Phys. Lett.*, vol. **83**, p. 1283, 2003.
28. T. C. Lu, Member, J. R. Chen, S. W. Chen, H. C. Kuo, C. C. Kuo, C. C. Lee, and S. C. Wang, "Development of GaN-based vertical-cavity surface-emitting lasers," *IEEE J. Sel. Topics Quantum Electron.*, vol. **15**, p. 1201, 2009.
29. Tien-Chang Lu, Chih-Chiang Kao, Hao-Chung Kuo, Gen-Sheng Huang and Shing-Chung Wang,"CW lasing of current injection blue GaN-based vertical cavity surface emitting laser" *Appl. Phys. Lett.*, vol. **92**, p. 141102, 2008.
30. Y. Higuchi, K. Omae, H. Matsumura and T. Mukai, "Room-temperature CW lasing of a GaN-based certical-cavity surface-emitting laser by current injection", *Appl. Phys. Express*, vol. **1**, p. 121102, 2008.

31. H. Yu, J. Yu, F. Sun, Z. Li, and S. Chen, "Systematic considerations for the patterning of photonic crystal devices by electron beam lithography," *Opt. Commun.*, vol. **271**, p. 241, 2007.

32. L. Vogelaar, W. Nijdam, H. A. G. M. van Wolferen, R. M. de Ridder, F. B. Segerink, E. Fluck, L. Kuipers, and N. F. vanHust, "Large area photonic crystal slabs for visible light with waveguiding defect structures: Fabrication with focused ion beam assisted laser interference lithography," *Adv. Mater.*, vol. **13**, p. 1551, 2001.

33. J. F. Carlin, J. Dorsaz, E. Feltin, R. Butté, N. Grandjean, M. Ilegems, and M. Laügt, "Crack-free fully epitaxial nitride microcavity using highly reflective AlInN/GaN Bragg mirrors," *Appl. Phys. Lett.*, vol. **86**, p. 031107, 2005.

34. M. Notomi, "Theory of light propagation in strongly modulated photonic crystals: Refractionlike behavior in the vicinity of the photonic band gap," *Phys. Rev. B*, vol. **62**, p. 10696, 2000.

35. S. Kako, T. Someya, and Y. Arakawa, "Observation of enhanced spontaneous emission coupling factor in nitride-based vertical-cavity surface-emitting laser," *Appl. Phys. Lett.*, vol. **80**, p. 722, 2002.

36. Honda, H. Kawanishi, T. Sakaguchi, F. Koyama, and K. Iga, "Characteristic temperature estimation for GaN-based lasers," *MRS Internet J. Nitride Semicond.*, vol. **4S1**, pp. G6.2-1–G6.2-6, 1999.

37. O. Painter, R. K. Lee, A. Scherer, A. Yariv, J. D. O`Brien, P. D. Dapkus, I. Kim, "Two-dimensional photonic band-gap defect mode laser," *Science*, vol. **284**, p. 1819, 1999.

38. H. G. Park, S. H. Kim, S. H. Kwon, Y. G. Ju, J. K. Yang, J. H. Baek, S. B. Kim, Y. H. Lee, "Electrically driven single-cell photonic crystal laser," *Science*, vol. **305**, p. 1444, 2004.

39. Y. S. Choi, K. Hennessy, R. Sharma, E. Haberer, Y. Gao, S. P. DenBaars, C. Meier, "GaN blue photonic crystal membrane nanocavities," *Appl. Phys. Lett.*, vol. **87**, p. 243101, 2005.

40. M. Imada, S. Node, A. Chutinan, and T. Tokuda, "Coherent two-dimensional lasing action in surface-emitting laser with triangular-lattice photonic crystal structure," *Appl. Phys. Lett.*, vol. **75**, p. 316, 1999.

41. H. Matsubara, S. Yoshimoto, H. Saito, Y. Jianglin, Y. Tanaka, and S. Noda, "GaN Photonic-crystal surface-emitting laser at blue-violet wavelengths," *Science*, vol. **319**, p. 445, 2008.

42. R. J. Horowicz, H. Heitmann, Y. Kadota and Y. Yamamoto, "GaAs microcavity quantum well laser with enhanced coupling of spontaneous emission to the lasing mode," *Appl. Phys. Lett.*, vol. **61**, p. 393, 1992.

43. N. Nakada, M. Nakaji, H. Ishikawa, T. Egawa, M. Umeno, and T. Jimbo, "Improved characteristics of InGaN multiple-quantum-well light-emitting diode by GaN/AlGaN distributed Bragg reflector grown on sapphire," *Appl. Phys. Lett.*, vol. **76**, p. 1804, 2000.

44. N. Nakada, H. Ishikawa, T. Egawa, and T. Jimbo, "Comparison of p-side down and p-side up GaN light-emitting diodes fabricated by laser lift-off," *Jpn. J. Appl. Phys.*, vol. **42**, p. L144, 2003.

45. S. Y. Huang, R. H. Horng, W. K. Wang, and D. S. Wuu, "Fabrication and characterization of InGaN resonant-cavity light-emitting diodes on silicon substrates," *Phys. Stat. Sol. (c)*, vol. **3**, p. 2137, 2006.

46. T. Wang, R. J. Lynch, P. J. Parbrook, R. Butté, A. Alyamani, D. Sanvitto, D. M. Whittaker, and M. S. Skolnick, "High-reflectivity $Al_xGa_{1-x}N/Al_yGa_{1-y}N$ distributed Bragg reflectors with peak wavelength around 350 nm," *Appl. Phys. Lett.*, vol. **85**, p. 43, 2004.

47. H. M. Ng, T. D. Moustakas, and S. N. G. Chu, "High reflectivity and broad bandwidth AlN/GaN distributed Bragg reflectors grown by molecular-beam epitaxy," *Appl. Phys. Lett.*, vol. **76**, p. 2818, 2000.

48. S. Adachi, *Physical Properties of III-V Semiconductor Compounds*, New York: Wiley, 1992.

49. T. Peng, and J. Piprek, "Refractive index of AlGaInN alloys," *Electron. Lett.*, vol. **32**, p. 2285, 1996.

50. D. Brunner, H. Angerer, E. Bustarret, F. Freudenberg, R. Höpler, R. Dimitrov, O. Ambacher, and M. Stutzmann, "Optical constants of epitaxial AlGaN films and their temperature dependence," *J. Appl. Phys.*, vol. **82**, p. 5090, 1997.

51. G. M. Laws, E. C. Larkins, I. Harrison, C. Molloy, and D. Somerford, "Improved refractive index formulas for the $Al_xGa_{1-x}N$ and $In_yGa_{1-y}N$ alloys," *J. Appl. Phys.*, vol. **89**, p. 1108, 2001.

52. G. S. Huang, T. C. Lu, H. H. Yao, H. C. Kuo, and S. C. Wang, "Crack-free GaN/AlN distributed Bragg reflectors incorporated with GaN/AlN superlattices grown by metalorganic chemical vapor deposition," *Appl. Phys. Lett.*, vol. **88**, p. 061904, 2006.

53. J. R. Chen, S. C. Ling, C. T. Hung, T. S. Ko, T. C. Lu, H. C. Kuo, and S. C. Wang, "High-reflectivity ultraviolet AlN/AlGaN distributed Bragg reflectors grown by metalorganic chemical vapor deposition," *J. Crystal Growth*, vol. **310**, p. 4871, 2008.

54. A. C. Tamboli, E. D. Haberer, R. Sharma, K. H. Lee, S. Nakamura, and E. L. Hu, "Room-temperature continuous-wave lasing in GaN/InGaN microdisks," *Nature Photonics*, vol. **1**, p. 61, 2006.

55. T. Peng, and J. Piprek, "Refractive index of AlGaInN alloys," *Electron. Lett.*, vol. **32**, p. 2285, 1996.
56. A. D. Rakic, A. B. Djurisic, J. M. Elazar, and M. L. Majewski, "Optical properties of metallic films for vertical-cavity optoelectronic devices," *Appl. Opt.*, vol. **37**, p. 5271, 1998.
57. M. T. Hill, M. Marell, E. S. P. Leong, B. Smallbrugge, Y. Zhu, M. Sun, P. J. van Veldhoven, E. J. Geluk, F. Karouta, Y. S. Oei, R. Notzel, C. Z. Ning, and M. K. Smit, "Lasing in metal-insulator-metal sub-wavelength plasmonic waveguides," *Opt. Exp.*, vol. **17**, p. 11107, 2009.
58. W. M. Rohsenow and H. Choi, *Heat Mass and Momentum Transfer*, New York: Prentice Hall, 1961.
59. C. Y. Lu, S. W. Chang, S. L. Chuang, T. D. Germann, and D. Bimberg, "Metal-cavity surface-emitting microlaser at room temperature," *Appl. Phys. Lett.*, vol. **96**, p. 251101, 2010.
60. M. T. Hill, "Status and prospects for metallic and plasmonic nano-lasers," *J. Opt. Soc. Am. B*, vol. **27**, p. B36, 2010.

Chapter 4

(Al,Ga)N Microdisk Cavities

Sylvain Sergent and Fabrice Semond

Centre de Recherche sur l'Hétéro-Epitaxie et ses Applications–CNRS, Rue Bernard Grégory, Sophia Antipolis, 06560 Valbonne, France

fs@crhea.cnrs.fr

4.1 Introduction

Microdisk cavities are optical resonators presenting resonant modes that are commonly called "whispering gallery modes". Their characteristics have been thoroughly studied since the beginning of the 20th century with the works of Mie and Rayleigh [1,2], and they have been under the scope of a particular attention since the early 1990s with the demonstration of a microdisk-based laser [3]. In the last two decades, such cavities have been widely used in the field of cavity quantum electrodynamics with the demonstration of the Purcell effect [4–6] and the strong-coupling regime [7]. Also, thresholdless lasers [8] and single photon sources [9] operating at cryogenic temperatures have been demonstrated using semiconductor microdisks. So far most of these studies have been carried out at long wavelength, in the near-infrared, using III-V arsenides. In III-V nitride materials, however, a few papers report

Handbook of Optical Microcavities
Edited by Anthony H. W. Choi

ISBN 978-981-4463-24-9 (Hardcover), 978-981-4463-25-6 (eBook)
www.panstanford.com

on the use of microdisk resonators for the fabrication of microlasers operating in the blue and near-UV range [10–14], but no cavity quantum electrodynamics effects have been demonstrated so far. Nevertheless, coupling optical modes of microdisks with quantum dots (QDs) in the III-V nitrides is an attractive target because, on the contrary to their group-III arsenide counterparts, such systems would build the path toward the demonstration of quantum electrodynamics effects at high temperature. This is essentially due to the large exciton binding energy in the (Al,Ga)N system and to their large band offsets allowing for efficient carrier localization up to room-temperature. As a result, GaN/AlN QDs have already been demonstrated to operate as single photon emitter at temperature up to 200 K [15].

In this chapter, we will first introduce the theoretical framework of resonant modes in microdisks. We will then address the processing and epitaxial challenges that one faces for the fabrication of (Al,Ga)N-based microdisks embedding GaN QDs. Finally, optical properties of state-of-the-art structures will be presented in order to discuss the possibility to demonstrate quantum electrodynamics effects in such structures.

4.2 Resonant Modes in Microdisks

Resonant modes in microdisks are commonly described through the whispering gallery mode approximation that allows separating the transverse, radial, and azimuthal components when calculating the electric field distribution. This approximation will be discuss in details. But to have a preliminary understanding of the confinement of light in microdisks, let us first consider a purely geometric approach. The geometry of microdisks allows a 3D confinement of light. In both transverse and in-plane directions, the confinement is obtained from the refractive index contrast between the slab material and the surrounding air. In the transverse direction, the microdisk behaves as a planar waveguide. The propagation of light in such a 2D waveguide can be represented by the effective index $n_{eff} = \beta\lambda_0/2\pi$, where λ_0 and β are the wavelength of light in vacuum and the propagation constant of the guided mode. In the in-plane direction, light can be seen as confined by total internal reflection at the disk edges (see Fig. 4.1(a)).

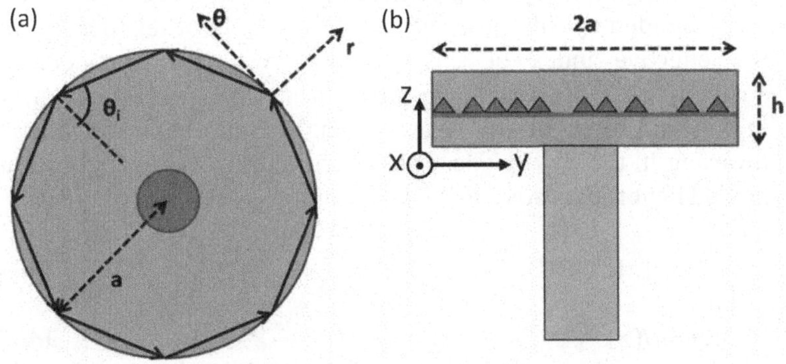

Figure 4.1 (a) In-plane and (b) transverse representation of a microdisk. Light reflected at the disk edges is represented by arrows in (a). "a" Represents the disk radius, h the slab thickness, and θ_i the angle of incidence of the ray of light at the disk edges. θ, r, and z are the azimuthal, radial, and vertical directions.

This condition of total internal reflection writes $\theta_i > \theta_c = \arcsin(n_{air}/n_{eff})$, where θ_i is the angle of incidence of the ray of light. In the case $\theta_i > \theta_c$, the light escapes the cavity and no resonant mode can be observed. If the condition of total internal reflection is verified, the reflection of light on the disk edges can lead to constructive interferences providing that:

$$2\pi a n_{eff} = p\lambda_0 \tag{4.1}$$

Where, $p \in \mathbb{N}$ and a is the disk radius. As a result the spectral spacing between two modes is written as:

$$\Delta\lambda = \frac{\lambda_0^2}{\pi a n_{eff}} \tag{4.2}$$

This geometric approach is simplified and the resonant modes is described more accurately in the following, using the whispering gallery mode approximation. First, let us calculate the transverse component of the electric field and the effective index.

4.2.1 Effective Index

We consider a microdisk constituted of an AlN layer of thickness h surrounded by infinite air layers (Fig. 4.1). In order to calculate the effective index of this structure, we first consider the structure as a 2D waveguide. In the Cartesian coordinates given in Fig. 4.1(b), light propagates along y and the waveguide is invariant in the (x, y) plane. The electric and magnetic fields of the guided light thus reads:

$$E = E(z)e^{j(\omega t - \beta y)} \tag{4.3}$$

$$H = H(z)e^{j(\omega t - \beta y)} \tag{4.4}$$

Where, $\omega = 2\pi c/\lambda_0$ is the radial frequency, c the velocity of light and λ the wavelength. In the AlN and air layers, the solutions of the system follow the Helmotz equations:

$$\Delta E + k^2 E = 0 \tag{4.5}$$

$$\Delta H + k^2 H = 0 \tag{4.6}$$

where k is the wavevector. The solutions of the system can be expressed as a combination of modes having orthogonal polarizations: transverse electric (TE) and transverse magnetic (TM) modes.

The emitters that we wish to couple to microdisk resonant modes are GaN QDs. Since GaN QDs present an electric dipole lying in the (x, y) plane, from now on we will only consider TE modes. For such TE modes, the electric field has a single component, so the Helmotz equation can be rewritten as follows:

$$\frac{\partial^2 E_{xi}(z)}{\partial^2 z^2} - (k_i^2 - \beta)E_{x,i}(z) = 0 \tag{4.7}$$

With i = AlN, air$_1$, air$_1$, and k_i is the wavevector in each layer. The general solution of the Helmotz equation writes:

$$E_{x,i}(z) = Ae^{jz\sqrt{k_i^2 - \beta}} + Be^{-jz\sqrt{k_i^2 - \beta}} \tag{4.8}$$

In order to keep the energy of the field finite in the semi-infinite air layers, $\sqrt{k_i^2 - \beta}$ has to be purely imaginary, so that the electric field is evanescent and reads in each air layer:

$$E_{x,\text{air}_1}(z) = A_{\text{air}} e^{-\frac{2\pi}{\lambda_0} z \sqrt{n_{\text{eff}}^2 - n_{\text{air}}^2}} \tag{4.9}$$

$$E_{x,\text{air}_2}(z) = A_{\text{air}} e^{\frac{2\pi}{\lambda_0} z \sqrt{n_{\text{eff}}^2 - n_{\text{air}}^2}} \tag{4.10}$$

For the AlN layer, we have:

$$E_{x,\text{air}_1}(z) = A_{\text{AlN}} e^{j\frac{2\pi}{\lambda_0} z \sqrt{n_{\text{AlN}}^2 - n_{\text{eff}}^2}} + B_{\text{AlN}} e^{-j\frac{2\pi}{\lambda_0} z \sqrt{n_{\text{AlN}}^2 - n_{\text{eff}}^2}} \tag{4.11}$$

The four unknown constants A_{AlN}, B_{AlN}, A_{air}, and B_{air} are then calculated from the boundary conditions implying both E_x and $\partial E_x/\partial_z$ to be continuous. The solution of the four linear equations expressing those boundary conditions is non-zero only if the determinant of the coefficients is zero. This can be written as follows:

$$\tan\left(h\frac{2\pi}{\lambda_0}\sqrt{n_{\text{AlN}}^2 - n_{\text{eff}}^2}\right) = \frac{2\dfrac{n_{\text{air}}^2}{n_{\text{AlN}}^2}\sqrt{n_{\text{AlN}}^2 - n_{\text{eff}}^2}\sqrt{n_{\text{eff}}^2 - n_{\text{air}}^2}}{2\dfrac{n_{\text{air}}^4}{n_{\text{AlN}}^4}\sqrt{n_{\text{AlN}}^2 - n_{\text{eff}}^2}\sqrt{n_{\text{eff}}^2 - n_{\text{air}}^2}} \tag{4.12}$$

There is no analytical solution to this equation, so that we must rely on numerical methods to find the value of the effective index. To illustrate the propagation of light in a waveguide, we consider a symmetrical slab constituted of a central AlN layer surrounded by air. We represent in Fig. 4.2(a) the variation of the effective index of TE modes as a function of the normalized slab thickness h/λ_0. For the thinnest slabs, Eq. 4.12 has only one solution called the fundamental mode. Increasing the thickness, the effective index of the fundamental mode tends to its upper limit n_{AlN} and new modes of smaller effective index appear. The modes are indexed in accordance with their order of appearance: the fundamental mode is also called TE_0 and the ith order mode, TE_i. We see that the number of modes allowed to propagate in the slab will be larger as the slab is thicker and the wavelength is shorter. As an example, for $h/\lambda_0 = 0.8$, the waveguide allows three TE modes to propagate; their profiles are represented in Fig. 4.2(b).

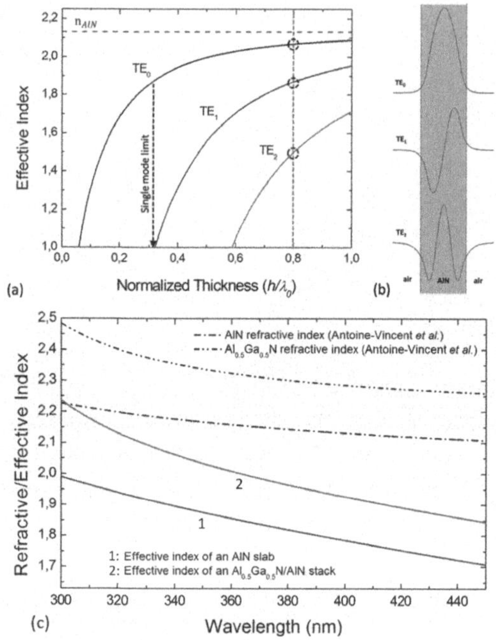

Figure 4.2 (a) Representation of the effective index of TE modes in an AlN slab surrounded by air as a function of its normalized thickness h/λ_0. The horizontal dashed line represents the refractive index of AlN at λ_0 = 400 nm. (b) Representation of the electric field profiles of TE_0, TE_1, and TE_2 for h/λ_0 = 0.8. (c) Calculated effective indices taking into account the spectral dependence of refractive indices for a 100 nm AlN slab surrounded by air and *a* (85 nm $Al_{0.5}Ga_{0.5}N$)/(15 nm AlN) stack surrounded by air. Refractive indices of AlN and $Al_{0.5}Ga_{0.5}N$ reported by Antoine-Vincent et al. [16] are also displayed.

In order to optimize the coupling coefficient of resonant modes with emitters, we want to minimize the spectral density of modes allowed to propagate in the cavity. As can be seen on Fig. 4.2(a), reducing the number of modes can be readily achieved by adjusting the slab thickness compared to the operation wavelength. Typically, at λ_0 = 400 nm the slab limit for single mode propagation is h = 125 nm (Fig. 4.2(a)). Here lies the necessity to fabricate the very thin group-III nitride membranes in order to obtain the optimized cavities. All the structures studied in the following are designed for a single mode operation and the resonant mode calculations will be limited to the fundamental TE mode.

If the graph displayed in Fig. 4.2(a) gives a good understanding of mode propagation in AlN waveguides, it should be noted that the refractive index of AlN is here considered constant and equal to $n_{\mathrm{AlN}}(\lambda_0 = 400\ \mathrm{nm})$. However, it does not accurately describe such structures because the refractive indices of AlN, GaN, and their alloys actually depend on the wavelength [16]. In order to model free-standing cavities, it is necessary to take into account this spectral dependence in calculations. Thus, we calculate the effective index for a typical waveguide thickness h = 100 nm, a wavelength ranging from 300 nm to 450 nm and two different stacks. On one hand, a pure AlN slab surrounded by air and on the other hand *a* (85 nm $Al_{0.5}Ga_{0.5}N$)/(15 nm AlN) stack surrounded by air. The effective indices of the waveguides represented in Fig. 4.2(c) present a significant decrease for the increasing wavelengths. Besides, in the whole spectral range, the $Al_{0.5}Ga_{0.5}N$/AlN stack exhibit a larger effective index. It shows that in order to maximize the effective index and thus, the light confinement in the cavity, one should consider pushing the operation wavelength toward the UV range and using (Al,Ga)N waveguides. We will see in details later, however, this actually raises difficulties due to an increase of absorption and scattering losses as we reduce the resonant wavelength.

4.2.2 Whispering Gallery Modes

As calculated the effective index of group-III nitride membranes, we can now calculate the radial (r) and azimuthal (θ) components of the electric field in the microdisk cavity. From Maxwell equations, the electric field follows:

$$\Delta E + \left(\frac{2\pi}{\lambda_0} n_{\mathrm{eff}}\right)^2 E = 0 \tag{4.13}$$

If one separates r and θ variables, the general solution writes:

$$E_r = \psi_r(r) e^{jm\theta} \tag{4.14}$$

$$E_\theta = \psi_\theta(r) e^{jm\theta} \tag{4.15}$$

where m is called the azimuthal number of the whispering gallery mode. The azimuthal number defines the periodicity of the electric field in the azimuthal direction. It can be related to the number of reflections on the disk edges in the geometric approach. Equations describing the electric field can then be formulated as follows:

$$\frac{d^2\psi r}{dr^2}+\frac{1d\psi r}{rdr}+\left(\left[\frac{2\pi}{\lambda_0}n_{\text{eff}}\right]^2-\frac{m^2}{r^2}\right)\psi_r=0 \tag{4.16}$$

$$\nabla E=0 \tag{4.17}$$

Inside the disk, for $r \le a$, the solutions writes:

$$E_r(r,\theta)=AJ_m\left(\frac{2\pi}{\lambda_0}n_{\text{eff}}r\right)e^{jm\theta} \tag{4.18}$$

$$E_r(r,\theta)=\frac{Bj}{m}\left(\frac{2\pi}{\lambda_0}n_{\text{eff}}r+1\right)J_m'\left(\frac{2\pi}{\lambda_0}n_{\text{eff}}r\right)e^{jm\theta} \tag{4.19}$$

where J_m is the Bessel function of first kind and of order m and A and B are constants. Outside the disk, because the field energy has to be finite, the field is rather described by Hankel functions that can be approximated as exponentially scaled Bessel functions. We thus have for $r > a$:

$$E_r(r,\theta)=AJ_m\left(\frac{2\pi}{\lambda_0}n_{\text{eff}}r\right)e^{-\frac{2\pi}{\lambda_0}\sqrt{n_{\text{eff}}^2-n_{\text{air}}^2}(r-\alpha)}e^{jm\theta} \tag{4.20}$$

In the so-called whispering gallery mode approximation, one considers the boundary condition $E_r(r, \theta) = 0$ for $r > a$. However, such an approximation does not take into account the outcoupling of light from the cavity. To understand this outcoupling, let us make the function substitution $\varphi_r(r)=\sqrt{r\psi(r)}$. Equation 4.16 then becomes:

$$-\frac{d^2\psi r}{dr^2}+V(r,\lambda_0)\psi_r=\left(\frac{2\pi}{\lambda_0}\right)^2\varphi_r \tag{4.21}$$

One can here recognize a similarity with the Schrödinger equation where the potential $V(r, \lambda_0)$ writes:

$$V(r,\lambda_0)=\left(\frac{2\pi}{\lambda_0}\right)^2\left[1-n^2(r)\right]+\frac{4m^2+3}{4r^2} \tag{4.22}$$

Here, the resonant mode can be considered as an eigenstate of a particle placed in a potential well, as illustrated in Fig. 4.3(a) for a given value of λ_0. It is then possible to describe the coupling between confined modes and the cavity environment [3]: light tunnels through the well barrier, that is, from $r = a$ to $r = n_{\text{eff}}a$, the field being evanescent; from $r = n_{\text{eff}}a$, light propagates in free space. As a result, the boundary condition $E(a) = 0$ gives a good approximation of these modes. Getting back to the geometric approach of the problem, we can see that the larger the value of m, the larger is the angle of incidence on the disk edges. For small values of m, $\theta_{\text{i}} > \arcsin(n_{\text{air}}/n_{\text{eff}})$, the radial component of the light wavevector prevails and the light cannot be confined into the disk [17]. Conversely, if m is large enough, $\theta_{\text{i}} < \arcsin(n_{\text{air}}/n_{\text{eff}})$, the wavevector is mainly azimuthal and we have: $E_\theta \ll E_r$ that is, $E(r, \theta) \approx E_{\text{r}}(r, \theta)$. We thus write the boundary condition:

$$E(a)=0 => J_m\left(\frac{2\pi}{\lambda_0}n_{\text{eff}}a\right)=0 \tag{4.23}$$

As illustrated in Fig. 4.3, for a given azimuthal number m, there exists a set of solutions to the latter equation. A number called the radial number n is attributed to each of them and $n = 1$ corresponds to the lowest energy solution (see Fig. 4.3(b)). As a result, a mode propagating in the disk can be fully described by its azimuthal and radial numbers m and n: it is thus written $\text{TE}_{m;n}$. The field profiles of the whispering gallery modes $\text{TE}_{22,n}$ are represented in Fig. 4.3(c) along the radius and for various values of n. For $n = 1$ the mode is localized at the periphery of the disk but for increasing n, one can see that the modes tend to propagate farther from the disk edges. Finally, the dependence of the resonant modes on θ can be evidenced by the representation of the in-plane field intensity that exhibits 2 m maxima in the azimuthal direction (Fig. 4.3(d)). It is important to mention that calculations represented in Fig. 4.3 do not take into account the spectral dependence of the effective index highlighted in the previous section. In practice, in order to identify confined modes within the whispering gallery mode approximation, the effective index will have to be calculated for each mode.

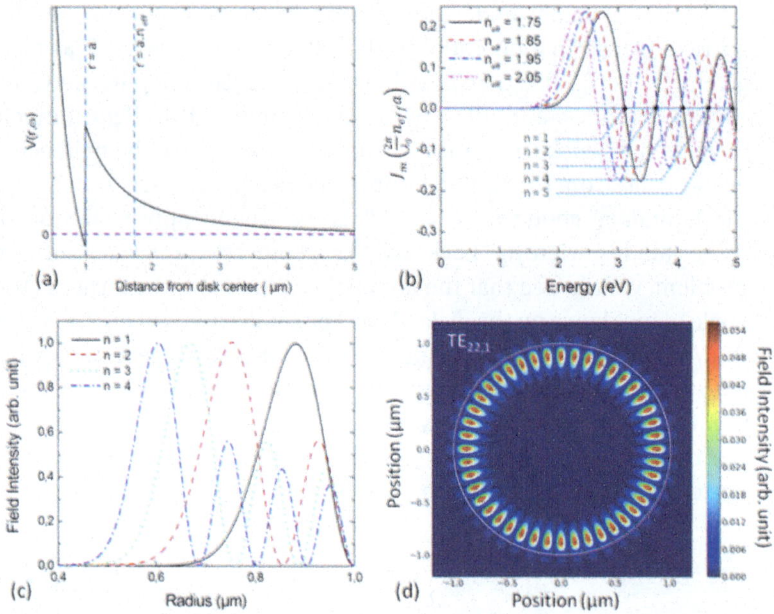

Figure 4.3 (a) Value of the potential defined in Eq. 4.22 along the radial direction for E = 3.2 eV, a = 1 μm, and n_{eff} = 1.75. (b) Representation of the Bessel function $J_m\left(\frac{2\pi}{\lambda_0}n_{\text{eff}}a\right)$ as a function of the mode energy for m = 22, a = 1 μm, and different values of n_{eff}. Black, red, blue, and pink lines are respectively for n_{eff} = 1.75, 1.85, 1.95, and 2.05. (c) Representation of the field intensity along the disk radius for m = 22, a = 1 μm, and various radial numbers n. Black, red, green, and blue lines are respectively for n = 1, 2, 3, and 4. (d) Representation of the field intensity of the mode $TE_{22,1}$ propagating in a microdisk of radius 1 μm and effective index 1.75.

Let us note that an accurate description of the resonant modes in microdisks would require numerical calculations, for example, by finite-difference time-domain analysis. However, the model proposed here gives a fair description of the modes for values of m sufficiently large (m > 6 according to [18]). It especially allows to identify with a good accuracy not only for the mode energy but also the mode volume and the intrinsic quality factor of each mode. This will be of great use to design the cavity and the emitter in order to optimize the Purcell factor.

4.2.3 Quality Factor and Origin of Losses

Since the first realization of a microdisk laser [3], microdisk quality factors have been drastically improved to reach several millions in cold cavities for most mature material systems like group-III arsenides [19] and silicon [20]. The quality factor of microdisks results from several phenomena such as losses Q_{tun}^{-1} due to the tunneling of light discussed in the previous section, slab absorption losses Q_{abs}^{-1} and losses Q_{scatt}^{-1} related to scattering of light at the disk interfaces with the surrounding air. As a result, the overall quality factor can be represented as:

$$Q^{-1} = Q_{\mathrm{tun}}^{-1} + Q_{\mathrm{abs}}^{-1} + Q_{\mathrm{scatt}}^{-1} \tag{4.24}$$

4.2.3.1 Intrinsic quality factor

The intrinsic quality factor of a mode is related to the loss originating from the tunneling of light mentioned before. It can be approximated as follows [3]:

$$Q_{\mathrm{tun}} = e^{2mJ} \tag{4.25}$$

where m is the azimuthal number and J is represented as:

$$J = \mathrm{argtanth}\left(\sqrt{1 - 1/n_{\mathrm{eff}}^2}\right) - \sqrt{1 - 1/n_{\mathrm{eff}}^2} \tag{4.26}$$

The value of the intrinsic quality factor is reported in Fig. 4.4(a) for $n_{\mathrm{eff}} = 1.75$. It is first observed that Q values above 10^4 are expected. Also, one can see a significant increase of the quality factor for increasing values of m. However, increasing m infinitely is not possible when considering light–matter coupling in microdisk resonators; in a given wavelength range, increasing the azimuthal number of a resonant mode requires to extend the disk radius but it also leads to a reduction of the spectral spacing between modes. If the spectral mode spacing is too small, a single emitter can couple to several modes, thus limiting the coupling coefficient of a specific mode. Such a mode spacing problem has been suspected to be the reason of high lasing threshold in nitride-based microdisks [12,14].

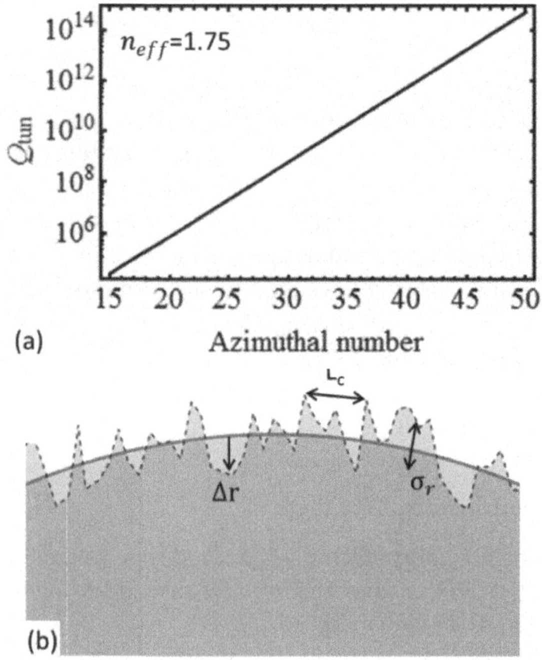

Figure 4.4 (a) Value of the intrinsic quality factor of a microdisk for $n_{\text{eff}} = 1.75$ as a function of m. (b) Schematics of the microdisk sidewalls showing the roughness amplitude Δr and its statistical parameters σ_{r} and L_{c}.

4.2.3.2 Absorption losses

Another loss mechanism comes from the slab absorption. It can be formulated as follows:

$$Q_{\text{abs}}^{-1} = \frac{\alpha\lambda}{2\pi n_{\text{eff}}} \tag{4.27}$$

where α is a spatially averaged absorption coefficient [21]. Such absorption losses directly depend on the active layer and its dimensionality. For example, the quality factors of microdisks studied by Simeonov et al. are shown to be limited by the absorption of the embedded InGaN quantum wells [14]. On the contrary, the use of GaN QDs presenting a low density of states will tend to minimize the absorption losses.

The absorption of the rest of the membrane slab, namely the (Al,Ga)N barriers in our case, also has to be taken into account. It has indeed been shown that the absorption tail below the barrier bandgap energy is significant, especially for highly defective crystals [22].

Last but not the least, the microdisk post also contributes to the absorption losses. In our case, because of the larger refractive index of the silicon post (see Section 4.2: The fabrication process of microdisks) as compared to AlGaN, light could couple with the post and could be absorbed by the silicon. For modes of the low radial number confined at the periphery, a very limited part of the field can couple with the substrate, so that its absorption can be neglected. However, the higher the radial number, the larger are both the overlap with the post area and the subsequent absorption. Now, let us note that adjusting the post diameter allows to "kill" parasitic modes of high radial number. This is especially useful in large microdisks where the spectral spacing between two modes of same radial number is small (see Eq. 4.2).

4.2.3.3 Scattering losses

The last loss mechanism occurring in microdisk resonators is related to the scattering of light due to surface irregularities. Scattering losses can be caused by the top and bottom face roughness of the membrane. However, they mostly originate from the roughness of the microdisk sidewalls created during the processing steps. Borselli et al. showed that it is possible to estimate scattering losses from the microdisk sidewall profile [23]. Let us introduce the statistical parameters of the roughness amplitude Δr such as its standard deviation σ_r and its correlation length L_c represented in Fig. 4.4(b). Taking into account the approximation $E_\theta \ll E_r$ mentioned in the previous section and provided that L_c is small compared to the considered wavelength, scattering losses of a resonant TE mode propagating at the periphery of an AlN microdisk (i.e., small radial number n) can be estimated as follows:

$$Q_{scatt}^{-1} = \frac{4\pi^{\frac{7}{2}} n_{air} n_{eff} (n_{AlN}^2 - n_{air}^2)^2}{3\lambda_0^3 n_{AlN} (n_{AlN}^2 - n_{air}^2)} \frac{L_c h \sigma_r}{\pi a} \tag{4.28}$$

According to Eq. 4.28, for a given fabrication process, that is, for a given roughness amplitude, it will not be possible to reduce the cavity radius or the emitter wavelength without increasing scattering losses. Such a limitation has a direct impact on the design of an optimal cavity. For instance, in order to observe the Purcell effect, one would indeed want to minimize the microdisk radius in order to limit the mode volume. However, we can see that it leads to a decrease of the quality factor. It is thus necessary to find a compromise in order to maximize the $Q{:}V$ ratio. Let us also note that, according to Eq. 4.28, pushing the operation wavelength far into the UV range would require very smooth sidewalls and thus, refined fabrication processes.

4.3 Fabrication of Microdisk Resonators

The fabrication of group-III nitride microdisk resonators faces challenges related to both material processing and epitaxial growth. First, this is due to the chemical inertness of group-III nitrides that makes difficult the definition of smooth sidewalls allowing for limited scattering losses. Second, because of the poor etching selectivity between nitrides, undercutting the microdisk membrane turns out to be quite difficult. So far, underetching has been realized through quite complex etching techniques that necessitate the fabrication of dedicated group-III nitride epilayer stacks and lead to significant fabrication imperfections of the microdisk resonators. For example, Tamboli et al. reported a photoelectrochemical etching technique for the fabrication of GaN microdisks embedding InGaN quantum wells and using an InGaN/InGaN superlattice as the sacrificial layer. With such a technique, the fabrication imperfections limit the Q factors to 4,600 [10], but it still allows for room-temperature lasing operation in the blue range [12]. Dorsaz et al. proposed to use an AlInN sacrificial layer to fabricate GaN microdisks embedding InGaN quantum wells. The underlying AlInN alloy is electrochemically oxidized and subsequently, wet-etched by a nitric acid solution [24]. This technique led to quality factors limited to 4,000 [13] but room-temperature lasing operation could be achieved at wavelengths as short as 408 nm [14]. As an alternative to these complex etching techniques, it has been proposed to use the good etching selectivity between the silicon substrate and group-III nitride epilayers to undercut the microdisk membrane. Using this technique low-temperature lasing operation could be achieved at the bulk GaN bandedge [11]. However, due to the lattice mismatch with the silicon substrate, GaN epilayers grown on silicon present a high density of threading dislocations that are centers of non-radiative recombinations and therefore, are very detrimental for room-temperature operation. This issue can be circumvented by using an emitter that efficiently confines charges and prevents non-radiative recombinations, for example, GaN QDs.

4.3.1 Growth of Thin GaN Quantum Dot Stacks on Silicon Substrates

GaN quantum dot stacks are grown by ammonia-based molecular beam epitaxy on Si(111) substrate. This growth technique enables the nucleation of GaN QDs on (Al,Ga)N buffer layers as thin as 20 nm and thus, to fabricate quantum dot stacks thinner than 100 nm. As calculated in Section 4.1, it enables the fabrication of single mode slabs. This is of significant importance in order to minimize the number of whispering gallery modes allowed to propagate and in order to optimize the coupling between GaN QDs and resonant modes. Besides, the ammonia-based molecular beam epitaxy of GaN QDs allows for a good control of the density [25] from 10^8 to 10^{11} cm^{-2} (Fig. 4.5(a)) that should lead to the fabrication of various optical systems ranging from low-threshold laser to robust single photon sources.

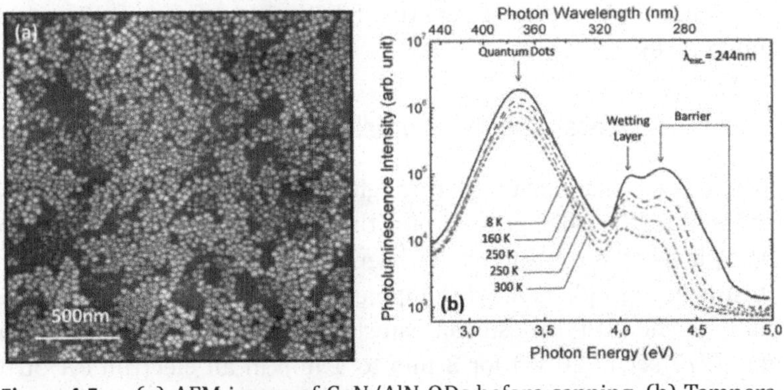

Figure 4.5 (a) AFM image of GaN/AlN QDs before capping. (b) Temperature-dependent photoluminescence spectra of GaN/$Al_{0.55}Ga_{0.45}N$ QDs for a 5.08 eV laser excitation.

Despite a high density of threading dislocation in group-III nitride membranes grown on silicon (10^{11} cm^{-2}), the temperature-dependent photoluminescence spectra of GaN/$Al_{0.55}Ga_{0.45}N$ QDs excited at 5.08 eV exhibit integrated intensity ratio of 0.3 between 300 K and 8 K [26], while the intensity of the GaN wetting layer and the barriers quenches by more than one order of magnitude (Fig. 4.5(b)). The intensity ratio of GaN QDs is even larger when using AlN barriers. This is a clear evidence of the effect of the large band offsets in such heterostructures that limit the escape of carriers through thermal activation and their diffusion to non-radiative recombination centers.

However, the photoluminescence (PL) ratios found for a 5.08 eV excitation do not only account for the carrier escape from GaN QDs, but also for carriers photocreated in the barrier (in the case of GaN/AlGaN QDs) or the wetting layer (in the case of GaN/AlN QDs) that diffuse toward non-radiative recombination centers, such as threading dislocations. To discriminate between mechanisms occurring inside and outside the QDs, we excite the QDs at 3.8 eV using an HeCd laser, so that no carrier is created neither in the barrier nor in the wetting layer. The resulting photoluminescence spectra present a 0.6 intensity ratio between 300 K and 8 K for $Al_{0.55}Ga_{0.45}N$ QDs and 0.8 for GaN/AlN QDs, which highlights the robustness of such emitters.

The optical properties of GaN QDs show good prospects for the development of optical devices operating at room-temperature and grown on silicon substrate. Moreover, such QD stacks grown on silicon are optimal for the fabrication of group-III nitride microdisks.

4.3.2 Processing of Microdisks and Structural Properties

Microdisks are fabricated through a classic top-down approach. After the NH_3-MBE growth of a thin (Al,Ga)N epilayer containing QD-planes, the top surface is covered with resist prior to electron-beam lithography and development steps. The epilayer is then fully etched to the silicon substrate by a $Cl_2/CH_4/Ar$ plasma (13/2/2 sccm, 20 W) for 3 min to 4 min in an electron cyclotron resonance-reactive ion etching system. The microdisk post is finally obtained by a $HF:HNO_3:CH_3COOH$ (HNA) wet under-etching of the silicon substrate, resulting in a free-standing microdisk. The three processing steps are illustrated in Fig. 4.6. As depicted in Fig. 4.6, the process allows to obtain circular cavities with smooth sidewalls. Atomic force microscopy measurements carried out on the top and the bottom face of the microdisk reveal root mean square roughnesses as low as 3 Å. It indicates that the surface and the interface of epitaxial layers are very smooth and abrupt respectively and also, it is an indication of the good selectivity of the HNA wet etching in between nitride materials and silicon.

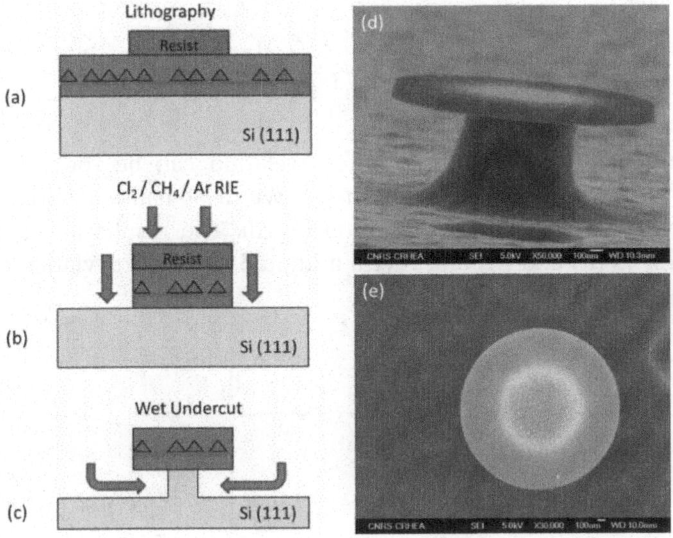

Figure 4.6 Schematics of the fabrication steps of an (Al,Ga)N-based microdisk containing GaN QDs. (a) E-beam lithography, (b) $Cl_2/CH_4/Ar$ dry etching of the group-III nitride material and (c) HNA wet underetching. (d) and (e) are scanning electron microscope (SEM) images of a resulting microdisk presenting a radius a = 1 μm. The lithography is performed at Institut d'Electronique Fonda-mentale, CNRS—Université Paris Sud, by D. Néel, S. David, X. Checoury and P. Boucaud.

It would be interesting to quantify the structural quality of such microdisks in order to address their impact on the quality factor as described in Eq. 4.28. From the top-view SEM image of a microdisk, it is possible to extract its in-plane profile and the relevant statistical parameters of the roughness amplitude especially the standard deviation and the correlation length L_c. The standard deviation:

$$\sigma_{\mathrm{r}} = \sqrt{\frac{1}{N}\sum_{i=1}^{N}\left[\Delta r(S_i) - \overline{\Delta r}\right]^2} \tag{4.29}$$

where N is the number of points extracted from the SEM image profile, $\Delta r(s)$ is the roughness amplitude of the microdisk as a function of the arc-length s (see Fig. 4.7a) and is the average value of $\Delta r(s)$. To estimate the correlation length, we first have to calculate the autocorrelation of the roughness amplitude:

$$A(S_i) = \sum_{j=1}^{N} \frac{\left[\Delta r(S_j)\right] - \overline{\Delta r}\left[\Delta r(s_{j+i}) - \overline{\Delta r}\right]}{\sigma_r^2} \tag{4.30}$$

The roughness amplitude autocorrelation can be roughly fitted by a Gaussian function, which half-width at half-maximum gives an estimation of the correlation length. The results are given in Figs. 4.7(a) and (b) and reveal a limited standard deviation of the

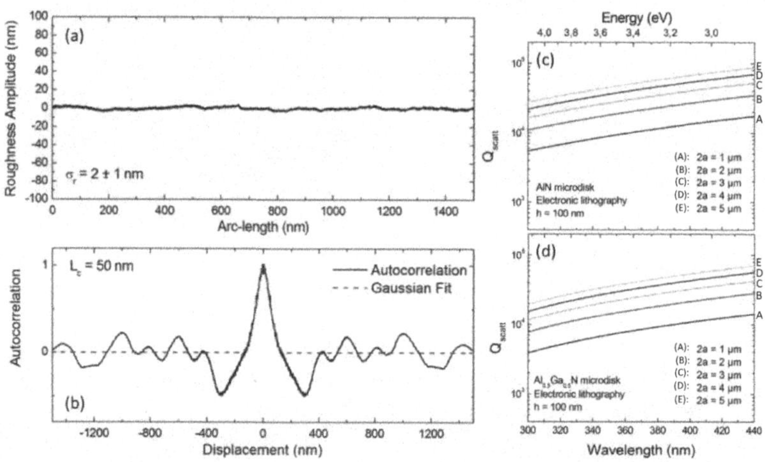

Figure 4.7 (a) Roughness amplitude Δr derived from an SEM image of a AlN microdisk. The calculated standard deviation is $\sigma_r = 2 \pm 1$ nm. (b) Autocorrelation of the roughness amplitude displayed in (a). The dashed line corresponds to a Gaussian fit. The calculated correlation length is $L_c \sim 50$ nm. (c) and (d) Value of the quality factor related to scattering losses derived from the roughness analysis of AlN and $Al_{0.5}Ga_{0.5}N$ microdisks. Calculations are made for microdisk diameters ranging from 1 µm to 5 µm and a 100 nm–thick membrane.

roughness amplitude σ_r = 2 ± 1 nm and a correlation length as short as $L_c \sim 50$ nm [27]. The error on σ_r is related to both the SEM image resolution and the noise induced by the profile extraction. According to Eq. 4.28, we calculate the losses related to scattering effects at the disk edges for AlGaN and AlN layers and various microdisk radii. The results represented in Figs. 4.7(c) and (d) give an estimation of Q_{scatt} as a function of the wavelength: in the near UV range, it turns out to be larger than 10^4 for microdisks presenting radii larger than 1 μm. One should also note that for identical geometries and residual roughnesses, AlGaN microdisks are more subject to scattering losses than AlN microdisks because of their larger refractive index (see Eq. 4.28).

4.4 Optical Properties

4.4.1 $Al_{0.5}Ga_{0.5}N$ Microdisks

Microdisks constituted of 100 nm-thick GaN/$Al_{0.5}Ga_{0.5}N$ QD stacks are characterized by microphotoluminescence. The microdisks are continuously excited from the top surface of the disk by a 266 nm laser through a microscope objective. The signal is collected by a charge-coupled device (CCD) camera either in a confocal geometry or through a lens placed on the side of the microdisk resonator. In confocal geometry, the microphotoluminescence spectra of 4 μm and 2 μm microdisks are dominated by the QD emission and do not exhibit any signature of confined modes. This is due to the smoothness and verticality of sidewalls that prevents any losses in the z-direction [28]. In side-collection geometry, however, one can observe whispering gallery modes with quality factors larger than 1,000 (see Fig. 4.8). Despite the very good structural quality of the microdisks, the resonators present significant losses. Furthermore, the quality factors do not vary much with the microdisk diameter and they abruptly decrease on the high-energy side of the 4 μm microdisk microphoto-luminescence spectrum (see Fig. 4.8(a)). This does not fit with the dependence of scattering losses that one would expect from Eq. 4.28. It can however, be explained by the exponential increase of the residual absorption in the $Al_{0.5}Ga_{0.5}N$ barriers. Typically, considering an absorption coefficient $\alpha \in$ 200 cm^{-1}; 200 cm^{-1} in the near UV range [22], the quality factor related to absorption losses would range between 800 and 1,600, according to Eq. 4.27. Therefore, in the near UV range, absorption is the main loss mechanism in $Al_{0.5}Ga_{0.5}N$ microdisk resonators.

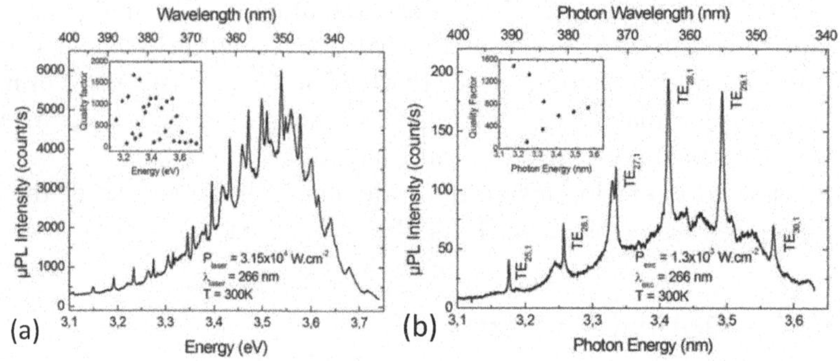

Figure 4.8 Microphotoluminescence spectra of (a) 4 μm and (b) 2 μm microdisks at room temperature. Insets present the quality factors of the whispering gallery modes as a function of the photon energy. (Data obtained by T. Guillet, C. Brimont and M. Mexis from Laboratoire Charles Coulomb, Université of Montpellier 2.)

Due to a large silicon post-diameter, only modes with radial numbers up to $n = 2, 3$ can propagate in the 4 μm microdisk resonator. As a result, the mode spacing is as large as 2.6 nm. Nevertheless, the spectral mode density is still too high to identify each whispering gallery mode although we can estimate the azimuthal number of $TE_{m,1}$ modes to range between $m = 40$ and $m = 60$. On the contrary, in 2 μm microdisks the mode spacing is sufficiently large to allow an identification of resonant modes. Whispering-gallery modes with radial numbers equal to $n = 2$ can propagate into the disk but the most intense modes correspond to $n = 1$ and azimuthal numbers range from $m = 25$ to $m = 30$.

4.4.2 AlN Microdisks

In order to reduce the influence of absorption on the microdisk quality factors, we propose to embed GaN QDs in 100 nm–thick AlN membranes. Because of the absence of alloy demixtion, the Urbach tail parameter of AlN $E_{AlN,Urb} = 90$ meV has indeed been found to be lower than the Urbach tail parameter of $Al_{0.5}Ga_{0.5}N$ $E_{Al_{0.5}Ga_{0.5}N,Urb} = 130$

meV [22], leading to a more abrupt absorption tail below the bandgap. Moreover, due to its larger bandgap, the absorption tail of AlN is pushed farther away in the UV range so that it has a lower impact than $Al_{0.5}Ga_{0.5}N$ in the near UV range.

Thus, the optical properties of 5 μm and 2 μm microdisks containing 4 GaN/AlN QD planes is studied (see Fig. 4.9). If compared to its $Al_{0:5}Ga_{0:5}N$ counterpart, the 2 μm microdisk resonator presents a significant improvement of the quality factor up to 5,000 at 2.88 eV and 3,000 at 3.14 eV. It is further improved when increasing the microdisk diameter: 5 μm microdisks present quality factors as high as 6,600 at 3.13 eV, in fair agreement with the linear dependence expected by Eq. 4.28. It is worth noting that quality factors as high as 7,300 at 3.013 eV have even been measured in similar AlN microdisks [29]. Quality factors are however, lower than expected by scattering losses calculations (see Fig. 4.7(c)). This is likely due to the substantial error on both the standard deviation and the Gaussian fit of the roughness autocorrelation. Owing to its large diameter the 5 μm microdisk exhibit a very high spectral density of modes leading to a mode spacing as small as $\Delta\lambda_0$ = 400 nm at λ_0 = 400 nm. It corresponds to the propagation of modes with radial numbers up to $n = 8$. On the contrary, the microphotoluminescence spectra of 2 μm microdisks reveal the propagation of a limited number of modes with radial numbers up to $n = 2$. As a result, it is possible to identify confined modes within the whispering gallery mode approximation: most intense peaks correspond to $TE_{m,1}$ modes presenting azimuthal numbers between $m = 17$ and $m = 26$ in the energy range (2.6 eV, 3.4 eV) (see Fig. 4.9(b)). Such modes correspond to mode volumes as small as $V = 6(\lambda/n)^3$. We can finally highlight the potential of such cavities for the observation of cavity quantum electrodynamics effects by calculating the expected Purcell factor assuming perfect spectral matching between the cavity mode and the emitter, $F_p = (3/4\pi^2)(\lambda/n)^3 Q|E(r)|^2$, with $E(r)$ is the electric field at position r. For an emitter placed at the electric field maximum, Purcell factor as high as $F_p = 60$ (resp. 40) can be expected for a microdisk presenting a 5 μm (resp. 2 μm) diameter. Such values show the great potential of AlN microdisks containing GaN QDs for the study of quantum electrodynamics effects at room temperature and the achievement of related optical devices such as low-threshold lasers or robust single-photon sources.

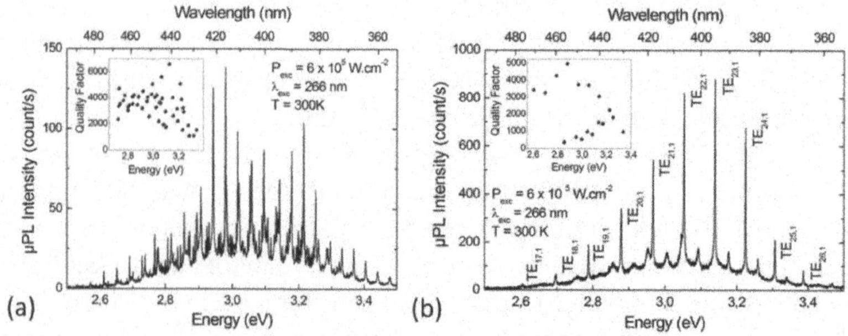

Figure 4.9 Room-temperature microphotoluminescence spectra of AlN microdisks containing four GaN QD planes and presenting (a) 5 μm and (b) 2 μm microdisk diameters. Insets show the quality factors of the whispering gallery modes as a function of the photon energy. (Data obtained by T. Guillet, C. Brimont and M. Mexis from Laboratoire Charles Coulomb, Université of Montpellier 2.)

4.5 Conclusion

Using thin (Al,Ga)N epilayers grown on silicon substrates and taking advantage of the high wet etching selectivity between nitrides and silicon, high quality (Al,Ga)N-based microdisk optical cavities are fabricated. The fabrication process is very simple, the main difficulty being the growth of a highly radiative active region in very thin epilayers directly grown on top of the silicon substrate surface. Using GaN QDs as an active layer, optical properties of microdisks are studied by photoluminescence in the UV range at room temperature. Whispering gallery modes are observed and well identified. Despite the difficulty to fabricate high quality factor optical resonators at short wavelength, Q factors ranging from 4,000 to 7,000 are easily achieved. For microdisks containing GaN/AlN QDs, quality factors are more than three times higher than those with GaN/AlGaN QDs. In the latter case, the quality factor is limited by the absorption in the AlGaN layer while for GaN/AlN QD-based microdisk, Q factors are limited by surface roughness scattering at the periphery of the disk. These nitride-based microdisks exhibit a higher Q factor than those achieved in any other kind of nitride-based optical microcavities, such as planar microcavities [30] and photonic crystal microcavites [31–33]. Because microdisk cavities exhibit high quality factors and are quite straightforward to fabricate, they offer a very interesting playground to study light-matter interactions in nitrides where room temperature cavity quantum electrodynamics effects are expected to be observed.

References

1. Mie, G. (1908). Beitrage zur optik trüber medien, speziell kolloïdaler metallosungen, *Ann. Phys.*, **25**, p. 337.
2. Rayleigh, L. (1912). The problem of the whispering gallery, *Scientic Papers*, **5**, p. 617.
3. McCall, S. L., Levi, A. F. J., Slusher, R. E., Pearton, S. J., and Logan, R. A. (1992). Whispering-gallery mode microdisk lasers, *Appl. Phys. Lett.*, **60**, p. 289.
4. Gérard, J.M. and Gayral, B. (1999). Strong Purcell effect for InAs quantum boxes in three-dimensional solid-state microcavities, *J. Lightw. Technol.*, **17**, p. 2089.
5. Kiraz, A., Michler, P., Becher, C., Gayral, B., Imamoglu, A., Zhang, L., Hu, E., Schoenfeld, W. V., and Petroff, P. M. (2001). Cavity-quantum electrodynamics using a single InAs quantum dot in a microdisk structure, *Appl. Phys. Lett.*, **78**, p. 3932.
6. Fang, W., Xu, J. Y., Yamilov, A., Cao, H., Ma, Y., Ho, S. T., and Solomon, G. S. (2002). Large enhancement of spontaneous emission rates of InAs quantum dots in GaAs microdisks, *Opt. Lett.*, **27**, p. 948.
7. Peter, E., Senellart, P., Martrou, D., Lemaître, A., Hours, J., Gérard, J.-M., and Bloch, J. (2005). Exciton-photon strong-coupling regime for a single quantum dot embedded in a microcavity, *Phys. Rev. Lett.*, **95**, p. 067401.
8. Michler, P., Kiraz, A., Zhang, L., Becher, C., Hu, E., and Imamoglu, A. (2000). Laser emission from quantum dots in microdisk structures, *Appl. Phys. Lett.*, **77**, p. 184.
9. Michler, P., Kiraz, A., Becher, C., Schoenfeld, W. V., Petroff, P. M., Zhang, L., Hu, E., and Imamoglu, A. (2000). A quantum dot single-photon turnstile device, *Science* **290**, p. 2282.
10. Haberer, E. D., Sharma, R., Meier, C., Stonas, A. R., Nakamura, S., DenBaars, S. P., and Hu, E. L. (2004). Comparison of ICl- and IBr-based plasma chemistries for inductively coupled plasma etching of GaN, InN and AlN, *Appl. Phys. Lett.*, **85**, p. 5179.

11. Choi, H. W., Hui, K. N., Lai, P. T., Chen, P., Zhang, X. H., Tripathy, S., Teng, J. H., and Chua, S. J. (2006). Lasing in GaN microdisks pivoted on Si, *Appl. Phys. Lett.*, **89**, p. 211101.

12. Tamboli, A. C., Haberer, E. D., Rajat, S., Lee, K. H., Nakamura, S., and Hu, E. L. (2007). Room-temperature continuous-wave lasing in GaN// InGaN microdisks, *Nat. Photonics*, **1**, p. 61.

13. Simeonov, D., Feltin, E., Bühlmann, H.-J., Zhu, T., Castiglia, A., Mosca, M., Carlin, J.-F., Butté, R., and Grandjean, N. (2011). Blue lasing at room temperature in high quality factor GaN/AlInN microdisks with InGaN quantum wells, *Appl. Phys. Lett.*, **90**, p. 061106.

14. Simeonov, D., Feltin, E., Altoukhov, A., Castiglia, A., Carlin, J. F., Butté, R., and Grandjean, N. (2008). High quality nitride based microdisks obtained via selective wet etching of AlInN sacrificial layers, *Appl. Phys. Lett.*, **92**, p. 171102.

15. Kako, S., Santori, C., Hoshino, K., Gotzinger, S., Yamamoto, Y., and Arakawa, Y. (2006). A gallium nitride single-photon source operating at 200 K, *Nat. Mater.*, **5**, p. 887.

16. Antoine-Vincent, N., Natali, F., Mihailovic, M., Vasson, A., Leymarie, J., Disseix, P., Byrne, D., Semond, F., and Massies, J. (2003). Determination of the refractive indices of AlN, GaN, and AlGaN grown on (111) Si substrates, *J. Appl. Phys.*, **93**, p. 5222.

17. Mair, R. A., Zeng, K. C., Lin, J. Y., Jiang, H. X., Zhang, B., Dai, L., Botchkarev, A., Kim, W., Morkoc, H., and Khan, M. A. (1998). Optical modes within III-nitride multiple quantum well microdisk cavities, *Appl. Phys. Lett.*, **72**, p. 1530.

18. Xu, Y., Lee, R. K., and Yariv, A. (2000). Finite-difference time-domain analysis of spontaneous emission in a microdisk cavity, *Phys. Rev. A*, **61**, p. 033808.

19. Srinivasan, K., and Painter, O. (2007). Mode coupling and cavity-quantum-dot interactions in a fiber-coupled microdisk cavity, *Phys. Rev. A*, **75**, p. 023814.

20. Soltani, M., Yegnanarayanan, S., and Adibi, A. (2007). Ultra-high Q planar silicon microdisk resonators for chip-scale silicon photonics, *Opt. Express*, **15**, p. 4694.

21. Slusher, R. E., Levi, A. F. J., Mohideen, U., McCall, S. L., Pearton, S. J., and Logan, R. A. (1993). Threshold characteristics of semiconductor microdisk lasers, *Appl. Phys. Lett.*, **63**, p. 1310.

22. Brunner, D., Angerer, H., Bustarret, E., Freudenberg, F., Hopler, R., Dimitrov, R., Ambacher, O., and Stutzmann, M. (1997). Optical

constants of epitaxial AlGaN films and their temperature dependence, *J. Appl. Phys.*, **82**, p. 5090.

23. Borselli, M., Johnson, T. J., and Painter, O. (2005). Beyond the Rayleigh scattering limit in high-Q silicon microdisks: theory and experiment, *Opt. Express*, **13**, p. 1515.
24. Dorsaz, J., Buhlmann, H. J., Carlin, J. F., Grandjean, N., and Ilegems, M. (2005), Selective oxidation of AlInN layers for current confinement in III-nitride devices, *Appl. Phys. Lett.*, **87**, p. 072102.
25. Sergent, S., Damilano, B., Huault, T., Brault, J., Korytov, M., Tottereau, O., Vennégués, P., Leroux, M., Semond, F., and Massies, J. (2011). Study of the growth mechanisms of GaN/(Al, Ga)N quantum dots: Correlation between structural and optical properties, *J. Appl. Phys.*, **109**, p. 053514.
26. Sergent, S., Moreno, J. C., Frayssinet, E., Chenot, S., Leroux, M., and Semond, F. (2009). GaN quantum dots grown on silicon for free-standing membrane photonic structures, *Appl. Phys. Express*, **2**, p. 051003.
27. Sergent, S., Moreno, J. C., Frayssinet, E., Laaroussi, Y., Chenot, S., Renard, J., Sam-Giao, D., Gayral, B., Néel, D., David, S., Boucaud, P., Leroux, M., and Semond, F. (2010). GaN quantum dots in (Al, Ga)N-based Microdisks, *J. Phys. Conf. Ser.*, **210**, p. 012005.
28. Peter, E., Dousse, A., Voisin, P., Lemaitre, A., Martrou, D., Cavanna, A., Bloch, J., and Senellart, P. (2007). Highly directional radiation pattern of microdisk cavities, *Appl. Phys. Lett.*, **91**, p. 151103.
29. Mexis, M., Sergent, S., Guillet, T., Brimont, C., Bretagnon, T., Gil, B., Semond, F., Leroux, M., Néel, D., David, S., Checoury, X., and Boucaud, P. (2011). High quality factor nitride-based optical cavities: microdisks with embedded GaN/Al(Ga)N quantum dots, *Opt. Lett.*, **36**, p. 2203.
30. Christmann, G., Simeonov, D., Butté, R., Feltin, E., Carlin, J.-F., and Grandjean, N. (2006). Impact of disorder on high quality factor III-V nitride microcavities, *Appl. Phys. Lett.*, **89**, p. 261101.
31. Vico Triviño, N., Rossbach, G., Dharanipathy, U., Levrat, J., Castiglia, A., Carlin, J.-F., Atlasov, K. A., Butté, R., Houdré, R., and Grandjean, N. (2012). High quality factor two dimensional GaN photonic crystal cavity membrane grown on silicon substrate, *Appl. Phys. Lett.*, **100**, p. 071103.
32. Sam-Giao, D., Néel, D., Sergent, S., Gayral, B., Rashid, M. J., Semond, F., Duboz, J. Y., Mexis, M., Guillet, T., Brimont, C., David, S., Checoury, X., and Boucaud, P. (2012). High quality factor AlN nanocavities

embedded in a photonic crystal waveguide, *Appl. Phys. Lett.*, **100**, p. 191104.

33. Sergent, S., Arita, M., Kako, S., Iwamoto, S., and Arakawa, Y. (2012). High-Q (>5000) AlN nanobeam photonic crystal cavity embedding GaN quantum dots, *Appl. Phys. Lett.*, **100**, p. 121103.

Chapter 5

MOCVD Growth of Nitride DBRs for Optoelectronics

Tao Wang
Department of Electronic and Electrical Engineering, University of Sheffield, Mappin Street, Sheffield, S1 3JD, UK
t.wang@sheffield.ac.uk

5.1 Introduction

The concept of vertical cavity surface emitting laser (VCSEL) was first proposed by Professor Iga in 1970s.[1,2] VCSEL requires both top and bottom distributed Bragg reflector (DBR) with close-to 100% reflectivity, between which is a planar Fabry–Pérot cavity with only a few λ optical thickness, that is, microcavity (MC). The DBR is normally formed from multiple repeats of two alternating layers each with a thickness of a quarter wavelength ($\lambda/4$), where a high contrast in reflective index between the two different layers is ideal. Emitting layers, such as quantum wells, are embedded within the cavity at the antinodes of the optical field in order to increase the coupling between photons and carriers. One of the key requirements for VCSEL is to achieve DBR with a high reflectivity.

Compared with an edge-emitting laser diode, VCSEL possesses a number of major distinctive advantages: (1) ultralow threshold

Handbook of Optical Microcavities
Edited by Anthony H. W. Choi

ISBN 978-981-4463-24-9 (Hardcover), 978-981-4463-25-6 (eBook)
www.panstanford.com

as a result of small cavity volume due to a cavity length typically with only a few micrometers; (2) completely circular field pattern; (3) single mode emission; (4) small beam divergence; (5) high speed modulation (>10 GHz) due to large relaxation frequency; (6) enhanced non-linear conversion; (7) facilitating fabrication of 2D laser array. As a result of the above superior properties compared with conventional edge-emitting lasers, VCSEL is a highly favorable lighting source, in particular, for application of high density information storage and high resolution laser printing.

The first room-temperature continuous wave (CW) operation of VCSEL was reported in 1989, which were based on a thick GaAs as an emitting region grown by metalorganic chemical vapour deposition (MOCVD), where both bottom and top DBRs were fabricated using dielectric materials.[3] Taking advantage of nearly matched lattice between GaAs and AlAs, GaAs/AlAs epitaxial DBR with a high reflectivity of >99.9% and large stop-band can be realized through simply increasing the number of repeats of $\lambda/4$ layers without concerning about the generation of extra defects. GaAs/AlAs DBR has been widely used in growth of GaAs-based VCSEL since 1992, one of the major achievements in the long wavelength spectral region.

The pioneering works from Amano and Akasaki in late 1980s[4,5] and Nakamura in the early 1990s[6] have led to major breakthrough in fabrication of III-nitride-based optoelectronics, mainly in the visible spectral region. Since then, the nitride optoelectronics have significantly changed the life of human being in many aspects. So far, GaN-based laser diodes (LDs) have been commercially available, although they suffer from high threshold current in the kiloampere/square centimeter range mainly due to the high carrier densities of states resulting from the high masses in III-nitrides. Therefore, VCSEL is becoming particularly important for fabrication of III-nitride-based LDs in terms of a reduced threshold.

So far, major achievements on III-nitrides are limited to the violet–blue spectral region where InGaN alloy is employed as an active region. In a remarkable contrast, the progress on III-nitride-based optoelectronics in the ultraviolet (UV) spectral region is much less impressive, where GaN or AlGaN alloy is required as an active region. Unlike InGaN alloy, the optical performance of the UV emitters is very sensitive to dislocations, thus leading to much higher threshold than that for the InGaN-based LDs. Recently, the shortest wavelength reported on electrical injection UV LDs is 336 nm, and the LDs can be operated only in a pulsed mode due to an extremely high threshold.[7] As mentioned above, VCSEL also provide one of best approaches in terms of a reduced threshold.

Unlike other III–V semiconductors such as GaAs or InP, growth of III-nitrides is mainly based on "lattice-mismatched heteroepitaxy" due to lack of affordable substrates. Both large lattice-mismatch and thermal co-efficient mismatch between III-nitride and foreign substrate, such as widely used sapphire, result in a tremendously high stress. This potentially generates both high dislocation density and serious cracking issues. This poses an even greater challenge in growth of III-nitride-based VCSEL, mainly in preparation of epitaxial DBR. Generally, bottom DBR employs epitaxial layers, such as $Al_xGa_{1-x}N/Al_yGa_{1-y}N$-based DBR.

The last decade has seen good progress in growth and fabrication of III-nitride VCSEL, although it is still a long way to obtain III-nitride VCSEL with performance, which is equivalent to GaAs-based VCSEL. In 1999, the joint efforts among three research teams in Japan, Germany, and Italy led to realization of the first nitride-based VCSELs optically pumped at 399 nm.[8] However, it is not until 2008 that the first electrically injected nitride-based-VCSEL has been demonstrated but at a low temperature (77 K), where an InGaN/GaN multiple quantum well structure (MQWS) is used as an active region and thus, the lasing is at 462 nm.[9] In later 2008, the first room temperature CW electrically injected InGaN-based VCSELs at 414 nm was reported, where both bottom and top DBR are made from dielectric materials.[10,11] In 2010, the first electrically injected InGaN-based VCSEL at 412 nm using an epitaxial DBR was successfully demonstrated.[12] Very recently, Nichia has demonstrated room temperature electrically injected VCSELs in the blue and green spectral region, where both bottom and top DBR are still made from dielectric materials. So far, the longest wavelength III-nitride-based VCSEL is at 503 nm.[13] At the opposite end in the spectral region, that is, the UV-based VCSEL, the progress lags far too behind the violet, or blue, or green VCSELs. Nevertheless, a breakthrough has been achieved through the joint efforts among the Sheffield team, Hong Kong team, and Singapore team, leading to demonstration of an optically pumped lasing at 343.7 nm at room temperature and the shortest wavelength so far.[14]

This chapter discusses a recent progress on development of DBR for III-nitride-based optoelectronics. Section 5.2 will describe a number of growth technologies developed for nitride DBR. These technological approaches that led to remarkable achievements in nitride DBR will be are described in detail, mainly based on lattice-mismatched $Al_xGa_{1-x}N/Al_yGa_{1-y}N$ DBR given in Section 5.2.1, and $Al_xIn_{1-x}N/Al_yGa_{1-y}N$ lattice-matched DBR presented in Section 5.2.2. A strain management is one of the key issues to achieve the lattice-mismatched $Al_xGa_{1-x}N/Al_yGa_{1-y}N$ DBR with a high reflectivity, which will be addressed in detail. Recently, a technique for air gap–based nitride DBR has been developed, although it is still the beginning. This can be achieved through combination of both epitaxial growth and chemically wetting-etching techniques, which will be discussed in Section 5.2.3. Section 5.3 will describe the research efforts on employing nitride-based DBR for fabrication of LDs. In addition to being applied in fabrication of VCSEL, DBR can also substantially improve optical performance of resonant cavity light emitting diodes (RCLEDs) and photo-detectors. Recent advance in this area will be introduced in Section 5.4. Finally, Section 5.5 will give a summary of the chapter and discuss perspectives on further improvements of DBR for III-nitride optoelectronics.

5.2 Nitride DBRs

Growth of III-nitrides is mainly based on "large lattice-mismatched heteroepitaxy" technique due to lack of suitable and affordable substrates for homoepitaxy. The research activity on lattice-mismatched heteroepitaxy actually started a long time ago, initially driven by an idea on integration of III–V compound semiconductors into silicon technology, represented by GaP grown on silicon and GaAs on silicon in early 1980s. Unfortunately, the progress on growth of the conventional III–V semiconductors on silicon has not been that impressive, as the density of defects generated is too high as a result of the large lattice-mismatch.

Although the lattice-mismatch between GaN and sapphire is as high as up to 16%, a huge success has been achieved on III-nitrides, following the invention of so-called two-step growth technology by Amano and Akasaki in 1986,[4,5] and Nakamura later on.[6] For the first time, the dislocation density of GaN was then reduced to a level of 10^8~10^9/cm^2, and thus the residual carrier density of GaN to 10^{16}~10^{17}/cm^3. This also made it possible to obtain *p*-GaN, another long-awaited major breakthrough. Although the dislocation density is still very high, surprisingly, blue LEDs with good performance were obtained for the first time in 1994. Right now, the standard two-step technology has been widely used for growth of III-nitrides the on sapphire: (1) a thin (~25 nm) GaN or AlN is grown as a nucleation layer at a low temperature and (2) the temperature is increased to more than 1,000°C to grow a thick GaN layer buffer prior to any device structure. This is a typical procedure for GaN grown on sapphire. Typically, the thickness of the high temperature (HT) GaN buffer is at least 1 μm in order to obtain reasonably good quality. Therefore, III-nitride DBR has to be grown on the above structure, namely, III-nitride DBR is basically grown on a thick GaN buffer layer on sapphire. DBR with high performance requires: (1) high reflectivity; (2) large stop-band; (3) free-cracking; (4) sharp interface between two alternating layers in each pair; (5) high crystalline quality, allowing for maintaining high crystalline quality of the active layer grown on the top of DBR.

Generally, the reflectivity (*R*) of a DBR structure with *N* pairs can be simply described by an equation below:[15,16]

$$R = \left[\frac{n_0 (n_\mathrm{h})^{2N} - n_\mathrm{s} (n_\mathrm{l})^{2N}}{n_0 (n_\mathrm{h})^{2N} + n_\mathrm{s} (n_\mathrm{l})^{2N}} \right]^2 \tag{5.1}$$

where n_0, n_h, n_l, and n_s are the respective refractive indices of the surround medium, and the two alternating materials (i.e., high refractive index layer and low refractive index layer) in each pair, and the substrate. Normally, n_0 is 1 for air. In this case, Eq. 5.1 can be simplified as:

$$R = \left[\frac{(n_\mathrm{h}/n_\mathrm{l})^{2N} - n_\mathrm{s}}{(n_\mathrm{h}/n_\mathrm{l})^{2N} + n_\mathrm{s}} \right]^2 \tag{5.2}$$

Therefore, both a high contrast in refractive index and a large number of pairs can lead to a high reflectivity.

Another important parameter to describe a DBR structure is the bandwidth of the photonic stop-band, which can be expressed by Eq. 5.3:[16,17]

$$\Delta\lambda = \frac{4\lambda_B}{\pi}\arcsin\left(\frac{n_h - n_l}{n_h + n_l}\right) \tag{5.3}$$

where λ_B is the central wavelength of the DBR. Equation 5.3 indicates that a large stop-band requires a high contrast in refractive index as well.

5.2.1 Lattice-Mismatched DBRs

The research on growth of III-nitride DBR started with the lattice-mismatched system, that is, AGaN–GaN system. The large lattice mismatch generates a number of issues. The most serious issue is due to cracking, and thus considerable efforts have been devoted to study the strain management, for which a number of approaches have been developed since early 1990s. When the wavelength moves down to the UV spectral region, $Al_xGa_{1-x}N$–$Al_yGa_{1-y}N$ system has to be employed for growth of DBR. In the meantime, AlGaN also has to be used for an emitting layer. Therefore, a careful design is necessary in order to eliminate the internal absorption issue.

5.2.2 Crack Issues in Growth of Lattice-Mismatched DBRs

It is a straightforward approach to use $Al_xGa_{1-x}N$–$Al_yGa_{1-y}N$ material system for growth of III-nitride DBRs. However, compared with AlAs/GaAs DBR, there exist a number of great challenges. The lattice mismatch between AlN and GaN is as high as 2.4%, and thus an overlying AlGaN grown on GaN suffers from a high tensile strain, potentially leading to a cracking issue. Furthermore, the thermal co-efficient mismatch between them is higher than 1%. Both GaN and AlN with a reasonably good quality need to be grown at a HT of more than 1,000°C. The large thermal stress makes the cracking issue become even worse. Another greater concern is that the contrast in refractive index between GaN and AlN is relatively small ($\Delta n/n \sim 0.15$). It means that a large number of repeats of $\lambda/4$ layers are required to maximize both reflectivity and stop-band. Finally, the large lattice mismatch between AlN and GaN also generates extra defects such as dislocations during growth of the DBR, thus potentially causing a degradation in optical performance of the active region grown on the bottom epitaxial DBR.

The research activity on III-nitride-based DBR started in early 1990s, initially with a low Al content layer such as $Al_{0.2}Ga_{0.8}N$/GaN DBR on sapphire.[18-21] As it is obvious that the contrast in reflective index is not so high, one cannot expect impressive data on both reflectivity and stop-band. Furthermore, the DBR was grown directly on a thin AlN layer as a buffer or nucleation layer, namely, the standard GaN buffer grown at a HT was not prepared, which I believe is specially designed in order to avoid the cracking issue. Such a growth approach may generate serious concerns about crystalline quality. Nevertheless, a clear abruptness of the interface between GaN and AlGaN has been observed, indicating that an epitaxial growth of AlGaN/GaN DBR could be possible.

As mentioned above, in order to obtain reasonably good quality, the thin GaN or AlN nucleation followed by a HT GaN buffer is necessary prior to growth of DBRs. In 1996 Redwing et al. followed the standard two-step approach to grow a thick III-nitride buffer prior to preparation of DBRs.[20] Instead of using a thick HT GaN buffer, they grew a 2 μm HT AlGaN with a low Al content of 12% as a buffer layer in order to avoid the cracking issue. This was followed by growth of 30 pairs of $Al_{0.12}Ga_{0.88}N/Al_{0.40}Ga_{0.60}N$ DBRs. Unfortunately, extensive cracks were still clearly observed, although a peak reflectivity of 93% has been estimated. Similarly, in order to suppress the formation of cracks, Krestnikov et al. modified the above two-step growth procedure, and then used a low temperature AlGaN nucleation layer, followed by a 1.1 μm $Al_{0.08}Ga_{0.92}N$ buffer layer instead of the standard GaN buffer layer prior to the growth of their DBR.[22] Although the modification may lead to degradation in crystalline quality, they obtained a cracking-free $Al_{0.15}Ga_{0.85}N$/GaN DBR with 37 pairs. The Al content is only 15%, which is quite low and thus may not be enough to obtain a high contrast in reflective index to achieve a DBR with a high reflectivity.

Ishikawa et al. adopted a compromised approach to eliminate the cracking issue.[23] Their study indicated that the cracks of the overlying AlGaN layer can be effectively suppressed if the crystal quality of the GaN underneath is deliberately decreased. It is fairly easy to understand it, as dislocation is also one of the channels for strain relaxation. Based on this idea, they obtained a 15-pair cracking-free AlGaN/GaN DBR with a peak reflectivity of 75% at 435 nm. However, it is obvious that it is not the best option to sacrifice crystalline quality to obtain free-cracks.

As stated above, a sharp interface is important for the growth of DBR with a high reflectivity. Shirasawa et al. employed X-ray diffraction (XRD) and atomic force microscopy (AFM) to study the quality of the interface between AlGaN and GaN on an AlGaN/GaN MQWS grown under different conditions.[21] They found that the sharp interface and smooth surface morphology can be obtained by means of taking a growth interruption approach, namely, there is a growth pause between GaN and AlGaN. During the growth interruption, NH_3 is still supplied. They studied the surface morphology as a function of interruption time (Fig. 5.1) indicating that the surface roughness can be significantly reduced as long as the interruption time is longer than 2 min. Based on the study, the 14-pair AlN/GaN DBR exhibit a peak reflectivity of 88% at 456 nm, while the reflectivity is reduced to 81% for the same structure but grown under non-interruption conditions.

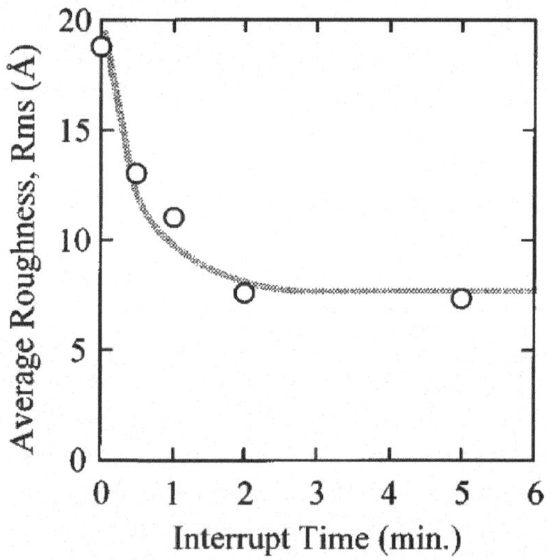

Figure 5.1 Surface roughness as a function of growth interruption time. Increasing interrupt time leads to a significant decrease in surface roughness. Reprinted from Ref. [21], Copyright (1998), with permission from Elsevier.

5.2.3 Strain Management in Growth of Lattice-Mismatched DBRs

Theoretically, the strain in a heterostructure depends on the thickness of the overlying layer and the layer underneath. If the optimal thickness can be found, it may be possible to minimize the strain, and thus possibly obtain a cracking-free heterostructure. Someya et al. studied the issue in order to find the optimal thickness of the GaN buffer layer for growth of an AlGaN/GaN DBR.[8,24,25] They found that a cracking-free DBR can be obtained on the top of the GaN buffer, as long as the thickness of the GaN buffer layer is less than 400 nm, which is less than the requirement for a normal growth of III-nitrides on sapphire. Based on the study, they achieved a 35-pair $Al_{0.34}Ga_{0.66}N$/GaN DBR with a peak reflectivity of 96% at 390 nm. The stop-band width is 14 nm. They predicted that the peak reflectivity can be further increased to 99% when the number of pairs is increased up to 50 (Fig. 5.2). However, the 42-pair DBR shows a slightly lower reflectivity than that of the 35-pair DBR.

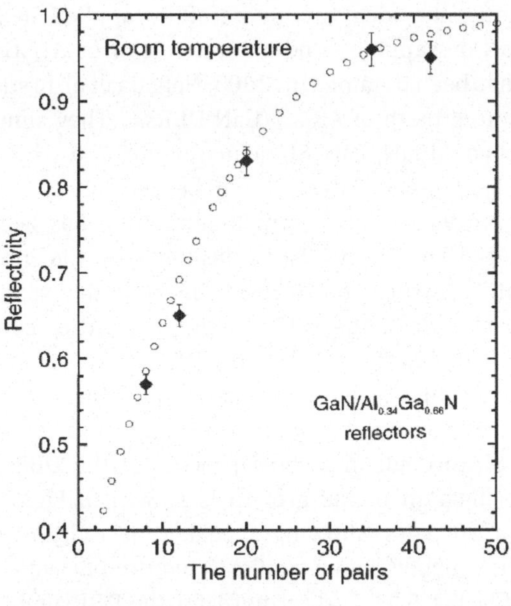

Figure 5.2 Measured peak reflectivity (solid squares) and theoretical reflectivity (open squares) as a function of the number of pairs. Reprinted with permission from Ref. [25], Copyright [1998], AIP Publishing LLC.

They attributed the degradation to an accumulation of interface roughness, optical absorption, and optical scattering due to dislocations and impurities. Another major reason, which they did not mention, may be still due to an accumulation of the strain, which is enhanced with increasing the number of pairs.

High crystalline quality is crucial for achieving III-nitride-based VCESL, and thus, a GaN or AlGaN buffer with reasonable thickness is necessary. Looking back at the history in developing III-nitride emitters, a GaN or AlGaN buffer with a thickness of more than 2 μm on sapphire may be necessary. In the last decade, a number of approaches to strain management have been developed in order to suppress the generation of cracks. Some of the approaches were originally designed for tackling the crack issue in growth of III-nitrides on silicon substrates, where the cracking issue is more serious than that for III-nitrides on sapphire.

An introduction of AlGaN/GaN superlattice (SL) to suppress crack generation was originally proposed for growth of GaN on silicon, leading to achieving a thick crack-free GaN film on silicon.[26] Subsequently, the idea was successfully applied in growth of a thick AlGaN on sapphire, and a 3 μm crack-free $Al_{0.2}Ga_{0.8}N$ film on sapphire has been obtained. In 2003, Nakada et al. firstly introduced the idea into growth of AlGaN/GaN DBRs.[27] They simply prepared a 100-period AlGaN/GaN SL structure prior to the growth of a 30-pair $Al_{0.41}Ga_{0.59}N$/GaN DBR.[27] The reflectivity was increased to 98% from 93%, and the crack generation has been effectively suppressed. Their study also indicated that the introduction of AlGaN/GaN SL structure can result in the highly compressed GaN layer in each pair of the DBR, making the lattice constant of the GaN close to that of the AlGaN in the DBR. Huang et al. modified the idea further.[28] Instead of a simple introduction of an AlGaN/GaN SL structure prior to growth of DBR, they inserted three sets of AlN/GaN SL through the whole process of the DBR growth. The schematic diagram of the growth is shown in Fig. 5.3. The DBR consists of 20 pairs, showing a peak reflectivity of 97% at 399 nm, which is higher than a similar structure but without using the AlN/GaN SL (Fig. 5.4). Furthermore, for the DBR aiming at 464 nm, they obtained a peak reflectivity of 99.4% and a wide stop-band of 29 nm. High-resolution transmission electron microscopy (TEM) study (Fig. 5.5) provides interesting results: many V-shape defects have been observed in the DBR and the crack lines are observed around the V-defects. The inserted AlN/GaN SLs can suppress the generation of cracks, but cannot reduce the number of the V-defects. Having looked at their growth conditions, they chose 980°C for the growth of their DBR instead of 1,050°C, while 1,050°C is their standard temperature for the growth of their HT GaN buffer. (Normally, growth of AlN layer needs to be performed at a higher temperature than that for growth of GaN). They also deliberately change N_2 as a carrier gas instead of H_2. Both the reduced growth temperature and the N_2 carrier gas employed certainly enhance the formation of the V-defects, which to some degree facilitate suppressing the generation of cracks. For this modified method, a careful design may be necessary, as the inserted AlN/GaN SLs has to account for part of the DBRs.

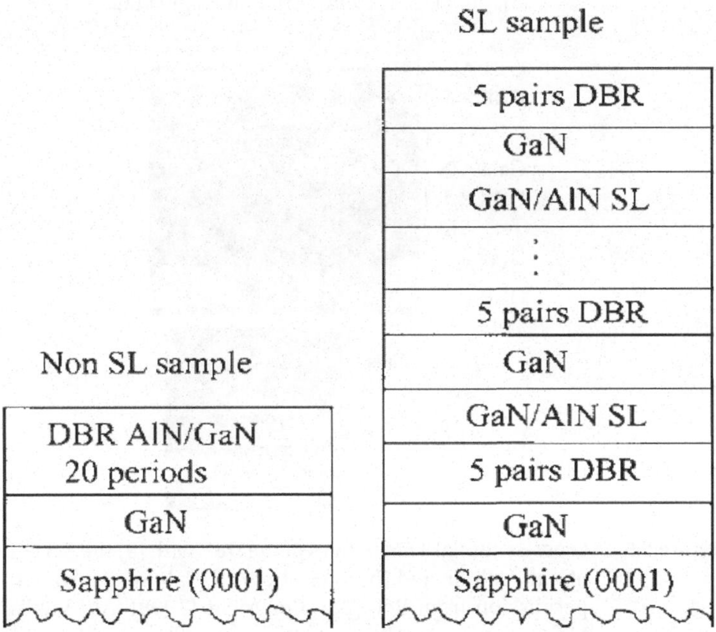

Figure 5.3 Schematic diagram for growth of a 20-pair GaN/AlN DBR using non-SL and SL (5.5 periods of GaN/AlN SL). Reprinted with permission from Ref. [28], Copyright [1998], AIP Publishing LLC.

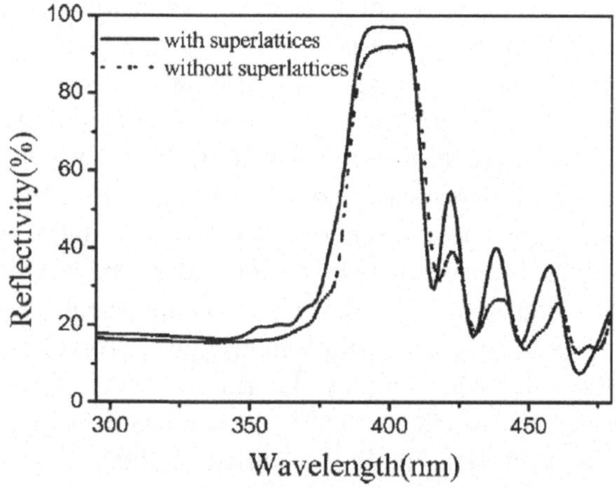

Figure 5.4 Reflectivity spectra of the 20-pair DBRs with and without the 5.5 periods of GaN/AlN SL. Reprinted with permission from Ref. [28], Copyright [1998], AIP Publishing LLC.

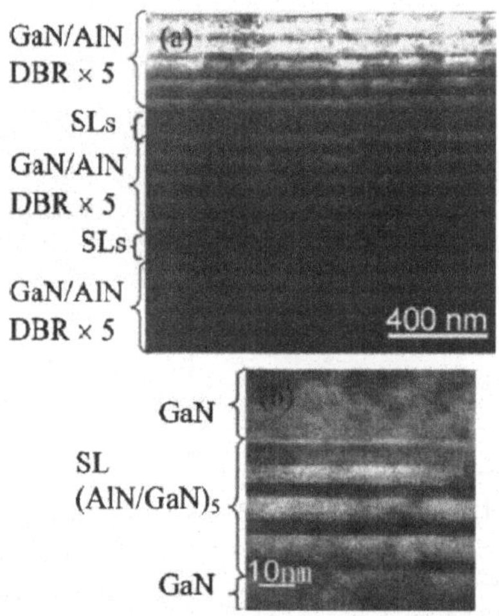

Figure 5.5 Cross-sectional TEM images of the DBR with (a) the 5.5 periods of GaN/AlN SL and (b) one set of AlN/GaN SL. Reprinted with permission from Ref. [28], Copyright [1998], AIP Publishing LLC.

Another simple idea for the strain management is to use a thin LT AlN layer. Theoretically, a tensile stress can potentially lead to the formation of cracks relatively easily. If a tensile stress can be converted into a compressive stress, the formation of cracks can be effectively suppressed. In 1999, Amano et al. introduced a thin LT AlN interlayer for growth of a HT AlGaN on GaN, and then successfully obtained a thick and crack-free AlGaN film.[29-31] Actually, the thin LT AlN interlayer can also further improve the crystalline quality of the overlying layer.

Although the LT AlN interlayer can effectively suppress the crack generation, solid evidences are required to prove the mechanism due to a change in stress. In 2000, Han et al. performed a detailed investigation by means of a special in situ technique, called multibeam optical stress sensor.[32] The equipment can in situ monitor any change in stress during a growth. A systematic change in stress of the overlying AlGaN during the growth has been found when the Al content of the LT $Al_xGa_{1-x}N$ interlayer is increased from 0 to 34%, 42%, 62%, and finally up to 100% (i.e., pure AlN). The data are given in Fig. 5.6, indicate that the tensile stress applied on the overlying AlGaN decreases with increase in the Al content of the LT $Al_xGa_{1-x}N$ interlayer. Finally, when the LT interlayer becomes pure AlN, the stress that the overlying AlGaN suffers from has been changed from tensile to compressive. They also used Griffith's criterion for crack propagation to calculate the critical thickness for the formation of cracks (h_g) of the overlying AlGaN as a function of the Al content of the LT AlGaN interlayer.[33] The calculation indicated that h_g can be increased as a result of increasing the Al content of the LT AlGaN interlayer. Figure 5.6 indicates that there still exists a tensile stress in the overlying AlGaN layer even using the 62% Al-content LT AlGaN interlayer. It needs to be pointed out here that the overlying AlGaN layer with a 20% Al content was used for the particular study. If the Al content of the overlying AlGaN increases, still there may exist a risk in crack generation assuming the overlying AlGaN is thick enough. Fortunately, Fig. 5.6 also indicates that the LT AlN interlayer can lead to a compressive stress on the overlying AlGaN layer, and thus the crack generation can be basically eliminated. Nomarski microscope measurements confirmed that a thick and crack-free AlGaN film has been obtained as a result of utilizing a pure LT AlN interlayer. Based on this study, the same group employed multiple LT AlN interlayers to successfully obtain crack-free $Al_{0.20}Ga_{0.80}N$/GaN DBRs even with 60 pairs. The peak reflectivity of 99% at 380 nm over an entire 2 inch wafer has been achieved.[34]

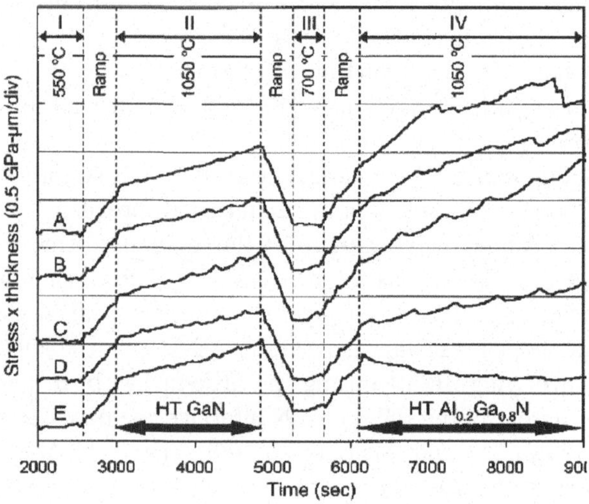

Figure 5.6 Stress × Thickness versus Time during growth of samples A–E. (A) x_{AlN} = 0.00 (i.e, GaN); (B) x_{AlN} = 0.34; (C) x_{AlN} = 0.42; (D) x_{AlN} = 0.62; and (E) x_{AlN} = 1.00 (AlN). Reprinted with Ref. [32], Copyright [2001], AIP Publishing LLC.

Selective area growth is another effective approach to eliminating cracks, which has been successfully used for the growth of GaN on silicon. Selective area growth of GaN on a patterned silicon substrate can effectively release the tensile stress between GaN and silicon. Ni et al. successfully applied this approach in the growth of nitride DBR, and then obtained a crack-free $Al_{0.46}Ga_{0.54}N$/GaN DBR with 40.5 pairs, showing a peak reflectivity of 98%.[35] They deposited a SiO_2 film and then fabricated it into square-shaped windows on an AlN layer on sapphire by means of standard lithography and subsequent chemical wetting-etching. The windows are of sizes 150 × 150 μm^2, 100 × 100 μm^2, 75 × 75 μm^2, and 50 × 50 μm^2 in a group, which form an array across an entire wafer. The growth of the DBR was performed selectively on the window areas. For this approach, the size of the SiO_2 window needs to be designed carefully, as the growth rate of GaN or AlGaN sensitively depends on the size. Normally, a small size leads to a higher growth rate. For examples, under identical growth conditions, the peak wavelength of the DBR grown on a 150 × 150 μm^2 window is around 391 nm, while it shifts to 410 nm for the DBR grown on a 50 × 50 μm^2 window. This can be seen clearly in Fig. 5.7.

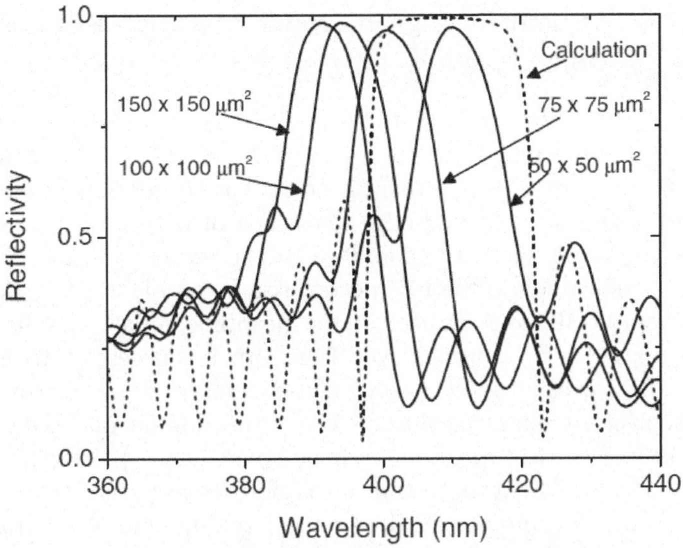

Figure 5.7 Reflectivity data from the DBR samples as a function of pattern sizes. Reprinted with permission from Ref. [35], Copyright [2009], John Wiley & Sons, Inc.

As stated above, a number of approaches to strain management have been reviewed. The LT AlN interlayer or the Al(Ga)N–GaN SL approach has been widely applied in growth of lattice-mismatched nitride DBR. So far, a peak reflectivity of over than 99% has been achieved on sapphire substrates. Due to an increasing interest in the development of III-nitride emitters on silicon substrates, the growth of III-nitride DBR on silicon has also been well progressed, but the growth approaches adopted for the strain-control are basically identical to those on sapphire addressed above.[36] Recently, the research on non-polar or semi-polar III-nitride emitters is becoming more and more intensive, as current III-nitride emitters suffer from self-built-in electric fields, leading to a reduced overlap of electron and hole wave-functions and thus, a low optical quantum efficiency. This is becoming even worse when the emitters are moving toward longer wavelength, such as in the green spectral region. Development of non-polar or semi-polar III-nitride emitters is aiming to challenge the so-called "green gap" issue. In the meantime, the research on growth of III-nitride DBR on non-polar substrates has started, but basically at a very preliminary stage.[37]

5.2.4 Structure Design in Growth of Lattice-Mismatched DBRs

So far, the DBRs discussed above are mainly for the applications in the violet–blue spectral region, that is, InGaN is used as an emitting layer. Basically, there is no serious concern about internal absorption, as the band-edge emissions of both GaN and AlGaN are in the UV spectra region. If a DBR structure is designed for a peak reflectivity in the UV spectral region, an $Al_xGa_{1-x}N$-$Al_yGa_{1-y}N$ ($x, y > 0$) DBR has to be used. In the meantime, AlGaN alloy also has to be used as an emitting layer. Therefore, it is necessary to design the DBR carefully, which not only maximizes the reflectivity but also avoids the internal absorption.[38,39] In addition, due to the self-built-in electrical fields induced by spontaneous polarizations in AlGaN-based MQW as an emitting region, there exists a large Stoke shift, which will make the design more complicated. The refractive index contrast in the AlGaN-based DBR system is quite small ($\Delta n \sim 0.4$; $x = 0$ and 1), while a DBR with a high peak reflectivity and wide band-width requires a high refractive index contrast. Therefore, it is required that a contrast in Al concentration in each pair have to be as high as possible. As a result, the Al concentration for the low refractive-index layer should be as high as possible, while the Al concentration for the high refractive-index layer as low as possible.

Here, a DBR structure with a peak reflectivity around 350 nm is used for the discussion. The DBR was grown on (0001) sapphire substrates by means of using a LT AlN interlayer on a thick GaN buffer in order to suppress the generation of cracks as discussed above.[38,39] The DBR consists of 36.5 ± 0.5 nm $Al_{0.49}Ga_{0.51}N$ and 34.5 ± 0.5 nm $Al_{0.16}Ga_{0.84}N$ in each pair, designed for a peak reflectivity at 353 nm (DBR I) (Fig. 5.8a, the cross-sectional scanning electron microscopy (SEM) image). For the DBRs with a peak reflectivity of 347 nm (DBR II), the layer thicknesses were decreased to 35.4 ± 0.5 nm and 33.5 ± 0.5 nm, respectively, with the same Al mole fractions.

Figure 5.8b shows the 300 K reflectivity spectrum of DBR I. It exhibits a symmetric, flat-topped stop-band, indicative of high sample quality. The peak reflectivity at 353 nm is 91 ± 2%, and the stop-band width is 17 nm. A simulation using transfer matrix theory of the full structure (DBR, interlayer, buffer layer, and substrate) using real and imaginary refractive index values from Brunner et al. was carried out and provided a good fit to the measured spectrum (Fig. 5.8b).[40] The band-gap energy of the $Al_{0.16}Ga_{0.84}N$ layers for the simulations was obtained from the PL measurements below. Figure 5.9 shows the reflectivity spectrum of DBR II: the peak wavelength is shifted to 347 nm due to the reduction in layer thicknesses. The peak reflectivity has decreased to around 88 ± 2% and the stop-band width to 15 nm. Furthermore, compared to DBR I, the stop-band is asymmetric, with the reflectivity decreasing more strongly on the short wavelength side. In order to explore the reason for this, 300 K PL measurements were carried on both DBRs. They exhibit almost identical spectra, the spectrum from DBR II (Fig. 5.9). An emission peak at 336.5 nm is observed, due to recombination in the $Al_{0.16}Ga_{0.84}N$ layers. This wavelength is only 10.5 nm shorter than the stop-band peak. Reflectivity simulations for this shorter wavelength DBR are shown in Fig. 5.9. It is seen that the asymmetry in the stop-band spectrum and somewhat lower peak reflectivity are reproduced well, suggesting strongly that absorption in the $Al_{0.16}Ga_{0.84}N$ layers leads to the degradation in DBR properties at shorter wavelengths. On the other hand, for DBR I with peak reflectivity at 353 nm, absorption in $Al_{0.16}Ga_{0.84}N$ layers has very little influence on the shape of the reflectivity spectrum. Hence, DBR I shows a symmetrical spectrum, while DBR II does not.

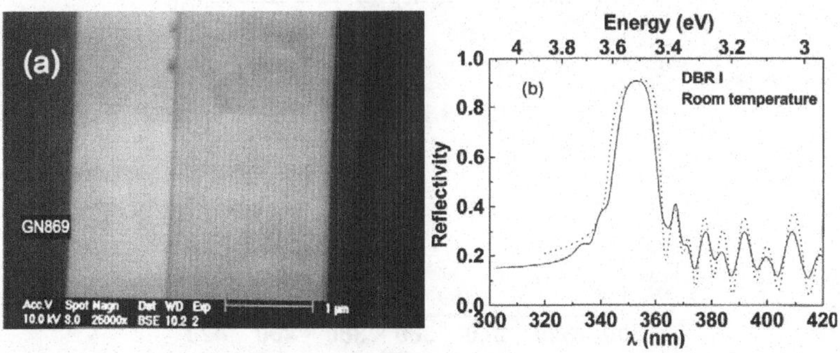

Figure 5.8 (a) Cross-sectional SEM image of DBR I, from which the average bilayer thickness is estimated to be 71 nm. (b) Measured (full) and simulated (dashed) 300 K reflectivity spectra of DBR I, showing a symmetric flat-topped stop-band. Reprinted with permission from Ref. [38], Copyright [2010], AIP Publishing LLC.

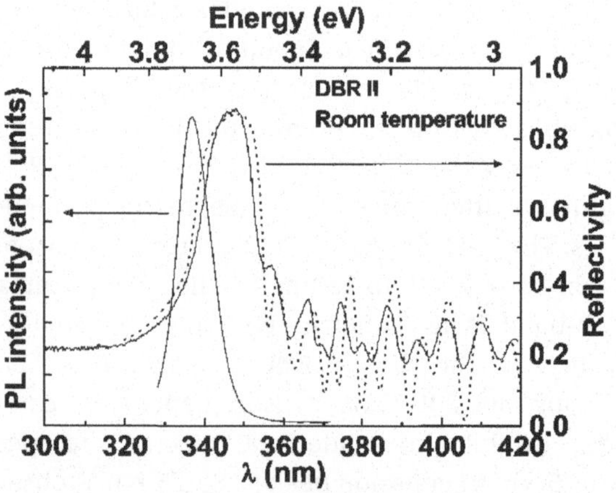

Figure 5.9 Measured (full) and simulated (dashed) 300 K reflectivity spectra of DBR II at a peak wavelength of 347 nm, showing asymmetric shape and lower peak reflectivity compared to DBR I. The 300 K PL spectrum shows an emission peakat 336.5 nm from the $Al_{0.16}Ga_{0.84}N$ layers. Reprinted with permission from Ref. [38], Copyright [2010], AIP Publishing LLC.

These conclusions are further confirmed by the low temperature (15 K) reflectivity and PL spectra of DBR II (Fig. 5.10). The spectrum now recovers a symmetrical shape, the stop-band

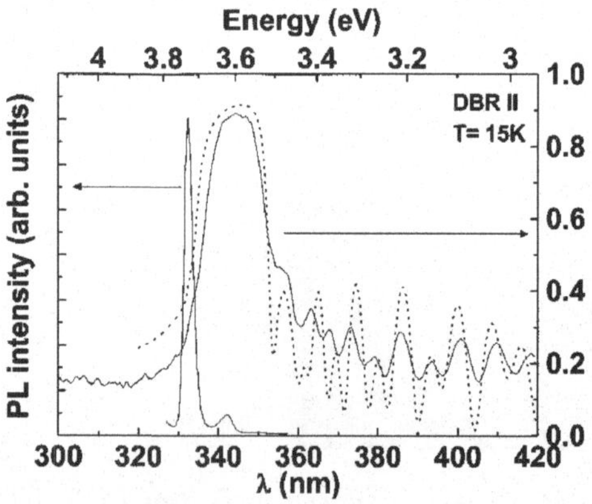

Figure 5.10 15 K reflectivity spectra (full experiment, dashed-theory) of DBR II, showing a symmetric stop-band and increased peak reflectivity compared to 300 K. The 10 K PL spectrum shows that the emission peak from the $Al_{0.16}Ga_{0.84}N$ layer shifts to around 331 nm. Reprinted with permission from Ref. [38], Copyright [2010], AIP Publishing LLC.

width increases to 17 nm, as for DBR I, and the peak reflectivity increases to 89%. The emission peak from the $Al_{0.16}Ga_{0.84}N$ layer becomes narrower and is shifted to 331 nm due to the temperature-induced band-gap increase, and is remote from the 345 nm stop-band compared to the situation at room temperature. Hence, the influence on the reflectivity is reduced, and the stop-band recovers a symmetrical shape. Once again these conclusions are reproduced well by the simulations shown in Fig. 5.10, where the parameters of Fig. 5.9 are employed, but with the $Al_{0.16}Ga_{0.84}N$ band-edge shifted by 6 nm to shorter wavelength. Therefore, such internal absorption should be avoided in designing a high quality DBR for short wavelengths, particularly below 360 nm.

5.3 Lattice-Matched DBRs

Ideally, the growth of lattice-matched nitride DBR is the best approach to eliminating any strain. Having looked at the whole III-nitride material system, one can find that a simple possibility is to incorporate InN into AlN for formation of AlInN alloy, which can be lattice-matched to GaN or AlGaN through tuning Al and In compositions (Fig. 5.11).[41] For examples, $Al_{0.82}In_{0.18}N$ is lattice-matched to GaN, $Al_{0.85}In_{0.15}N$ is lattice-matched to $Al_{0.20}In_{0.80}N$, etc. Figure 5.11 also indicates that for any lattice-matched AlInN–Al(Ga)N system, the AlInN alloy exhibits higher band gap than the Al(Ga)N.

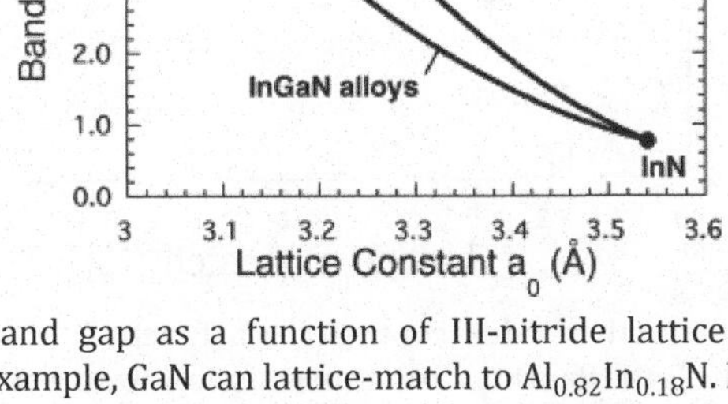

Figure 5.11 Band gap as a function of III-nitride lattice constant. For example, GaN can lattice-match to $Al_{0.82}In_{0.18}N$. Reprinted with permission from Ref. [41], Copyright [2005], John Wiley & Sons, Inc.

Carlin et al. confirmed that AlInN exhibits a lower optical index than GaN by means of an in situ reflectivity measurement during the growth, although they may be slightly different compared with that at room temperature.[42] A detailed investigation has been carried out by Carlin et al. and an equation has been obtained to describe the optical index contrast between $Al_{1-x}In_xN$ as a function of indium content within a range from 6% to 21%.

$$\frac{\Delta n}{n}(\mathrm{Al}_{1-x}\mathrm{In}_x\mathrm{N/GaN}) = -0.127 + 0.35x \tag{5.4}$$

Equation 5.4 indicates the optical index contrast is around −12.7% when x is zero, that is, the optical index contrast for AlN/GaN. In this case, the lattice-mismatch is 2.4%. Figure 5.12 gives the optical index contrast as a function of lattice-mismatch. The most interesting point in Fig. 5.12 is on the region where the lattice-mismatch remains within ±0.5%, as it is expected that such a small lattice-mismatch cannot lead to a serious strain issue on the DBR. In this case, another major advantage of AlInN/GaN DBRs can be clearly demonstrated in Fig. 5.12, showing the maximal index contrast increases from 3% for the AlGaN/GaN DBR to 8% for the AlInN/GaN DBR if the lattice-mismatch is limited within ±0.5%.

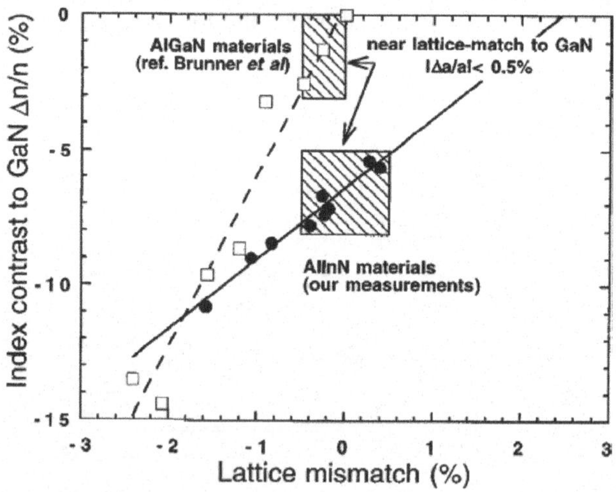

Figure 5.12 Refractive index contrast as a function of lattice-mismatch to GaN. Reprinted with permission from Ref. [42], Copyright [2003], AIP Publishing LLC.

However, the major challenge lies in growth of AlInN with a high crystalline quality. Firstly, InN has to be grown at a low temperature around 600°C, while AlN is normally grown at a temperature which has to be even higher than that for GaN, that is, above 1,100°C. The previous experience on growth of InGaN indicates that high indium content can be obtained only by lowering the growth temperature, but at the expense of a reduced crystalline quality. Therefore, it is expected that the growth of AlInN will be extremely difficult. Secondly, we have learnt that the large difference in interatomic spacing between GaN and InN has led to a severe miscibility gap, generating the well-known indium segregation. This leads to a severe issue on phase separation and thus non-uniformity. The issue becomes more serious for the growth of AlInN alloy, as the difference in interatomic spacing between AlN and InN is even larger, compared with GaN/InN. Thirdly, indium-contained alloy has to be grown at a very slow rate. The large difference in growth temperature for AlInN and GaN requires a number of growth interruptions, allowing for many temperature ramping processes. Both lead to a substantial increase in the total growth time. Typically, growth of a 30-pair AlInN/GaN DBR takes 16.5 h. Lastly, the involved temperature-ramping processes for growth of GaN may cause degradation in desorption and the crystal quality of the AlInN layer grown just before the GaN.

A research team led by Professor Grandjean at EPFL has performed a large amount of excellent work on the growth of AlInN alloy.[43–51] They deposited AlInN on a thick GaN buffer at a temperature between 800 and 850°C. In order to enhance the incorporation of InN into AlN, N_2 was used a carrier gas, which is similar to growth of InGaN. As the indium composition is highly sensitive to the growth temperature, it is crucial to control the temperature precisely and uniformly across a wafer. This is not so easy due to non-uniformity in temperature induced by the wafer-bowing. Despite the above growth challenges, they have obtained a 100 nm $Al_{0.83}In_{0.17}N$ film with a flat surface. Generally, the quality degradation in AlInN film becomes worse with increasing the thickness. Fortunately, for the applications in the UV or the visible spectral region, the thickness of AlInN required in each pair is around 30–60 nm, and the subsequently grown GaN layer at a HT can help to smooth the surface to some degree. Therefore, although the surface morphology of the $Al_{0.83}In_{0.17}N$ film is less impressive compared with a standard GaN, it may not be necessary to concern about it.

From the previous experience on the growth of Al-contained III-nitride alloys, we have learnt that a high flow rate of carrier gas is helpful for reducing the pre-reaction issue. It has been found that increasing the flow-rate of the carrier gas can significantly improve optical properties of the AlInN (Fig. 5.13), where the 500 nm AlInN layer becomes more transparent as a result of increasing the flow-rate of the carrier gas.[41]

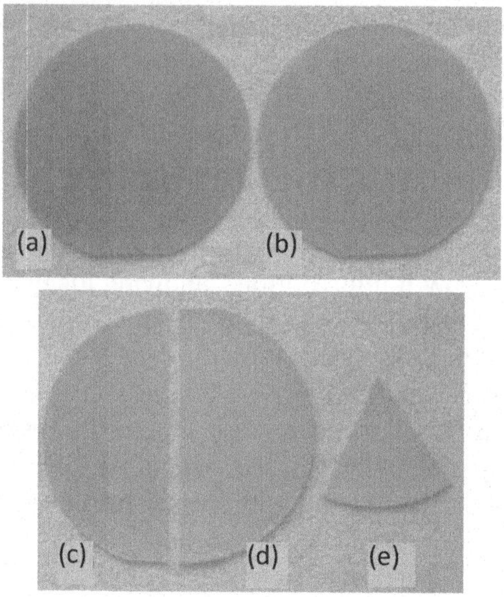

Figure 5.13 Photographs of $Al_{0.84}In_{0.16}N$ layers grown under an increasing flow-rate of carrier gas from left to right. Under a high flow-rate, the layer becomes transparent. Reprinted with permission from Ref. [41], Copyright [2005], John Wiley & Sons, Inc.

Considering a large difference in growth temperature between AlInN and GaN, a two-temperature approach with growth interruption has been applied in preparing AlInN/GaN DBRs. The EPFL has reported their cracking-free $Al_{0.82}In_{0.18}N$/GaN DBR with 50 pairs, showing excellent results: peak reflectivity of 99.7% at 410 nm and ~30 nm stop-band width (Fig. 5.14).[51]

The degradation in AlInN normally takes place during the process of temperature-ramp, mainly exhibited in the form of indium desorption. This will lead to an enhanced surface roughness, and a change in the AlInN refractive index close to the interface. In response to this issue, Sadler et al. further modified the growth–interruption approach discussed above; the growth interruption does not take place straight after the deposition of the AlInN.[52,53] Instead, a thin GaN is initially capped at the AlInN growth temperature before the growth interruption for ramping the temperature to ≥1,000°C for growth of GaN. This prevents desorption of indium from the surface of the AlInN, and thus, minimizes non-uniformity and unnecessary strain in the structure due to the indium desorption from the AlInN layer.

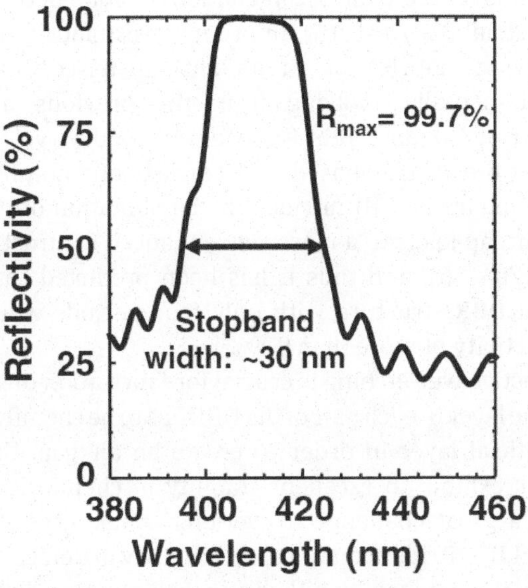

Figure 5.14 Reflectivity spectra of a 50 pair $Al_{0.82}In_{0.18}N/GaN$ DBR Reprinted with permission from Ref. [51], Copyright [2007], IOP Publishing. All rights reserved.

So far, we have addressed the issue on growth of the lattice-matched AlInN/GaN DBRs for the applications mainly in the blue spectral region. For the UV applications, an AlInN/AlGaN DBR structure has to be used instead of the AlInN/GaN DBR. In that case, if the AlInN/AlGaN DBR is directly grown on the GaN buffer, a large strain will be built up, and thus generate cracking eventually. At the moment, there is not an easy solution. Currently, a short period GaN/AlGaN SL structure mentioned is inserted before the growth of the AlInN/AlGaN DBR in order to avoid the formation of cracks (Section 5.1.2), which may be one of the efficient options. In long term, it may be necessary to develop a new buffer technology. Nevertheless, some good results have been obtained, leading to a 35-pair DBR with a peak reflectivity of 99.5% and a stop-band width of about 20 nm at 343 nm.[51]

5.3.1 Air Gap–Based DBRs

Fundamentally, a further increase in both reflectivity and stop-band width requires a higher contrast in refractive index. Unfortunately, as stated above, the contrast in refractive index between GaN and AlN is so small ($\Delta n/n \sim 0.15$). In order to respond to this issue, the technology for fabrication of an air gap–based DBR has been developed, actually evolving from the previous approach for fabrication of InP-air DBRs for near infrared wavelengths.[54,55] It means that the air gap–based DBR consists of a number of repeats of alternating air and III-nitride. The combination of the III-nitride and the air gap leads to a massively enhanced contrast in refractive index ($\Delta n/n \sim 0.6$), and thus it has been predicted that a GaN/air gap–based DBR structure with only four periods would lead to a peak reflectivity of more than 99%.[56]

A selective wet-etching is crucial for fabrication of such GaN–air DBR, as one layer in each pair of the DBR has to be chemically removed as a sacrificial layer in order to create an air gap. Unfortunately, III-nitrides exhibit an excellent stability in chemical solution, and thus it is a great challenge to wet-etch single crystal III-nitrides. However, if III-nitrides contain a high density of defects, which could potentially be attached by either an alkali or an acid, this may lead to an etching rate, which is practically usable. Compared with the growth of GaN or AlGaN alloy, indium-contained III-nitrides, such as InGaN or AlInN, are grown at a relatively lower temperature. Consequently, the density of defects in the indium-contained alloys is relatively high. In addition, the strength of chemical bonds in the indium-contained alloy is not as strong as that in either GaN or AlGaN alloy. Therefore, the indium-contained alloys could be used as an efficiently sacrificial layer. The experiments have confirmed that either InGaN or AlInN can be effectively employed as a sacrificial layer for fabrication of the GaN/air-based DBR.[57–62] So far, a number of methods for chemical etching of III-nitrides have been developed, among which so-called photo-enhanced electrochemical (PEC) wet-etching is the most efficient approach.[63] The PEC etching process combines the initial oxidation of the semiconductor surface and the subsequent dissolution of the resulting oxides, leading to an efficient wet-etching. The mechanism can be expressed by the equation below:[63]

$$2GaN + 6h^{+} + 6OH^{-} \rightarrow Ga_2O_3 + 3H_2O + H_2\uparrow$$

$$Ga_2 + 6OH^{-} \rightarrow 2GaO_3^{3-} + 3H_2O.$$

The equation requires that it be necessary to generate holes in order to promote the oxidation process. The holes can be conveniently generated under an illumination of a lighting source with photon energy larger than the band gap. The above equations uses GaN as an example, but it is actually suitable for AlGaN or InGaN as long as the holes can be provided.

The equation also implies that a selective wet-etching can be achieved through employing a lighting source with a proper wavelength. For example, for an AlGaN–InGaN system, a selective wet-etching can be achieved under an illumination of a lighting source whose photon energy is higher than the band gap of the InGaN, but lower than that of the AlGaN. In that case, the InGaN layer can be chemically etched, while the AlGaN layer remains intact. For an AlGaN–GaN system, the photon energy of the chosen lighting source needs to be higher than the band gap of GaN, but lower than that of the AlGaN.

The PEC wet-etching allows us to obtain a significantly enhanced etching rate, which also depends on the excitation power of the illuminated lighting source. For example, using a 45% KOH solution as an electrolyte, an etching rate of 400 nm per minute has been obtained under an illumination of 6 mW 325 nm He–Cd laser (a power density of about 0.57 W/cm^2), whose photon energy is higher than the band gap of GaN (≈362 nm).[64] One of the popular lighting sources for the PEC wet-etching is Xenon lamp. A simple set-up called electrochemical cell is typically required for the PEC wet-etching. A Ti/Au or Cr/Au-coated glass microscope slide as a contact is normally made to the front surface of the GaN.[63,65] No bias is required for the PEC wet-etching process. Both KOH and H_3PO_4 are widely used for the PEC wet-etching of III-nitrides, but the etching rate of III-nitrides in KOH is normally higher than those in H_3PO_4, of course depending on their concentrations.[63,65]

Using the PEC wet-etching approach, successful fabrication of a three-period $Al_{0.08}Ga_{0.92}N$/air gap–based DBR structure was reported in 2005.[66] The device is shown in Fig. 5.15. For the fabrication, an InGaN-based SL structure is used as the sacrificial layer, and consists of a 2.5-period 20 nm $In_{0.04}Ga_{0.96}N$/20 nm $In_{0.07}Ga_{0.93}N$. The utilization of the InGaN-based SL aims to produce an efficient carrier confinement. This can effectively prevent the photo-generated carriers diffusing and then recombining before reacting with the electrolyte. This design is very important, as otherwise the etching rate is extremely slow. In order to achieve a best selectivity, an unintentionally doped GaN film was used to filter out any part of the illuminated Xenon lamp light with photon energy higher than the band gap of $Al_{0.08}Ga_{0.92}N$. In 2007, the UCSB team reported a 4.5 periods $Al_{0.08}Ga_{0.92}N$/air DBR structure with a peak reflectivity of ~70%,[67] although in the theory the peak reflectivity is predicted to be above 99%.

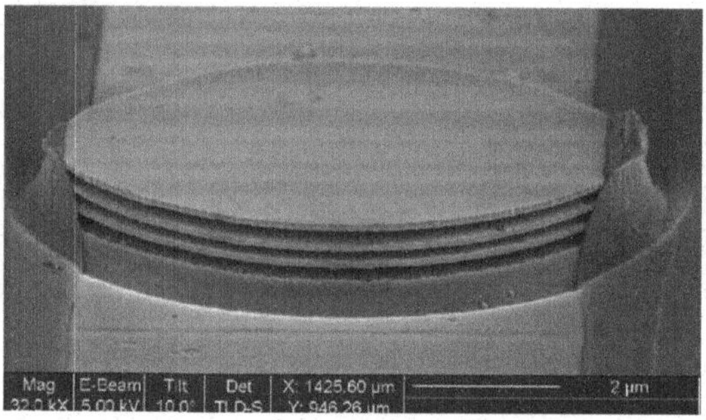

Figure 5.15 SEM image of an air-gap DBR structure fabricated by using the PEC wet-etching approach. Reprinted with permission from Ref. [66], Copyright [2005], AIP Publishing LLC.

Considerable efforts have been devoted to the fabrication of (Al)GaN/air gap–based DBRs using the lattice-mismatched (Al)GaN–AlInN system, where the AlInN is chosen as a sacrificial layer. An obvious advantage of using this system lies in its lattice-mismatch, as there is no any concern about strain relaxation after being fabricated into (Al)GaN/air gap–based DBRs. More importantly, the crystal quality of AlInN is far worse than that of InGaN due to a number of growth challenges (Section 5.2.2). Consequently, the density of defects in AlInN alloy is expected to be much higher than that in InGaN, potentially leading to a higher etching rate in comparison with InGaN as a sacrificial layer. The experiments have indicated that the wet-etching of the AlInN layer may not require the PEC process discussed above. Either nitric acid or KOH is supposed to be good enough to obtain an excellent selective wet-etching.[59] As discussed above, the PEC is more complicated as both the UV light with a specific wavelength and a metal contact as an electrode have to be carefully chosen.

Bellanger et al. optimized the conditions for using hot nitric acid to selectively etch lattice-mismatched GaN/$Al_{0.83}In_{0.17}N$ pairs, which were grown by molecular beam epitaxy (MBE).[60] It is worth pointing out that their AlInN layer was grown at 600°C, which is much lower than the typical temperature for the growth of the AlInN using MOCVD (at ~800°C). Detailed study indicates that the nitric acid with a 5 M concentration provides a best compromise between etching time and selectivity, leading to an etching rate of AlInN at ~100 nm/h and an excellent etching selectivity to GaN of ~500. In order to further enhance the etching rate, the nitric acid solution temperature is set to be ~120°C, just below its boiling point. This requirement is made in order to avoid generating a large number of bubbles, as the appearance of the bubbles could potentially cause the air-gap structure to collapse, the typical problem in fabricating the air gap–based DBRs.

Another great challenge for the fabrication of the (Al)GaN/air-gap DBR is to maintain a good mechanical property in order to reduce or ideally eliminate bending of the left (Al)GaN layers after the removal of the AlInN layers. Therefore, a careful design in both the device structure and the wafer structure has to be made. Bellanger et al. designed a microbridge configuration with a dimension of $10 \times 8\ \mu m^2$, which can be fabricated using a standard lithograph and subsequent dry-etching technique (Fig. 5.16).[60] The wet-etching was performed using the hot nitric acid under the conditions stated above. The final DBR is shown in Fig. 5.16, a four pair GaN/air-gap DBR structure. In addition, in order to make the mechanical performance strong enough, the GaN layer in each pair is deliberately designed to be 100 nm, which is equivalent to an optical length of ~$5/4\lambda$. The ~$5/4\lambda$ optical length leads to two stop-bands of high reflectivity in the reflectivity spectra, as shown in Fig. 5.17, one centered at 400 nm (~74% peak reflectivity and ~50 nm stop-band width) and another at 590 nm (>99% peak reflectivity, and 170 nm stop-band width), respectively. However, the peak reflectivity of the stop-band at 400 nm is far below the predicted value, which has to be above 99%. The major reason is due to the non-uniformity across the radial direction, as the GaN layer has been slightly etched away, which cannot be completely negligible. Please bear in mind that the total etching time for fabricating the structure shown in Fig. 5.16 is around 100 h, given that the etching rate of AlInN is ~100 nm/h. This non-uniformity becomes a severe issue at short wavelength, as the reflectivity is more sensitive to the fluctuation in thickness of the air gap at the short wavelength than that at the long wavelength.

Figure 5.16 SEM images of a four period nitride air-gap DBR (a) before and (b) after, etching using hot nitric acid. Reprinted from Ref. [60], Copyright [2009], The Japan Society of Applied Physics.

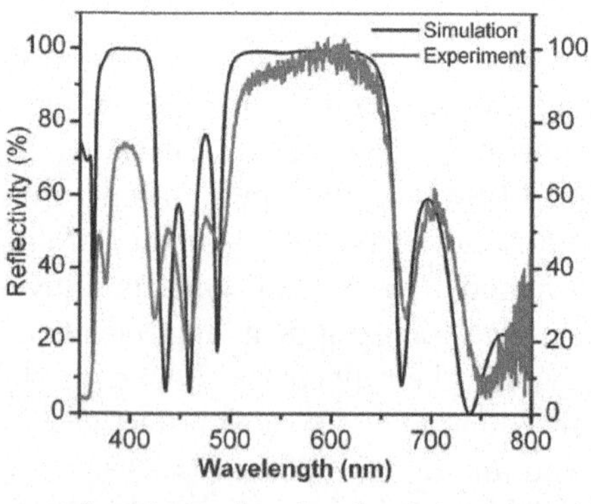

Figure 5.17 Measured and simulated reflectivity spectra for the air-gap DBR microbridge. Reprinted from Ref. [60], Copyright [2009], The Japan Society of Applied Physics.

A different approach to the etching of AlInN is called anodic oxidation process, which involves the dissolution of nitrilotriacetic acid (NTA) under an injection current.[58] The approach does not require any illumination. It can be carried out even at room temperature. The oxidation rate can be controlled by the injection current. This process can provide a high selectivity, meaning that the GaN layer was not attacked at all; however, it is complicated. Furthermore, the subsequent dissolution of the resulting oxides needs to be performed in a separate process, which makes it less impressive.

Another simple approach has been reported, where neither an illumination required for the PEC wet-etch nor the pre-oxidation process for the anodic process is required. Either hot KOH or hot nitric acid can be used to wet-etch AlInN, while the GaN in the solution remains intact. In the case of the hot KOH, the highest etch rate of 240 nm/min has been reported if the concentration is 6 M at 60°C.[68] In the case of using the hot nitric acid, a lateral etching rate of 140–210 nm/h has been reported.[59] In both cases, the etching rate sensitively depends on the crystal quality of AlInN, as basically either the hot KOH or the hot nitric acid is used to actually attack the defects in the AlInN in order to achieve the lateral etching. However, it can be expected that such chemical etching of AlInN could become very difficult only by using either hot KOH or hot nitric acid once the crystal quality of AlInN is significantly improved in the future.

Although some promising results have been achieved, it is still a long way to use the wet-etching in a practical application. A number of great challenges need to be met: (1) the fabrication process is extremely time-consuming, typically 10–100 h; (2) it can be fabricated based a small size, typically on a micrometer scale; (3) the uniformity issue is also severe, as a 100% selectivity in such chemical etching cannot be obtained, given that a long time is required for a whole etching process; (4) the most concerned issue is due to wafer bending or structure collapse, and thus, a careful design is crucial for maintaining a good mechanical property; (5) finally, the small size and non-uniformity of the devices present a great difficulty in performing measurements and thus, obtaining a reliable reflectivity spectrum.

5.4 DBRs for LDs

Major applications in utilizing DBR are for fabrication of VCSEL. It has been predicted that the threshold for lasing can be reduced if the cavity length is on a scale of the emission wavelength.[69,70] The research on GaN-based VCSEL started in 1990s. Compared with the GaN edge-emitting LDs, there exist a number of great challenges in device fabrication in addition to the difficulties in epitaxial growth. As expected, the first GaN-based VCSEL is obtained based on InGaN/GaN MQW as an emitting region under an optical pumping.

5.4.1 InGaN/GaN-Based Visible VCSELs

A joint effort led by Arakawa's team at University of Tokyo led to the first room-temperature optically pumped GaN-based VCSEL at 399 nm grown on sapphire substrate, reported in 1999.[8] The epitaxial AlGaN/GaN-based DBR is used as the bottom DBR, consisting of 43 pairs of 40 nm $Al_{0.34}Ga_{0.66}N$/38 nm GaN. A 15 pairs of ZrO_2/SiO_2-based oxide DBRs are used as the top DBRs. The cavity with an optical thickness of 2.5λ contains 195 nm GaN and 26 periods of 3 nm $In_{0.1}Ga_{0.9}N$ quantum well and 5 nm $In_{0.01}Ga_{0.99}N$ barrier. In order to eliminate the cracking, the GaN buffer thickness has been carefully designed, and a 400 nm thickness has been chosen. A 2D array has been fabricated using a standard reactive ion etching, consisting of an array of disk-shaped VCSEL structures 18 μm in diameter each separated in a 22 μm spacing, as shown in Fig. 5.18. Based on such structure, a Q factor of 500 has been achieved, corresponding to a MC with 98% reflecting mirrors. This is in a very reasonable agreement with the measured reflectivities from the bottom DBR and the oxide DBR, which are 98% and 99.5%, respectively. For their optical pumping measurements, a high power pulsed nitrogen laser was used to pump a dye laser at 367 nm as an excitation source. The optical characteristics are shown in Fig. 5.19, exhibiting the emission intensity as a function of optical power and the emission spectra below and above the optical threshold. The optical threshold is estimated to be 43 nJ/pulse, and the emission peak width is less than 0.1 nm above the optical threshold, indicating a clear lasing action.

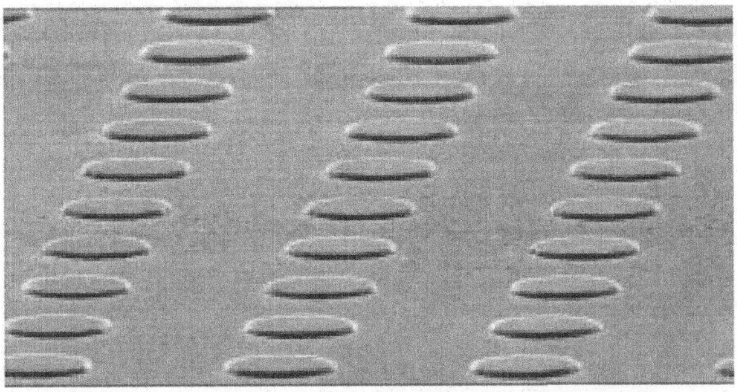

Figure 5.18 SEM image of a 2D array of GaN-based VCSELs. Reprinted with permission from Ref. [8], Copyright [1999], AAAS.

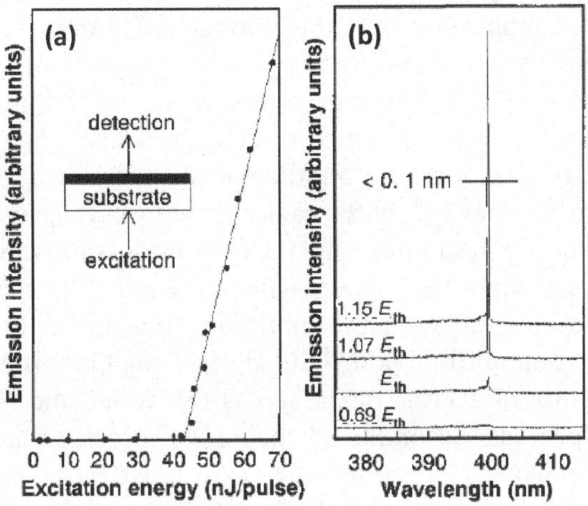

Figure 5.19 (a) Emission intensity as a function of excitation energy per pulse. (b) Emission spectra from the VCSEL as a function of excitation power. Reprinted with permission from Ref. [8], Copyright [1999], AAAS.

In parallel with developing the VCESL using epitaxial DBRs, considerable efforts have been devoted to the research on GaN-based VCSEL using oxides as both top and bottom DBRs. This approach can easily eliminate the cracking issue. As the contrast in refractive index in dielectric DBR is much higher compared with GaN–AlGaN system, both a close-to 100% reflectivity and a broad stop-band can be easily obtained. As the cavity layer are sandwiched between both top and bottom dielectric DBRs, the foreign substrate needs to be removed so that a so-called laser lift-off (LLO) technique is required. For a standard LLO process, a high power, UV (λ = 193 nm or

248 nm) excimer laser is irradiated from the backside of a GaN/sapphire (i.e., Al_2O_3). The UV photons can go through Al_2O_3 (e.g., ~9.9 eV or 125 nm) as a window with essentially no attenuation. Most of the UV power is absorbed and concentrated within the first 50~100 nm of GaN at the GaN–Al_2O_3 interface. The localized heating due to a high UV flux causes the thermal decomposition of GaN into N_2 and Ga droplets, accompanied by a build-up of explosive shockwave that can "de-bond" the GaN layer.

Actually, the LLO process has achieved a great success in fabricating III-nitride-based optoelectronics on sapphire substrate. It has also been successfully applied in fabricating GaN-based VCSEL. Basically, a cavity layer containing InGaN/GaN MQWs is grown on sapphire or any other foreign substrate, and then a top DBR structure using the oxides such as SiO_2/SiN or ZrO_2/SiO_2 is deposited on the top surface of the cavity layer. Subsequently, a so-called wafer bonding process is required, allowing the separate structure to be mounted in another substrate but in a reversed way. Therefore, another side of the cavity structure, which used to be the interface between sapphire and the cavity structure, can be exposed to the air for deposition of the second dielectric DBR structure.

However, there are a few limitations in using the LLO approach for fabrication of III-nitride VCSEL. Firstly, the LLO process involves the scanning of a laser beam across the wafer and is inherently a serial process. Secondly, as the LLO process creates localized thermal and mechanical stresses, care must be taken in surface preparation, wafer bonding, and maintenance of the laser beam profile. Thirdly, the LLO technique relies on a window-absorber relation between substrate and epilayer (GaN), and therefore cannot be extended to other potentially useful substrates such as Si, SiC, or GaN. Furthermore, a great concern is that the first 50~100 nm of GaN at the GaN–Al_2O_3 interface after the LLO becomes very rough, leading to a severe non-uniformity in cavity length. In addition, an additional wet-etching process is also required in order to remove the remained Ga metals.

In order to respond to the above issue, Tawara et al. reported a different approach for fabricating dielectric DBR for VCSEL.[70] Basically, two individual dielectric DBRs were prepared on sapphire separately, and a cavity structure containing InGaN/GaN MQWs with a high quality was grown on SiC. They took a conventional dry-etching approach to remove the SiC substrate in order to make the interface at the epi-film and SiC remain atomically flat after the removal, which has been confirmed by scanning electron microscope. This is extremely important for the subsequent wafer bonding process. Finally, using the wafer-bonding technique, the separated cavity structure and the two dielectric DBRs are bonded together to form the VCSEL in a sandwiched manner (schematic diagram, Fig. 5.20).[70] Taking the advantage of the high contrast in refractive index in the dielectric DBRs, both the dielectric DBRs show a high reflectivity of 99% at 400 nm and a very wide stop-band of ~100 nm. In order to obtain a best crystal quality, the $Al_{0.07}Ga_{0.093}N$ cavity with a 4λ optical thickness was grown on *n*-type 6H-SiC substrates by MOCVD. The cavity also contains three periods of 5 nm–thick $In_{0.02}Ga_{0.98}N$ barrier and a 2.5 nm–thick $In_{0.15}Ga_{0.85}N$ quantum well layer, which has been designed in order to ensure that the MQW structure is placed at the antinode position of the resonant optical wave in the cavity. Such VCSEL exhibits a few advantages over the VCSEL fabricated by the above hybrid VCSELs grown on sapphire: (1) the utilization of both dielectric DBRs leads to an improved reflectivity, as both DBRs exhibit ~99% reflectivity and (2) the growth

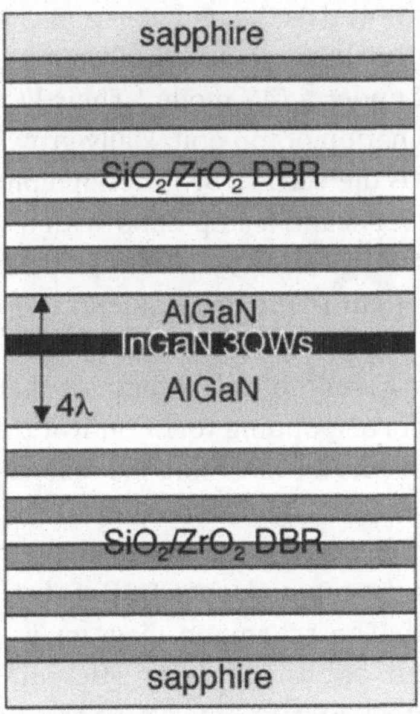

Figure 5.20 Schematic diagram of InGaN VCSEL fabricated using both top and bottom dielectric DBRs. Reprinted with permission from Ref. [70], Copyright [2003], AIP Publishing LLC.

of the cavity structure including InGaN MQWs on SiC results in better crystal quality due to less mismatch between AlGaN and SiC than that between GaN and sapphire. As a result the optical performance has been significantly improved, compared with the above hybrid VCSELs grown on sapphire. They have achieved a reduced threshold for an optical pumped lasing. The optical threshold at 400 nm has been reduced to 17.5 nJ/pulse. The optical gain of InGaN/GaN MQWs at the threshold has been significantly increased to 9,000 cm^{-1}, while it is only 890 cm^{-1} reported from the above hybrid VCSELs grown on sapphire.

Although the optical pumped GaN-based VCSEL was successfully fabricated in 1999, it is not until the year 2008 that the first electrically pumped GaN-based VCSEL was reported.[9] The conventional fabrication of VCSEL, such as GaAs-based VCSEL, normally employs a standard annular *p*-contact. However, it is difficult to apply such method in fabrication of nitride VCSEL, as they lead to a serious problem in current spreading due to the high resistivity of *p*-type nitrides. Therefore, an intra-cavity current injection structure has been proposed, where a transparent and conductive indium tin oxide (ITO) is employed as the *p*-contact.

Lu et al. reported the first electrically pumped GaN-based VCSEL operated at 77 K under a CW mode.[9] Their VCSEL structure is still based on a combination of top epitaxially grown AlN/GaN DBR and bottom Ta_2O_5/SiO_2 dielectric DBR. The top epitaxial DBR exhibits a reflectivity of ~99.4% and a stop-band with a width of 25 nm. The fabrication starts with a SiN layer; firstly, deposited and then a small window with less than 10 μm in a diameter is opened for fabrication of the *p*-contact. This is followed by deposition of a layer of ITO with a thickness of 240 nm as a *p*-contact. This ITO is also used as part of the MC (corresponding to 1λ optical thickness) in order to meet the requirements for matching the resonance phase condition. Finally, an aperture for light extraction of diameter of 10 μm or slightly larger is opened on the top of the *p*-contact, followed by the deposition of the top dielectric DBR. Fabrication of a *n*-contact is fairly standard. The schematic diagram is given in Fig. 5.21.[9] The electrical and optical characteristics at 77 K are given in Fig. 5.22, showing a 1.4 V turn-on voltage and a threshold current of 1.4 mA for lasing, equivalent to 1.8 KA/cm^2. Figure 5.23 shows the emission spectrum as a function of injection current, where a very narrow emission peak width of 0.15 nm at 462.8 nm has been observed when the injection current density is above the threshold. Furthermore, the spontaneous emission coupling factor β has been estimated to be 7.5×10^{-2}, which is three order magnitudes higher than that of the typical edge emitting LDs.

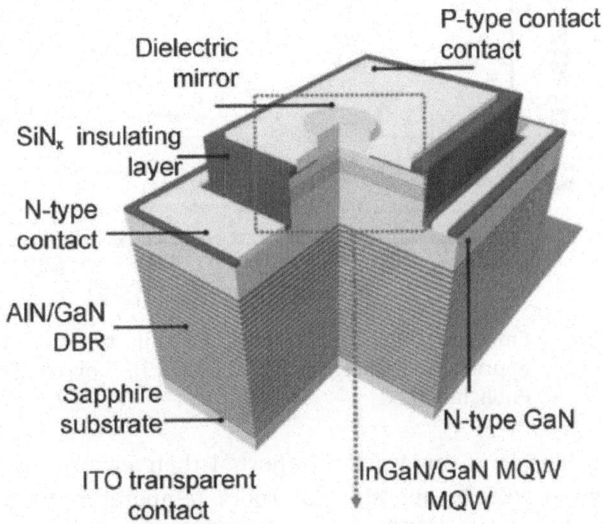

Figure 5.21 Schematic diagram of the GaN VCSEL. Reprinted with permission from Ref. [9], Copyright [2008], AIP Publishing LLC.

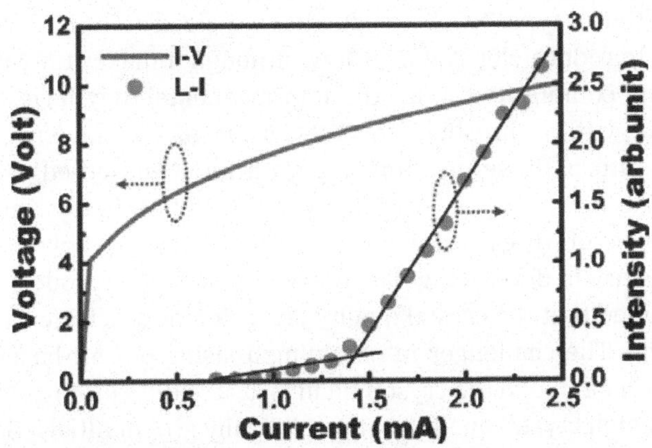

Figure 5.22 Optical and electrical characteristics (I–L and I–V curves) of the GaN VCSEL measured under the CW condition at 77 K. Reprinted with permission from Ref. [9], Copyright [2008], AIP Publishing LLC.

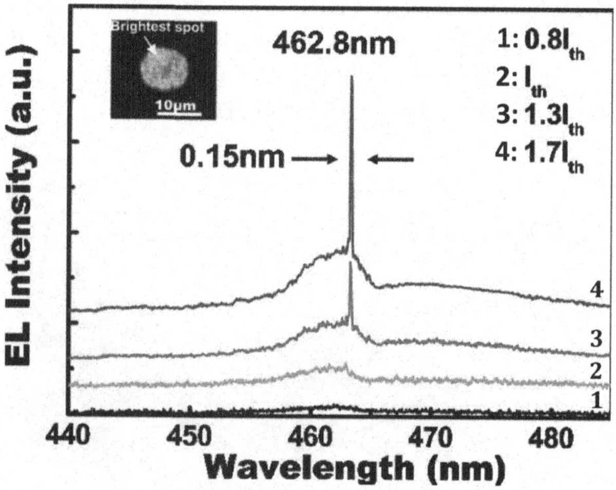

Figure 5.23 Emission spectrum as a function of injection current at 77 K. Reprinted with permission from Ref. [9], Copyright [2008], AIP Publishing LLC.

In 2010, the same group reported their electrically pumped GaN-based VCSEL, operating at room temperature.[12] Compared with their previous VCSEL operated at 77 K, a few new features have been added: (1) a heavily doped *p*-InGaN with a thickness of 2 nm is added as a contact layer in order to reduce the series resistance of the *p*-contact, although it leads to a slight increase in absorption; (2) A thin *p*-type AlGaN is added as an electron blocking layer in order to reduce electron overflow, although it makes the MC design become complicated. With the improvements, they have achieved a lasing with a threshold current density of 12.4 kA/cm^2 at room temperature. However, compared with InGaN-based edge emitting LDs, the performance of the GaN-based VCSEL is less impressive in terms of threshold current density. Having looked back at the history in developing GaN-based edge emitting diodes, it can be concluded that the crystal quality of epi-film is crucial for achieving a lasing. The challenges in obtaining a lasing of GaN-based VCSEL may still be due to the crystal quality, as it is difficult to obtain a MC with a high crystal quality on an epitaxially grown AlGaN/GaN DBR.

As stated above, the fabrication of GaN-based VCSEL with an MC sandwiched between two dielectric DBRs may be another good alternative option, as the current epitaxial growth technology can give rise to a GaN epi-film with an excellent quality even on sapphire. Therefore, we do not need to worry about the crystal quality of the MC on a single GaN epilayer. Using a combination of the wafer-bonding and the subsequent LLO techniques, a GaN-based VCSEL with a reasonably good performance can be fabricated. Actually, the first room-temperature electrically pumped GaN-based VCSEL was successfully fabricated based on this strategy, reported by Nichia Chemical Co. in 2008.[10]

The epi-layers used in Nichia's VCSEL were grown on c-plane sapphire, where two-period InGaN (9 nm)/GaN (13 nm) MQWs grown on a single GaN layer are used as the emitting region. Similar to the above procedure, a SiO_2 layer was first deposited, followed by opening a current aperture with a diameter of 8 μm for subsequent deposition of a 50 nm ITO as a *p*-type contact. On the top of the ITO, an extra Nb_2O_5 layer with an optical thickness of $\sim\lambda/8$ was prepared prior to the further growth of the dielectric DBRs, consisting of 11.5-pair SiO_2/Nb_2O_5. This extra Nb_2O_5 layer ensures that both the ITO layer and the InGaN/GaN MQWs are located at a node of an optical standing wave and an antinode, respectively, giving rise to an enhanced coupling between photons and carriers. The wafer-bonding on highly conductive Si and the subsequent LLO process were then performed in order to remove the sapphire substrate, allowing for deposition of the second DBR on the opposite side of the MC. Finally, the second DBR consisting of seven-pair SiO_2/Nb_2O_5 is prepared. The thickness of the MC is about 1.1 μm in total, being equivalent to the optical thickness of 7λ.

Figure 5.24 shows the I–L and I–V characteristics of the VCSEL under CW operation at RT, where the threshold for lasing has been estimated to be 13.9 kA/cm^2. The emission spectra below and above the threshold current are given in Fig. 5.25, where a narrow emission peak with a line-width of 0.03 nm has been observed.[10]

It is worth pointing out that the near-field patterns of the GaN-based VCSEL show a special feature below and above the threshold current, as demonstrated in Fig. 5.26. Below the threshold current, one can observe a uniform light emission across the current aperture, while a strong localization in the lasing spot has been observed and the lasing spot has been found to concentrates on a small area in a diameter of ~2 μm within the 8 μm-diameter current aperture.

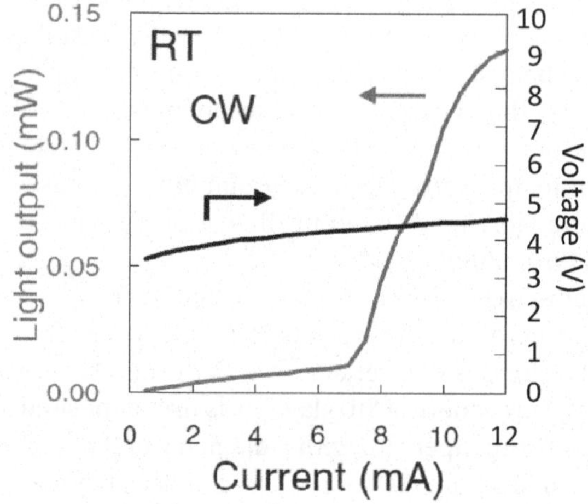

Figure 5.24 I–L and I–V characteristics of the VCSEL as a function of injection current measured under a CW operation mode at RT. Reprinted with permission from Ref. [10], Copyright [2009], The Japan Society of Applied Physics.

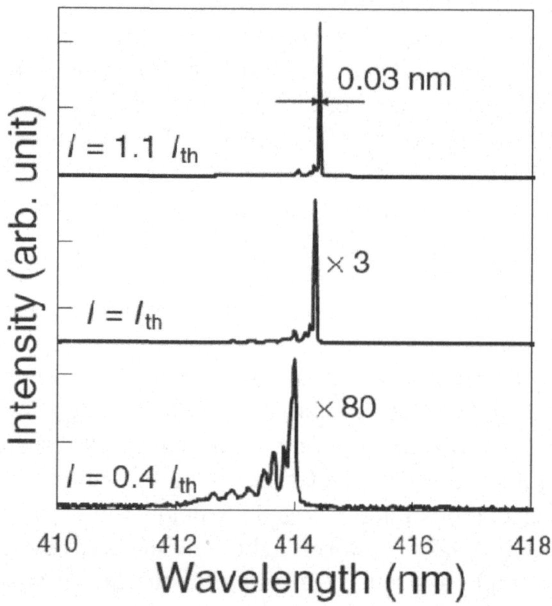

Figure 5.25 Emission spectra of the VCSEL below, at, and above the threshold current. Reprinted with permission from Ref. [10], Copyright [2009], The Japan Society of Applied Physics.

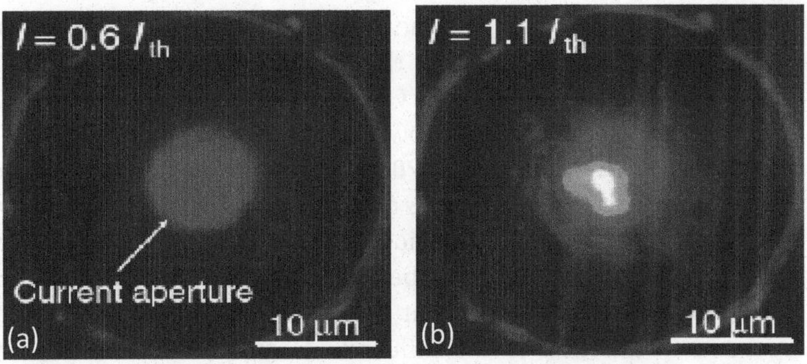

Figure 5.26 Near-field patterns of the VCSEL (a) below and (b) above the threshold current. Reprinted with permission from Ref. [10], Copyright [2009], The Japan Society of Applied Physics.

This similar phenomenon has been observed in the first electrically pumped GaN-based VCSELs, operating at 77 K. So far, the origin is unclear, but it is likely to be related to inhomogeneity on a micrometer scale. The GaN-based VCSEL fabricated using either epitaxially grown DBRs or pure dielectric DBRs show a similar behavior. Furthermore, it seems to have nothing to do with whether the LLO process is used or not. Therefore, it may be more likely related to indium segregation rather than surface morphology or variation in MC length.

Very recently, Nichia has reported that they have been successful in fabricating GaN-based VCSEL in the blue and green spectral region.[13] They still used the bottom and top dielectric DBRs, similar to the above description. However, their MC was grown on c-plane GaN substrate instead of sapphire, namely, the MC was grown based on homoepitaxy technology, which can lead to the InGaN/GaN MQWs with a massively improved quality. As the LLO process cannot be applied in removing the GaN substrate, a combination of mechanical polishing and subsequent chemical–mechanical polishing is used to remove the GaN substrate for the subsequent wafer-bonding process. The rest of the procedures remained identical. For the blue VCSEL at 451 nm, the threshold current density has been reduced to 3.0 kA/cm^2 under a CW operation mode. For the green VCSELs at 503 nm, the VCSEL has to be operated only under a pulse mode, as the threshold current density is very high, ~28 kA/cm^2.

Figure 5.27 summaries the threshold current density of all Nichia's GaN-based VCSELs in the spectral region from violet to green, showing a clear boundary at 471 nm. The GaN-based VCSEL at an emission wavelength below 470 nm can be operated under a CW mode, while those above 470 nm can work only under a pulsed mode. A very interesting result is that all the GaN-based VCSEL below 460 nm show a similar threshold current density. The increased threshold current density has been found when the wavelength is above 460 nm. This means that there may be some other unclear reasons, in addition to the influence of quantum-confined Stark effect (QCSE) on the optical performance, which they concluded.

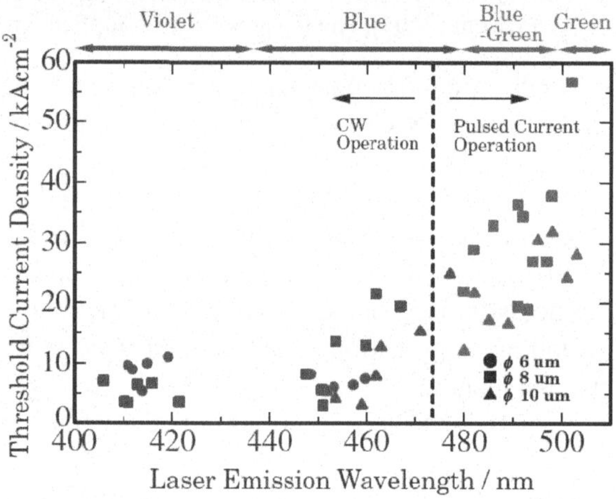

Figure 5.27 Current status of development of the GaN-based VCSELs in terms of wavelength and threshold current. Reprinted with permission from Ref. [13], Copyright [2011], The Japan Society of Applied Physics.

5.4.2 AlGaN/GaN-Based UV VCSELs

Compared with the GaN-based VCSEL in the visible spectral region, it is much more difficult to grow and fabricate III-nitride-based VCSEL in the UV spectral region. A number of great challenges need to be met before it becomes possible. Firstly, when the wavelength moves down toward 363 nm (~GaN band gap) or less, Al-contained MQWs have to be used as the emitting region. Consequently, a high crystal quality is crucial, as the performance of the UV emitters is much more sensitive to dislocations than InGaN-based violet, or blue, or green emitters. Secondly, it is difficult to apply the classic GaN buffer technique in the growth of the UV emitters on sapphire, as it leads to a severe cracking problem. Therefore, neither a conventional GaN buffer nor a free-standing GaN substrate (an extremely expensive option) is applicable to the growth of UV emitters. Thirdly, it is difficult to apply the LLO technique in removing substrate in the case of UV emitters, and thus the approach using both bottom and top dielectric DBRs may not work efficiently. In that case, an epitaxially grown DBR will have to be used as the bottom DBRs, and thus Al(Ga)N/$Al_xGa_{1-x}N$ instead of Al(Ga)N/GaN has to be used for growth of bottom DBRs. This causes the contrast in refractive index ($\Delta n/n \sim 0.15$ for AlN/GaN) even smaller, making the growth of the epitaxial DBR even more difficult.

The joint efforts among the teams in Sheffield, Hong Kong, and Singapore have led to the first 343 nm optically pumped VCSEL, operated at room temperature, which is so far the shortest wavelength VCSEL.[14] Unlike the conventional VCSEL fabricated in a form of an array on a micrometer scale, they adopted a nanostructure, namely, an array of close-packed nanopillars with diameters of around 500 nm, which is close to the emission wavelength. This aims to obtain an enhanced coupling strength.

As discussed in Section 5.2.4, the Sheffield team developed a thin AlN interlayer approach to achieve a high reflectivity $Al_xGa_{1-x}N/Al_yGa_{1-y}N$ DBR on a standard GaN buffer layer, leading to a crack-free 25-pair $Al_{0.2}Ga_{0.8}N/Al_{0.4}Ga_{0.6}N$ DBR with a reflectivity of up to 94% at 340–350 nm.[38,39] An AlInGaN quaternary alloy instead of GaN or AlGaN is used as a quantum well, aiming to enhance the optical performance of Al-contained MQWs grown on the DBRs as a result of indium segregation-induced localization. The previous experiments have confirmed that the utilization of the AlInGaN quaternary alloy has significantly improved the optical performance of the 350 nm UV LEDs.[71,72] In their structure, an AlGaN/GaN SL is used for the MC and the active region consisting of 10-pair of 2.5 nm $Al_{0.05}In_{0.045}Ga_{0.905}N$ and 5.0 nm $Al_{0.2}In_{0.002}Ga_{0.798}N$ MQWs is located at the center of the MC. A four pair of SiO_2/HfO_2 is used as the top dielectric DBR. The nanopillar structure was fabricated by means of a nanosphere lithography using silica nanoparticles.

Figure 5.28 shows the emission spectra of the VCSELs fabricated in a form of nanopillar structure measured under different optical pumping powers at room temperature, showing a clear lasing action at 343.7 nm when the optical pumping is above 0.52 MW/cm^2, the optical threshold for lasing.

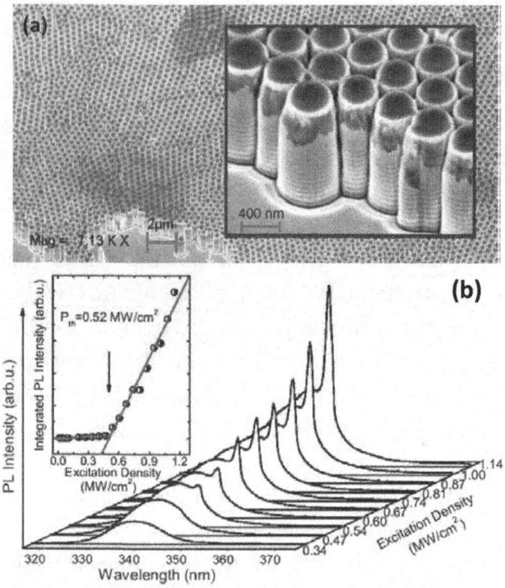

Figure 5.28 (a) SEM images of the VCSEL nano-pillar arrays, and (b) emission spectra and integrated emission intensity as a function of optical pumping density. Reprinted with permission from Ref. [14], Copyright [2010], AIP Publishing LLC.

5.4.3 III-Nitride-Based Polariton Laser

An MC containing MQWs provides a good confinement for both exciton and photon, possibly leading to a coupling between exciton and photon. This can possibly form cavity polaritons which are part exciton and part photon. The exciton binding energies of GaN and its alloy are very high (~50 meV), making them stable at room temperature. Therefore, it is possible to achieve a room temperature polariton laser-based on GaN MC.[73–76]

Another major advantage of the polariton laser is that the threshold is extremely low in theory, which is based on inversionless lasing. This is particularly important for GaN-based LDs, as they suffer from a high threshold current. This high threshold current is to a large degree due to the high carrier densities of states resulting from the high masses. However, the GaN-based polariton mass has a very small mass of order 10^{-4} m_e.

A strong coupling between exciton and photon is required for realization of polariton laser, meaning that the splitting between the two polariton branches (i.e., upper polariton branch (UPB) and lower polariton branch (LPB)) has to be greater than the line-widths of the exciton and cavity in an uncoupled mode, where the coupling strength determines the splitting between the two polariton branches (i.e., vacuum Rabi splitting). Therefore, both MQW and DBR structure require a high quality in terms of optical properties. Basically, the structure of the polariton laser is similar to that of VCSEL.

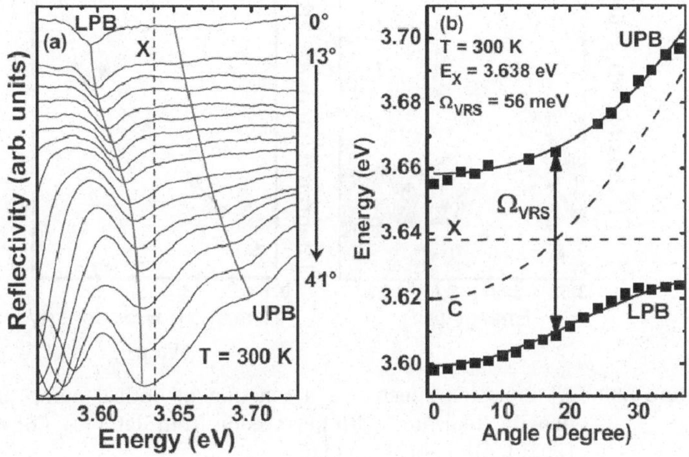

Figure 5.29 (a) RT reflectivity spectra as a function of angle and (b) RT experimental dispersion curve deduced from PL spectra (black squares) and fits of the LPB and the UPB (black lines). Reprinted with permission from Ref. [14], Copyright [2008], AIP Publishing LLC.

Christmann et al. reported a room temperature optically pumped polariton lasing, based on a GaN/AlGaN MQW MC.[75] The lattice-matched AlInN/AlGaN bottom DBR and the dielectric top DBRs are used as mirrors. They chose the AlGaN/GaN MQWs instead of InGaN/GaN MQWs in order to avoid a broadening line-width of the exciton induced by indium segregation. In order to enhance the oscillator strength of the exciton, 67-period GaN (1.2 nm)/$Al_{0.2}Ga_{0.8}N$ (3.6 nm) MQWs are used in the 3λ cavity structure. Figure 5.29 shows an RT angular-resolved reflectivity and an RT experimental dispersion curve deduced from PL spectra, respectively, showing a large vacuum Rabi splitting of 56 meV. Figure 5.30 shows both RT emission spectra as a function of optical pumping power density and integrated intensity as a function of optical pumping power density. The measurements were performed at room temperature when $K_{||}$ (momentum) is zero. A clear lasing has been observed at a relatively low-threshold pump power density of 17.7 W/cm^2.

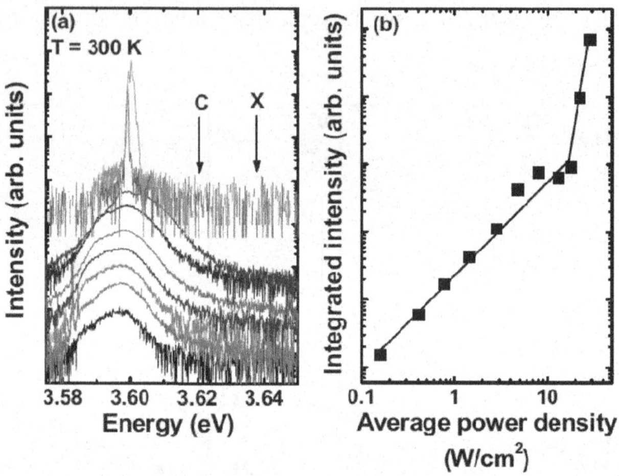

Figure 5.30 RT emission spectra as a function of optical pump power density. Reprinted with permission from Ref. [14], Copyright [2010], AIP Publishing LLC.

Very recently, Das et al. reported an ultralow threshold GaN nanowire polariton laser by optical pumping, where a single GaN nanowire is located in SiO_2 cavity with both top and bottom DBRs consisting of seven pairs of SiO_2/TiO_2.[76] The threshold carrier density is three order magnitudes lower than that of the photon lasing in the same device.

5.5 DBRs for Other GaN-Based Optoelectronics

The DBR structure can also be applied in fabrication of LEDs and photo-detectors with enhanced performance. In particular, the resonant-cavity (RC) LEDs (RCLEDs) exhibit superior performance to the conventional LED, such as a narrower spectrum line-width, a larger bandwidth, an excellent light-emitting directionality, and a better fiber-coupling efficiency. The most impressive feature is that light extraction efficiency can be significantly enhanced. This is very important for fabricating UV LED or any III-nitride LED on a silicon substrate, which can eliminate the internal absorption issue. Even with a DBR structure simply as a reflector, the optical performance has been observed to be significantly improved.

Lin et al. reported their 448 nm RCLED with an increase in light output intensity with a factor of 2.45 and a significant reduction in emission line width by 10 nm.[77] The fabrication is quite simple: a standard blue LED structure with both top and bottom dielectric DBRs, and the bottom dielectric DBR was prepared after sapphire was polished and then lapped. The angle-dependent EL measurements indicated that the emission line width increases from 13 nm to 36 nm when the emission angle changes from 90° to 45°. In the meantime, a small blue shift has been observed at a larger emission angle. Both are the typical characteristics of MC effect.

Using an AlInN/GaN lattice-mismatched DBR, Ishikawa et al. have been successful in eliminating the internal absorption issue of blue LEDs grown on silicon substrates, significantly improving the optical performance.[78] Their experiment has demonstrated that the light output power of the LED can be enhanced by up to a factor of 3.6 with increasing the pair-number of the DBR to 14.5, compared with the non-DBR LED. A clear line narrowing has been observed with increasing the number of DBR pairs.

An AlGaN/AlGaN DBR simply used as a reflector was incorporated below the light-emitting region in an UV LED structure grown by using a LT AlN interlayer on GaN buffer layer on sapphire, where the AlN inter-layer was used in order to prevent the formation of cracks for the growth of any further device structure. The AlGaN/AlGaN DBR structure employed for the 340 nm UV LED aims to eliminate any internal absorption issue due to using the GaN buffer. The output power has been increased by a factor of ~2.3, compared with the similar UV LED but without using the AlGaN/AlGaN DBR structure.[79] This is the first DBR-enhanced UV-LED reported so far.

Yonemaru et al. reported resonant-cavity enhanced UV metal–semiconductor–metal photo-detectors.[80] The structure is pretty simple, as shown in Fig. 5.31, where a 40 nm GaN as an absorption layer is sandwiched by two AlGaN layers. The cavity length is

964 nm, equivalent to an optical thickness of 7λ. As a result, two resonant peaks have been observed in the spectral region. Based on this design, an enhanced responsivity by a factor of about two has been observed due to the resonant cavity effect, selectively at wavelengths of 364 and 357 nm, respectively.

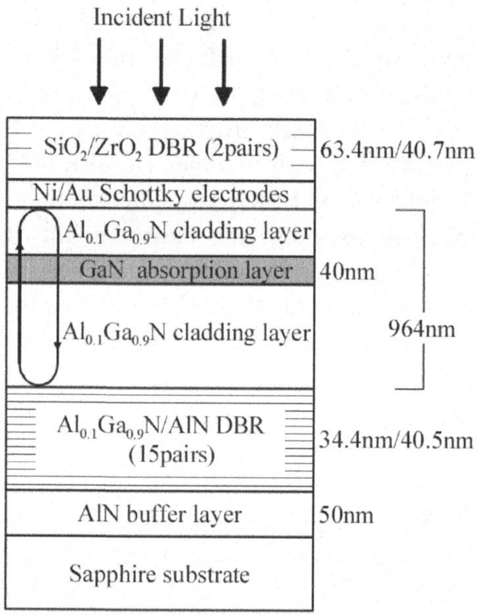

Figure 5.31 Schematic diagram of the resonant cavity-enhanced UV metal–semiconductor–metal photo-detector. Reproduced with permission from Ref. [80], Copyright [2005], John Wiley & Sons, Inc.

5.6 Summary

The growth of III-nitrides is mainly based on "lattice mismatched heteroepitaxy", which started a long time ago. Compared with the previous research on GaP grown silicon and GaAs on silicon, the advance of III-nitride-based optoelectronics on sapphire is a miracle. However, so far, except the edging emitting LDs, the development of III-nitride-based VCSELs and distribute feedback Bragg (DFB) lasers is far behind the other III–V semiconductor system. Furthermore, the major achievements are limited to the "violet/blue" spectral region. Further, efforts have to be put on development of new technologies for the fabrication of UV emitters and long wavelength emitters with high output power, like green emitters. In this paper, we reviewed the recent progress in the development of both epitaxial DBR and dielectric DBR for III-nitride-based optoelectronics. Further, efforts might be on how to improve strain management, even for the growth on an expensive GaN substrate. In this case, a new buffer technology may need to be developed. One of the most impressive material properties in III-nitrides lies in its large exciton energy, allowing the excitons to stay stably at room temperature. This provides us possibly with a unique opportunity for developing room temperature polariton lasers with an extremely low threshold using III–V semiconductors. Although an optically pumped III-nitride polariton laser has been reported, it may still be a long way for achieving an electrically injected polariton laser due to a number of challenges in both epitaxial growth and device fabrication. Another interesting trend toward developing III-nitride VCSEL may be represented by the growth along the non-polar or semi-polar direction.

Acknowledgments

The author is grateful to Michael Y. H. Wang for his contribution.

References

1. K. Iga (2000). Surface-emitting laser: Its birth and generation of new optoelectronics field, *IEEE J. Quant. Electron.* **6**, 1201.
2. H. Soda, K. Iga, C. Kitahara, and Y. Suematsu (1979). GaInAsP/InP surface emitting injection lasers, *Jpn. J. Appl. Phys.* **18**, 2329.
3. F. Koyama, S. Kinoshita, and K. Iga (1989). Room-temperature continuous wave lasing characteristics of GaAs vertical cavity surface emitting laser, *Appl. Phys. Lett.* **55**, 221.
4. H. Amano, N. Sawaki, I. Akasaki, and Y. Toyoda (1986). Metal-organic vapor phase epitaxial growth of a high quality GaN film using an AlN buffer layer, *Appl. Phys. Lett.* **48**, 354.
5. H. Amano, M. Kito, K. Hiramatsu, and I. Akasaki (1989). P-Type conduction in Mg-doped GaN treated with low-energy electron beam irradiation (LEEBI), *Jpn. J. Appl. Phys.* **28**, L2112.
6. S. Nakamura, T. Mukai, and M. Senoh (1994). Candela-class high-brightness InGaN/AlGaN double-heterostructure blue-light-emitting diodes, *Appl. Phys. Lett.* **64**, 1687.

7. H. Yoshida, Y. Yamashita, M. Kuwabara, and H. Ka (2008). Demonstration of an ultraviolet 336 nm AlGaN multiple-quantum-well laser diode, *Appl. Phys. Lett.* **93**, 241106.
8. T. Someya, R. Werner, A. Forchel, M. Catalano, R. Cingolani, and Y. Arakawa (1999). Room temperature lasing at blue wavelengths in gallium nitride microcavities, *Science* **285**, 1905.
9. T. C. Lu, C. C. Kao, H. C. Kuo, G. S. Huang, and S. C. Wang (2008). CW lasing of current injection blue GaN-based vertical cavity surface emitting laser, *Appl. Phys. Lett.* **92**, 141102.
10. Y. Higuchi, K. Omae, H. Matsumura, and T. Mukai (2008). Room-temperature CW lasing of a GaN-based vertical-cavity surface-emitting laser by current injection, *Appl. Phys. Express* **1**, 121102.
11. K. Omae, Y. Higuchi, K. Nakagawa, H. Matsumura, and T. Mukai (2009). Improvement in lasing characteristics of GaN-based vertical-cavity surface-emitting lasers fabricated using a GaN substrate, *Appl. Phys. Express* **2**, 052101.
12. T. C. Lu, S. W. Chen, T. T. Wu, P. M. Tu et al. (2010). Continuous wave operation of current injected GaN vertical cavity surface emitting lasers at room temperature, *Appl. Phys. Lett.* **97**, 071114.
13. D. Kasahara, D. Morita, T. Kosugi, K. Nakagawa et al. (2011). Demonstration of blue and green GaN-based vertical-cavity surface-emitting lasers by current injection at room temperature, *Appl. Phys. Express* **4**, 072103.
14. R. Chen, H. D. Sun, T. Wang, K. N. Hui, and H. W. Choi (2010). Optically-pumped ultraviolet lasing from nitride nanopillars at room temperature, *Appl. Phys. Lett.*, **96**, 241101.
15. T. E. Sale (1995). *Vertical Cavity Surface Emitting Lasers*. Taunton, UK: Research Studies Press.
16. C. J. R. Sheppard (1995). Approximate calculation of the reflection coefficient from a stratified medium, *Pure Appl. Opt.* **4**, 665.
17. A. M. Green, D. G. Gevaux, and C. C. Phillips (2004). Analytical method for the optical modelling of resonant-cavity-enhanced photodiodes, *Semicond. Sci. Technol.* **19**, 673.
18. M. A. Khan, J. N. Kuznia, J. M. Van Hove, and D. T. Olson (1991). Reflective filters based on single-crystal GaN/$Al_xGa_{1-x}N$ multilayers deposited using low-pressure metalorganic chemical vapor deposition, *Appl. Phys. Lett.* **59**, 1449.
19. I. J. Fritz, T. J. Drummond (1995). AlN-GaN quarter-wave reflector stack grown by gas-source MBE on (100) GaAs, *Electron. Lett.* **31**, 68.

20. J. M. Redwing, D. A. S. Loeber, N. G. Anderson, M. A. Tiscler, and J. S. Flynn (1996). An optically pumped GaN–AlGaN vertical cavity surface emitting laser, *Appl. Phys. Lett.* **69**, 1.

21. T. Shirasawa, N. Mochida, A. Inoue, T. Honda, T. Sakaguchi, F. Koyama, and K. Iga (1998). Interface control of GaN/AlGaN quantum well structures in MOVPE growth, *J. Cryst. Growth* **189**, 124.

22. I. L. Krestnikov, W. V. Lundin, A. V. Sakharov, V. A. Semenov et al. (1999). Photopumped InGaN/GaN/AlGaN vertical cavity surface emitting laser operating at room temperature, *Phys. Status Solidi B* **216**, 511.

23. H. Ishikawa, N. Nakada, M. Nakaji, G. Y. Zhao, T. Egawa, T. Jimbo, and M. Umeno (2000). Investigations on strained AlGaN/GaN/sapphire and GaInN multi-quantum-well surface LEDs using AlGaN/GaN Bragg reflectors, *IEICE Trans. Electron.* **E83C**, 591.

24. T. Someya, K. Tachibana, J. Lee, T. Kamiya, and Y. Arakawa (1998). Lasing emission from an $In_{0.1}Ga_{0.9}N$ vertical cavity surface emitting laser, *Jpn. J. Appl. Phys.* **37**, L1424.

25. T. Someya, and Y. Arakawa (1998). Highly reflective $GaN/Al_{0.34}Ga_{0.66}N$ quarter-wave reflectors grown by metal organic chemical vapor deposition, *Appl. Phys. Lett.* **73**, 3653.

26. H. Ishikawa, G.-Y. Zhao, N. Nakada, T. Egawa, T. Jimbo, and M. Umeno (1999). GaN on Si substrate with AlGaN/AlN intermediate layer, *Jpn. J. Appl. Phys.* **38**, L492.

27. N. Nakada, H. Ishikawa, T. Egawa, and T. Jimbo (2003). Suppression of crack generation in GaN/AlGaN distributed Bragg reflector on sapphire by the insertion of GaN/AlGaN superlattice grown by metal-organic chemical vapor deposition, *Jpn. J. Appl. Phys. 2,* **42**, L144.

28. G. S. Huang, T. C. Lu, H. H. Yao, H. C. Kuo, and S. C. Wang, C. W. Lin, and L. Chang (2006). Crack-free GaN/AlN distributed Bragg reflectors incorporated with GaN/AlN superlattices grown by metalorganic chemical vapor deposition, *Appl. Phys. Lett.* **88**, 061904.

29. H. Amano, M. Iwaya, T. Kashima, M. Katsuragawa et al. (1998). Tress and defect control in GaN using low temperature interlayers, *Jpn. J. Appl. Phys.* **37**, L1540.

30. H. Amano, M. Iwaya, N. Hayashi, T. Kashima, S. Nitta, C. Wetzel, and I. Akasaki (1999). Control of dislocations and stress in AlGaN on sapphire using a low temperature interlayer, *Phys. Status Solidi B* **216**, 683.

31. S. Yamaguchi, M. Kosaki, Y. Watanabe, Y. Yukawa, S. Nitta, H. Amano, and I. Akasaki (2001). Metalorganic vapor phase epitaxy growth of crack-free AlN on GaN and its application to high-mobility AlN/GaN superlattices, *Appl. Phys. Lett.* **79**, 3062.
32. J. Han, K. E. Waldrip, S. R. Lee, J. J. Figiel, S. J. Hearne, G. A. Petersen, and S. M. Myers (2001). Control and elimination of cracking of AlGaN using low-temperature AlGaN interlayers, *Appl. Phys. Lett.* **78**, 67.
33. S. J. Hearne, J. Han, S. R. Lee, J. A. Floro, D. M. Follstaedt, E. Chason, and I. S. T. Tsong (2000). Brittle-ductile relaxation kinetics of strained AlGaN/GaN heterostructures, *Appl. Phys. Lett.* **76**, 1534.
34. K. E. Waldrip, J. Han, J. J. Figiel, H. Zhou, E. Makarona, and A. V. Nurmikko (2001). Stress engineering during metalorganic chemical vapor deposition of AlGaN/GaN distributed Bragg reflectors *Appl. Phys. Lett.* **78**, 3205.
35. X. Ni, R. Shimada, T.D. Kang, J. H. Leach, and U. Ozgur (2009). GaN-based vertical cavities with crack-free high-reflectivity patterned AlGaN/GaN distributed Bragg reflectors, *Phys. Status Solidi A* **206**, 367.
36. H. Ishikawa, K. Asano, B. Zhang, T. Egawa, and T. Jimbo (2004), Improved characteristics of GaN-based light-emitting diodes by distributed Bragg reflector grown on Si, *Phys. Status Solidi A* **201**, 2653.
37. T. Zhu, A. Dussaigne, G. Christmann, C. Pinquier, E. Feltin, D. Martin, R. Butté, and N. Grandjean (2008). Nonpolar GaN-based microcavity using AlN/GaN distributed Bragg reflector, *Appl. Phys. Lett.* **92**, 061114.
38. T. Wang, R. J. Lynch, P. J. Parbrook, R. Butté, A. Alyamani, D. Sanvitto, D. M. Whittaker, and M. S. Skolnick (2004). High-reflectivity $Al_xGa_{1-x}N/Al_yGa_{1-y}N$ distributed Bragg reflectors with peak wavelength around 350 nm, *Appl. Phys. Lett.* **85**, 43.
39. A. Alyamani, D. Sanvitto, A. A. Khalifa, M. S. Skolnick et al. (2007). GaN hybrid microcavities in the strong coupling regime grown by metal-organic chemical vapor deposition on sapphire substrates, *J. Appl. Phys.* **101**, 093110.
40. D. Brunner, H. Angerer, E. Bustarret, F. Freudenberg, R. Hoepler, R. Dimitrov, O. Ambacher, and M. Stutzmann (1997). Optical constants of epitaxial AlGaN films and their temperature dependence, *J. Appl. Phys.* **82**, 5090.
41. J.-F. Carlin, C. Zellweger, J. Dorsaz, S. Nicolay, G. Christmann, E. Feltin, R. Butté, and N. Grandjean (2005). Progresses in III-nitride distributed

Bragg reflectors and microcavities using AlInN/GaN materials, *Phys. Status Solidi B* **242**, 2326.

42. J.-F. Carlin, and M. Ilegems (2003). High-quality AlInN for high index contrast Bragg mirrors lattice matched to GaN, *Appl. Phys. Lett.* **83**, 668.

43. S. Nicolay, J.-F. Carlin, E. Feltin, R. Butté et al. (2005). Midinfrared intersubband absorption in lattice-matched AlInN/GaN multiple quantum wells, *Appl. Phys. Lett.* **87**, 111106.

44. E. Feltin, R. Butte, J. F. Carlin, J. Dorsaz, N. Grandjean, M. Ilegems (2005). Lattice-matched distributed Bragg reflectors for nitride-based vertical cavity surface emitting lasers, *Electron. Lett.* **41**, 94.

45. J.-F. Carlin, J. Dorsaz, E. Feltin, R. Butté, N. Grandjean, M. Ilegems, and M. Laügt (2005). Crack-free fully epitaxial nitride microcavity using highly reflective AlInN/GaN Bragg mirrors, *Appl. Phys. Lett.* **86**, 031107.

46. F. Medjdoub, J. F. Carlin, M. Gonschorek, N. Grandjeanm et al. (December 11–13, 2006). Can InAlN/GaN be an alternative to high power/high temperature AlGaN/GaN devices? *IEEE International Electron Devices Meeting*. San Francisco, CA.

47. M. Gonschorek, J.-F. Carlin, E. Feltin, M. A. Py, and N. Grandjean (2006). High electron mobility lattice-matched AlInN/GaN field-effect transistor heterostructures, *Appl. Phys. Lett.* **89**, 062106.

48. F. Medjdoub, J. F. Carlin, M. Gonschorek, N. Grandjeanm et al. (2006). Small-signal characteristics of AlInN/GaN HEMTs, *Electron. Lett.* **42**, 779.

49. C. Gaquiere, F. Medjdoub, J. F. Carlin, N. Grandjeanm et al. (June 3–8, 2007). AlInN/GaN a suitable HEMT device for extremely high power high frequency applications. *IEEE/MTT-S International Microwave Symposium*. Honolulu, HI.

50. A. Lupu, F. H. Julien, S. Golka, N. Grandjeanm et al. (2008). Lattice-matched GaN-InAlN waveguides at lambda = 1.55 micrometer grown by metal-organic vapor phase epitaxy, *IEEE Photonics Technol. Lett.* **20**, 102.

51. R. Butte, J.-F. Carlin, E. Feltin, M. Gonschorek et al. (2007). Current status of AlInN layers lattice-matched to GaN for photonics and electronics, *J. Phys. D Appl. Phys.* **40** 6328.

52. T. C. Sadler, M. J. Kappers, and R. A. Oliver (2009). The effect of temperature and ammonia flux on the surface morphology and composition of $In_xAl_{1-x}N$ epitaxial layers, *J. Cryst. Growth* **311**, 3380.

53. T. C. Sadler, M. J. Kappers, and R. A. Oliver (2009). Optimisation of GaN

overgrowth of InAlN for DBRs, *Phys. Status Solidi C* **6**, S666.

54. K. Streubel, S. Rapp, J. Andre, and N. Chitica (1997). Fabrication of InP/air-gap distributed Bragg, *Mater. Sci. Eng. B* **44**, 364.
55. A. Hasse, S. Irmer, J. Daleiden, N. Dharmarasu, S. Hansmann, and H. Hillmer (2006). Monolithic integration of a tunable photodetector based on InP/air-gap Fabry-Pérot filters, *Electron. Lett.* **42**, 974.
56. E. Cho, D. Pavlidis, and E. Sillero (2007). GaN/air gap based micro-opto-electro-mechanical (MOEM) Fabry-Pérot filters, *Phys. Status Solidi C* **4**, 2767.
57. I. M. Watson, C. Liu, E. Gu, M. D. Dawson, P. R. Edwards, and R. W. Martin (2005). Use of AlInN layers in optical monitoring of growth of GaN-based structures on free-standing GaN substrates, *Appl. Phys. Lett.* **87**, 151901.
58. J. Dorsaz, H. J. Buhlmann, J. F. Carlin, N. Grandjean, and M. Ilegems (2005). Selective oxidation of AlInN layers for current confinement in III-nitride devices, *Appl. Phys. Lett.* **87**, 072102.
59. C. Xiong, P. R. Edwards, G. Christmann, E. Gu, M. D. Dawson, J. J. Baumberg, R. W. Martin, and I. M. Watson (2010). High-reflectivity GaN/air verticaldistributed Bragg reflectors fabricated by wet etching of sacrificial AlInN layers, *Semicond. Sci. Technol.* **25**, 032001.
60. M. Bellanger, V. Bousquet, G. Christmann, J. Baumberg, and M. Kauer (2009). Highly reflective GaN-based air-gap distributed Bragg reflectors fabricated using AlInN wet etching, *Appl. Phys. Express* **2**, 121003.
61. E. D. Haberer, R. Sharma, A. R. Stonas, S. Nakamura, S. P. DenBaars, and E. L. Hu (2004). Removal of thick (>100 nm) InGaN layers for optical devices using band-gap-selective photoelectrochemical etching, *Appl. Phys. Lett.* **85**, 762.
62. A. R. Stonas, T. Margalith, S. P. DenBaars, L. A. Coldren, and E. L. Hu (2001). Development of selective lateral photoelectrochemical etching of InGaN/GaN for lift-off applications, *Appl. Phys. Lett.* **78**, 1945.
63. V. Cimalla, J. Pezoldt, and O. Ambacher (2007). Group III nitride and SiC-based MEMS and NEMS: Materials properties, technology and applications, *J. Phys. D Appl. Phys.* **40**, 6386.
64. M. S. Minsky, M. White, and E. L. Hu (1996). Room temperature photoenhanced wet etching of GaN, *Appl. Phys. Lett.* **68**, 1531.
65. L.-H. Peng, C.-W. Chuang, J.-K. Ho, C.-N. Huang, and C.-Y. Chen (1998).

Deep ultraviolet enhanced wet chemical etching of gallium nitride, *Appl. Phys. Lett.* **72**, 939.

66. R. Sharma, E. D. Haberer, C. Meier, E. L. Hu, and S. Nakamura (2005). Vertically oriented GaN-based air-gap distributed Bragg reflector structure fabricated using band-gap-selective photoelectrochemical etching, *Appl. Phys. Lett.* **87**, 051107.
67. R. Sharma, Y.-S. Choi, C.-F. Wang, A. David, C. Weisbuch, S. Nakamura, and E. L. Hu (2007). Gallium-nitride-based microcavity light-emitting diodes with air-gap distributed Bragg reflectors, *Appl. Phys. Lett.* **91**, 211108.
68. E. Sillero, D. López-Romero, F. Calle, M. Eickhoff, J. F. Carlin, N. Grandjean, and M. Ilegems (2007). Selective etching of AlInN/GaN heterostructures for MEMS technology, *Microelectron. Eng.* **84**, 1152.
69. H. Yokoyama, and S. D. Broson (1989). Rate equation analysis of microcavity lasers, *J. Appl. Phys.* **66**, 4801.
70. T. Tawara, H. Gotoh, T. Akasaka, N. Kobayashi, and T. Saitoh (2003). Low-threshold lasing of InGaN vertical-cavity surface-emitting lasers with dielectric distributed Bragg reflectors, *Appl. Phys. Lett.* **83**, 830.
71. T. Wang, Y. H. Liu, Y. B. Lee, J. P. Ao, J. Bai, and S. Sakai (2002). 1 mW AlInGaN-based ultraviolet light-emitting diode with an emission wavelength of 348 nm grown on sapphire substrate, *Appl. Phys. Lett.* **81**, 2508.
72. T. Wang, P. J. Parbrook, and M. A. Whitehead (2003). MOCVD growth of AlInGaN quaternary system and its optical investigation, *Phys. Status Solidi C* **0**, 2019.
73. M. S. Skolnick, T. A. Fisher, and D. M. Whittaker (1998). Strong coupling phenomena in quantum microcavity structures, *Semicond. Sci. Technol.* **13**, 645.
74. D. Bajoni, P. Senellart, E. Wertz, I. Sagnes, A. Miard, A. Lemaître, and J. Bloch (2008). Polariton laser using single micropillar GaAs-GaAlAs semiconductor cavities, *Phys. Rev. Lett.* **100**, 047401.
75. G. Christmann, R. Butté, E. Feltin, J.-F. Carlin, and N. Grandjean (2008). Room temperature polariton lasing in a GaN/AlGaN multiple quantum well Microcavity, *Appl. Phys. Lett.* **93**, 051102.
76. A. Das, J. Heo, M. Jankowski, W. Guo, L. Zhang, H. Deng, and P. Bhattacharya (2011). Room temperature ultralow threshold GaN nanowire polariton laser, *Phys. Rev. Lett.* **107**, 066405.
77. C.-C. Lin, and C.-T. Lee (2010). GaN-based resonant-cavity light-emitting

diodes with top and bottom dielectric distributed Bragg reflectors, *IEEE Photonics Technol. Lett.* **22**, 1291.

78. H. Ishikawa, T. Jimbo, and T. Egawa (2008). GaInN light emitting diodes with AlInN/GaN distributed Bragg reflector on Si, *Phys. Status Solidi C* **5**, 2086–2088.
79. T. Wang, P. J. Parbrook, C. N. Harrison, J. P. Ao, and Y. Ohno (2004). Highly improved performance of a 350 nm ultraviolet light-emitting diode containing $Al_xGa_{1-x}N/Al_yGa_{1-y}N$ distributed Bragg reflectors, *J. Cryst. Growth* **267**, 583.
80. M. Yonemaru, A. Kikuchi, and K. Kishino (2002). Improved responsivity of AlGaN-based resonant cavity-enhanced UV photodetectors grown on sapphire by RF-MBE, *Phys. Status Solidi A* **192**, 292.

Chapter 6

III-Nitride Photonic Crystal LEDs for Lighting Applications

T. N. Oder,[a] J. Y. Lin,[b] and H. X. Jiang[b]

[a]*Department of Physics and Astronomy, Youngstown State University, Youngstown, OH 44555, USA*
[b]*Department of Electrical and Computer Engineering, Texas Tech University, Lubbock, TX 79409, USA*

tnoder@ysu.edu

Realizing lighting sources that are efficient, affordable, and environmentally friendly is one of the major goals of research and development of III-nitride light-emitting diodes (LEDs). Only about 5% of light produced in these materials escapes out in to air, the remaining is mostly guided laterally and eventually absorbed. Photonic crystals (PCs) have attracted intensive interest in scientific research and technical applications, because they offer approaches to manipulate photons, allowing the control of light propagation. Of the many other potential practical applications, using PC to dramatically increase light extraction efficiency of LEDs is very promising. The proceeding work reviews some of the pioneering work undertaken in incorporating PCs in the III-nitride LEDs for the purpose of enhancing light extraction. The incorporation of PC into both blue and ultraviolet (UV) nitride LEDs resulted in enhancement in extraction efficiency, light emission direction,

Handbook of Optical Microcavities
Edited by Anthony H. W. Choi

ISBN 978-981-4463-24-9 (Hardcover), 978-981-4463-25-6 (eBook)
www.panstanford.com

and pattern control and modulation speed control. More recent developments have showed that embedding 2D PCs in LEDs further enhances extraction efficiency because of improved interaction with guided modes. PC LEDs have become commercially available for high-end products. In addition several groups have started to explore III-nitride PC surface-emitting lasers, PC micro- and nanocavities, and microdisk lasers.

6.1 Introduction

III-nitride-based devices offer a great potential for applications such as UV/blue lasers, solar-blind UV detectors, and high-power electronics. Researchers in this field have made extremely rapid progress toward materials growth as well as device fabrication. The successes of super-bright blue light-emitting diodes (LEDs) and UV/blue laser diodes (LDs) based on the III-nitride system are a clear indication of the great potential of this material system. The recent success of the III-nitride edge-emitters and detectors is encouraging for the study of micro- and nanophotonic structures and devices. These micro- and nanophotonic devices range from arrays of emitters, detectors, and waveguides to optical switches and PCs. New physical phenomena and properties begin to dominate as the device size scale approaches the wavelength of light, including modified spontaneous emission, enhanced quantum efficiency (QE), and lasing in microcavities, all of which warrant fundamental investigations. The micro- and nanophotonic technologies, such as the integration of Si transistors in the 1960s, are eventually expected to provide the ability to integrate arrays of thousands of cheaply fabricated optical circuit elements such as waveguides, switches, and resonators. on a single chip. Together with their potential for optical circuit elements integration, these micro- and nanophotonic devices open many important applications such as optical communications, signal and image processing, optical interconnects, computing, enhanced energy conversion and storage, chemical, biohazard substance, and disease detection [1].

III-nitride optoelectronic devices offer special benefits including wide tunable bandgap covering the entire visible spectral range to deep UV (for InN at about 0.7 eV to AlN at about 6.1 eV), the ability to operate at very high temperatures and power levels is due to their mechanical hardness and larger bandgaps, large band offset for GaN/AlGaN or InGaN/AlGaN heterostructures, and exceptionally large exciton binding energies allowing novel quantum device designs, and high emission efficiencies. These unique properties may allow the creation of optoelectronic and photonic devices with unprecedented properties and functions. However, these unique properties also make the device physics and architectures of the nitrides different and much more challenging than other conventional semiconductors. For example, most of the nitride emitter structures are grown on sapphire substrates, which are insulating and cause extra difficulties in Ohmic contacts fabrication. Additionally, the conductivity of *p*-type materials is much lower for the nitrides, which makes hole-injection very difficult and requires innovative methods to improve the *p*-type conductivity as well as the current spreading.

In recent years, there have been dramatic advances in the utilization of PCs for enhancing the functionality of III-nitride photonic devices. PCs are artificially created ordered micro- and nanostructures in which media with different refractive indices are periodically arranged. In an ideal 3D PC, a photonic energy bandgap is created, in which no electromagnetic modes of a range of frequencies can exist [2–4]. Similar to the effect of energy bandgap in semiconductors, the photonic bandgap (PBG) dramatically changes the photon propagation properties. For example, if light is generated in the PC, it cannot propagate in certain directions; if light is sent from outside, it will be completely reflected. PCs have attracted intensive interest in scientific research and technical applications, because they offer approaches to manipulate photons, allowing the control of light propagation. These properties of PCs have been utilized in applications such as PC fibers for wave-guiding and enhancing nonlinearity, and PC sensors.

Of the many other potential practical applications, using PC to dramatically increase light extraction efficiency of LEDs is very promising. For LEDs, light extraction is a big concern. LED materials typically have a large refractive index, leading to a very small light extraction coefficient with most generated light entrapped in the LED [2]. The enhancement of external efficiency of LEDs using PCs has been studied previously in the infrared (IR) wavelength regions, where the demonstrations have been carried out using optical pumping on semiconductor materials not yet fabricated into electrically pumped LED devices [2,5–11]. A maximum of 30-fold enhancement in photoluminescence (PL) emission intensity from a freestanding InGaAs PC slab was reported at a wavelength of 1100 nm [11]. A sixfold enhancement in emission intensity at 925 nm has been obtained from an LED structure upon PC formation under optical pumping [10].

Nitride PCs that function in the visible and UV regions have been studied only in recent years. In particular, the refractive index is about 2.4 for III-nitride LEDs, the light extraction angle (or the critical angle for light to escape) is only ~23°, this amounts to only about 5%–10% of the light extracted from the LEDs. The need for improvement in the extraction efficiency in LEDs is exceptionally great, especially for deep UV LEDs ($\lambda < 340$ nm) based upon III-nitrides, which presently have very low QE. All lateral guided modes are lost due to parasitic absorption in UV LEDs. Chip-level innovations, such as the incorporation of PC, are key to achieving a significant enhancement in QE of deep UV photonic devices based upon III-nitrides. The effects of PC band structure and reduced group velocity were first observed in GaN [7,8]. Several research groups have since exploited PCs to enhance the extraction efficiency of III-nitride LEDs early on, and some of the results will be highlighted in this review.

Ideal PBG is achieved by periodicity in 3D. For extraction of light in LEDs, it is sufficient to eliminate light propa-gation only in the horizontal plane with the use of 2D PC. It was shown that the magnitude of Purcell enhancement factor for spatially extended band modes in 2D PC is similar to that for a tiny mode in a 3D nanocavity and that light extraction enhancement that arises from Zone folding or Bragg scattering of the photonic bands is the more important effect [9]. For this purpose, triangle lattice of air holes in a dielectric background are shown to be one of the most prominent 2D structures to have in-plane PBG effect [12]. The incorporation of PCs into III-nitride UV/blue emitters offers a solution for improving the extraction efficiency, but it is challenging. This is due in part to the difficulty in fabrication associated with the nanometer-scale features required for blue and UV wavelengths. Because PC structures must be built at the same scale as the wavelength to be suppressed, the difficulty of producing PC increases with shorter wavelengths. Additionally, nitride photonic devices are generally grown on insulating sapphire substrates, which make electron-beam (e-beam) lithography patterning of nanoscale features difficult (due to charge accumulation). One other challenge is the ability to transfer the designed patterns from e-beam resist to the samples. Because the III-nitrides are hard materials, the most effective method to achieve this is by high-density plasma etching. This requires an optimized resist or a lift-off material that will withstand the plasma etching as well as maintain the small features of the PC. Despite these technological challenges, tremendous progresses have been made.

In Section 6.2, we briefly highlight the major conventional lighting sources and discuss how they contrast with LEDs. Section 6.3 is an overview of the pioneering work the authors performed in fabricating and characterizing current-driven nitride PC LEDs. Section 6.4 reviews some of the more recent advances in using embedded 2D PCs in the nitride LEDs. Our concluding remarks in Section 6.4 highlight the need to develop methods for mass-scale patterning of PCs for industrial adaption as well as some of the more recent interest in surface plasmons (SPs) for enhancement of light extraction.

6.2 Lighting Sources and LEDs

6.2.1 Light Source in Human History

The importance of light has long been recognized in human history to the point that some societies have worshiped the sun. The *Book of Genesis* opens with a statement of God creating light to dispel the darkness and disorder. Thus, the invention of fire by early humans is regarded as a key game changer in the development of human society because it was a form of energy that provided heating and lighting. Over the years, humans have developed artificial lighting sources that has enabled independence from the natural light and thereby "extended" the day so that humans can increase their productivity [6,13]. No wonder then that the demand for artificial light has been increasing through the centuries and lighting technologies that deliver efficient, bright, and durable lighting source are in great demand. The major lighting sources that have been developed over the years include the following:

(a) Filament-based incandescent lighting sources

The incandescent bulb, invented by Heinrich Goebel in 1854 and made practical by Thomas Edison in 1879 relies on light produced from a heated object. When electrical current passes through the tungsten filament (resistor), the filament heats up to a very high temperature (>750°C), emitting electromagnetic radiation which has a significant amount of visible light. When it was first made commercially available by General Electric in 1909, the incandescent bulb had an efficacy of only 10 lm/W [14]. A standard 100 watt light bulb in today's use has a rated light output of 1,750 lumens, and an average efficiency of 17.5 lm/W. With the use of a small quantity of halogen gas and quartz glass, the temperature of the filament can be increased, yielding brighter light of efficacy of 15–20 lm/W. Note that an ideal white light would have an output of about 243 lm/W or 683 lm/W for yellowish-green light [15].

(b) Fluorescent lamp

Thomas Edison is also named as the inventor of the fluorescent lamp based on a patent filed in 1896 and granted in 1907 for the X-ray fluorescent lamp [14]. Again, it was General Electric that made this technology commercially available in 1938. The fluorescent lamp is made of a phosphor-coated tube filled with a low-pressure inert gas with a small amount of mercury. Applied voltage accelerates electrons in the tube which collide with gaseous mercury atoms causing excitation. UV radiation emitted from the de-excitation is then absorbed by the phosphor which "fluoresces", converting the UV radiation to visible light. Fluorescent lamps are used in many commercial lighting applications because of their higher efficacies that range from 60 to 100 lm/W. Compact fluorescent bulbs are now being used in the place of incandescent bulbs in residential lighting applications because they utilize less energy and produce less heat.

(c) Light-emitting diode

A light-emitting diode (LED) is a device made from a solid-state material (inorganic or organic semiconductor) that emits light when an electric current is applied to it, a phenomenon known as electroluminescence (EL). The British radio engineer H. J. Round was the first to report observation of light from a solid-state diode in 1907 [15,16]. The observed light was produced from a silicon carbide (SiC) Schottky diode, which can produce light either under strong forward-bias conditions or by avalanche multiplication under reverse-bias conditions. Today's LEDs are made from *p-n* junction diodes where light is emitted following radiative recombination of electrons and holes when the electrons are injected from the *n*-side to the *p*-side of the *p-n* junction diode. The blue SiC LEDs were sold commercially in the early 1990s but never became a viable lighting source due to the fact that SiC is indirect bandgap material. Currently, 20% of electrical energy generated worldwide is used for lighting and 90% of this is for residential and commercial lighting applications, which predominantly employ either incandescent or fluorescent light sources [14,17].

6.2.2 LEDs versus Other Lighting Sources

Solid-state LEDs have rapidly emerged as a viable source of lighting that is poised to replace the incandescent and fluorescent lighting sources. It is necessary to lay out the comparison between the traditional lighting sources against the emerging LED for lighting and a few of these are outlined here. When comparing brightness with the best R&D devices reported at 100 lm/W, the efficacies of incandescent bulbs are 10–20 lm/W, and that of fluorescent tubes is 50–100 lm/W and for LEDs, it is 30–50 lm/W [18]. A cool white LED made by Lumileds (LXML-PWC1-0120) reached 124 lm/W in 2010. Cool White (5650 K) Luxeon Rebel LED has an efficacy of 135 lm/W [19]. In 2006, Nichia developed a lamp-type LED with a luminous efficacy of 150 lm/W [20]. This implies that the incandescent bulb is not an efficient source of lighting as most of the energy is instead converted to heat. Furthermore, LEDs provide directional lighting, allowing for less "light pollution" [21]. LEDs with proper current regulation have an indefinite lifetime, which is a huge difference from other lighting sources. An LED from Cree can replace a 65 watt light bulb using only 10.5 watts [22]. This LED would not need to be replaced during its 50,000 hour lifetime and would consume about 525 kilowatt-hours for a cost of $63; the 65 watt bulb on the other hand would last for 2,000 hours, requiring 25 replacements, would consume 3,250 kilowatt-hours of energy and cost $465 [22,23]. This amounts to 85% less energy when an incandescent bulb is replaced by the Cree LED light bulb. When comparing between environmental pollution effects, incandescent light is not efficient source of lighting, implying that it consumes enormous amount of energy, whose production has a detrimental effect on the environment. Many countries have banned or in the process of banning incandescent bulbs all together because of the enormous energy waste. Fluorescent tubes on the other hand contain toxic mercury which is harmful to the environment, and several states in the United States have laws banning disposal of fluorescent tubes as general waste [20]. LED lights however do not contain materials harmful to the environment. LEDs thus render promise for applications such as general lighting, display technology, fiber optic systems, and machine vision, and they no wonder have generated a tremendous interest in research and developing with the primary goals of developing LEDs that are brighter, cheaper, and more energy efficient to meet the ever rising market demand.

LEDs that emit light in the blue-UV wavelength range are now based on GaN. The history of the development of GaN LEDs can be traced to the work of P. H. Maruska in 1968 at the Radio Corporation of America that was propelled by the need to develop blue light needed for full color imaging [24]. This led to the first development of the blue GaN LED by Pankove and Miller in 1971 with an emission peak of 475 nm [25]. Even though Maruska later incorporated Mg in the GaN and got a violet emission at 430 nm, this was based on impact excitation of filled Mg acceptors by hot electrons, followed by recombination and yielded inefficient LEDs of no commercial value [26,27]. No wonder then that effort on GaN was pretty much abandoned by the scientific community, except for Isamu Akasaki who refused to give up and his efforts finally paid off in 1989 when he succeeded in developing *p*-type GaN through electron beam annealing of Mg-doped GaN [28]. Further improvements of the GaN LED came through the work of Nakamura in 1995, when he developed blue and green GaN heterostructure LEDs with efficiencies exceeding 10% and thereby launched a revolution in the development of LED for lighting applications [29].

6.3 First III-Nitride PC-LED Work

6.3.1 Introduction

In semiconductor LEDs, about $1/(4n^2)$ of the light emitted radiates through the top and bottom without counting the effect of photon recycling, where n is the semiconductor refractive index [9]. For the III-nitride LEDs ($n \sim 2.4$), this amounts to a maximum of 12% of the total light created in the quantum well (QW) is emitted to air from both the top and bottom surfaces and the remaining 88% is guided in GaN (66%) and sapphire (22%) layers [30]. To improve the amount of light extracted from LEDs, different schemes have been employed, including using the flip-chip technology [31], thin film [32], and roughing the surface [33]. Surface roughening is among the most widely used approaches and relies on randomizing the light reflected at the interfaces destroying the light propagation in straight paths and leading to multiple attempts to escape the semiconductor slab through the critical cone. Although surface roughening offers high extraction efficiency in LEDs, it provides little control on the direction of the light emission in the device, resulting in Lambertian radiation patterns.

From his early work on PCs, Eli Yablonovitch recognized PCs as an alternative method of improving the light output from LEDs [4]. Work done by various researches since then have shown progressively improved outcome when PCs are incorporated in LEDs for the purpose of improving light extraction. Ideal PGB is achieved by periodicity in 3D, but its realization met practical difficulties. For extraction of light in LEDs, it is sufficient to eliminate light propagation only in the horizontal plane with the use of 2D PCs. David et al. showed that for GaN LEDs, triangular lattice PC is suitable for light extraction and favors omnidirectional behavior [34].

Research and development efforts on PCs initially concentrated on materials for applications in the IR wavelength regions for communication with quite impressive and rapid progress [9,10,35]. However, very little work was done on fabricating PCs on the group III-nitride materials for the blue/green and UV wavelengths applications prior to the first III-nitride PC-LED work reported by the authors who were then based at Kansas State University [36].

This lack was due in part to the difficulty in fabrication associated with the nanometer scale periodicity required. It has been reported that for incident light at λ = 488 nm in GaN, the PBG is likely to be observed at lattice periodicity a = 221 nm [37].

6.3.2 Optically Pumped III-Nitride PC LEDs

Initial investigations performed by the authors involved the nanofabrication of triangular lattice PCs with diameter/periodicity as small as 100/180 nm on an InGaN/GaN multiple quantum well (MQW). The optical pumping and characterization were carried using near-field scanning optical microscopy (NSOM) uniquely configured for UV wavelengths. A 20-fold enhancement of light intensity at wavelength of 475 nm by the PCs was observed at room temperature.

The III-nitride materials used for the research were grown by metalorganic chemical vapor deposition (MOCVD) where a 30 nm low temperature AlN buffer layer, followed by a 3 μm thick GaN layer, was deposited on double side polished sapphire substrates. Five periods of $In_{0.2}Ga_{0.8}N$/GaN (3 nm/20 nm) MQWs were then grown between a pair of 20 nm thick GaN cladding layers. Thus, the distance from the surface to the bottom of the MQW active region was about 160 nm. The nanofabrication of the PCs was accomplished by first spin-coating the III-nitride samples with polymethylmethacrylate (PMMA) and using e-beam lithography to write circles in a triangular lattice patterns. Triangle lattice of air holes in a dielectric background has been shown to be one of the most prominent 2D structures to present PBGs [38,39]. The lithography was performed with a LEO-440 scanning electron microscopy (SEM) system operated at 40 kV. The samples were then developed for 70 s in a solution of methyl isobutyl ketone and isopropyl alcohol and subsequently etched for 30 s using inductively coupled plasma (ICP) etching in a BCl_3/Cl_2 plasma. The residual PMMA resist was removed in warm acryl stripper by ultrasonic cleaning. Circular holes in the triangular lattice with diameters (d) from 60 to 140 nm and periodicity (a) from 120 to 300 nm were defined in an area of about 12 μm × 12 μm. The resulting structure is illustrated in Fig. 6.1. The surface morphology of the etched nitride PCs was characterized using SEM and atomic force microscopy (AFM). Figures 6.2(a) and 6.2(b) show SEM images of the nitride PCs after the patterns were etched into the nitride materials with hole diameter/periodicity of 100/180 and 120/300 nm, respectively [36]. The later structure with hole diameter/periodicity of 120/300 nm and an air-fill factor of 29% was extensively characterized. Figure 6.2(c) is an AFM line profile across the nitride PCs, showing that most of them were etched to a depth of about 200 nm, which was sufficiently deep enough to penetrate through the MQW layers of our structure.

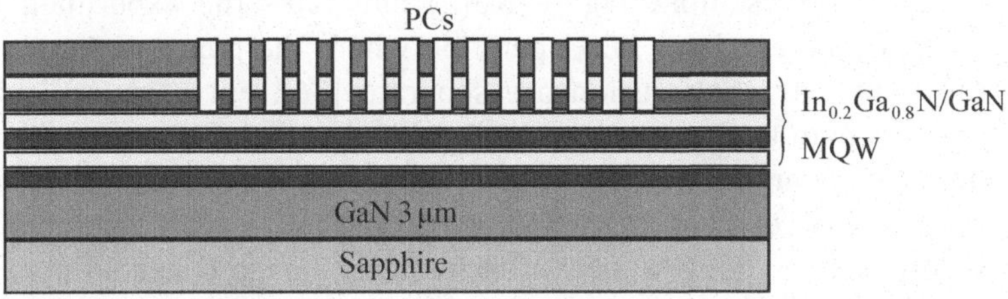

Figure 6.1 Structure of the III-nitride materials on which the PCs were fabricated.

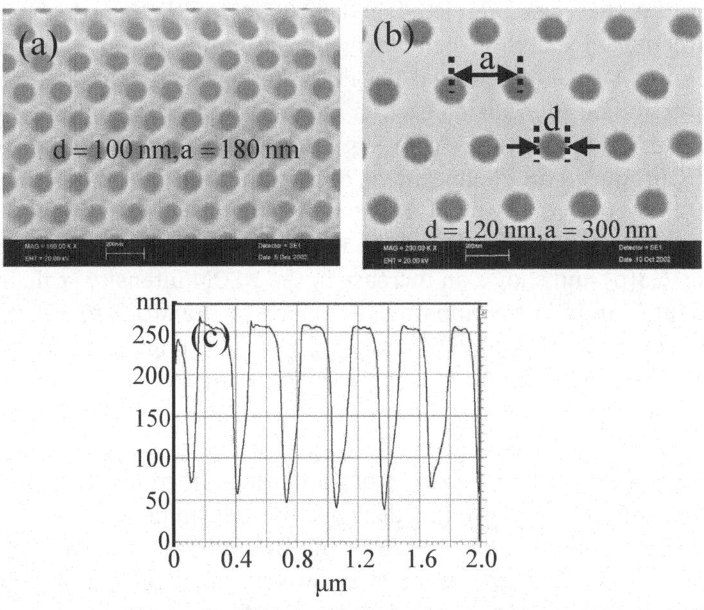

Figure 6.2 SEM images of the nitride PCs with hole diameter/periodicity of (a) 100/180 and (b) 120/300 nm. The AFM line profile across the PCs with an etch depth of about 200 nm is shown in (c). Reproduced from Ref. [36] with permission by AIP.

The optical characterization of the sample was accomplished using the NSOM system specially configured for UV wavelengths. Laser light of wavelength 266 nm was focused onto a small spot of about 3 μm on the unpatterned part of the sample at about 10 μm away from the edge of the PCs region and the emission intensity collected by the NSOM tip placed directly above the center of the PCs region. The emission intensity was also collected at an equal distance (30 μm) from the same pump spot from an unpatterned region of the sample. Figure 6.3(a) illustrates the experimental setup which emulates LED operation where the pump region and the light extraction region are separate [36]. Furthermore, this setup is akin to the one used by Boroditsky et al. [9] and enabled direct comparison between the emission light intensity collected in the MQW region patterned with PCs and an unpatterned region. Figure 6.3(b) shows a 3D NSOM intensity image collected above the 12 μm × 12 μm region patterned with PCs. The laser pump spot was located about 30 μm from the PCs in the unpatterned region of the sample. Figure 6.3(c) shows the PL spectra at T = 300 K for the two cases depicted in Fig. 6.3(a), with the propagation direction of emission light from the laser pump spot parallel to the Γ–K lattice direction of the PCs. The emission peak at 475 nm is attributed to localized exciton recombination in the well regions of the MQW structure. From Fig. 6.3(a), the region patterned with PCs produced an enhancement of the emission intensity of about 20-fold compared to the unpatterned region of the sample. The line scan across the 3D NSOM image in Fig. 6.3(b) is plotted in Fig. 6.3(d) and shows an increase in the NSOM intensity scale from about 5 units in the unpatterned region of the MQW to 135 units in the region of the MQW with PCs, giving an enhancement factor consistent with the PL data shown in Fig. 6.3(c).

By visual inspection with the naked eye, it was possible to observe a significant change in the light reflected from the samples with the PCs when viewed from different directions. A measurement setup was therefore designed to determine the amount of light emitted from the MQW as the orientation of the PC lattice was varied. The excitation laser was focused to a point about 100 μm from the PC region and the NSOM intensity was collected above the PCs in a 20 μm × 20 μm region enclosing the PCs. The sample was then manually turned about the vertical axis to vary the orientation angle θ (0° to 120°), defined as an angle between laterally propagating emitted light in the MQW and Γ–M direction of the PC lattice. Figures 6.4(a), 6.4(b), and 6.4(c) show an illustration of the setup and the scanned NSOM intensity image across the PC region for three different orientations, of 0°, 15°, and 30°, respectively [36]. A plot of the NSOM intensity versus orientation θ is shown in Fig. 6.4(d), which shows that the NSOM intensity is minimum for orientation of 0°, 60°, and 120° corresponding to the Γ–*M* direction. The NSOM intensity, however, increases threefold for orientation of 30° or 90° corresponding to the Γ–*K* direction of the PCs lattice. When compared to the NSOM intensity from an area without the PCs, the intensity increases by a factor 7 along the Γ–*M* direction, but increases by a factor of 20 along the Γ–*K* direction. The dispersion diagram of transverse electric modes in a triangular lattice structure of PCs indicates a larger PBG along the Γ–*K* direction is larger than that along the Γ–*M* direction [39]. This possibly translates into a large change in enhancement factor for light traveling along different directions as our data show.

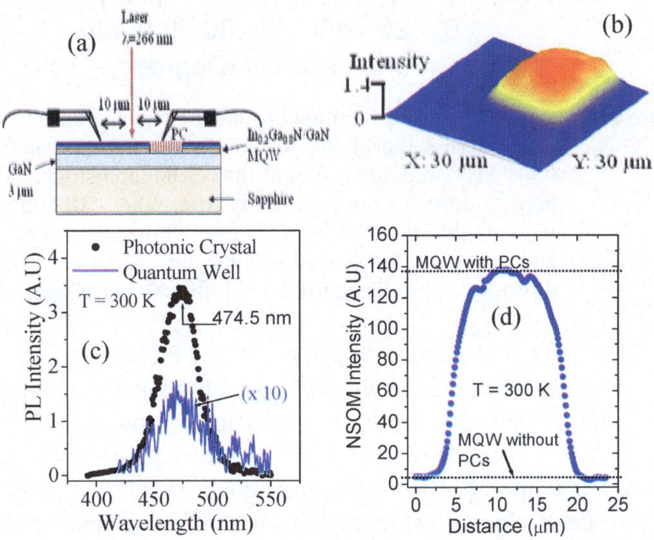

Figure 6.3 (a) Schematic setup for NSOM measurement of the PCs. (b) 3D NSOM intensity image collected above the 12 μm × 12 μm region patterned with PCs. (c) PL spectra at T = 300 K for the two cases in (a), measured by NSOM in collection mode. (d) Line scan across the NSOM image showing an increase in intensity from about 5 units in the unpatterned region of the MQW to about 135 units in the region of the MQW with PCs. Reproduced from Ref. [36] with permission by AIP.

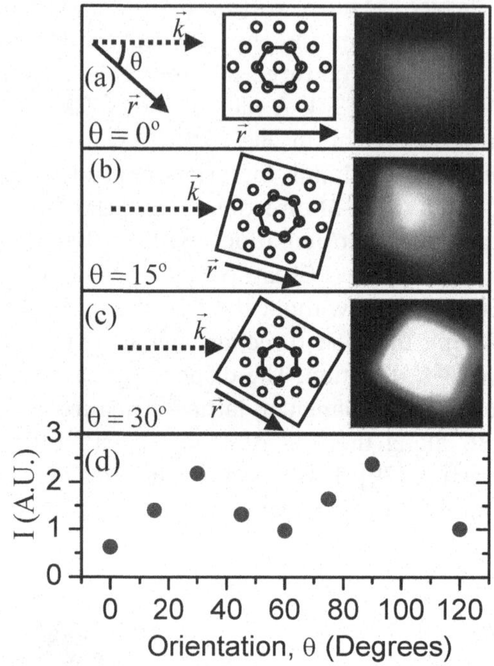

Figure 6.4 (a)–(c) Illustration of the setup and the scanned NSOM intensity image across the PC region for three different orientations: $\theta = 0°$, 15°, and 30°. A plot of the NSOM intensity vs. θ shown in (d) indicates a 60° periodicity where the NSOM intensity is minimum for $\theta = 0°$ or 60° corresponding to the Γ–*M* direction and maximum for $\theta = 30°$ or 90° corresponding to the Γ–*K* direction. Reproduced from Ref. [36] with permission by AIP.

There are two possible ways by which PCs can improve LED output power. The first is through Purcell spontaneous emission rate enhancement due to the band structure and nanocavities of the PCs and is directly related to the internal QE. E. M. Purcell first showed that modification of spontaneous emission occurs when the source is placed inside a cavity with dimensions comparable to the wavelength, and the emission is enhanced by a factor f given by the following equation [5,9,40]

$$f = \frac{3Qg(\lambda/n)^3}{2\pi V_{\text{cav}}} \tag{6.1}$$

where Q is the quality factor, g is mode degeneracy, λ is the wavelength, n is the refractive index, and V_{cav} is the cavity volume. A large Q and a small V_{cav} are required for Purcell enhancement that would significantly affect the dynamics of the emitter. Q is limited to 10–30 at room temperature by the material properties of the semiconductor. For nitride materials, using $n \sim 2.4$, $\lambda = 475$ nm, and $V_{cav} \sim 0.0024\ \mu m^3$ lead to a maximum f value of about 5.7 for single mode degeneracy, implying that overall improvement of LED operation by the Purcell effect may be limited.

The second method by which PCs can improve LED output power is by Bragg scattering, which is directly related to the extraction efficiency. Light extraction in this case is enhanced in two possible ways. First, because multiple scattering of photons by lattices of periodically varying refractive indices in the PCs acts to form PBGs in which lateral propagation of the Bloch-guided modes is prohibited, light generated in the bandgap region can couple only to radiation modes and is radiated outward. The lattice dimensions of the PCs would have to be tuned to match this bandgap region. Second, the refractive index periodicity creates a cut-off frequency for guided modes. Guided modes are folded by the PCs at the Brillouin zone boundaries, allowing phase matching to the radiation modes that lie above this cut-off frequency [10]. The guided modes that phase match to the radiation modes become leaky resonances of the PCs which Bragg scatter the light emitted from the active region. The result is significant enhancement of light extraction due to coupling to free space modes in the two cases above. The center of a PBG in GaN with dielectric constant $\varepsilon = 8$ is estimated to be at a normalized frequency (a/λ) value of about 0.5 for a triangular air hole lattice structure [36]. The normalized frequency that corresponds to our structure is 0.63 for a wavelength of 475 nm and a periodicity of 300 nm, which implies that photon modes in our structure lie above the cut-off frequency of the guided modes. Thus, the observed enhancement in our measurement is mainly due to coupling to leaky modes above the cut-off frequency of the Bloch-guided modes in our PCs. However, the angle dependence in Fig. 6.4 could be indicative of the PBG effect of PCs in our investigations.

6.3.3 Electrically Pumped III-Nitride PC LEDs

Based on the promising results obtained from the optically pumped PC LEDs, we fabricated PCs on III-nitride LEDs and investigated the enhancement of light when the LEDs were electrically pumped. The schematic diagrams of the 460 nm blue and 340 nm UV LED structures are shown in Fig. 6.5(a) and 6.5(b) [41]. The III-nitride materials were grown by MOCVD on sapphire substrates using trimethylgallium for Ga, trimethylaluminum for Al, trimethylindium for In, and ammonia for N_2. For Mg doping, bis-cyclopentadienyl-magnesium was transported into the growth chamber during growth while SiH_4 was used for Si doping. The active layers for the blue LEDs were $In_{0.2}Ga_{0.8}N/GaN$ MQWs, while for UV LEDs, these were $Al_{0.11}In_{0.03}Ga_{0.86}N/Al_{0.2}In_{0.03}Ga_{0.77}N$ MQWs. Out of these materials, conventional broad area LEDs of 300 × 300 μm^2 were fabricated by standard photolithography and ICP dry etching. Additional patterns consisting of ring-shaped LEDs of inner/outer diameters of 80/140 μm and 90/180 μm and square shaped LEDs (200 μm × 200 μm), fabricated from the UV materials as described previously. PCs with triangular lattice patterns of circular holes with diameter/periodicity of 300/700 nm were fabricated on the LEDs using e-beam lithography and ICP dry etching as described above. *I–V*, *L–I*, and electroluminescence (E-L) measurements on specific LEDs were taken before and after the fabrication of PCs and the data compared. Typical E-L data for the respectively [41]. The inset in Fig. 6.6(b) shows the SEM image of the nitride PCs after the triangular array of holes were etched 460 nm blue and 340 nm UV are given in Figs. 6.6(a) and 6.6(b),

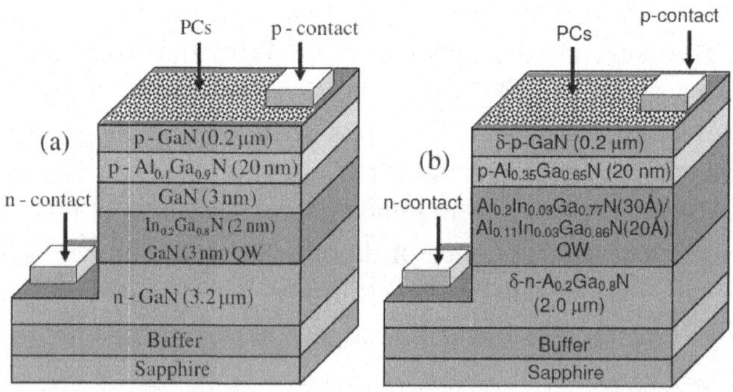

Figure 6.5 Schematic diagrams of (a) 460 nm blue and (b) 340 nm UV LED structures. Reproduced from Ref. [41] with permission by AIP.

in the nitride LEDs. Optical microscope images shown in Fig. 6.7 are for (a) 300 × 300 μm^2 blue LEDs, (b) 340 nm UV square LEDs, and (c) 340 nm UV ring LEDs with inner/outer diameters of 80/140 μm [41]. The LEDs without the PCs are shown on the left and the LEDs with the PCs shown on the right, with the corresponding LEDs under the same electrical current injection. The images were taken with a charge-coupled device (CCD) camera under the same settings for accurate comparison. The LEDs patterned with PCs become much brighter under the same current injection.

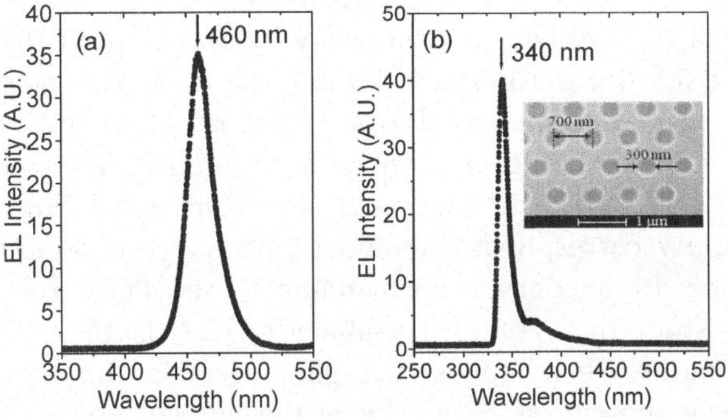

Figure 6.6 (a) and (b) Typical EL data from 460 nm blue and 340 nm UV LEDs, respectively. The inset in (b) shows the SEM image of the nitride PCs after the triangular array of holes were etched in the nitride LEDs. Reproduced from Ref. [41] with permission by AIP.

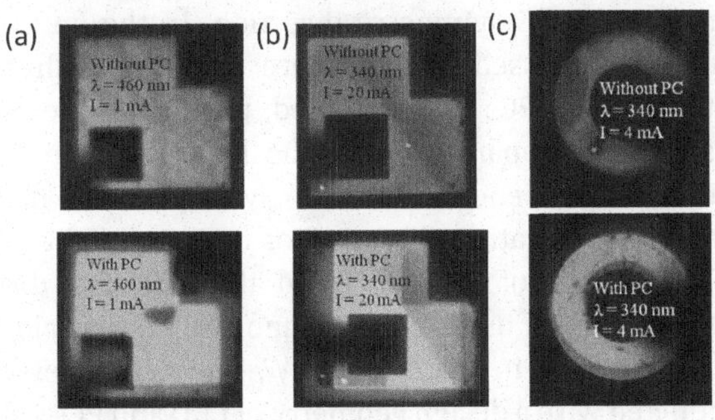

Figure 6.7 (a) Optical microscopy images of 300 × 300 μm^2 blue LED without (top) and with (bottom) PCs, both under 1 mA electrical current injection. (b) and (c) show optical microscopy images of 340 nm UV square and ring LEDs, respectively. The ring-LEDs have inner/outer diameters of 80/140 μm. Reproduced from Ref. [41] with permission by AIP.

The *I–V* and *L–I* data for blue LEDs are shown in Figs. 6.8(a) and 6.8(b), while those for UV LEDs are shown in Figs. 6.8(c) and 6.8(d) [41]. The forward voltage (V_F) shows an increase from 4.0 V to 4.2 V for blue LEDs (Fig. 6.8(a)) while that for UV LEDs shows decrease by a much larger value from 7.6 V to 6.0 V (Fig. 6.8(c)) following patterning with PCs. The slight increase in V_F for blue LEDs is attributable to reduction in total active area and *p*-type contact region. The reason for the much larger drop in V_F for UV LEDs is unclear but may be related to the enhanced current density as well as current spreading as a result of fabrication of the PCs. The total power shown in Figs. 6.8(b) and 6.8(d) was obtained by averaging the intensity measured from the top and bottom sides of the LEDs. At 20 mA, the integrated power of the 460 nm blue LED increased from 1.37 mW to 2.24 mW as a result of PCs corresponding to an increase in intensity by 63%. For the 340 nm UV LEDs, the integrated power increased from 44.5 μW to 86.6 μW, corresponding to about 95% increase in intensity. The normalized frequency corresponding to the PCs structure used in this study (a = 700 nm) is about 1.5/2.1 for the 460/340 nm wavelengths. This implies that photon modes in our structure lie above the cut-off frequency of the guided modes. Therefore, enhancement in our LEDs is mainly due to coupling to leaky modes as discussed above. Note here that in this PC-LED work, the 2D PCs were formed everywhere including transparent metal layer, except on the contact pads of the 300 × 300 μm^2 shaped LEDs, neither was the LED designed to optimize the extraction direction in relation to the PC symmetry direction. In the optical pumping configuration discussed above where a 20-fold enhancement of light extraction by PCs was obtained, the light generation region (pump spot) was separated from the light extraction region and the extraction efficiency was strongly dependent on the angle between the propagation direction of emission light and the PCs lattice. The amount of light extracted along the Γ–*K* direction was found to be as much as three times more than that along the Γ–*M* direction of the PCs in the nitride QW. The next investigation was thus conducted with a design geometry of LED and PCs with different lattice constants and hole sizes. The III-nitride LEDs structure used were similarly grown by MOCVD on sapphire substrates. The schematic diagrams of the 333 nm UV LED structure with LED mesa and contact pads are shown in Fig. 6.9(a), where the active region

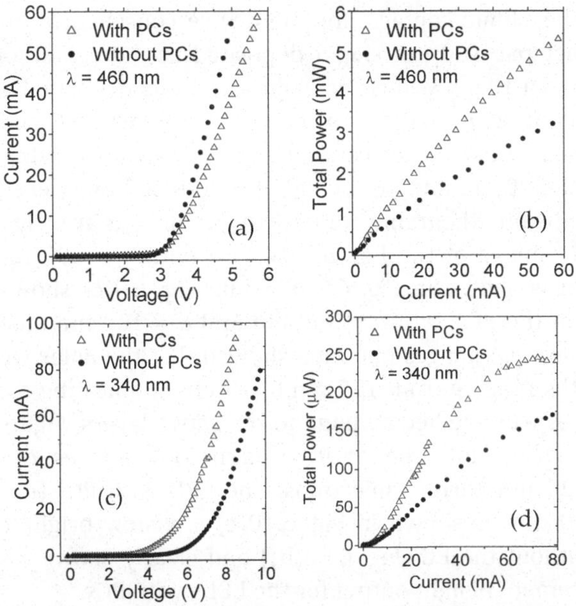

Figure 6.8 The *I–V* and *L–I* characteristics of square (300 × 300 μm^2) LEDs. (a) and (b) are results for 460 nm blue LEDs, while (c) and (d) are results for 340 nm UV LEDs. The optical power shown in (b) and (d) is obtained by averaging the values measured from the top and bottom side of the unpackaged bare LED chips. For 460 nm blue LEDs, the forward voltage (V_F) shows an increase by about 0.2 V and the optical power enhanced by 63% at 20 mA following patterning of PCs. For 340 nm UV LEDs, the forward voltage (V_F) shows a decrease by about 1.6 V following patterning of PCs while the power shows an enhancement by 95% at 20 mA. While the optical power level is still low for the UV LEDs studied here, comparison is being made on the overall intensity enhancement using PCs. Most recently, our MOCVD growth endeavors have since yielded UV LED materials with improved qualities and higher optical power output exceeding 1 mW at 120 mA dc at 300 nm. Reproduced from Ref. [41] with permission by AIP.

for the LEDs was an $Al_{0.11}In_{0.03}Ga_{0.86}N/Al_{0.2}In_{0.03}Ga_{0.77}N$ double QW [42]. The hexagonal mesa of side length 120 μm, as shown in Fig. 6.9(a), was defined by e-beam lithography and etched by ICP dry etching. For efficient extraction of the guided light, the PCs were designed such that the Γ–*K* direction of the PCs was perpendicular to the sides of the mesa, as depicted in Fig. 6.10(a)

[42]. The ohmic contacts for the device consisted of a hexagonal *p*-contact pad (60 μm sides) deposited at the center of the LED mesa and a 10 μm wide *n*-type ohmic contact deposited around the mesa along with a 100 μm × 100 μm *n*-contact pad. The PCs with triangular lattice patterns of circular holes with varying diameter d = 100–200 nm and periodicity a = 300–600 nm were fabricated using e-beam lithography and ICP dry etching as already described. An AFM image of the PCs on UV LED with a = 600 nm and d = 200 nm is shown in Fig. 6.9(b) while Fig. 6.9(c) shows the SEM image of the PCs with a = 300 nm and d = 100 nm etched on UV LED. The AFM images obtained showed that the holes were etched to depths ranging from 175 nm to 190 nm, indicating that most of the holes were etched through to the active layers. Figures 6.10(b) and 6.10(c) show the optical microscopy images of UV LEDs under 20 mA current injections. The LED with PCs (a = 600 nm, d = 200 nm) is shown in Fig. 6.10(c) is visibly brighter than the LED without the PC (Fig. 6.10(b)) and clearly shows a significant enhancement of light output for the LEDs with PCs.

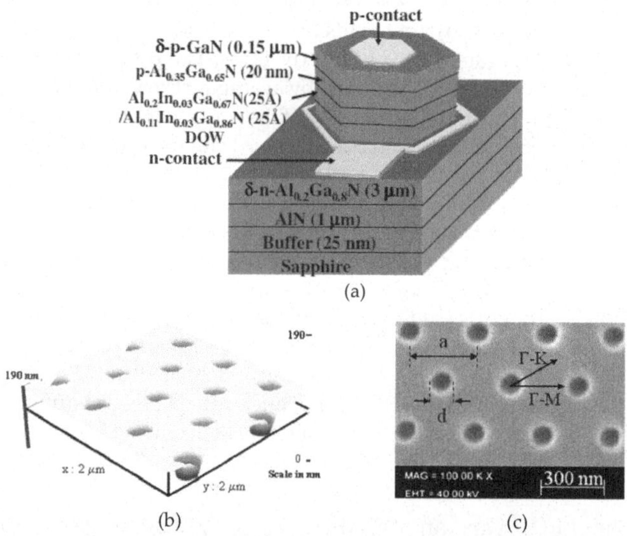

Figure 6.9 (a) Schematic diagram of 333 nm UV LED structure showing mesa and contact pads. (b) AFM image of the PCs on UV LED with a = 600 nm and d = 200 nm. The etch depth is around 190 nm. (c) The SEM image of the PCs on UV LED with lattice constant a = 300 nm and hole diameter d = 100 nm. Reproduced from Ref. [42] with permission by AIP.

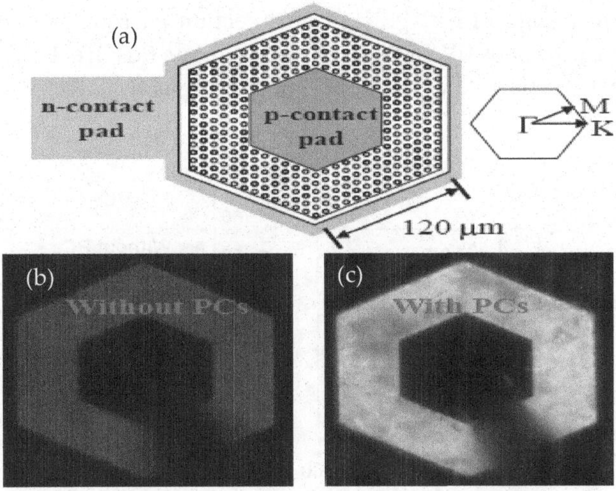

Figure 6.10 (a) Schematic top view diagram of the hexagonal mesa. The p-contact pad side length is 60 mm. Triangular array of holes were etched on top of mesa all around p-contact pad with Γ–K direction of PCs perpendicular to the side of mesa. The change coupled device images of UV LEDs at 20 mA current injections (b) without PCs and (c) with PCs (a = 600 nm, d = 200 nm). The darker area in the right bottom side of both of the images is due to the light obstruction by the p-contact probe. Reproduced from Ref. [42] with permission by AIP.

A typical EL spectrum of the 333 nm UV LEDs, shown in Fig. 6.11(a), reveals no change in the peak position as well as the line width of the spectrum due to PC formation implying that the spontaneous emission is not altered by the formation of PCs [42]. This is not surprising because the lattice constant of the PCs (600 nm) is larger than the normally required value ($\sim\lambda/2$ = 167 nm) to tune into the PBG of spontaneous emission regions. The I–V characteristics of the UV LEDs with PCs (a = 600 nm, d = 200 nm) and without PCs are shown in Fig. 6.11(b) from which it can be seen that the LEDs with or without the PCs have equal forward voltage at 20 mA (V_f) of about 5.7 V. The LED with PC indicates a more rapid increase in forward current, which could be caused by increase in the surface recombination of the injected electrons and holes as a result of increased surface around etched holes. The output power, measured using an integrating sphere, versus the injection current (L–I) for LEDs without and with PCs (a = 600 nm, d = 200 nm) are shown in Fig. 6.12(a) [42]. At an injection current of 20 mA, the output power of UV LEDs with and without PCs are 147 mW and 60 mW, respectively, giving an enhancement by a factor of about 2.5 with PCs.

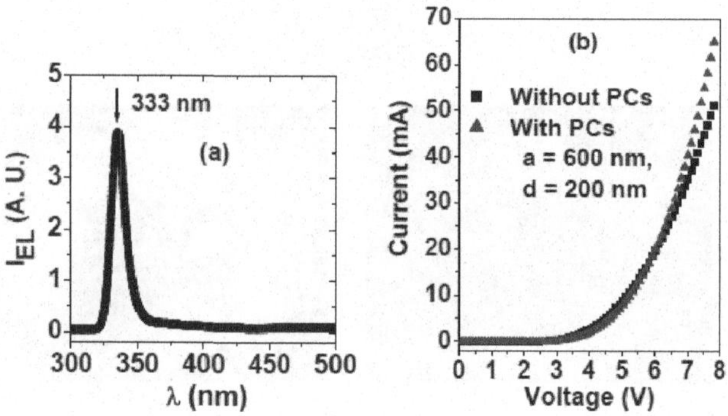

Figure 6.11 (a) Typical EL spectrum of the 333 nm UV LEDs. (b) The I-V characteristics of UV LEDs with PCs (a = 600 nm, d = 200 nm) and without PCs. Reproduced from Ref. [42] with permission by AIP.

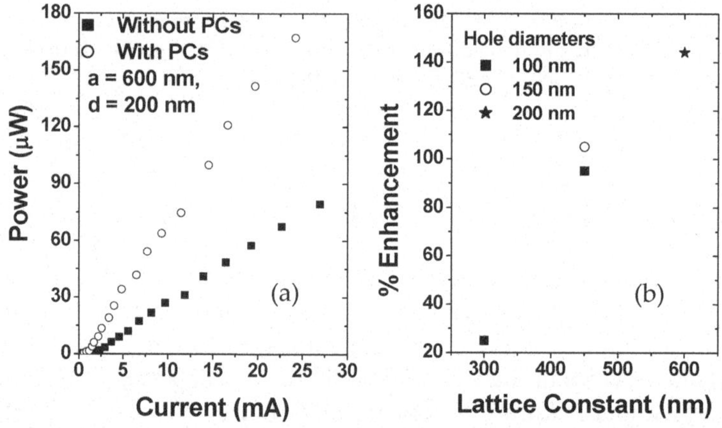

Figure 6.12 (a) The *L–I* characteristics of LEDs without PCs and with PCs, measured using an integrating sphere. At 20 mA, the output power was enhanced by a factor of 2.5 in the PC-LED. (b) The PC lattice constant and the air hole diameter dependence of the enhancement factor in the UV LEDs. While the optical power level is still low for the UV LEDs studied here, comparison is being made on the overall intensity enhancement using PCs. Reproduced from Ref. [42] with permission by AIP.

As explained above, we believe the observed enhancement is predominantly the result of Bragg scattering, where periodicity of the PC folds the higher momentum guided modes above the air light line, phase-matching them to the radiation mode and leak out as Bragg scattered light [34,43]. This conclusion is reached, based on the wavelength of our LED (λ = 333 nm) and the lattice constant of the PCs (a = 300–600 nm), yielding a/λ = 0.9–1.8, which is above the estimated normalized frequency ($\omega a/2\pi c \sim 0.5$) for triangular PCs in GaN. Figure 6.12(b) shows the variation of the output power enhancement factor in the UV LEDs with the lattice constant (a) and the air hole diameter (d) of the PC structure. In contrast to the expectations, we got surprisingly higher enhancement for larger dimensions of lattices [44]. This implies that the light extraction was dominated by the process of Bragg scattering instead of the effect of the PBG creation. For the same air-filling factor f = 0.1 (a/d = 3), the enhancement increases with the lattice dimension. However, a decrease in the enhancement factor was observed for a > 600 nm. For the same value of a = 450 nm, the enhancement is larger for a larger hole diameter. This may be due partly to the fact that the etching is not perfectly vertical and the actual hole size near the active layer is smaller than the targeted. In any case, the bigger targeted holes seem to perform better than the smaller holes in terms of light extraction. We believe that further enhancement can be achieved by improving vertical etching as well as increasing the air-filling factor. Figure 6.13 shows the PCs lattice constant and the air hole diameter dependence of the power output enhancement factor in the UV LEDs. An enhancement factor ~2.9 was observed for 299 nm UV PC-LEDs with PCs (a = 500 nm, d = 300 nm). In contrast to expectations, we got surprisingly higher enhancement for moderately larger dimensions of lattices. This implies that the light extraction was dominated by the process of Bragg scattering instead of the effect of the PBG creation. Enhancement factor increases with d up to d = 300 nm then decreases with d. This may be due partly to the fact that the etching is not perfectly vertical and the actual hole size near the active layer is smaller than the targeted. Also, the much bigger size holes effectively remove a large portion of the active region, thus reducing the power. We believe that further enhancement can be achieved by improving vertical etching.

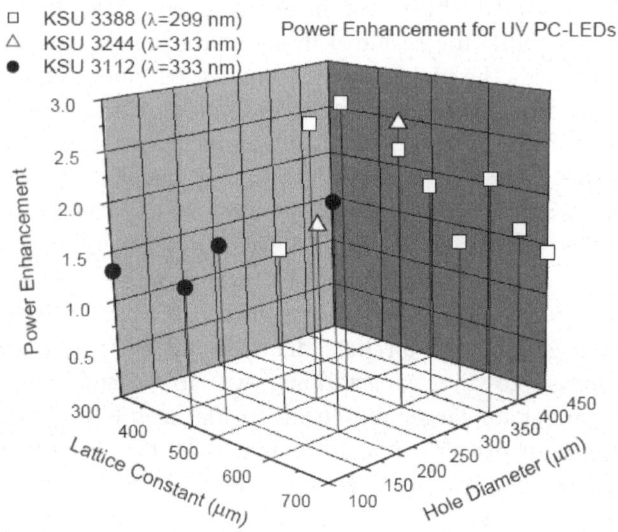

Figure 6.13 3D plot of the PCs lattice constant and the air hole diameter dependence of the enhancement factor in the UV LEDs. Three different sets of UV PC-LEDs were studied. A highest enhancement factor of about 2.9 was observed with PCs parameters $a = 500$ nm, $d = 300$ nm. Reproduced from Ref. [44] with permission by SPIE.

6.4 Current Approaches Using Embedded PCs

As mentioned earlier, PCs can be used to enhance light extraction through the PBG effect or through the Bragg reflection effect, both of which come with practical challenges. The PBG approach requires tuning the PC dimensions to yield a PBG that would inhibit laterally guided modes. The relatively low refractive index and the kind of epilayer thickness encountered in GaN as well as the small PC lattice constants required make the PBG approach for light extraction enhancement in GaN LEDs a challenging task. In the approach of using PCs as Bragg diffraction gratings, the light-guided modes are out-coupled to leaky modes by the diffractive properties of the PCs. This approach accommodates larger PC lattice constants than the PBG approach and thus relaxes some of the practical challenges in the fabrication [45]. However, the underlying challenge is met in diffracting all the guided modes because each guided mode propagates in GaN at different angles.

Wierer et al. showed that light extraction with 2D PCs fabricated on the surface of the III-nitride thin film was comparable to that of unencapsulated rough surface LED [45]. Better light extraction efficiency has been achieved when the LED and PC parameters were designed to optimize as much interaction of the guided modes with the PhCs as possible. This leads to reduction in the extraction lengths of the guided modes compared to other competing dissipation processes such as absorption the QWs, doped layers, and contacts. Several research conducted by David et al. [34,46], Matioli et al. [47–49], and Shields et al. [50] showed that PCs fabricated on the surface have limited extraction efficiency due to low interaction of the PCs with some of the optical-guided modes. The spatial modulation of the refractive index of the PCs causes a change in the in-plane wave vector of the guided modes of light from LEDs. The guided modes acquire an imaginary component (k_{im}) of the wave vector, causing it to diffract outward into air; the larger this component, the more likely this mode is diffracted into air with a characteristic extraction length of $1/2k_{im}$. Matioli et al. calculated the extraction length of the guided modes in GaN LED structure and for emitted light of wavelength 460 nm and 150 nm deep surface PCs, the extraction lengths of high order modes vary up to about 150 μm, but for low order modes, these could be as large as 400 μm [30]. The smaller extraction lengths for high order modes means they can be extracted by surface PCs. Low order modes, however, are poorly extracted by surface PCs. Further improvement in the extraction efficiency was realized by embedding the PCs in the LED structure. In this technique, air gaps are embedded to obtain high contrast refractive index that boosts the diffraction strength while leaving the surface unaffected.

The embedded PCs yielded a 3.5-fold enhancement of vertically emitted power and 94% extraction efficiency for encapsulated LEDs because of improved interaction with low- and high-order light modes. In a subsequent investigation, Matioli et al. determined the extraction lengths in surface PCs to be in the range of 55–120 μm, while for embedded PCs, this was in the range of 64–80 μm [47]. The shorter extraction lengths of the guided modes in embedded PCs indicated that this is a very promising approach in improving light extraction in III-nitride LEDs. In addition to permitting enhanced interaction of the guided modes with the PCs for maximum extraction, another advantage of the embedded PC is that the top surface of the LED remains planar, allowing easy formation of contacts, and there is no need for etching that would otherwise introduce damage to the device surface [30]. It needs to be pointed, however, that the fabrication of embedded 2D-PC LED structures does not come without its challenges [50]. Some of the challenges come from achieving the small dimensions of a few hundreds of nanometers required for embedded PCs structures. Furthermore, the growth of very thin high-quality layers on top of the PCs is required and it is more challenging to ensure that the intermediate processing step will not introduce contamination close to the critical active layers.

6.5 Concluding Remarks

Because of the enormous potential for energy savings, as well as environmental benefits, there is currently a great interest world-wide in developing III-nitride LED technologies for solid-state lighting (SSL). However, significantly improvement in the wall-plug efficiency has to be accomplished before the SSL mission could be achieved. PC incorporation is one of the most promising approaches. To realize the true potential of PC for boosting the LED efficiencies for practical uses, however, the prospects for industrial mass-scale adoption must be established. A number of groups have already demonstrated the patterning of PC features by using techniques such as laser holography, nanoimprint, one-shot exposure patterning using laser interference, self-organized micro- and nano-scale patterning, lateral epitaxial growth on Si mold-substrate with 2D grooves, and growth on sapphire substrate with 2D PC features [51–56]. Earlier nitride PC LEDs were fabricated by standard e-beam lithographic patterning together with pattern transfer by dry etching techniques. Overall, the incorporation of PC into nitride LEDs results in three dominant effects: (a) enhancement in extraction efficiency [36,41,42,57]; (b) light emission direction and pattern control [57]; and (c) modulation speed control [57]. PC LEDs have become commercially available for high-end products [59].

In addition to the interest in PCs for enhanced light extraction efficiency in LEDs, several groups have started to explore III-nitride

PC surface-emitting lasers, PC micro- and nanocavities, microdisk lasers [60,61]. Most notably, a current-driven blue-violet PC surface-emitting laser based on the band-edge effect has been realized [61]. These devices open the door to a range of applications such as super-high-resolution laser sources, which can be focused to spot sizes smaller than blue-violet wavelengths and optical tweezers for ultrafine manipulation [61]. As the efficiencies of III-nitride LEDs continue to rise with the utilization of sophisticated extraction strategies, the goal of III-nitride LEDs for general lighting gets closer to being realized.

Another very interesting and promising approach is the incorporation of metal SPs [62]. In theory, with its high density of states, SPs can enhance the electron-hole recombination rate in QWs by coupling the emission into the SP mode of thin metal film, and visible light can be efficiently extracted by SP mode scattering into air. Time-resolved PL measurements have confirmed that the recombination rate in GaN LED QWs can be significantly increased and a 14-fold enhancement in peak PL intensity has been observed from silver-coated InGaN/GaN QW materials [63,64]. On the other hand, there is so far no reported work on current-injected SP-coupled GaN LED devices. The fundamental difficult involved in obtaining current-injected SP-coupled GaN LEDs originates from the III-nitride LED structural design. For example, the coupling spacing of SPs with excitons in QWs is limited to less than 50 nm, while for GaN LED this spacing is *p*-GaN thickness, which has to be much more than 50 nm in order to achieve optimal performance. There are also technical difficulties related with fabrications of both PCs and SPs. To achieve the optimal functions of PCs and SPs, patterned features will have a size in sub-100 nm range, and there are currently no existing cost-effective approaches for the manufacturing of such nanostructures in the LED industry. Thus, future research will need to address manufacturing approaches to incorporate PC and SP technologies into GaN LED products for SSL applications.

Recently, there has also been intensive R&D effort in the area of UV and deep UV optoelectronic materials and devices. High efficiency deep UV light sources (below 280 nm) have a broad spectrum of applications of direct benefit to societal concerns, ranging from information storage, environmental protection, medical diagnostic and treatment, water and air purification, to biological threat detection. The applications of present deep UV systems are limited by cost, size, weight, power requirement, and performance capability. The realization of chip-scale devices based on AlGaN semiconductors would overcome limitations of current deep UV systems, as well as allow for integration with other functional photonic devices. As we discussed in previous sections, a large fraction of the light generated in the semiconductors is trapped by total internal reflection as guided modes due to the higher refractive index of semiconductor and eventually lost due to parasitic absorption within the semiconductor. Moreover, UV LEDs using AlGaN as active layers possess unique polarization properties compared to more conventional semiconductors. In Al-rich AlGaN alloys, the recombination between the conduction band electrons and holes in the top-most valance band is prohibited for the polarization configuration of $\mathbf{E} \perp \mathbf{c}$, while the allowed $\mathbf{E} // \mathbf{c}$ emission component cannot easily get out of the escaping cone [65,66]. Here, **E** denotes the electric field component of the light and **c** the c-axis of the epilayers. Thus, the need of finding effective methods and schemes to enhance the extraction efficiencies is exceptionally great for deep UV LED. 2D AlN PCs with varying periodicity/diameter down to 150 nm/75 nm were fabricated [67]. With PCs formation, a 20-fold enhancement in the band edge emission intensity at 208 nm over unpatterned AlN epilayer has been observed. The emission intensity increases with decreasing lattice constant of AlN PCs. However, the spectral peak energy was observed to decrease with decreasing lattice constant, indicating a possible release of compressive stress as a result of PC formation. Successful fabrication of AlN PCs opens up new opportunities for exploring novel physical phenomena in the artificially structured PBG material system and their applications in the deep UV spectral region.

Acknowledgments

We would like to acknowledge the contributions from our former group members and collaborators to the work summarized in this review: Dr. J. Li, Dr. Z. Y. Fan, Dr. J. Shakya, Dr. K. H. Kim, Dr. K. B. Nam, Dr. M. L. Nakarmi, Dr. N. Nepal, and Dr. A. Sedhain. The research was supported by NSF and DOE. Jiang and Lin are grateful to the AT&T Foundation for the support of Edward Whitacre and Linda Whitacre Endowed Chairs.

References

1. Jiang, H. X., and Lin, J. Y. (2007). *Wide Bandgap Light-Emitting Materials and Devices*, eds. Neumark, G. F., Kuskovsky, I., and Jiang H. X., Chapter 2 "III-nitride micro-cavity light-emitters," (Wiley–VCH Verlag GmbH).
2. Boroditsky, M., and Yablonovitch, E. (1997). A light extraction efficiency from light-emitting diodes, *Proc. SPIE.*, **3002**, pp. 119–122.
3. John, S. (1987). Strong localization of photons in certain disordered dielectric superlattices, *Phys. Rev. Lett.*, **58**, pp. 2486–2489.
4. Yablonovitch, E. (1987). Inhibited spontaneous emission in solid-state physics and electronics, *Phys. Rev. Lett.*, **58**, pp. 2059–2062.
5. Barra, A., Cassagne, D., and Jouanin, C. (1999). Visible light control by GaN photonic band gaps, *Phys. Status Solidi A*, **176**, pp. 747–749.
6. Bowers, B. (1998). *Lengthening the Day: A History of Lighting Technology* (Oxford University Press, New York).
7. Coquillat, D., Ribayrol, A., De La Rue R. M., Le Vassor d'Yerville, M., Cassagne, D., and Jouanin, C. (2001). First observation of 2D photonic crystal band structure in InGaN-sapphire epitaxial material, *Physica Status Solidi A*, **183**, pp. 135–138.
8. Coquillat, D., Ribayrol, A., de la Rue R. M., Vassor D'yerville M. L., Cassagne, D., and Albert, J. P. (2001). Observations of band structure and reduced group velocity in epitaxial GaN–sapphire 2D photonic crystals, *Appl. Phys. B*, **73**, pp. 591–593.
9. Boroditsky, M., Vrijen, R., Krauss, T. F., Coccioli, R., Bhat, R., and Yablonovitch E. (1999). Enhancement from thin-film 2-D photonic crystals, *J. Lightwave Technol.*, **17**, pp. 2096–2112.
10. Erchak, A. A., Ripin, D. J., Fan, S., Rakich, P., Joannopoulos, J. D., Ippen, E. P., Petrich, G. S., and Kolodziejski, L. A. (2001). Enhanced coupling to vertical radiation using a two-dimensional photonic crystal in a semiconductor light-emitting diode, *Appl. Phys. Lett.*, **78**, pp. 563–565.
11. Ryu, H. Y., Lee, Y. H., Sellin, R. L., and Bimberg, D. (2001). Over 30-fold enhancement of light extraction from free-standing photonic crystal slabs with InGaAs quantum dots at low temperature, *Appl. Phys. Lett.*, **79**, pp. 3573–3575.
12. Lem, H. V. D. and Moroz, A. (2000). Towards two-dimensional complete photonic bandgap structures below infrared wavelengths, *J. Opt. A: Pure Appl. Opt.*, **2**, pp. 395–399.
13. Tsao, J. Y., Saunders, H. D., Creighton, J. R., Coltrin, M. E., and Simmons, J. A., (2010). Solid-state lighting: An energy-economics Perspective, *J. Phys. D: Appl. Phys.*, **43**, pp. 354001–354018.

14. Liu, Y. S. (2008). *III-Nitride Devices and Nanoengineering*, Chapter One, "An overview of the development of major light sources: From light bulb to solid state lighting", (Imperial College Press, London).
15. "Lighting the Way to the Future with LED," Redneck USA, accessed November 1, 2011, http://redneckusa.wordpress.com/2009/11/07/lighting-the-way-to-the-future.
16. Schubert E. F. (2006). *Light Emitting Diodes* (Cambridge University Press, Cambridge).
17. "U.S. Lighting Market Characterization. Volume I: National Lighting Inventory and Energy Consumption Estimate", accessed October 13, 2011, http://apps1.eere.energy.gov/buildings/publications/pdfs/ssl/lmc_vol1_final.pdf.
18. Craford, M. G. (2005). LEDs for solid state lighting and other emerging applications: Status, trends, and challenges. *Proc. SPIE*, **5941**, pp. 1–10.
19. "Luxeon Rebel ES Datasheet DS61", Philips, Lumileds, accessed November 1, 2011, http://www.luxeon.hu/docs/DS61.pdf.
20. "Nichia Develops Record 150 lm/W White LED" *Semiconductor Today*, December 22, 2006 issue, accessed November 1, 2011, http://www.semiconductor-today.com/news_items/DEC_06/NICH_221206.htm.
21. Scoyk D. V., "LED Lighting: Is It Time to Change From Fluorescent and Incandescent Lighting?", accessed, November 1, 2011, http://www.superiorlight.com/docs/LED-whitepaper-3-11-10.pdf.
22. "Cree Reports 131 lm/W from Prototype White LED at 20 mA" LEDs Magazine, June 21, 2006 issue, accessed November 1, 2011, http://www.ledsmagazine.com/news/3/6/19.
23. "Ultra-Efficient Lighting", Cree, Inc, accessed October 13, 2011, http://www.creells.com/The-Cree-Difference/Efficiency.aspx.
24. Maruska, H. P. "A Brief History of GaN Blue Light Emitting Diodes," accessed November 1, 2011, http://www.sslighting.net/news/features/maruska_blue_led_history.pdf.
25. Pankove, J. I., Miller, E. A., Richman, D., and Berkeyheiser, J. E. (1971). Electroluminescence in GaN, *J. Luminescence*, **4**, pp. 63–66.
26. Maruska, H. P., Anderson, L. J., and Stevenson, D. A. (1974). Microstructural observations on gallium nitride light-emitting diodes, *J. Electrochem. Soc.*, **121**, pp. 1202–1207.
27. Pankove, J. I., and Lampert, M. A. (1974). Model for electroluminescence in GaN, *Phys. Rev. Lett.*, **33**, pp. 361–365.

28. Amano, H., Kito, M., Hiramatsu, K., and Akasaki, I. (1989). P-type conduction in Mg-doped GaN treated with low-energy electron beam irradiation (LEEBI), *Jpn. J. Appl. Phys.*, **28**, pp. L2112–L2114.

29. Nakamura, S., Senoh, M., Iwasa, N., and Nagahama, S. (1995). High-brightness InGaN blue, green and yellow light-emitting diodes with quantum well structures, *Jpn. J. Appl. Phys.*, **34**, pp. L797–L799.

30. Matioli, E., Iza, M., Choi, Y.-S., Wu, F., Keller, S., Masui, H., Hu, E., Speck, J., and Weisbuch, C. (2009). GaN-based embedded 2D photonic crystal LEDs: Numerical optimization and device characterization, *Phys. Status Solidi C*, **6**, pp. 675–679.

31. Wierer, J. J., Steigerwald, D. A., Krames, M. R., O'Shea, J. J., Ludowise, M. J., Christenson, G., Shen, Y.-C., Lowery, C., Martin, P. S., Subramanya, S., Götz, W., Gardner, N. F., Kern, R. S., and Stockman, S. A. (2001). High-power AlGaInN flip-chip light-emitting diodes *Appl, Phys. Lett.*, **78**, pp. 3379–3381.

32. Fujii, T., David, A., Schwach, C., Pattison, P. M., Sharma, R., Fujito, K., Margalith, T., Denbaars, S. P., Weisbuch, C., and Nakamura, S. (2004). Microcavity effect in GaN-based light-emitting diodes formed by laser lift-off and etch-back technique, *Jpn. J. Appl. Phys.*, **43**, pp. L411–L413.

33. Haerle, V., Hahn, B., Kaiser, S., Weimar, A., Bader, S., Eberhard, F., Plössl, A., and Eisert, D. (2004). High brightness LEDs for general lighting applications using the new Thin GaN^{TM} Technology, *Phys. Status Solidi A*, **201**, pp. 2736–2739.

34. David, A., Benisty, H., and Weisbuch, C. (2007). Optimization of light-diffracting photonic-crystals for high extraction efficiency LEDs, *J. Display Technol.*, **3**(2), pp. 133–148.

35. Ryu, H. Y., Hwang, J. K., Lee, Y. J., and Lee Y. H. (2002). Enhancement of light extraction from two-dimensional photonic crystal slab structures, *IEEE J. Sel. Top. Quantum Electron.*, **8**, pp. 231–237.

36. Oder, T. N., Shakya, J., Lin, J. Y., and Jiang, H. X. (2003). III-Nitride Photonic Crystals, *Appl. Phys. Lett.*, **83**, pp. 1231–1233.

37. Meade, R. D., Brommer, K. D., Rappe, A. M., and Joannopoulos, J. D. (1992). Existence of a photonic band gap in 2-D, *Appl. Phys. Lett.*, **61**, pp. 495–497.

38. Purcell, E. M. (1946). Spontaneous emission probabilities at radio frequencies, *Phys. Rev.*, **69**, p. 674.

39. Joannopoulos J. D., Meade, R. D., and Winn, J. N. (1995). *Photonic Crystals* (Princeton University Press, Princeton, NJ).

40. Todorov, Y., Sagnes, I., Abram, I., and Minot, C. (2007). Purcell enhancement of spontaneous emission from quantum cascades inside mirror-grating metal cavities at THz frequencies, *Phy Rev. Lett.*, **99**, pp. 223603–223607.

41. Oder, T. N., Kim, H. S., Lin, J. Y., and Jiang, H. X. (2004). III-Nitride blue and ultraviolet photonic crystal light emitting diodes, *Appl. Phys. Lett.* **84**, pp. 466–468.

42. Shakya, J., Kim, K. H., Lin, J. Y., and Jiang, H. X. (2004). Enhanced light extraction in III-nitride ultraviolet photonic crystal light-emitting diodes, *Appl. Phys. Lett.*, **85**, pp. 142–144.

43. Boroditsky, M., Krauss, T. F., Coccioli R., Vrijen R., Bhat, R., and Yablonovitch, E. (1999). Light extraction from optically pumped light-emitting diode by thin-slab photonic crystals, *Appl. Phys. Lett.*, **75**, pp. 1036–1038.

44. Shakya, J., Kim, K. H., Oder, T. N., Lin, J. Y., and Jiang, H. X. (2004). III-Nitride blue and UV photonic crystal light-emitting diodes, *Proceedings of SPIE*, **5530**, pp. 241–250.

45. Wierer, J. J., David, A., and Megens, M. M., (2009). III-nitride photonic-crystal light-emitting diodes with high extraction efficiency, *Nat. Photon.*, **3**, pp. 163–169.

46. Matioli, E., Fleury, B., Rangel, E., Melo, T., Hu, E., Speck, J. S., and Weisbuch, C. (2010). High extraction efficiency GaN-based photonic-crystal light-emitting diodes: Comparison of extraction lengths between surface and embedded photonic-crystals, *Appl. Phys. Express*, **3**, pp. 032103-1–032103-3.

47. Matioli, E., Fleury, B., Rangel, E., Melo, T., Hu, E., Speck, J. S., and Weisbuch, C. (2010). High extraction efficiency GaN-based photonic-crystal light-emitting diodes: Comparison of extraction lengths between surface and embedded photonic-crystals, *Appl. Phys. Express*, **3**, pp. 032103-1–032103-3.

48. Matioli, E., Rangel, E., Iza, M., Fleury, B., Pfaff, N., Speck, J., Hu, E., and Weisbuch, C. (2010). High extraction efficiency light-emitting diodes based on embedded air-gap photonic-crystals, *Appl. Phys. Lett.*, **96**, pp. 031108-1–031108-3.

49. Matioli, E., and Weisbuch, C. (2010). Impact of photonic crystals on LED light extraction efficiency: Approaches and limits to vertical structure designs, *J. Phys. D: Appl. Phys.*, **43**, pp. 354005–354026.

50. Shields, P. A., Charlton, M., Lee, T., Zoorob, M. E., Allsopp, D. W. E., and Wang, W. N. (2009). Enhanced light extraction by photonic quasi-

crystals in GaN blue LEDs. *IEEE J. Sel. Top. Quantum Electron.*, **15**(4), pp. 1269–1274.

51. Hsieh, M. L., Lo, K. K., Lan, Y. S., Yang, S. Y., Lin, C. H., Liu, H. M., and Kuo, H. C. (2008). One-shot exposure for patterning two-dimensional photonic crystals to enhance light extraction of InGaN-based green LEDs, *IEEE Photon. Techn. Lett.*, **20**, pp. 141–143.
52. Ng, W. N., Leung, C. H., Lai, P. T., and Choi, H. W. (2006). Photonic crystal light-emitting diodes fabricated by microsphere lithography, *Nanotechnology*, **19**, pp. 255302–255306.
53. Orita, K., Takase, Y., Fukushima, Y., Usuda, M., Ueda, T., Takigawa, S., Tanaka, T., Ueda, D., and Egawa, T. (2008). Integration of photonic crystals on GaN-based blue LEDs using silicon mold substrates, *IEEE J. Quant. Electron.* **44**, pp. 984–989.
54. Park, J., Oh, J. K., Kwon, K. W., Kim, Y. H., Jo, S. S., Lee, J. K., and Ryu, S. W. (2008). Improved light output of photonic crystal light-emitting diode fabricated by anodized aluminum oxide nano-patterns, *IEEE Photon. Techn. Lett.*, **20**, pp. 321–323.
55. Su, Y. K., Chen, J. J., Lin, C. L., Chen, S. M., Li, W. L., and Kao, C. C. (2008), GaN-based light-emitting diodes grown on photonic crystal-patterned sapphire substrates by nanosphere lithography, *Jpn. J. Appl. Phys.*, **47**, pp. 6706–6708.
56. Truong, T. A., Campos, L. M., Matioli, E., Meinel, I., Hawker, C. J., Weisbuch, C., and Petroff, P. M., (2009). Light extraction from GaN-based light emitting diode structures with a noninvasive two-dimensional photonic crystal, *Appl. Phys. Lett.*, **94**, pp. 023101-1–023101-3.
57. Shakya, J., Lin, J. Y., and Jiang, H. X. (2004). Time-resolved electroluminescence studies of III-nitride ultraviolet photonic-crystal light-emitting diodes, *Appl. Phys. Lett.*, **85**, pp. 2104–2106.
58. McGroddy, K., David, A., Matioli, E., Iza, M., Nakamura, S., DenBaars, S., Speck, J. S., Weisbuch, C., and Hu, E. L. (2008). Directional emission control and increased light extraction in GaN photonic crystal light emitting diodes, *Appl. Phys. Lett.*, **93**, pp. 103502-1–103502-3.
59. Hodapp, M. W. (1997). Applications for high-brightness light-emitting diodes, *Semiconductors and Semimetals*, **48**, pp. 227–356.
60. Lee, J., Ahn, S., Kim, S., Kim, D. U., Jeon, H., Lee, S. J., and Baek, J. H. (2009). GaN light-emitting diode with monolithically integrated photonic crystals and angled sidewall deflectors for efficient surface emission, *Appl. Phys. Lett.*, **94**, pp. 101105-1–101105-3.

61. Matsubara, H., Yoshimoto, S., Saito, H., Jianglin, Y., Tanaka, Y., and Noda, S. (2008). GaN photonic-crystal surface-emitting laser at blue-violet wavelengths, *Science*, **319**, pp. 445–447.

62. Ritchie, R. H. (1957). Plasma losses by fast electrons in thin films, *Phys. Rev.*, **106**, pp. 874–881.

62. Neogi, A., Lee, C-W., Everitt, H. O., Kuroda, T., Tackeuchi, A., and Yablonovitch, E. (2002). Enhancement of spontaneous recombination rate in a quantum well by resonant surface plasmon coupling, *Phys. Rev. B*, **66**, pp. 153305-1–153305-4.

64. Okanoto, K., Niki, I., Shvartser, A., Narukawa, Y., Mukai, T., and Scherer, A. (2004). Surface-plasmon-enhanced light emitters based on InGaN quantum wells, *Nat. Mater.*, **201**(3), pp. 601–605.

65. Li, J., Nam, K. B., Nakarmi, M. L., Li, J., Lin, J. Y., Jiang, H. X., Carrier, P., and Wei, S. H. (2003). Band structure and fundamental optical transitions in wurtzite AlN, *Appl. Phys. Lett.*, **83**, pp. 5163–5165.

66. Nam, K. B., Li, J., Nakarmi, M. L., Lin, J. Y., and Jiang, H. X. (2004). Unique optical properties of AlGaN alloys and related ultraviolet emitters, *Appl. Phys. Lett.*, **84**, pp. 5264–52646.

67. Nepal, N., Shakya, J., Nakarmi, M. L., Lin, J. Y., and Jiang, H. X. (2006). Deep ultraviolet photoluminescence studies of AlN photonic crystals, *Appl. Phys. Lett.*, **88**, pp. 133113-1–133113-3.

Chapter 7

Nanostructures in the III-Nitride/Silicon Material System

Yongjin Wang
Institute of Communication Technology, Nanjing University of Posts and Telecommunications, Nanjing, Jiang-Su 210003, People's Republic of China

wangyj@njupt.edu.cn

III-nitride compounds, especially gallium nitride (GaN), are promising direct band-gap semiconductor materials for optoelectronic applications. In principle, it is easy to tune the emission wavelength of the III-nitride system: add indium to shift emission toward longer wavelength, or add aluminum to move into the UV region. Thus, one can fabricate III-nitride light emitting devices that cover the entire visible and (near-) UV spectrum [1,2].

From the fabrication point of view, the substrate used for III-nitride growth is still one bottleneck for developing III-nitride optical devices. Sapphire and SiC are the most common substrates of choice due to the fact that bulk lattice-matched substrates are not readily available. However, both sapphire and SiC are difficult to manufacture. The laser lift-off from sapphire substrates is hardly suitable for mass production. SiC substrates are too expensive for low-cost applications. In recent years, a high-quality III-nitride-

Handbook of Optical Microcavities
Edited by Anthony H. W. Choi

ISBN 978-981-4463-24-9 (Hardcover), 978-981-4463-25-6 (eBook)
www.panstanford.com

on-silicon system has been developed by using an intermediate AlGaN/AlN buffer layer to reduce the crack density [3–6]. Silicon micromachining is a mature technique for removing the silicon substrate to generate freestanding III-nitride slabs. It is thus feasible to fabricate optical components on thin freestanding III-nitride slabs.

The design, fabrication and characterization of III-nitride nanostructures are demonstrated in this chapter. The overall objective is to study the light matter interaction in III-nitride nanostructures. Fundamental studies as well as applied research are carried out with the aim of ultimately offering small, easy-to-manufacture, and efficient III-nitride nanostructures. The strong light-matter interaction is thus investigated in III-nitride nanostructures, and the subsequent realization of III-nitride photonic devices are developed for practical use.

This chapter is divided into two parts. Part 1 is focused on III-nitride resonant structures, and is organized as: freestanding III-nitride resonant gratings [7,8], III-nitride resonant structures fabricated by a front-side process [9], and III-nitride resonant structures grown on freestanding HfO_2 structures [10]. Part 2 describes the light extraction efficiency improved by III-nitride nanostructures, and includes freestanding circular III-nitride gratings [11], freestanding III-nitride subwavelength nanostructures [12], patterned growth of InGaN/GaN multiple quantum wells (MQWs) [13,14], and freestanding III-nitride nanocolumn slab with bottom subwavelength nanostructures [15,16].

7.1 Freestanding III-Nitride Resonant Gratings

Figure 7.1(a) shows the cross sectional schematic diagram of the proposed III-nitride resonant gratings. The silicon handle layer under the resonant grating region is removed, and the freestanding grating is comprised of 200 nm AlN layer (t_1) and 650 nm GaN layer (t_2). The grating period, grating width and grating thickness are expressed by P, W and h, respectively. The duty ratio $D(= W/P)$ is defined as the ratio of the grating width with respect to the grating period. The optical responses of the freestanding III-nitride resonant gratings are simulated by using the rigorous coupled-wave analysis (RCWA) method. Actually, these simulations are just referenced for fabricating III-nitride resonant devices. In reality, AlGaN buffer layer is used to grow GaN layer, and the Al mole fraction is gradually changed. It is often a high secret for company. Correspondingly, the refractive indices of GaN and AlN layers are difficult to be precisely obtained from commercial III-nitride/Silicon wafer. For simplicity, the III-nitride grating is supposed to be comprised of GaN and AlN layers, and the grating height h is 200 nm. The refractive indices of GaN and AlN used in the calculation are 2.325 and 2.2, respectively [17].

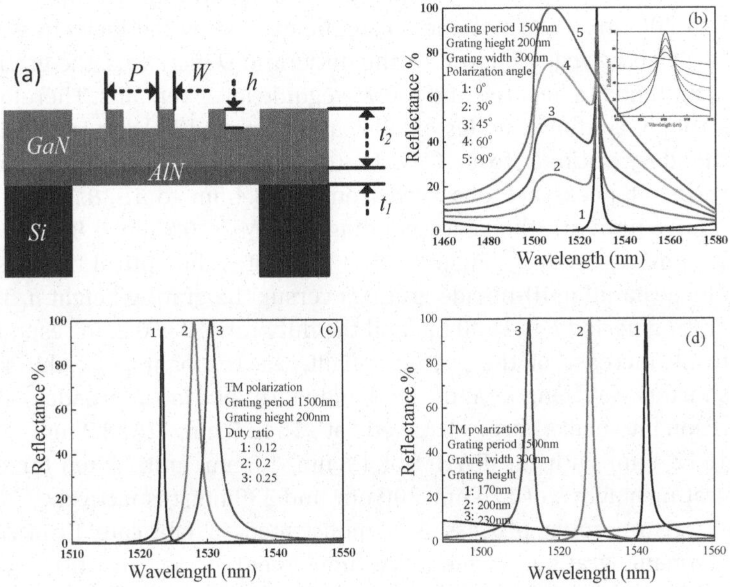

Figure 7.1 (a) The schematic diagram of the freestanding III-nitride resonant gratings; (b), (c), and (d) Reflectance spectra of freestanding III-nitride gratings calculated by RCWA: (b) versus polarization angle; (c) versus duty ratio; (d) versus grating height.

Figure 7.1(b) shows a set of reflectance spectra of the freestanding III-nitride gratings of various polarization angles from 0° (transverse magnetic (TM) polarization, namely, TM is polarized in the plane of the grating and perpendicular to the grating lines) to 90° (transverse electric (TE) polarization, namely, TE is polarized in the plane of the grating and parallel to the grating lines). The grating period P is 1500 nm, the duty ratio D is 0.2, and the grating height h is 200 nm. The resonances occur at different parametric locations for TE and TM polarization due to the different eigenvalue expressions for TE and TM modes [18]. The resonant peak is observed at 1528.2 nm with a full-width at half-maximum (FWHM) of 1.8 nm under TM polarization. The resonant peak is gradually suppressed with increasing the polarization angle from 0° to 90°, while another broad peak around 1508.2 nm is gradually enhanced.

The optical responses of freestanding III-nitride gratings also show a clear dependence on the duty ratio *D*, as shown in Fig. 7.1(c). The grating period *P* is 1500 nm and the grating height *h* is 200 nm. The resonances can't be excited by the incident wave in a slab waveguide layer. As the duty ratio *D* increases, the grating layer tends to be large and the waveguide layer remains. Therefore, strongly confined modes exhibit larger linewidth. With increasing the duty ratio *D* from 0.12 to 0.2 and 0.25, the shifts of the resonant peaks are observed from 1523.4 nm to 1528.2 nm and 1530.6 nm with the corresponding FWHM of 0.5 nm, 1.8 nm, and 2.2 nm, respectively. Figure 7.1(d) illustrates the optical responses of freestanding III-nitride gratings versus the grating height *h*. The grating period *P* is 1500 nm and the duty ratio *D* is 0.2. In response to an increase of the grating height, the resonant peak shifts to shorter wavelength and the bandwidth becomes broader. The resonant peaks are observed at 1542.5 nm, 1528.2 nm, and 1512.4 nm with the FWHM of 1.3 nm, 1.8 nm, and 2.2 nm for the grating height *h* of 170 nm, 200 nm, and 230 nm, respectively.

A top-down process is demonstrated to fabricate III-nitride resonant gratings, which are implemented on an III-nitride/silicon wafer, consisting of 200 nm GaN layer, 450 nm $Al_xGa_{1-x}N$ layer (0.70~0.20 Al mole fraction), 200 nm AlN buffer layer, and 200 μm silicon handle layer. The freestanding gratings use the large index contrast between the III-nitride layer and the surrounding air to confine light in the vertical direction, improving the optical performance of the III-nitride gratings. Fast atom beam (FAB) etching is developed for processing grating structures on III-nitride membrane. FAB etching is used as a well-controlled, low-damage etching technique since it uses neutral particles or atoms for dry etching. The silicon handle layer below the III-nitride grating region is completely removed by deep reactive ion etching (DRIE).

Figure 7.2 illustrates the schematic fabrication process of freestanding III-nitride gratings. A positive electron beam (EB) ZEP520A resist is first spin-coated onto the III-nitride device layer, and nanoscale circular gratings are patterned in ZEP520A resist using EB lithography (steps a–b). The patterns are then transferred to III-nitride membrane by FAB etching (step c). FAB etching is generated by the neutralization of ions extracted from direct-current Cl_2 plasma. FAB etching is performed with an etching rate of about 11.5 nm/min at a high voltage of 3.0 KV and accelerated current of 10 mA. The processed patterns are protected by thick photoresist after removing the residual resist and then, the silicon handle layer below the III-nitride grating region is patterned from the backside by photolithography and etched down to the AlN layer by DRIE (steps d–e), which makes the III-nitride grating suspend in space. The freestanding III-nitride gratings are finally generated by removing the residual photoresist (step f).

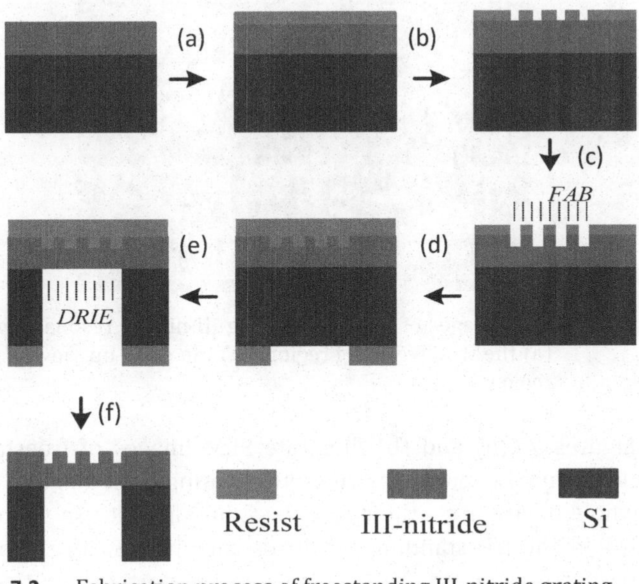

Figure 7.2 Fabrication process of freestanding III-nitride grating.

Figure 7.3 illustrates scanning electron microscope (SEM) images of freestanding III-nitride gratings. The thin membrane can guarantee sufficient stiffness for the designed resonant gratings and the grating arrays are well fabricated. There are 80-period grating lines in each grating array, as shown in Fig. 7.3(a). Each grating line has a length of 60 μm, the grating period P is 1500 nm, and the grating height h is approximately 230 nm. The magnified view is illustrated in Fig. 7.3(b).

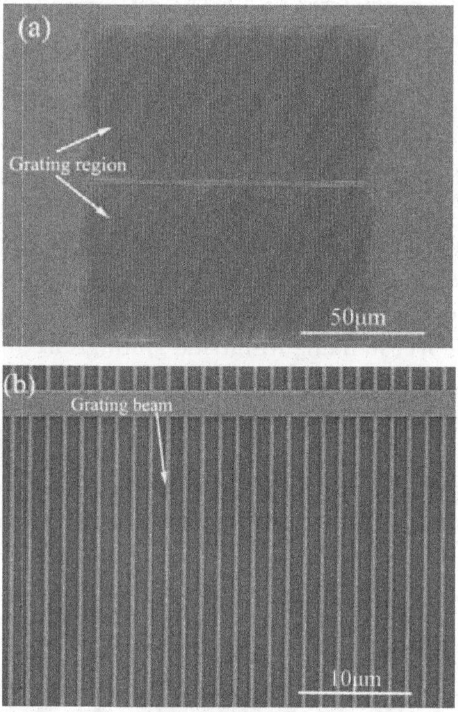

Figure 7.3 SEM images of the freestanding III-nitride resonant gratings. (a) the whole grating region; (b) the close-up view of grating region.

Figures 7.4(a) and (b) illustrate SEM images of freestanding circular III-nitride gratings, where the grating period is 850 nm. In the whole device area, the circular grating doesn't use supporting arms, and the freestanding III-nitride membrane can sustain the residual stress during DRIE process to support the grating. The top-down view of the freestanding III-nitride gratings exhibits their well-defined circular symmetry, providing the polarization-independent optical performance. Figure 7.4(c) shows a close-up view of freestanding circular III-nitride gratings, where the grating period is 1500 nm and the grating width is 200 nm. Figure 7.4(d) illustrates the tilting-view of III-nitride gratings with grating period of 1500 nm. The grating profile imperfections can be observed in Fig. 7.4(d), and the grating width generated in reality deviates slightly from the designed value 200 nm due to the fabrication issues. The III-nitride crystal defects, which introduce undesirable scattering losses, are also observed.

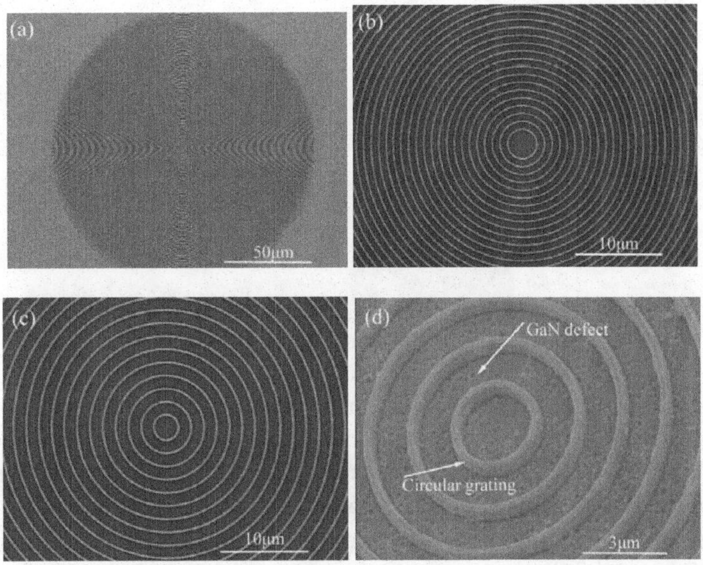

Figure 7.4 SEM images of the freestanding circular III-nitride gratings. (a) Whole grating region for 850 nm-period circular grating; (b) and (c) Close-up views of grating region for 850 nm- and 1500 nm-period circular grating, respectively. (d) Tilting-view of 1500 nm period circular grating.

The reflectance measurements are performed to characterize the optical responses of fabricated gratings. The optical measurement configuration is schematically illustrated in Fig. 7.5. A tunable laser (Agilent 81682A) operating at 1460~1580 nm range is used as the light source to characterize the resonant responses of the freestanding III-nitride gratings. The $\lambda/4$ plate is installed to get the circular polarization and the linearly polarized light is generated by the polarizer. The polarized light beam is then incident onto the III-nitride gratings by an infrared objective lens. In order to acquire sample images, one light source and an infrared charge-coupled device camera are installed on the setup. The reflected light is collected and sent to an infrared spectrometer. The experimental spectra are normalized to those of a commercial gold mirror.

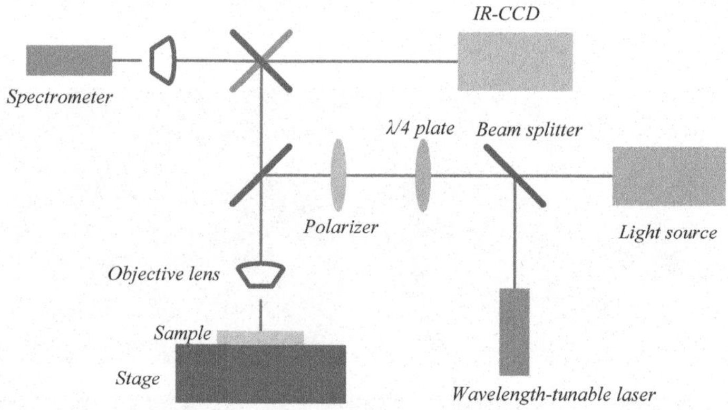

Figure 7.5 The optical measurement configuration.

The optical behavior of linear III-nitride grating is sensitive to polarization due to its 1D configuration. Figure 7.6(a) shows the measured optical responses of linear III-nitride grating versus polarization angles at a numerical aperture (NA) of 0.25. The grating period is 1500 nm, the grating height is approximately 230 nm, and the grating width is approximately 300 nm. The strong resonant peak is observed at 1522.1 nm with a FWHM of 5.5 nm under TM polarization (0°). This resonant peak is gradually suppressed with increasing the polarization angle from 0° to 90° (namely the transition from TM polarization to TE polarization), while another broad peak around 1490 nm is found under TE polarization. The III-nitride crystal defects and the grating

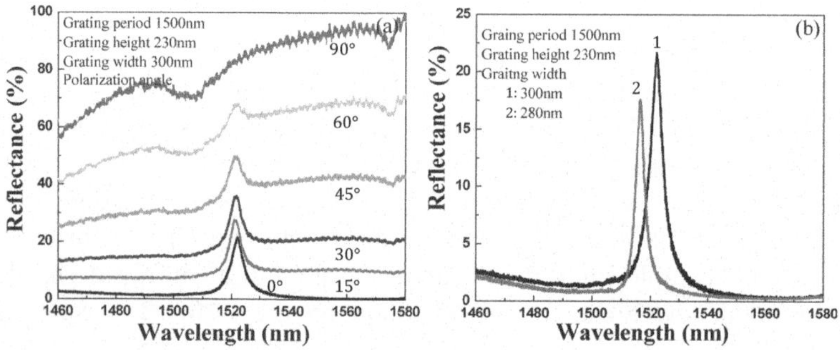

Figure 7.6 Reflectance spectra of the freestanding III-nitride resonant gratings as a function of wavelength. (a) Reflectance spectra versus polarization angle. (b) Reflectance spectra versus duty ratio.

grating height generated in reality and the ideal element result in profile imperfections introduce the undesirable scattering losses. Moreover, the deviations between material system, duty ratio and the reduction of the reflected intensities. The resonant peak of 280 nm wide III-nitride grating is observed at 1516.4 nm with a FWHM of 4 nm under TM polarization, as shown in Fig. 7.6(b). Given the possible strain, fabrication deviations, and other loss mechanisms, it's a very promising result of the fabricated III-nitride resonant gratings.

Owing to their circular rotation symmetry, the fabricated circular III-nitride gratings are expected to demonstrate the polarization-independent responses, which are measured by rotating the sample with angle from 0° to 90° with respect to initial measurement. Figure 7.7 illustrates the measured reflectance spectra versus polarization angle. The grating period is 850 nm and the grating width is approximately 120 nm. The circular III-nitride grating consists of 230 nm grating layer and 620 nm waveguide layer. The polarization-independent responses are experimentally demonstrated although a small shift is still observed, which may be caused by alignment error in the measurement. For example, the reflectance peaks are measured at 1489.1 nm and 1486.75 nm with a rotation angle of 15° and 90°, respectively, with almost same FWHM of about 6 nm. These results indicate that the top-down process is promising for fabricating III-nitride resonant devices.

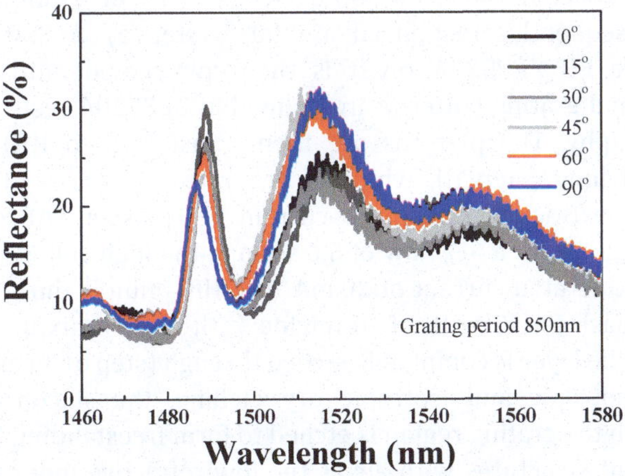

Figure 7.7 Reflectance spectra versus rotation angle of circular III-nitride grating.

7.1.1 III-Nitride Resonant Structures Fabricated by a Front-side Process

Based on the silicon backside technology, guided mode resonant III-nitride gratings are designed by using RCWA method and fabricated by FAB etching, where EB resist acts as an etching mask. Since EB resist serves as an etching mask, the etching depth of III-nitride is limited. Therefore, the applications are constrained for tunable optical components and the incorporation into a surface emitting devices.

To fabricate guided mode III-nitride resonant devices on III-nitride/silicon material system, another competitive method is to etch III-nitride layer through and remove silicon substrate by a front-side process. Hafnium oxide (HfO_2) film is introduced as an etching mask, and the III-nitride layer is completely etched into the silicon substrate by inductively coupled plasma reactive ion etching (ICP-RIE). The silicon substrate underneath the structures is etched using a front-side process to generate the freestanding III-nitride structures. Air serves as the low refractive index materials on the top and bottom, and the freestanding III-nitride structures present good optical confinement in the vertical direction due to the high refractive index difference.

Figure 7.8 schematically illustrates the fabrication process of freestanding III-nitride resonant structures. A 200 nm thick HfO_2 film is deposited onto the top device layer by EB evaporation and is subsequently crystallized at 800°C (step a). A 350 nm thick positive EB ZEP520A resist is then spin-coated onto the HfO_2 layer, and grating patterns are defined in ZEP520A resist using EB lithography. The patterns are then transferred to HfO_2 film by FAB etching (step d), which is generated by the neutralization of ions extracted from direct-current SF_6 plasma. FAB etching is performed with a SF_6 flow of 5.6 sccm at the high voltage of 2.0 KV and accelerated current of 20 mA. The HfO_2 film is thus used for a hard mask for ICP-RIE of III-nitride with Cl_2/N_2 mixture, and the III-nitride layer is completely etched through (step e). In association of anisotropic and isotropic dry etching, the silicon substrate beneath the grating region is etched to form freestanding III-nitride resonant structures with air as the low refractive index materials on the top and bottom (step f). The HfO_2 cladding layer can be further removed by buffered HF etching (step g).

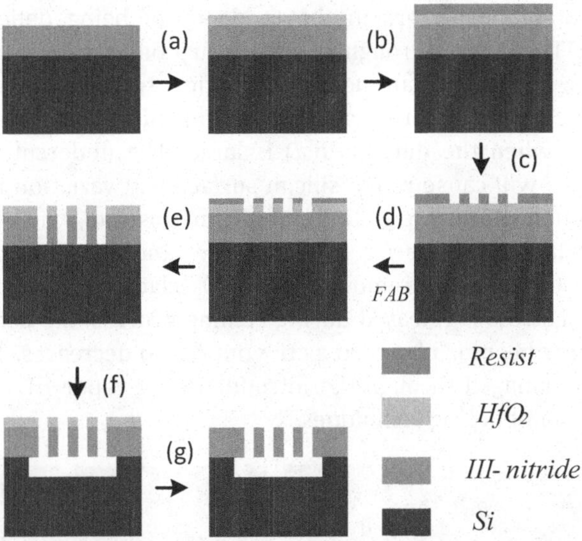

Figure 7.8 The fabrication process of freestanding III-nitride resonant structures.

Figure 7.9(a) illustrates freestanding III-nitride grating that consists of 60-period grating with the grating length of 60 μm. The HfO_2 film is completely etched into the III-nitride layer and an etching window is formed for further ICP-RIE etching of silicon, where EB resist serves as an etching mask. From the fabrication point of view, the ridge shape of HfO_2 part tends to be trapezoidal when we make it rectangular by EB lithography, and the sidewall of the III-nitride part is generated by ICP-RIE. This brings about variations in grating widths on top and bottom. Figure 7.9(b) shows the zoom-in SEM image of the freestanding III-nitride gratings, where the grating period is 1100 nm. The corresponding bottom and top grating widths W_b and W_t are measured ~716 nm and ~224 nm, respectively. For simplicity, the duty ratio D is defined here as the ratio of the bottom grating width W_b to the grating period P. Figure 7.9(c) shows freestanding circular III-nitride grating, and the inset is the zoom-in SEM image of circular grating with the grating period of 700 nm, where cross arms with the widths of ~200 nm are connected to the freestanding circular gratings. Suspended linear grating with the grating length of 30 μm is illustrated in Fig. 7.9(d). The grating period and width are 600 nm and ~100 nm, respectively. As the duty ratio D decreases, the grating beams are in the tendency of being deflected and fragile. The sidewall roughness and warp on the gratings become serious, resulting in an increase in optical scattering and loss. On the other hand, the release of the III-nitride grating tends to be difficult when the duty ratio *D* is large. The undercut of silicon substrate will cause rough silicon surface and variation in air gap between the III-nitride grating and silicon substrate, which introduce additional optical interferences. The freestanding III-nitride grating can be obtained by removing the HfO_2 cladding layer. Since the residual stress is released during etching process, the deformation of grating beam is observed as the duty ratio decreases. This issue can be managed through the introduction of vapor HF etching or critical point drying techniques.

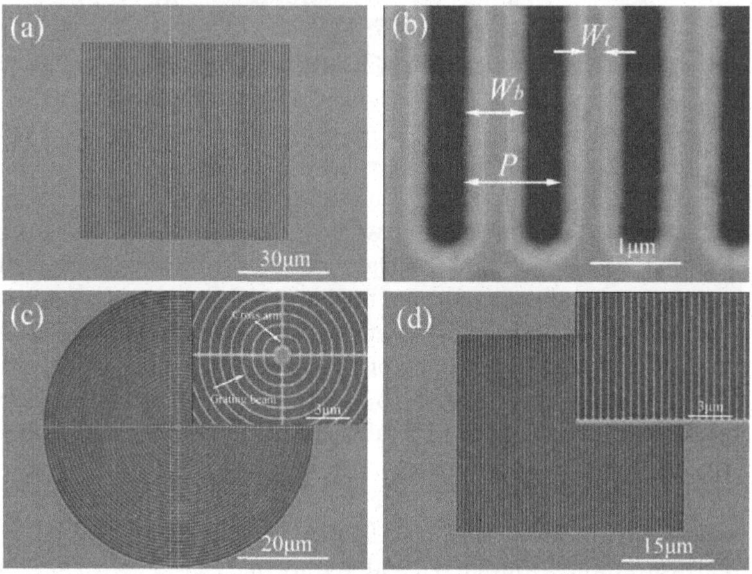

Figure 7.9 SEM images of III-nitride gratings with HfO_2 cladding layer. (a) The whole grating region for 1100 nm period III-nitride grating. (b) The zoom-in view of III-nitride grating. (c) Freestanding 700 nm period circular III-nitride grating and the zoom-in view. (d) Freestanding 600 nm period III-nitride grating and the zoom-in view.

The III-nitride gratings are separate and serve as the guiding layer. When one light beam is incident, guided modes are excited through phase matching by the III-nitride grating, and the broad reflection is determined by the refractive index contrast. Since fabricated III-nitride gratings are configured with 1D symmetry, their optical responses are sensitive to polarization. Figure 7.10 illustrates the reflectance spectra versus polarization. The grating period is 1100 nm, and the duty ratio D is about 0.632. The gratings are measured by rotating the sample with an angle of 90° with respect to the initial measurement. Under TM polarization, one sharp reflection dip of ~43.5% is clearly observed around 1538.8 nm, and reflectance over 99% is measured in the range of 1461.2 m~1507.5 nm. Two distinct reflection peaks under TE polarization are found at 1477 nm and 1550.6 nm with measured reflectance of 70.8% and 70%, respectively.

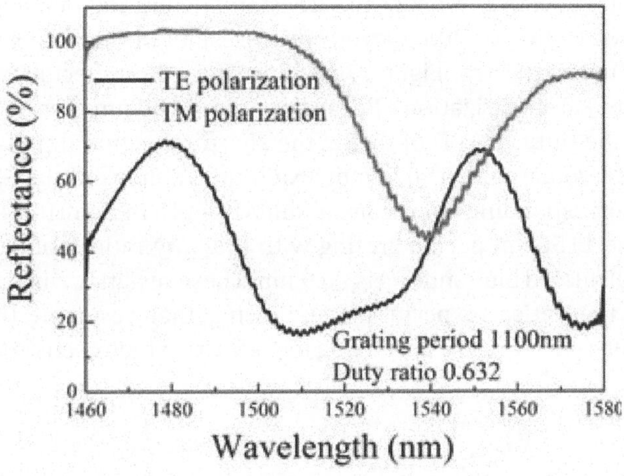

Figure 7.10 Reflectance spectra versus polarization for 1100 nm period III-nitride grating.

Figure 7.11 shows the reflectance spectra versus period and duty ratio under TM polarization. For the 1100 nm period grating, the duty ratio D are 0.632 and 0.488, and the 1150 nm period gratings are with the duty ratio D of 0.636 and 0.519, respectively. The measured reflectance spectra are thus comparable. It is clearly shown that the broad reflection band shifts to longer wavelength as the corresponding grating period increases and the reflection band changes as the duty ratio varies.

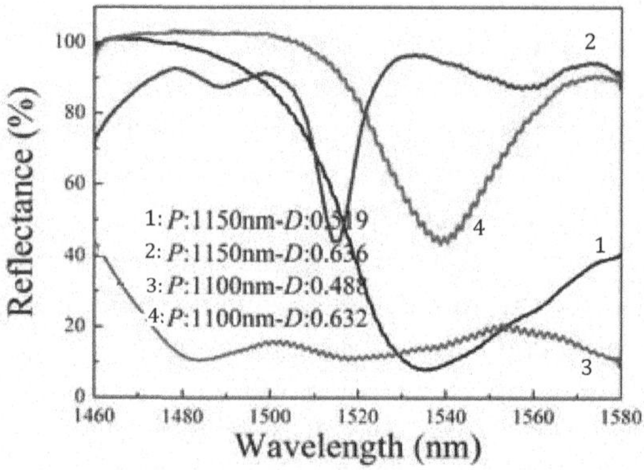

Figure 7.11 Reflectance spectra versus grating period and duty ratio.

The effective index of the freestanding grating decreases after removing HfO_2 cladding layer, which results in the blue shifting of reflectance spectra. Figure 7.12 illustrates the reflectance spectra under TM polarization. Regarding the 1100 nm period grating with the duty ratio *D* of 0.632, the sharp reflection dip of ~49.3% is observed around 1507.9 nm, exhibiting a blue-shift of ~30.9 nm. The corresponding reflection dip shifts from 1514.5 nm to 1487.6 nm for the 1150 nm period grating with the duty ratio *D* of 0.636, and the calculated blue shifting is 26.9 nm. These results are in consistent with theoretical expectation, indicating that resonant III-nitride grating reflectors are promising for refractive index sensing.

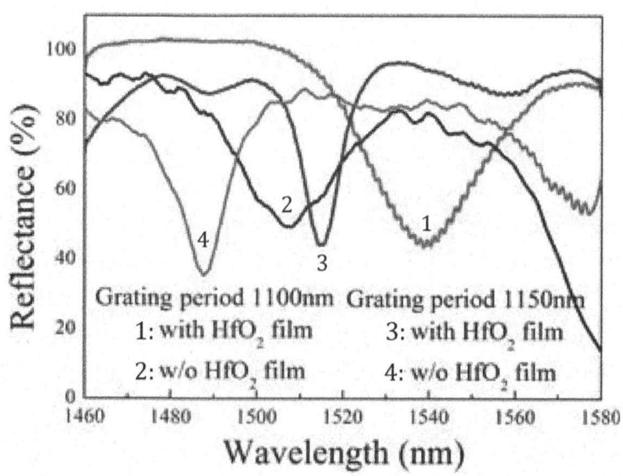

Figure 7.12 Reflectance spectra after removing HfO_2 cladding layer.

7.1.2 III-Nitride Resonant Structures Grown on Freestanding HfO_2 Structures

Among the approaches toward creating III-nitride resonant structures, growth of III-nitride on freestanding structured template is an emerging technology. Nanoscale structures locally change the growth conditions, and novel III-nitride structures are achieved since freestanding III-nitride structures are formed by growth method and free of the etching damage. Meanwhile, the as-grown III-nitride structures provide a natural optical cavity to support guided mode resonances, opening the possibility to achieve the interaction between the excited light and the epitaxial structures.

Freestanding structured HfO_2 film is chosen as a template for growing III-nitride materials. HfN surface is formed by nitrifying HfO_2 substrate during molecular beam epitaxy (MBE) growth. From the growth point of view, small material lattice mismatch between HfN and GaN crystals makes HfN film as a superior buffer layer for the growth epitaxial of GaN [19,20]. On the other hand, HfO_2 film is an excellent optical material with high thermal and chemical stability. Freestanding HfO_2 structures demonstrate their guided mode resonances due to the high refractive index contrast between HfO_2 film and air [21].

The whole fabrication process is schematically illustrated in Fig. 7.13. The process starts from the blank deposition of HfO_2 film on the silicon substrate with an EB evaporator (step a). A positive EB ZEP520A resist is then spin-coated onto the HfO_2 layer, and grating patterns are patterned in ZEP520A resist using EB lithography (step b). Subsequently, the patterns are transferred to HfO_2 layer by FAB etching (step c). FAB etching, which is generated by the neutralization of ions extracted from direct-current SF6 plasma, is performed with a SF6 gas of 5.6 sccm at the high voltage of 2.0 KV and accelerated current of 20 mA. The HfO_2 gratings are then released by a combination of anisotropic and isotropic dry etching of silicon, which makes the HfO_2 grating freely suspend (step d). The anisotropic etching of silicon is carried out to produce vertical silicon trenches and the isotropic etching is used to release the HfO_2 gratings laterally, where the remaining EB resist and HfO_2 film act as the etching mask. The freestanding HfO_2 gratings are generated by removing the residual resist (step e). Finally, the epitaxial growth of III-nitride material by MBE is conducted on the suspended HfO_2

gratings. After nitrifying HfO_2 substrate at the temperature of 680°C for 20 minutes, a 150-nm thick buffer layer is deposited at the temperature of 680°C, and a 450 nm high-temperature GaN layer is then grown at the temperature of 750°C. The six-pair 3 nm InGaN/12 nm GaN MQWs is subsequently deposited at the temperature of 660°C. Finally, a 20 nm GaN layer is grown at the temperature of 660°C.

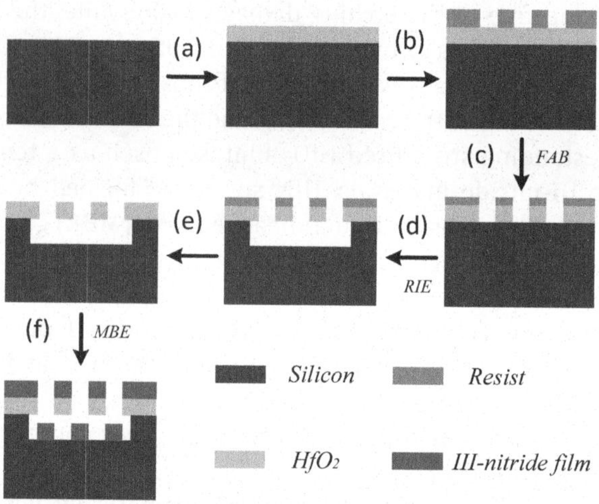

Figure 7.13 Schematic process of III-nitride grating grown on freestanding HfO_2 grating by MBE.

Figure 7.14(a) illustrates SEM image of freestanding HfO_2 grating template with a grating period of 1020 nm. One period grating is comprised of HfO_2 grating beam and air opening. Actually, suspended HfO_2 grating beam functions as a template for the epitaxial growth of III-nitride materials. For simplicity the air opening is used to describe the epitaxial III-nitride grating. Figure 7.14(b) shows the 60-period epitaxial grating grown on 1020 nm period suspended HfO_2 grating with the air opening of 400 nm, and the inset is the zoom-in image of freestanding III-nitride grating. The 60-μm long suspended HfO_2 grating beam can guarantee sufficient stiffness for MBE growth. III-nitride nanocolumns are grown on the top surface and the sidewall of the grating beam. The flat III-nitride grating surface can be achieved by improving the growth condition. Decreasing the original air opening to 200 nm, the resultant grating opening illustrated in the inset of Fig. 7.14(c) is easily filled with III-nitride nanocolumns, and the epitaxial III-nitride grating is similar to that grown on unpatterned area. Increasing the original air opening to 600 nm, the epitaxial grating beams shown in Fig. 7.14(d) are in the tendency of being deflected and fragile. The grating openings are easily filled with III-nitride nanocolumns as the grating period decreases.

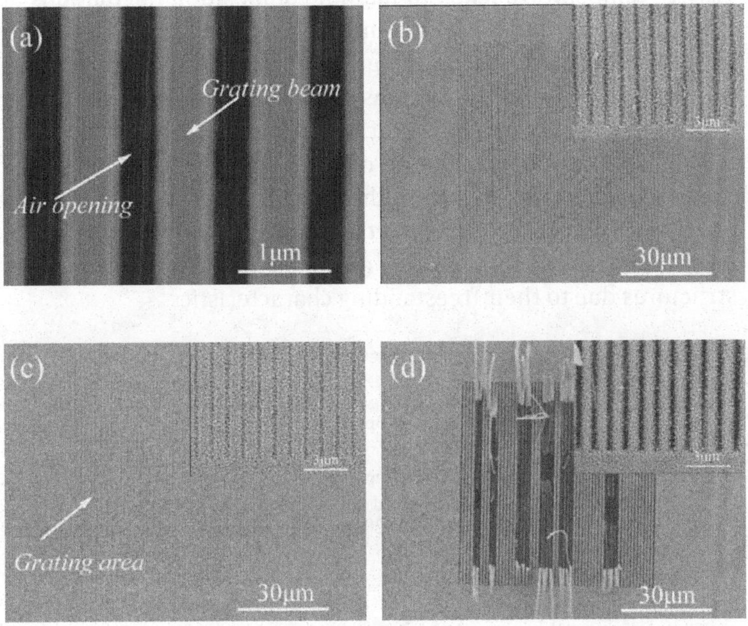

Figure 7.14 (a) SEM image of freestanding HfO_2 grating template. (b) Epitaxial III-nitride grating with air opening of 400 nm, and the inset is the zoom-in image. (c) Epitaxial III-nitride grating with air opening of 200 nm and the zoom-in image. (d) Epitaxial III-nitride grating with air opening of 600 nm and the zoom-in image.

The emission properties of the resultant epitaxial structures are characterized using a micro photoluminescence (PL) system at room temperature. The excitation source is a continuous wave 325 nm He-Cd laser source, and the pump light is focused on the sample through a UV-compatible objective lens (20× and numerical aperture: 0.36). The emitted light is collected by the same objective lens and measured using a multichannel analyzer system (Hamamatsu C10027). Figure 7.15 shows the PL spectra of

1020 nm period epitaxial gratings with various air openings. Regarding the excitation of InGaN/GaN MQWs grown on unpatterned area, two distinct PL peaks are observed around 441 nm and 620.5 nm, and a clear broad shoulder is found at ~748.7 nm in the PL spectra. The emission from epitaxial materials is dependent on indium content in the InGaN quantum well [22]. Variation in growth temperature leads to indium composition fluctuations and eventually results in the broad emission spectra [23–25]. Compared with unpatterned substrate, grating structures locally change the epitaxial growth conditions, giving an influence on indium composition in the epitaxial gratings. It can be seen that the tuning of the PL spectra is much more obvious as the original air opening increases. As to the grating with designed air opening of 400 nm, only two PL peaks are observed around 437.9 nm and 662.6 nm, respectively. The PL intensity is also improved for epitaxial grating structures due to their freestanding characteristics.

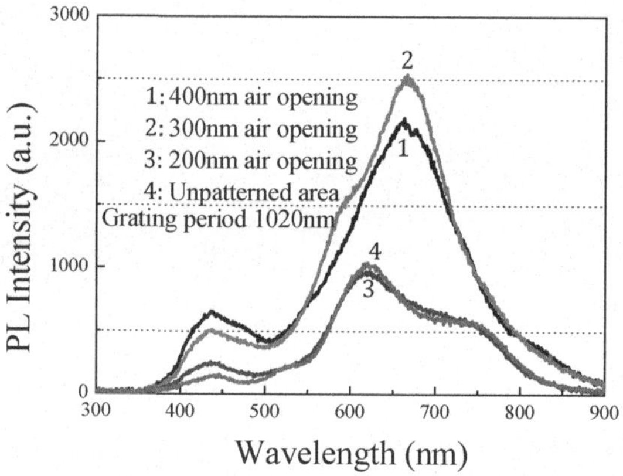

Figure 7.15 PL spectra of 1020 nm period epitaxial III-nitride grating versus air opening.

Since the epitaxial gratings are freely suspended with air as the low refractive index material on the top and bottom, the freestanding epitaxial gratings serve as an optical cavity to support guided mode resonances. Figure 7.16(a) shows the reflectance spectra versus polarization. Under TE polarization, one sharp reflection dip ~4% is clearly observed at 1500.5 nm for 1040 nm period epitaxial grating. Measured reflectances over 70% are with a broad stopband in the range of 1512.8 m ~1563.5 nm. The reflection band shifts and the shape changes under TM polarization, which is perpendicular to the grating lines. As the grating period decreases from 1040 nm to 1020 nm, the broad reflection band exhibits a distinct blue-shift, indicating that the guided mode resonances shift proportionally to the grating period. Compared with the original HfO_2 gratings, the effective index of the freestanding grating is increased after the epitaxial growth of III-nitride materials, resulting in the red shifting of reflectance spectra and the broadening of the reflection band. The reflection dip illustrated in Fig. 7.16(b) has a red-shift of ~15 nm, and the reflection band widens about 10 nm. These results suggest that such resonant gratings can be developed for refractive index sensing.

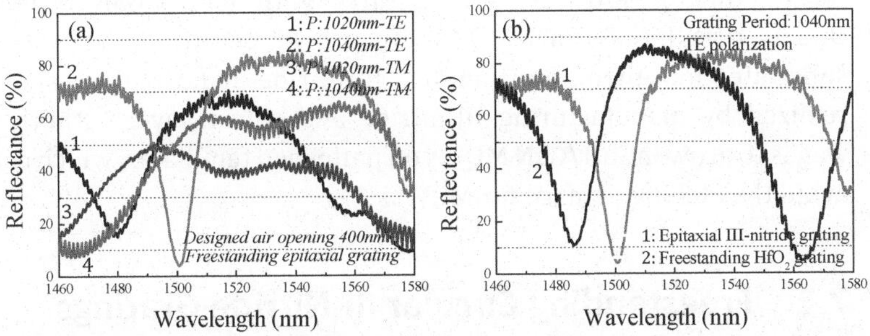

Figure 7.16 (a) Reflectance spectra of epitaxial III-nitride grating. (b) Comparison of reflectance spectra between HfO_2 grating and resultant III-nitride grating.

Although there is no clear signal to reflect the interaction between the excited light and the suspended structures, it is possible to optimize the grating parameters to tune the resonant locations corresponding to the PL spectra. Suspended III-nitride grating has a potential to realize the interaction between the excited light and the grating structure through guided mode resonance. In that case, guided mode resonant emission can be achieved in the epitaxial gratings when the excited light is coupled to one mode of the grating waveguide.

A thin freestanding III-nitride slab and control of the light propagation direction are essential ingredients to obtain III-nitride lighting devices with high light-extraction efficiency. One promising application of nanostructure is to change the total internal reflection condition at the GaN/air interface to improve the light extraction efficiency. Since the refractive index difference between the GaN (~2.5) and air is large, most of the emitted light from the InGaN/GaN MQWs is confined inside thick III-nitride layer due to the total internal reflection. The confined modes propagate within the III-nitride layer and their numbers increase with the III-nitride layer thickness. With the introduction of nanostructures on freestanding III-nitride slab, these challenging issues can be managed: (1) nanostructures break the total internal reflection condition at the GaN/air interface to enhance the light-extraction efficiency, and (2) freestanding slab provides a feasible way to thin III-nitride layer from the backside to reduce the confined modes.

Two possible methods are demonstrated here to fabricate freestanding III-nitride nanostructures with InGaN/GaN MQWs. One is that InGaN/GaN MQWs are grown on III-nitride/Silicon substrate and then, freestanding III-nitride nanostructures are realized by manufacturing III-nitride and silicon layers. Another way is to grow InGaN/GaN MQWs on patterned templates, which can avoid the etching damage for MQWs.

7.2 Freestanding Circular III-Nitride Gratings

EB lithography is a precise technique to define nanostructures, including gratings, photonic crystals, and microdisks. Circular grating structures, coupling the emitted radiation by multi-directional diffraction, are widely employed to achieve polarization-independent optical property, small-divergence output beam, and vertical emission perpendicular to the device plane [26–28]. FAB etching is able to etch III-nitride layer with high lateral resolution. Circular III-nitride gratings are realized by combining EB lithography and FAB etching. The silicon substrate under III-nitride grating region is completely removed to generate freestanding circular III-nitride gratings and then, FAB etching from the backside is developed to thin the freestanding III-nitride slab. This novel technique functions to reduce the number of optical modes inside III-nitride slab and to improve the light-extraction efficiency. The freestanding circular III-nitride gratings are implemented on the III-nitride/silicon substrate. Figure 7.17 shows the material structures.

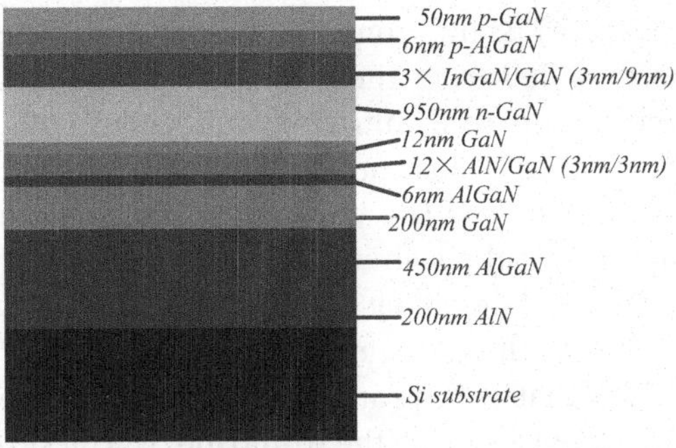

Figure 7.17 The schematic layer structure of material for making freestanding circular III-nitride grating.

Figure 7.18 illustrates the schematic fabrication process of freestanding circular III-nitride gratings. Circular gratings are first patterned in ZEP520A resist using EB lithography (steps a–b). The grating patterns are then transferred to III-nitride slab with a grating height of approximately 200 nm by FAB etching (step c). The processed patterns are protected by thick photoresist and then, the silicon substrate under the circular grating region is patterned from the backside by photolithography and etched down to the AlN layer by deep reactive ion etching (steps d–e). The freestanding circular III-nitride gratings are finally generated by removing the residual photoresist (step f). The freestanding III-nitride slab is thinned from the backside to a thickness of ~1000 nm (steps g–h).

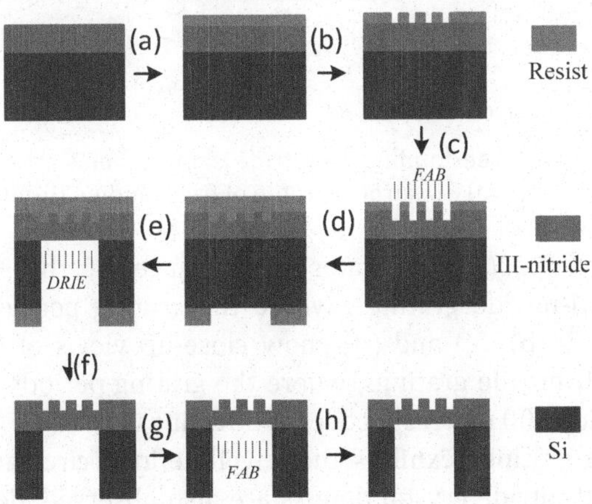

Figure 7.18 Fabrication process of freestanding circular III-nitride grating.

Figure 7.19(a) shows one SEM image of freestanding III-nitride slab obtained from the silicon substrate side. The size of the freestanding square III-nitride slab is about 300 μm × 300 μm. The 3D surface profile illustrated in Fig. 7.19(b) is measured by an optical interferometer. The surface deflection regarding the freestanding III-nitride slab is downward due to the residual stress, and the peak-to-valley value is measured around 3.25 μm at the center of the 300 μm × 300 μm III-nitride slab. These results indicate that it is promising to fabricate freestanding III-nitride lighting devices using the III-nitride/silicon system.

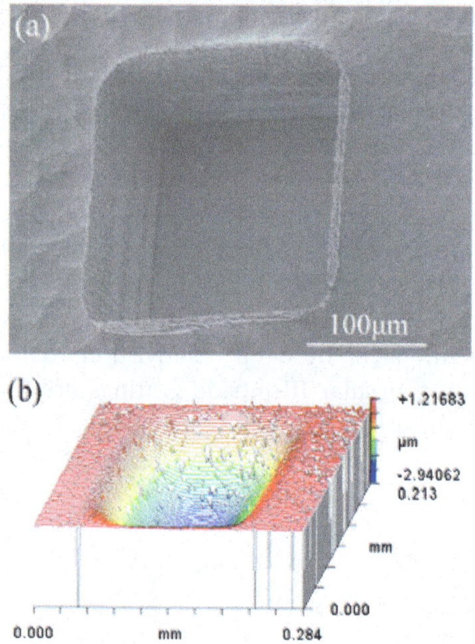

Figure 7.19 (a) Freestanding III-nitride slab obtained from Si substrate side. (b) 3D surface profile of freestanding III-nitride slab.

Figure 7.20(a) illustrates SEM image of the freestanding circular III-nitride gratings, where the grating period is 300 nm. Figures 7.20(b), (c) and (d) show close-up views of freestanding circular III-nitride gratings, where the grating periods are 300 nm, 600 nm, and 800 nm, respectively. The tilting view of the freestanding III-nitride gratings exhibits their well-defined circular symmetry, and the III-nitride crystal defects are also observed, degrading the internal quantum efficiency of the III-nitride material. In the whole device area, the circular gratings are well fabricated, and no broken grating beams are observed.

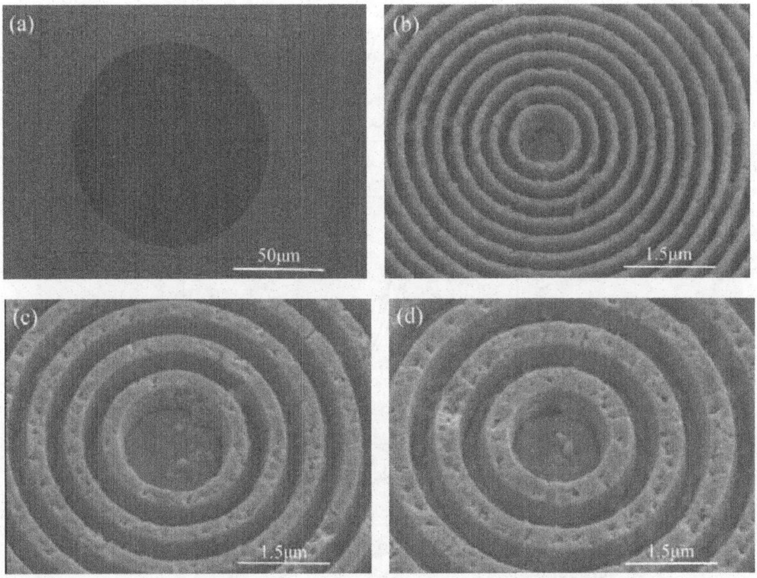

Figure 7.20 SEM images of the freestanding circular III-nitride gratings. (a) Whole grating region for 300 nm-period circular grating. (b), (c), and (d) Close-up views of circular grating with grating period of 300 nm, 600 nm, and 800 nm, respectively.

Figure 7.21(a) shows the micro-PL spectra of fabricated samples, where the pump light is focused on the sample from top surface. The micro-PL spectra of freestanding III-nitride slab are first compared with those of non-freestanding III-nitride slab. The luminescence peak of the non-freestanding III-nitride slab at 426 nm is normalized to 1, and a 2.5-fold enhancement in micro-PL peak intensity is observed from the freestanding III-nitride slab. The luminescence intensity integrated over the emission spectrum is increased by about 2.5 times. The measurements confirm the improvement in the light-extraction efficiency of freestanding III-nitride slab. The light loss into the silicon substrate is avoided in the freestanding III-nitride slab and thus, the light-extraction efficiency is improved.

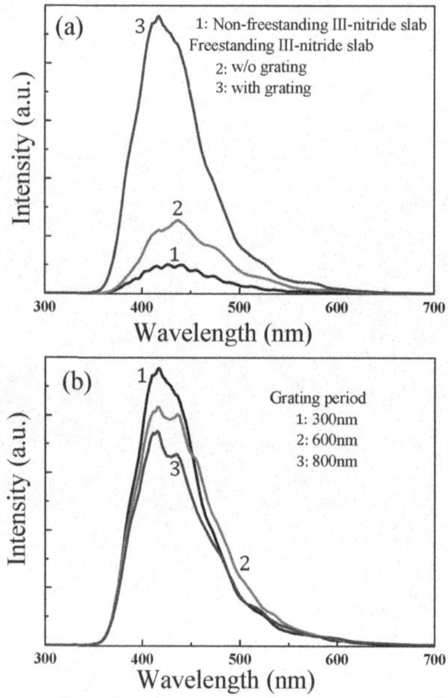

Figure 7.21 (a) PL spectra of fabricated samples obtained from top surface. (b) PL spectra versus grating period obtained from top surface.

Circular gratings on the top surface are useful for enhancing the light-extraction efficiency. The fabricated circular gratings have a periodic distribution of the refractive index. Therefore, a modification of the direction of light propagation occurs by multi-directional diffraction effect, which breaks the total internal reflection condition at the surface of the III-nitride slab. When the MQWs layer is excited by the pump laser source, the emitted light propagates in the circular grating. More photons can leave the device, and the micro-PL intensity is significantly improved. As illustrated in Fig. 7.21(a), the micro-PL intensities are significantly increased by the assistance of circular grating patterns, 10-fold peak intensity and 8-fold integrated intensity enhancements are obtained from 300 nm-period circular grating, respectively. The micro-PL spectra also show a clear dependence on the periodicity of circular gratings, as illustrated in Fig. 7.21(b). The luminescence peaks are observed around 416 nm, and the peak intensities are 7-fold, 8-fold, and 10-fold enhancements for circular grating with grating period of 800 nm, 600 nm, and 300 nm, respectively. For a proper nanoscale grating, the grating structures allow only specific wavelengths of light to be resonant inside the structure. When the round-trip gain is equal to one (and saturated), the light becomes a true laser beam for further III-nitride surface-emitting lasers.

7.2.1 Freestanding III-Nitride Subwavelength Nanostructures

As to freestanding III-nitride subwavelength nanostructures, the light extraction efficiency is improved by reducing the number of confined modes inside the GaN slab, eliminating silicon absorption of the emitted light, and decreasing the losses caused by the total internal reflection [29,30].

Subwavelength nanostructures can break the total internal reflection condition at the GaN surface to improve the light extraction efficiency [31–33]. Self-assemble technique is a very competitive approach to define subwavelength nanostructures [34,35]. Silica microspheres are self-assemble into a single layer hexagonal array on the GaN surface and then, III-nitride subwavelength nanostructures are achieved by FAB etching with self-assembled silica nanospheres acting as an etching mask. Such nanostructures modify the light propagation direction at the GaN surface. The reflection losses at the GaN/air interface are reduced, and the light extraction efficiency is thus increased.

Figure 7.22 illustrates the fabrication process of freestanding III-nitride subwavelength nanostructures. The top III-nitride layer is first protected by thick photoresist, and the silicon substrate is patterned from the backside by photolithography (step a). The silicon substrate is then etched down to the AlN layer using DRIE, where the AlN layer acts as a definite etch stop (step b). The freestanding III-nitride slab is thinned to a thickness of ~900 nm from the backside by FAB etching (step c). After removing the residual resist, silica nanospheres are spin-coated onto the III-nitride surface with the rotation speed of 3500 rpm for 40 sec, which makes the silica nanospheres self-assemble into a single layer close-packed array. In order to evaporate the solvent, the substrate is subsequently kept at 90° for 10 min and 145°C for 30 min, respectively (step d). With an etching mask of silica nanospheres, the nanostructure patterns are transferred to the III-nitride slab by FAB etching, and the residual silica particles are finally removed in buffered HF solution, generating III-nitride subwavelength nanostructures (step e).

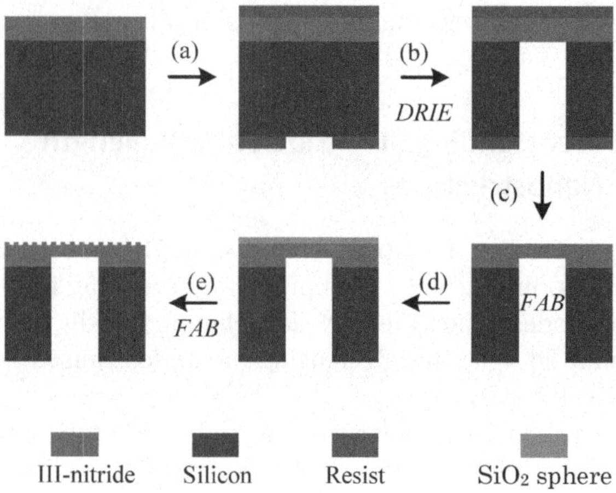

Figure 7.22 Fabrication processes of freestanding III-nitride subwavelength nanostructures.

The size of freestanding square III-nitride slab is about 300 μm × 300 μm. Figures 7.23(a) and (b) show the 30° tilt-view SEM images of silica nanospheres forming a close-packed single

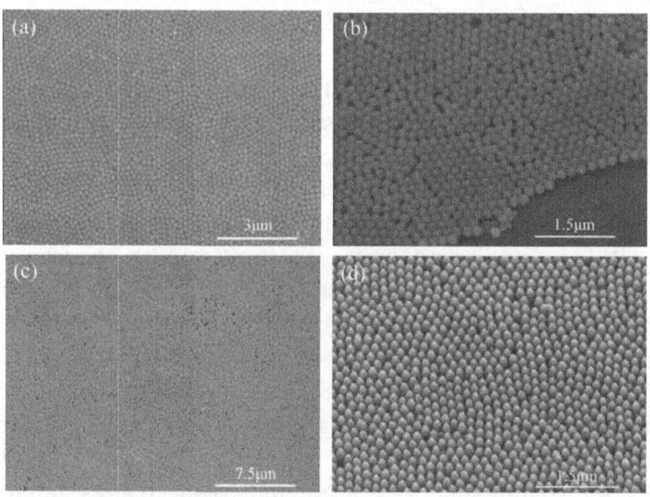

Figure 7.23 (a) and (b) Tilt-view SEM images of single-layer close-packed silica nanospheres. (c) and (d) Tilt-view SEM images of fabricated III-nitride subwavelength nanostructures.

layer on the III-nitride surface. These silica nanospheres have a uniform diameter size of about 170 nm and display short-range order. Subwavelength nanostructures are transferred to the III-nitride surface by FAB etching, as illustrated in Figs. 7.23(c) and (d). The tilt views exhibit the well-defined III-nitride subwavelength nanostructures using self-assemble technique.

Figure 7.24(a) shows the measured PL spectra of the III-nitride slabs. The original III-nitride slab on silicon substrate is measured as reference, and the luminescence peak at 428 nm is normalized to 1. A 3.3-fold enhancement in the peak intensity is observed at 415 nm for the III-nitride slab with surface texturing, and the luminescence intensity integrated over the emission spectrum is increased by 3.1 times. Subwavelength nanostructures have a periodic distribution of the refractive index and modify the light-propagation direction at the GaN surface. Therefore, the total internal reflection condition is broken and the reflection losses are suppressed, which is able to help more photons leave the device when the MQWs active layers are excited by the pump laser beam. Hence, the measured PL intensity is improved.

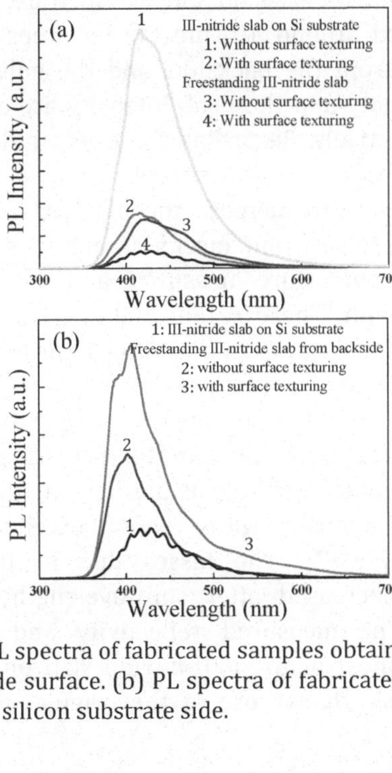

Figure 7.24 (a) PL spectra of fabricated samples obtained from the top III-nitride surface. (b) PL spectra of fabricated samples obtained from silicon substrate side.

After backside thinning of the freestanding III-nitride slab, silicon absorption of the emitted light is eliminated and the number of optical modes is reduced. Hence, the light extraction efficiency is increased. The measurements confirm the improvement in the light-extraction efficiency of the freestanding III-nitride slab. A 2.9-fold peak intensity enhancement and a 3.1-fold integrated intensity enhancement are obtained, and the intensity peak is observed at 423 nm. The freestanding III-nitride slab with surface texturing can take all the mentioned advantages, i.e., reducing the number of confined modes inside the III-nitride slab, eliminating silicon absorption of the emitted light, and modifying the light propagation direction at the III-nitride surface. The experimental results provide solid evidences on significantly improving the light extraction efficiency. The luminescence peak at 416 nm has a 14-fold enhancement in the peak intensity, and the integrated intensity is also increased by a factor of 14, as shown in Fig. 7.24(a).

Figure 7.24(b) illustrates a clear enhancement in PL intensity when the pump light is incident on the freestanding III-nitride slab from silicon substrate side. Compared with the PL intensities of the original III-nitride slab on silicon substrate, the luminescence peak is observed around 402 nm for the freestanding III-nitride slab measured from the backside, and the enhancements of the integrated intensity and the peak intensity are about 2 times and 2.5 times, respectively. Regarding the freestanding III-nitride slab with surface texturing, subwavelength nanostructures work as patterned substrate to increase the light extraction efficiency. A 4.6-fold peak intensity enhancement and a 4.3-fold integrated intensity enhancement are measured, and the PL intensity peak is found at 407 nm. These results indicate that the PL intensity form the backside is also significantly improved by introducing subwavelength texturing of the GaN surface.

Figure 7.25(a) shows the measured reflectivity from the top III-nitride surface, where the interference fringes are attributed to the multiple reflections of visible light at the GaN interfaces. Subwavelength nanostructures show excellent anti-reflective characteristics [36,37]. The losses due to the total internal reflection are decreased after subwavelength texturing of the GaN surface. The measured reflectivity and the interference peaks of the nanostructured III-nitride slab on silicon substrate are much smaller than those of the original III-nitride slab on silicon substrate. The subwavelength texturing technique can help more excited photon to escape out the III-nitride slab and thus enhance the extraction efficiency, as shown in Fig. 7.24(a). The blue curve illustrated in Fig. 7.25(a) represents the measured reflectivity of the freestanding nontextured III-nitride slab. The number of confined modes inside the III-nitride slab is decreased by backside thinning of the freestanding III-nitride slab, resulting in the reduction of the interference fringes. As to the freestanding nanostructured III-nitride slab, the reflectivity is decreased due to the reduction of the total internal reflection, and the number of confined modes is also reduced. These reflection measurements are in line with the above analysis on the improvement in the light extraction efficiency by reducing the number of optical modes inside the III-nitride slab, eliminating silicon absorption of the emitted light, and decreasing the reflection losses. Figure 7.25(b) illustrates the reflectivities of the freestanding III-nitride slab measured from the backside and it is shown that they behave very similarly to those obtained from the top side of the freestanding III-nitride slab.

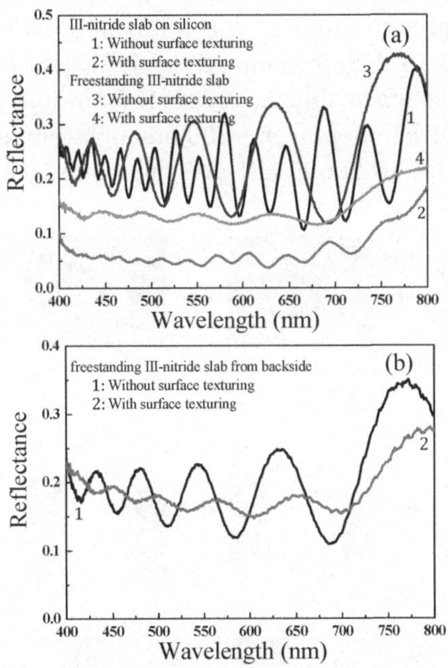

Figure 7.25 (a) Reflectivity of fabricated samples obtained from the top III-nitride surface. (b) Reflectivity of fabricated samples obtained from silicon substrate side.

7.2.2 Patterned Growth of InGaN/GaN MQWs

Compared to other fabrication techniques, novel epitaxial structures can be generated with smooth facets through patterned growth of III-nitride [38,39] and are free of etching damage. The patterned growth of InGaN/GaN MQWs on freestanding nanoscale III-nitride gratings are carried out by MBE. Through the introduction of nanoscale grating structures, the selective growth occurs and depends on the grating period and the grating width.

The fabrication process is schematically illustrated in Fig. 7.26. Nanoscale gratings are patterned in ZEP520A resist using EB lithography, and the resist structures act as a mask for FAB etching of GaN (steps a–b). The Cl_2 gas is used as the process gas, and the etching depth is about 200 nm (step c). Then the residual EB resist is stripped and the processed device layer is protected by thick photoresist (step d). Silicon substrate beneath the III-nitride grating region is patterned from the backside and etched down to the AlN layer by DRIE (step e). The freestanding III-nitride gratings are generated by removing the residual photoresist and cleaned for the epitaxial growth (step f). The epitaxial growth is conducted on the processed III-nitride template by MBE (step g). The epitaxial films with a designed thickness of ~420 nm incorporate ~140 nm low-temperature buffer layer, ~200 nm high-temperature GaN layer, six-pair 3 nm InGaN/9 nm GaN QWs layer and 10 nm GaN top layer.

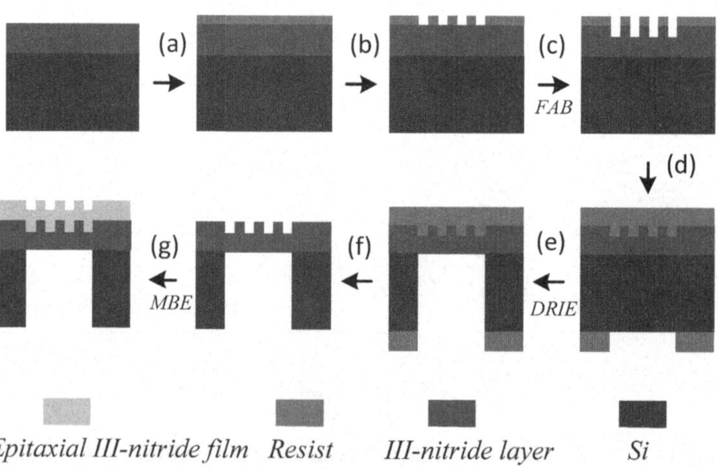

Figure 7.26 Schematical process of patterned growth on freestanding III-nitride grating by MBE.

Various freestanding III-nitride gratings are fabricated on an III-nitride/silicon substrate by a combination of EB lithography, FAB etching of III-nitride and DRIE of silicon. Figure 7.27 illustrates SEM images of freestanding GaN gratings. The grating period and the grating width are expressed by *P* and *W*, as shown in Fig. 7.27(a), where *P* is 500 nm and *W* is ~300 nm. One period grating consists of the grating ridge and the grating opening. The III-nitride gratings illustrated in Figs. 7.27(b), (c), and (d) have the same grating width of ~200 nm and have different grating periods of 500 nm, 450 nm, and 400 nm, respectively. The variation in the grating width *W* means the different distributions between the grating ridge and the grating opening, which plays an important role in the epitaxial growth.

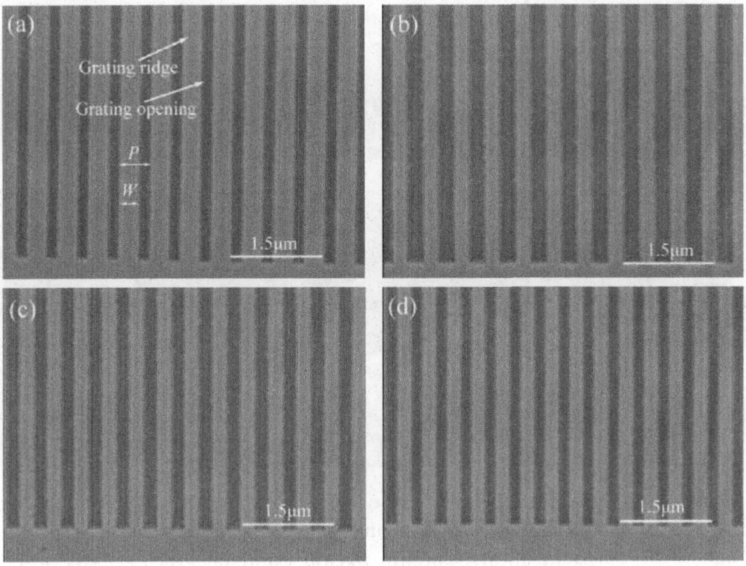

Figure 7.27 SEM images of III-nitride grating templates for the epitaxial growth. (a) 500 nm period, 300 nm wide grating; (b) 500 nm period, 200 nm wide grating; (c) 450 nm period, 200 nm wide grating; (d) 400 nm period, 200 nm wide grating.

The built-in residual stress in III-nitride thin film on silicon substrate, which is due to the lattice mismatch and the thermal expansion coefficient mismatch, can result in the deflection problems for freestanding III-nitride membrane. Although thin III-nitride membrane can guarantee sufficient stiffness for the fabrication of freestanding gratings during DRIE of silicon process, the fracture-related problems shown in Fig. 7.28(a) are evident in the freestanding III-nitride membrane after the epitaxial growth. These problems might be solved by adjusting the fabrication process. The crack networks, which are caused by the lattice mismatch in the epitaxial layers, are observed on unpatterned substrate, as illustrated in the inset of Fig. 7.28(a). The crack does not occur in the III-nitride grating region, indicating the III-nitride gratings can compensate the lattice mismatch.

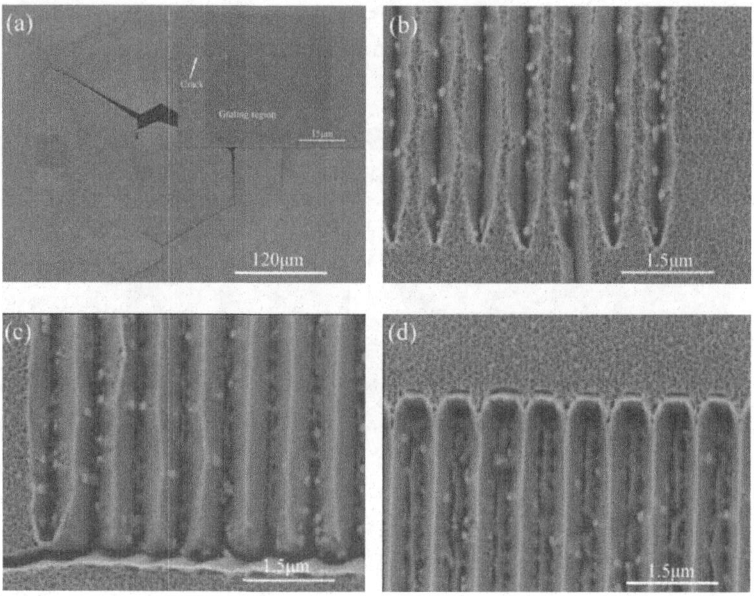

Figure 7.28 (a) Epitaxial grating on freestanding III-nitride membrane, and the inset is the zoom-in view of grating region. (b), (c), and (d) Resultant 700 nm period epitaxial gratings; (b) 500 nm wide grating; (c) 350 nm wide grating; (d) 250 nm wide grating.

Figures 7.28(b), (c), and (d) show the epitaxial structures on the 700 nm period III-nitride gratings with the grating width *W* of ~500 nm, ~350 nm, and ~250 nm, respectively. Compared with unpatterned substrate, grating structures locally change the diffusion conditions of adatoms from neighboring areas. Coherent growth is suppressed, and the selective growth takes place on the grating ridge with a preferential growth process on the low-energy side $\{10\bar{1}1\}$

facets. As the grating width *W* decreases, the area of the grating ridge is reduced. Thus, the surface diffusion can be sufficiently enhanced, resulting in complete coalescence between two side facets. Epitaxial gratings with smooth facets are observed in Figs. 7.28(c) and (d). Especially, Fig. 7.28(d) demonstrates that the selective growth can also occur in the grating openings. Compared with Fig. 7.28(b), it can be concluded that a critical growth area is needed for the selective growth. When the growth area is too small, the selective growth is suppressed. On the other hand, it's difficult to complete the selective growth if the growth area is too large. The critical growth area might be dependent on the surface diffusion, which could be improved by adjusting the grating parameters.

The shape and the cross section of the epitaxial films are shown in Fig. 7.29. Since the sample is used for the development of backside thinning technique by wet etching of Al-based compounds, some freestanding epitaxial slabs are damaged in the wet etching process. The measured thickness of epitaxial films is about 510 nm, a little larger than the estimated thickness of ~420 nm. The freestanding III-nitride slab is deflected due to the residual stress, and the slab is thinner than that on silicon substrate, as shown in Fig. 7.29(a). One cross section image of epitaxial grating is illustrated in Fig. 7.29(b). The inset is the zoom-in image of epitaxial grating, and the shape changes are clearly observed on different templates.

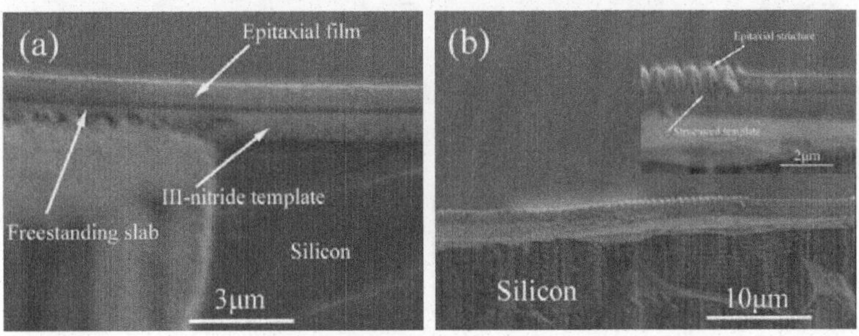

Figure 7.29 (a) The cross section of the epitaxial films. (b) Freestanding epitaxial grating structures, and the inset is the zoom-in view of grating region.

The PL of InGaN/GaN MQWs deposited on unpatterned area is shown in Fig. 7.30(a). Since the silicon substrate is removed and the slab is thinned by wet etching, the PL intensity is great for freestanding InGaN/GaN MQWs slab. Figure 7.30(b) shows the PL spectra of 700 nm period epitaxial gratings with various grating widths. The PL peaks at ~436.4 nm are associated with the excitation of the InGaN/GaN MQWs active layers. With decreasing the grating width *W* from ~500 nm to ~250 nm, the PL peak and the integrated intensity are significantly increased, corresponding to the improvement in the selective growth. The PL spectra of 500 nm period epitaxial gratings are shown in Fig. 7.30(c) and demonstrate the similar optical performances. The PL peaks are about 436.4 nm, and the corresponding PL intensities are improved, indicating that small grating period is helpful for the patterned growth. However, the PL spectra illustrated in Fig. 7.30(d) is different as the grating period decreases to 450 nm, where the number of III-nitride nanocolumns is gradually increased. However, the PL from 200 nm grating width sample is stronger than it is from 250 nm grating width sample for the 450 nm period epitaxial gratings. It might be explained by the formation of epitaxial nanocolumns. Both epitaxial grating and nanocolumns contribute to the PL excitation. The number of epitaxial nanocolumns is increased with

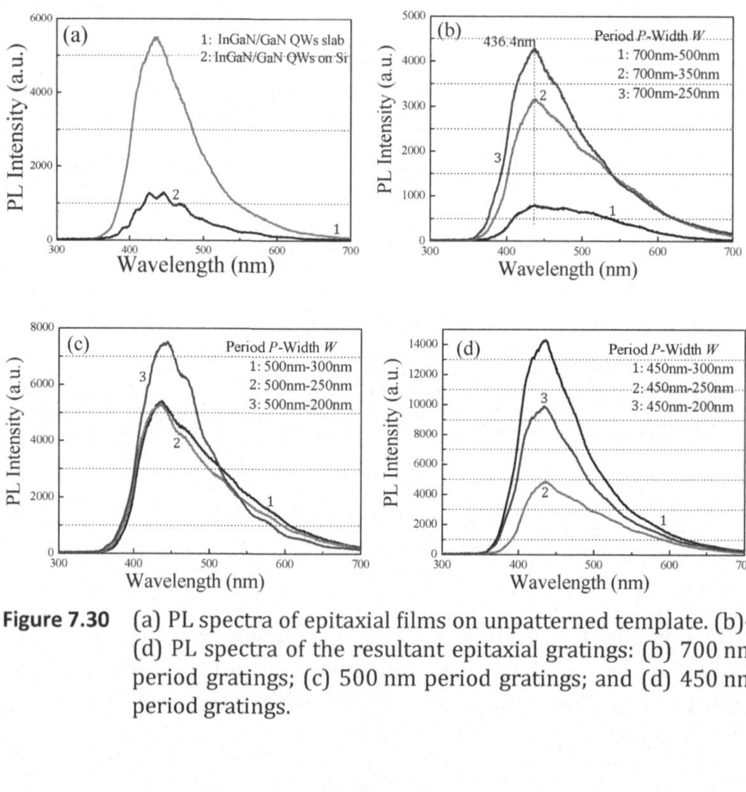

Figure 7.30 (a) PL spectra of epitaxial films on unpatterned template. (b)–(d) PL spectra of the resultant epitaxial gratings: (b) 700 nm period gratings; (c) 500 nm period gratings; and (d) 450 nm period gratings.

increasing the grating width, whereas the epitaxial gratings with smooth facets are easily formed with decreasing the grating width. Hence, the epitaxial structures generated in reality determine which one plays the dominant influence on the PL spectra. On the other hand, thin InGaN/GaN MQWs layers are incorporated in the upper part of the epitaxial gratings, the film structures beneath smooth side facets are rough, and the scattering losses are thus very large. Consequently, there is no clear signal to reflect the interaction between the excited light and the grating structures.

The selective growth of III-nitride is also validated in other nanostructured GaN templates. Figure 7.31(a) shows SEM image of fabricated 2D GaN nanocolumns in the whole device region. The period P is 1500 nm, and the diameter of GaN nanocolumn varies. Figures 7.31(b)–(d) show the magnified images of periodic GaN nanocolumns with the designed diameters of 900 nm, 800 nm, and 700 nm, respectively. EB resist acts as an etching mask during FAB etching, and the thickness of EB resist varies at the edge due to the proximity effect in EB lithography. FAB etching is a dry etching technique with high directionality of the beam flux. The generated GaN nanocolumns are thus tapered profiles with the etching depths of about 230 nm. The corresponding bottom diameters of GaN nanocolumns are measured ~896 nm, ~785 nm, and ~693 nm, respectively. The top diameters of GaN nanocolumns are about 684 nm, 619 nm, and 508 nm, respectively.

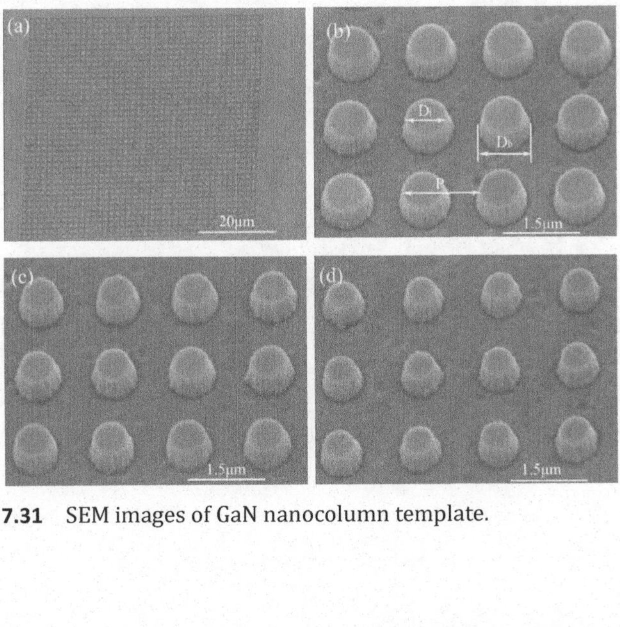

Figure 7.31 SEM images of GaN nanocolumn template.

Figure 7.32(a) shows SEM image of the resultant freestanding epitaxial III-nitride structures. The freestanding slab is easily broken due to large changes in the strain state during the epitaxial growth. The coherent growth takes place on unpatterned GaN substrate, and the crack networks are generated in the epitaxial layers due to the lattice mismatch. The inset in Fig. 7.32(a) shows the cross-sectional image of epitaxial structures in the GaN nanocolumn region. About 700 nm thick epitaxial films, at the upper part of which six-pair 3 nm InGaN/9 nm GaN MQWs active layers are integrated, are grown on the processed GaN substrate. Figures 7.32(b)–(d) demonstrate the epitaxial structures in the nanocolumn region with the corresponding top diameters of ~684 nm, ~619 nm, and ~508 nm, respectively. The selective growth is observed on the GaN nanocolumns, depending on the size of GaN nanocolumn. For GaN nanocolumn with the top diameter of ~684 nm, the circular shapes become the hexagonal shapes with six lateral {10$\bar{1}$1} facets and a rough {0001} facet on the top after MBE growth, and the neighboring areas are with the rough epitaxial structures due to the coherent growth. Full coalescence of 2D hexagonal

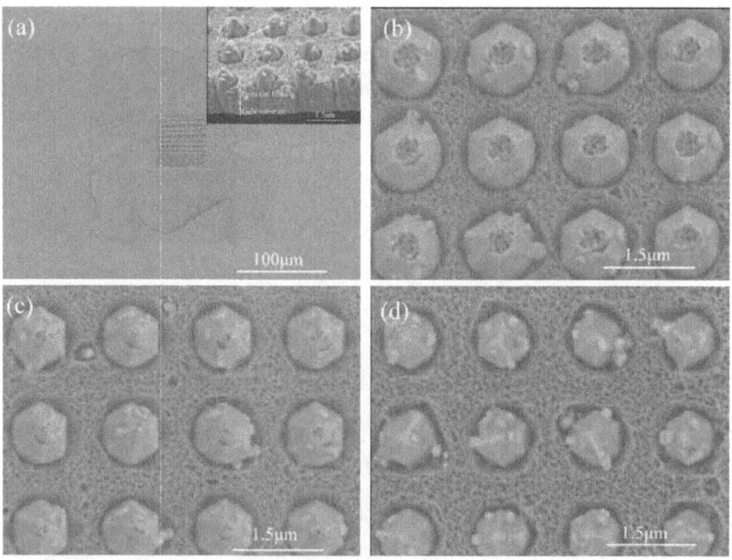

Figure 7.32 Epitaxial structures in GaN nanocolumn region.

III-nitride pyramids can be completed as the growth time extends. With decreasing the top diameter to ~619 nm, the area of top {0001} facet decreases, and the epitaxial structures have a tendency to be hexagonal pyramids with self-organized facets. Full coalescence is formed for the GaN nanocolumns with the top diameter of ~508 nm, and periodic hexagonal III-nitride pyramids are generated with random tips on side facets. These results demonstrate that the selective growth of III-nitride takes place with the assistance of GaN nanocolumns, and a preferential growth process takes place on $\{10\bar{1}1\}$ facets since the growth rate of the lateral $\{10\bar{1}1\}$ facets is faster than that of the top {0001} facet. This selective growth of III-nitride is also dependent on the size of GaN nanocolumns.

7.2.3 Freestanding III-Nitride Nanocolumn Slab with Bottom Subwavelength Nanostructures

For the light emission application, III-nitride nanocolumns are epitaxially grown on silicon substrate to improve the internal quantum efficiency and the light-extraction efficiency. However, only small part of the excited light is emitted in air when silicon substrate remains. Figure 7.33 shows possible paths of the extracted light for III-nitride nanocolumns grown on silicon substrate. A significant portion of the excited light is absorbed by silicon substrate, and the rest light is trapped in the GaN layer, as shown in Fig. 7.33(a). The freestanding III-nitride nanocolomun slab with bottom subwavelength nanostructures is improved in Fig. 7.33(b). The absorption of silicon substrate is eliminated for the freestanding III-nitride nanocolumn slab, and the extracted light from the topside can be slightly increased by the reflection of light from the bottom surface. Especially, the accessible paths from the bottom surface are greatly changed, and more light can escape in air with the assistance of the low-index buffer layer, which is necessary for the epitaxial growth of III-nitride. Subwavelength nanostructures are also able to break the total internal reflection condition at the bottom surface to enhance the light-extraction efficiency.

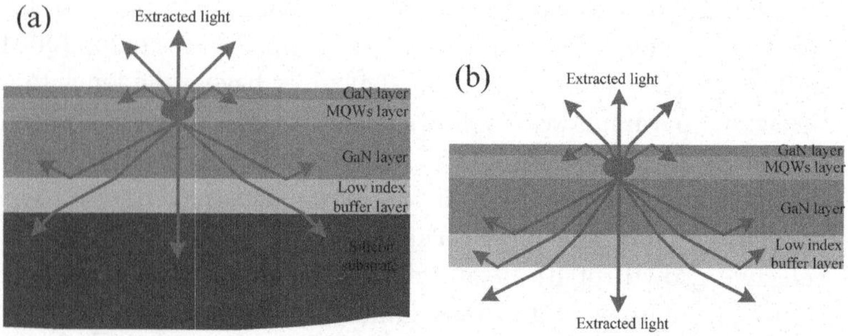

Figure 7.33 Possible paths of the extracted light for III-nitride nanocolumns grown on silicon substrate.

The 1200 μm-diameter freestanding III-nitride nanocolumn slabs with bottom subwavelength nanostructures are experimentally achieved by a combination of self-assemble technique, silicon-on-insulator (SOI) technology, manufacturing of silicon, and the growth of III-nitride. Figure 7.34 schematically illustrates the fabrication process of the freestanding III-nitride nanocolumn membrane with bottom subwavelength nanostructures. The starting 2 cm × 2 cm n-type (100) SOI substrate consists of 10 μm silicon top layer, 1 μm buried oxide layer, and 200 μm silicon substrate. The silicon substrate is first patterned from the backside by photolithography (step a). The silicon substrate is then etched down to the buried oxide layer using DRIE, where the buried oxide layer acts as an etching stop layer (step b). After removing the residual photoresist, the buried oxide layer is removed by buffered HF etching (step c). Subsequently, 170 nm-diameter silica nanospheres are spin-coated onto the silicon top surface, which makes silica nanospheres self-assemble into a close-packed monolayer. The prepared template is then kept in an oven at 90°C for 10 min and 145°C for 30 min, respectively, evaporating the solvent (step d). The subwavelength nanostructure patterns are transferred into the silicon top layer by FAB etching where silica nanospheres serve as an etching mask. FAB etching is conducted with the SF_6 flow of 5.6 sccm at the high voltage of 2.0 KV and the accelerated current of 20 mA. The residual silica particles are removed by BHF etching to generate subwavelength nanostructures on the silicon top layer (step e). The InGaN/GaN MQWs active layers are deposited on the prepared SOI template by MBE. The resultant III-nitride nanocolumn structure incorporates 200 nm low-temperature buffer-layer, 300 nm high-temperature GaN layer, six-pair 3 nm InGaN/9 nm GaN MQWs layer and 10 nm GaN top-layer (step f). Finally, the freestanding silicon membrane is removed by RIE of silicon from the backside, which makes the III-nitride nanocolumn membrane totally suspend in space (step g).

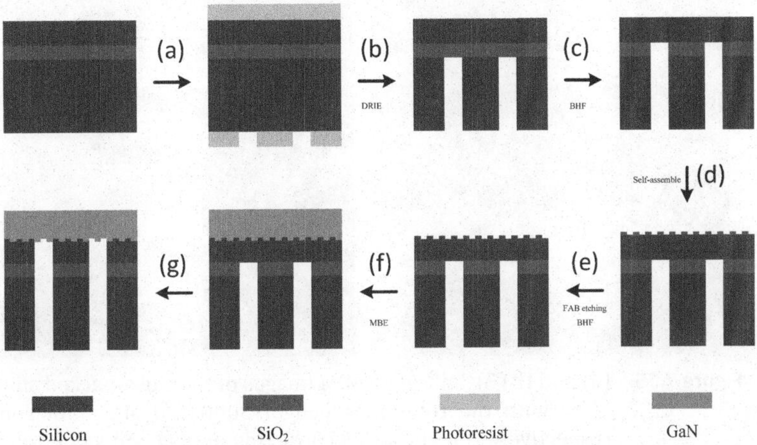

Figure 7.34 Schematic fabrication process of the freestanding III-nitride nanocolumn membrane with bottom subwavelength nanostructures.

Silica nanospheres are spin-coated onto the silicon top surface, and the rotation speed during the spin-coating step is crucial for forming closed-packed silica single layer. Figures 7.35(a) and (b) show the 30° tilt-view SEM images of closed-packed silica layer that are obtained with the rotation speed of 3000 rpm and 3500 rpm, respectively. Dual closed-packed silica layer is clearly observed at the rotation speed of 3000 rpm. Figure 7.35(c) illustrates the area consisting of dual and single silica layer. After FAB etching, the resultant silicon nanostructures are shown in Fig. 7.35(d). With silica monolayer acting as etching mask, subwavelength silicon nanostructures are gradually tapered and display short-range order. Silicon top layer is only partially engineered with the dual silica layer acting as etching mask. The following III-nitride nanocolumns are grown on nanostructured silicon substrate where silica monolayer serves as etching mask.

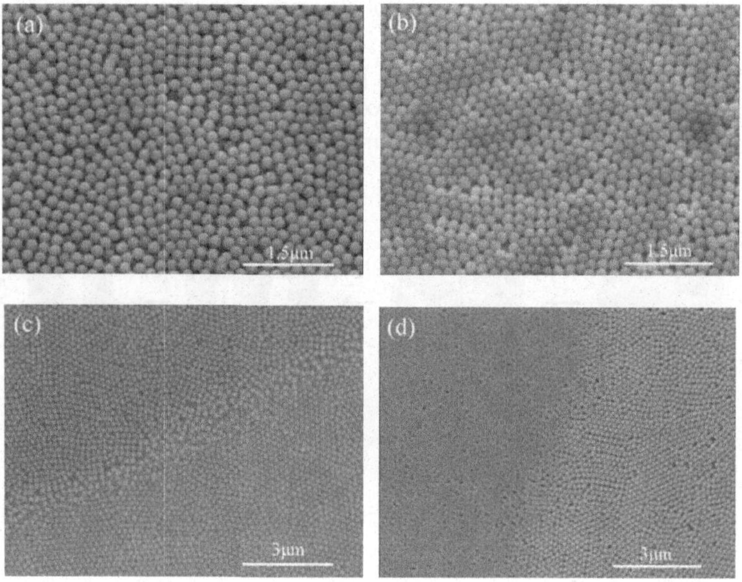

Figure 7.35 (a) and (b) 30° tilt-view SEM images of the close-packed silica layer with the rotation speed of 3000 rpm and 3500 rpm, respectively. (c) Silica area consisting of dual and single silica layer. (d) Fabricated subwavelength silicon nanostructure.

Figure 7.36(a) shows one photograph of the freestanding III-nitride nanocolumn slab with a diameter of 1200 μm. The inset is the 30° tilt-view SEM image of the III-nitride nanocolumns from the topside. The III-nitride nanocolumns are epitaxially grown on silicon substrate, and the cross-sectional profiles of the III-nitride nanocolumn slab are illustrated in Figs. 7.36(b) and (c). The total thickness is about 1.5 μm, and the resultant III-nitride nanocloumns incorporate ~550 nm low temperature buffer layer, ~850 nm high temperature III-nitride layer, six-pair 3 nm InGaN/9 nm GaN MQWs layer, and ~20 nm GaN top layer. The tapered silicon nanostructures are clearly observed with the heights of about 170 nm, and III-nitride nanocolumns are well grown, as illustrated in Fig. 7.36(c). RIE of silicon that is based on SF_6/C_4F_8 is carried out to remove silicon substrate from the backside. Etching/passivation cycle and gas flow rate are optimized to reduce surface damage due to over etching of silicon. The III-nitride nanocolumn slabs with a diameter of 1200 μm are totally suspended in space, and the bottom surfaces of the freestanding III-nitride nanocolumn slab are shown in

Fig. 7.36(d). Subwavelength III-nitride nanostructures are well achieved and surface profiles can be further improved by optimizing RIE of silicon.

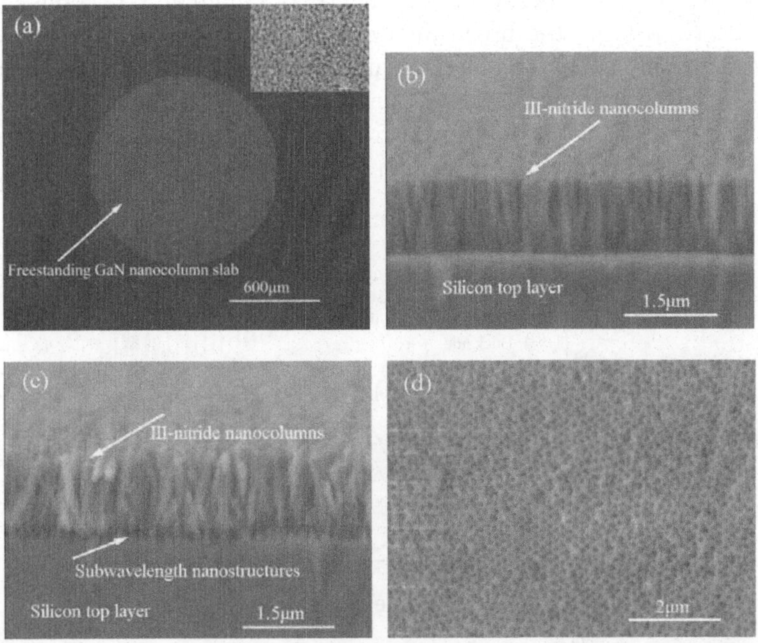

Figure 7.36 (a) Digital photography of the freestanding III-nitride nanocolumn slab, and the inset is the 30° tilt-view SEM image of nanocolumns from the topside; the cross-sectional profiles of nanolcolumns, (b) grown on the flat silicon substrate, (c) grown on nanostructured silicon substrate; (d) the bottom surface of the nanocolumn slab with bottom subwavelength nanostructures.

The III-nitride nanocolumns are experimentally demonstrated to act as an antireflective coating to reduce optical reflection over a broad wavelength range [40–43]. The reflectivity of a bare silicon surface is high, while optical reflection is greatly decreased by the introduction of the III-nitride nanocolumns, as shown in Fig. 7.37(a). Multiple reflections that lead to the broad interference fringes in the reflectance spectra are caused by the immediate change of the refractive index at the GaN/flat silicon interface. Regarding the freestanding III-nitride nanocolumn slab, surface profile plays an important role in optical reflection. The reflectivity obtained from the topside is lower than that obtained from the backside because more reflection is reduced at the III-nitride nanocolumn/air interface. A graded refractive index step is realized by introducing subwavelength nanostructures, hence, the multiple reflections are decreased and the broad interference fringes are suppressed, as shown in Fig. 7.37(b). The measured reflectivity is below 3% in the wavelength range of 500–750 nm. Optical reflection obtained from the backside is also low for the freestanding III-nitride nanocolumn slab with bottom subwavelength nanostructures.

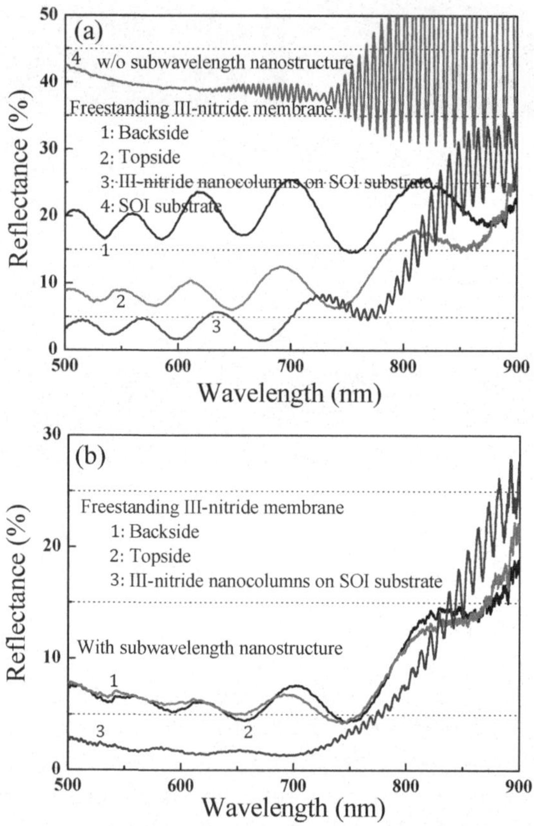

Figure 7.37 (a) Measured reflectivity of the III-nitride nanocolumn slab grown on the flat silicon substrate. (b) Measured reflectivity of the III-nitride nanocolumn slab grown on nanostructured silicon substrate.

The PL measurements are performed using a 325 nm He-Cd laser source. Figure 7.38(a) shows the PL spectra of the III-nitride nanocolumn slabs grown on the flat silicon substrate. The PL peak at ~365 nm is attributed to the excitation of the top III-nitride crystal, and the emission peak of the InGaN/GaN MQWs active layers is at ~471 nm for the non-freestanding III-nitride slab, which is normalized to 1. Silicon absorption of the emitted light is eliminated, and the reflection of light from the bottom surface can slightly increase the PL intensity for the freestanding III-nitride nanocolumn slab. The luminescence intensity integrated over the emission spectrum is increased by about 2.3 times, where the pump light is focused on the sample from top surface. The emission peak shifts to ~482 nm and a 2.3-fold enhancement in the PL peak intensity is measured. As aforementioned, the bottom surface can provide more paths accessible to emit light in air for the freestanding III-nitride nanocolumn slab. As a result, the emission peak of the III-nitride crystal disappears, and the III-nitride nanocolumns produce a strong blue emission peak at ~446 nm with a FWHM of about 88 nm, which is associated with the InGaN/GaN MQWs active layers. The integrated intensity and the peak intensity are greatly increased by a factor of 33 and 54, respectively. Regarding the III-nitride nanocolumn slab grown on nanostructured silicon substrate, the measured PL spectra shown in Fig. 7.38(b) are similar to those illustrated in Fig. 7.38(a). The PL intensity obtained from the bottom surface is significantly increased by introducing bottom subwavelength nanostructures. It should be noted that the relative emission peaks of the InGaN/GaN MQWs active layers exhibit blue shifts, which can be attributed to the occurrence of strain relaxation by growing III-nitride nanocolumns on nanostructured silicon substrate. The emission peak of the InGaN/GaN MQWs is observed at ~432 nm with a FWHM of about 64 nm, and the normalized enhancements of the integrated intensity and the peak intensity are about 91 times and 203.8 times, respectively. The measurements confirm the improvement in the light-extraction efficiency of the freestanding III-nitride nanocolumn slab with the assistance of the low-index buffer layer and bottom subwavelength nano-structures.

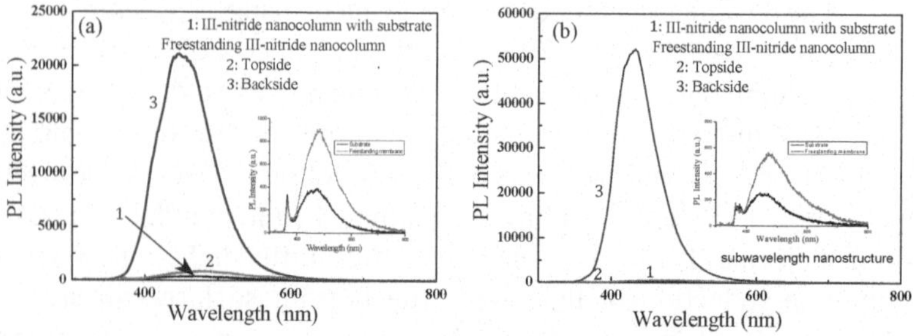

Figure 7.38 (a) The PL spectra of the III-nitride nanocolumn slab grown on the flat silicon substrate. (b) The PL spectra of the III-nitride nanocolumn slab grown on nanostructured silicon substrate.

References

1. U. Schwarz (2008). Ultraviolet laser diodes: Indium-free success, *Nature Photon*, **2**, 521.
2. A. Khan, *et al.* (2008). Ultraviolet light-emitting diodes based on group three nitrides, *Nature Photon*, **2**, 77.
3. J.-M. Bethoux, *et al.* (2003). Growth of high quality crack-free AlGaN films on GaN templates using plastic relaxation through buried cracks, *J. Appl. Phys.*, **94**, 6499.
4. F. Schulze, *et al.* (2006). Metalorganic vapor phase epitaxy grown InGaN/GaN light-emitting diodes on Si(001) substrate, *Appl. Phys. Lett.*, **88**, 121114.
5. H. W. Choi, *et al.* (2006). Lasing in GaN microdisks pivoted on Si, *Appl. Phys. Lett.*, **89**, 211101.
6. Nitronex. Application Note 11: Substrates for GaN RF devices, June 2008, http://www.nitronex.com/pdfs/AN-011%20GaN%20Substrates.pdf.
7. Y. J. Wang, *et al.* (2009). Frestanding circular GaN grating fabricated by fast atom beam etching, *Appl. Phys. A: Materials Science & Processing*, **97**, 39.
8. Y. J. Wang, *et al.* (2009). Freestanding GaN resonant gratings at telecommunication range, *IEEE Photon. Technol. Lett.*, **21**, 1184.
9. Y. J. Wang, *et al.* (2011). The resonant III-nitride grating reflector, *J. Micromech. Microeng.*, **21**, 105025.
10. Y. J. Wang, *et al.* (2011). III-nitride grating grown on freestanding HfO_2 gratings, *Nanoscale Res. Lett.*, **6**, 497.
11. Y. J. Wang, *et al.* (2010). Fabrication and characterization of freestanding circular GaN gratings, *Opt. Express*, **18**, 773.

12. Y. J. Wang, *et al.* (2010). Fabrication and characterization of subwavelength nanostructures on freestanding GaN slab, *Opt. Express*, **18**, 2940.
13. Y. J. Wang, *et al.* (2011). Patterned growth of InGaN/GaN wells on freestanding GaN grating by molecular beam epitaxy, *Nanoscale Res. Lett.*, **6**, 117.
14. Y. J. Wang, *et al.* (2011). Lateral epitaxial overgrowth of GaN on patterned GaN-on-silicon substrate by molecular beam epitaxy, *Semicond. Sci. Technol.*, **26**, 045015.
15. Y. J. Wang, *et al.* (2010). Large area, freestanding GaN nanocolumn membrane with bottom subwavelength nanostructure, *Opt. Express*, **18**, 5504.
16. Y. J. Wang, *et al.* (2010). Enhancement of light extraction from freestanding GaN nanocolumn slab with bottom subwavelength nanostructures, *Appl. Phys. B: Lasers and Optics*, **11**, 575.
17. D. Brunner, *et al.* (1997). Optical constants of epitaxial AlGaN films and their temperature dependence, *J. Appl. Phys.*, **82**, 5090.
18. S. S. Wang, *et al.* (1993). Theory and applications of guided mode resonance filters, *Appl. Opt.*, **32**, 2606.
19. X. Xu, *et al.* (2005). Epitaxial condition and polarity in GaN grown on a HfN-buffered Si(111) wafer, *Appl. Phys. Lett.*, **86**, 182104.
20. H. Sameshima, *et al.* (2009). A freestanding GaN/HfO_2 membrane grown by molecular beam epitaxy for GaN-Si hybrid MEMS, *IEEE J. Sel. Top. Quantum Electron.*, **15**, 1332.
21. Y. J. Wang, *et al.* (2011). Freestanding HfO_2 grating fabricated by fast atom beam etching, *Nanoscale Res. Lett.*, **6**, 397.
22. T. Kuykendall, *et al.* (2007). Complete composition tunability of InGaN nanowires using a combinatorial approach, *Nat. Mater.*, **6**, 951.
23. R. Houdrè, *et al.* (1996). Vacuum-field Rabi splitting in the presence of inhomogeneous broadening: Resolution of a homogeneous linewidth in an inhomogeneously broadened system, *Phys. Rev. A*, **53**, 2711.
24. L. C. Andreani, *et al.* (1998). Effect of inhomogeneous brodening on optical properties of excitons in quantum wells, *Phys. Rev. B*, **57**, 4670.
25. G. Christmann *et al.* (2006). Room temperature polariton luminescence a GaN/AlGaN quantum well microcavity, *Appl. Phys. Lett.*, **89**, 071107.
26. G. A. Turnbull, *et al.* (2005). Effect of gain localization in circular-grating distributed feedback lasers, *Appl. Phys. Lett.*, **87**, 201101.
27. Y. Chen, *et al.* (2007). Nanoimprinted circular grating distributed feedback dye laser, *Appl. Phys. Lett.*, **91**, 051109.

28. X. K. Sun, *et al.* (2008). Surface-emitting circular DFB, disk- and ring-Bragg resonator lasers with chirped gratings: a unified theory and comparative study, *Opt. Express*, **16**, 9155.

29. S. Noda, *et al.* (2009). Light-emitting diodes: photonic crystal efficiency boost, *Nature Photon.*, **3**, 129.

30. J. J. Wierer, *et al.* (2009). III-nitride photonic-crystal light-emitting diodes with high extraction efficiency, *Nature Photon.*, **3**, 163.

31. I. Schnitzer, *et al.* (1993). 30% external quantum efficiency from surface textured, thin-film light-emitting diodes, *Appl. Phys. Lett.*, **63**, 2174.

32. H. K. Cho, *et al.* (2006). Light extraction enhancement from nano-imprinted photonic crystal GaN-based blue light-emitting diodes, *Opt. Express*, **14**, 8654.

33. S. J. An, *et al.* (2008). Enhanced light output of GaN-based light-emitting diodes with ZnO nanorod arrays, *Appl. Phys. Lett.*, **92**, 121108.

34. W. N. Ng, *et al.* (2008). Photonic crystal light-emitting diodes fabricated by microsphere lithography, *Nanotechnology*, **19**, 255302.

35. C. H. Chan, *et al.* (2009). Patterning periodical Motif on substrates using monolayer of microspheres: Application in GaN light-emitting diodes, *Jpn. J. Appl. Phys.*, **48**, 020212.

36. T. X. Lee, *et al.* (2007). Light extraction analysis of GaN-based light-emitting diodes with surface texture and/or patterned substrate, *Opt. Express*, **15**, 6670.

37. Y. F. Huang, *et al.* (2007). Improved broadband and quasi-omnidirectional anti-reflection properties with biomimetic silicon nanostructures, *Nature Nanotechnology*, **2**, 770.

38. A. Kikuchi, *et al.* (2004). InGaN/GaN multiple quantum disk nanocolumn light-emitting diodes grown on (111) Si substrate, *Jpn. J. Appl. Phys.*, **43**, L1524.

39. K. Kishino, *et al.* (2009). Improved Ti-mask selective-area growth (SAG) by rf-plasma-assisted molecular beam epitaxy demonstrating extremely uniform GaN nanocolumn arrays, *J. Cryst. Growth*, **311**, 2063.

40. C. H. Chiu, *et al.* (2008). Broadband and omnidirectional antireflection employing disordered GaN nanopillars, *Opt. Express*, **16**, 8748.

41. S. L. Diedenhofen, *et al.* (2009). Broad-band and omnidirectional antireflection coatings based on semiconductor nanorods, *Adv. Mater.*, **21**, 973.

42. C. H. Chang, *et al.* (2009). Broadband and omnidirectional antireflection from conductive indium-tin-oxide nanocolumns prepared by glancing-angle deposition with nitrogen, *Appl. Phys. Lett.*, **94**, 051114.

43. H. Y. Chen, *et al.* (2008). Gallium nitride nanorod arrays as low-refractive-index transparent media in the entire visible spectral region, *Opt. Express*, **16**, 8106.

Chapter 8

Coupling of a Light Emitter with Surface Plasmon Induced on a Metal Nanostructure for Emission Enhancement

Yang Kuo, Horng-Shyang Chen, Dong-Mong Yeh, Yu-Lung Jung, Chia-Phen Chen, Yean-Woei Kiang, and Chih-Chung Yang

Institute of Photonics and Optoelectronics, National Taiwan University, 1, Roosevelt Road, Section 4, Taipei 10617, Taiwan

ccy@cc.ee.ntu.edu.tw

We first introduce the basic concept of surface plasmon (SP) coupling with a light emitter or radiating dipole for enhancing its emission efficiency and review some of the past implementations of SP-coupled GaN-based light-emitting diode (LED). Then, the SP resonance behaviors and their potential applications to InGaN/GaN quantum well (QW) coupling for enhancing LED emission efficiency of three metal nanostructures are discussed, including surface metal nanoparticles (NPs), embedded metal NPs in GaN, and metal nanodisks penetrating into GaN. With a surface metal NP distribution, we demonstrate the QW emission enhancement when the emission wavelength of the QWs beneath roughly coincides with the SP resonance wavelength. The SP resonance wavelength

Handbook of Optical Microcavities
Edited by Anthony H. W. Choi

ISBN 978-981-4463-24-9 (Hardcover), 978-981-4463-25-6 (eBook)
www.panstanford.com

can be controlled to fall into the green range by designing the NP size. Although the embedded metal NPs in GaN can become closer to the QWs for enhancing the coupling strength, their SP resonance wavelengths are always longer than the green range so that this technique is not suitable for the application to LEDs in the blue and green ranges. Finally, the experimental results of SP resonance features of metal nanodisks penetrating into GaN are illustrated. Certain features can be used for coupling with the green-emitting QWs to enhance their emission efficiency.

8.1 Introduction

Surface plasmon (SP) represents collective oscillation of excited electrons at the interface between a metal and a dielectric or semiconductor material [1]. An SP includes the electromagnetic field distribution of a skin depth in the metal and the evanescent or near-field distribution in the dielectric or semiconductor material. At an extended metal/dielectric interface, an SP can propagate along the interface and is called a surface plasmon polariton (SPP) [2]. On the other hand, it can exist as the local electron oscillation around a metal nanoparticle (NP) or nanostructure and is called a localized surface plasmon (LSP) [3]. Either SPP or LSP confines electromagnetic field around the metal structure.

The evanescent or near field distribution of an SP in the dielectric or semiconductor material surrounding the metal structure can extend a range of several tens nm from the metal/dielectric interface. Such a spatially localized energy distribution is similar to the behavior of a resonance cavity. As shown in Figs. 8.1(a) and (b) for SPP and LSP, the structures of metal/dielectric interface and metal NP for inducing SPP and LSP, respectively, can be regarded as the resonance cavities. Because of the effective radiation of an LSP or SPP (under a momentum matching condition) and significant metal dissipation, the energy loss of such a cavity is quite large. Therefore, its quality (Q) factor is usually small. However, due to the small range of strong field distribution near the metal structure, the small mode volume of the cavity can still lead to useful applications, including enhanced Raman spectroscopy, chemical sensing, biomedical imaging and therapy, optical signal guiding, and enhanced absorption and emission.

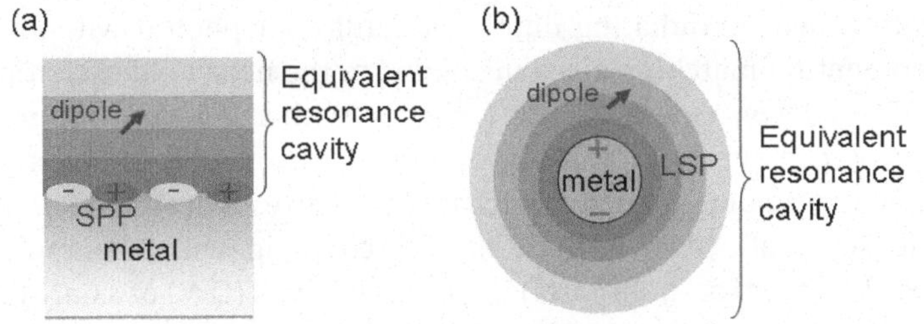

Figure 8.1 Evanescent or near field distributions of (a) a metal/dielectric interface and (b) a metal NP for inducing SPP and LSP, respectively.

An SP can interact with a nearby light emitter through its evanescent or near-field coverage for enhancing emission [4–9] (see Figs. 8.1(a) and (b)). From the viewpoint of resonance cavity, the emission enhancement through SP coupling with a radiating dipole can be understood as the Purcell effect, which says that the emission efficiency of a radiating dipole in a resonance cavity can be enhanced by the Purcell factor, F_p, given by [10].

$$F_p = \frac{3}{4\pi^2}\left(\frac{\lambda}{n}\right)^3\left(\frac{Q}{V}\right). \tag{8.1}$$

Here, λ and n are the wavelength in vacuum and the refractive index of the cavity medium, respectively. Q and V are the quality factor and mode volume of the cavity, respectively. As mentioned earlier, in a resonance cavity formed by an SP, Q is usually quite small. However, because V is also quite small, the Purcell factor in such a cavity can still be a large number.

The coupling process between an SP and a radiating dipole is analogous to the mechanism of stimulated or amplified spontaneous emission in a laser system. The interaction or coupling between an SP and a radiating dipole can be regarded as a process of amplifying the evanescent or near field of the SP by a gain medium, which includes the radiating dipole. In other words, the radiating dipole transfers energy into the SP for emission [11]. An SPP can effectively radiate when the momenta of the SPP and photon are matched that can usually be implemented through a rough metal interface or a metal-grating structure. Actually, unlike the SPP excited by a plane wave, the coupled SPP on a flat metal/dielectric interface with a nearby radiating dipole can still emit photon without any momentum matching mechanism due to the near-field excitation of SPP. However, a metal grating structure with an appropriate grating period can significantly enhance SPP emission. LSP can effectively radiate without any extra mechanism. Figures 8.2(a) and (b) show the schematic drawings of the SP coupling with the radiating dipoles (electron-hole pairs) in a quantum well (QW) of a GaN-based light-emitting diode (LED). In Fig. 8.2(a), a metal grating is formed on the top surface of an LED to induce SPP and/or LSP for coupling with the dipoles in the QW. For effective coupling, the thickness of the p-GaN layer must be thinner than several tens nm. In Figure 8.2(b), metal NPs are formed on the top surface to induce LSP for coupling. In these two figures, the curves represent the evanescent or near field distributions of the SPs.

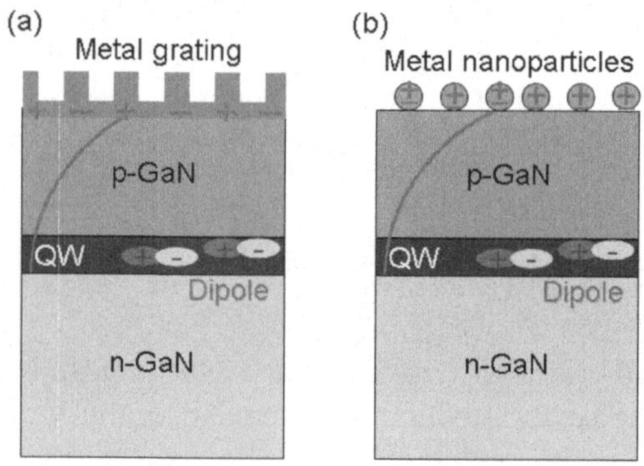

Figure 8.2 Schematic illustrations of the coupling mechanisms of SPs, induced on (a) a surface metal grating and (b) a surface metal NP array, with the radiating dipoles (electron-hole pairs) in a QW of a GaN-based LED.

Therefore, SP coupling with a light emitter can effectively create an alternative emission channel, i.e., SP radiation. As depicted in the floating chart shown in Fig. 8.3, the carriers in a QW can produce radiative recombination for light emission and non-radiative recombination to generate heat (energy loss). Their weightings depend on the internal quantum efficiency (IQE) of the QW. Through SP coupling, the carrier energy is transferred into SP for emission. Although metal dissipation can cause SP loss, as long as this energy loss is smaller than that due to the non-radiative recombination process in the active region, an emission enhancement of the device can be expected. In other words, SP coupling is particularly useful for increasing emission efficiency of a light-emitting device of inherently low IQE. For the application to an InGaN/GaN QW LED, a blue-emitting LED may not need such a technique for emission efficiency improvement because its IQE has been quite high (can be >80%). However, this technique is useful for a green-emitting LED because its IQE is still quite low (<30%) due to the difficulty of growing high-indium InGaN/GaN QW of high crystal quality.

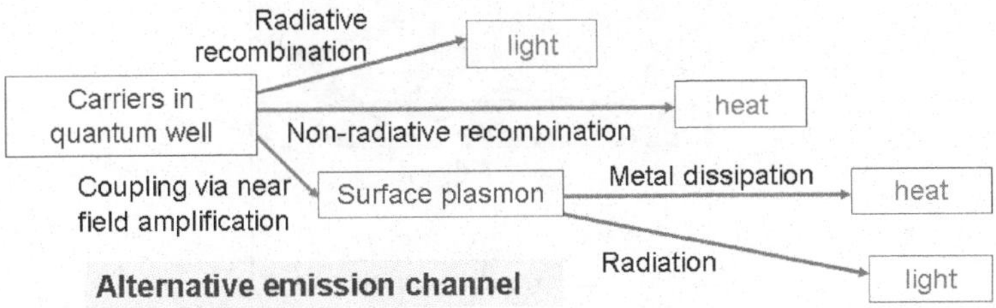

Figure 8.3 Floating chart showing the energy flow in a system with SP-QW coupling. Light can be emitted directly from the QW through radiative recombination of carriers or from SP radiation through SP-QW coupling.

SP coupling can lead to several advantages when it is applied to LED development. First, the IQE can be effectively increased by the generation of an alternative emission channel (SP emission). Second, by designing the metal structure properly, SP can radiate photons directly into air so that the light extraction efficiency of the device can also be enhanced. Third, due to the effective transfer of the energy of part of carriers into the SP mode, the carrier density left in the QW for radiative or non-radiative recombination is reduced so that the efficiency droop effect in an LED can be weakened [12]. Fourth, because SP has certain polarization nature, the SP radiation is polarized, which is different from the un-polarized emission of carrier recombination in the QW. An LED of polarized output is useful for the liquid-crystal-display backlighting application. Fifth, the energy loss for producing heat in the QW due to non-radiative recombination is reduced and the new energy loss is generated in the surface metal structure due to metal dissipation of SP. Because the heat in the surface metal structure can be more easily removed, device cooling becomes more effective. Finally, since an SP mode can coherently couple with all the nearby dipoles, as schematically depicted in Fig. 8.4 by considering an LSP case with a metal NP on the device surface, the coupling process can produce localized coherent emission from an LED. Localized coherent emission can result in stronger LED output intensity, when compared with the case of purely incoherent emission.

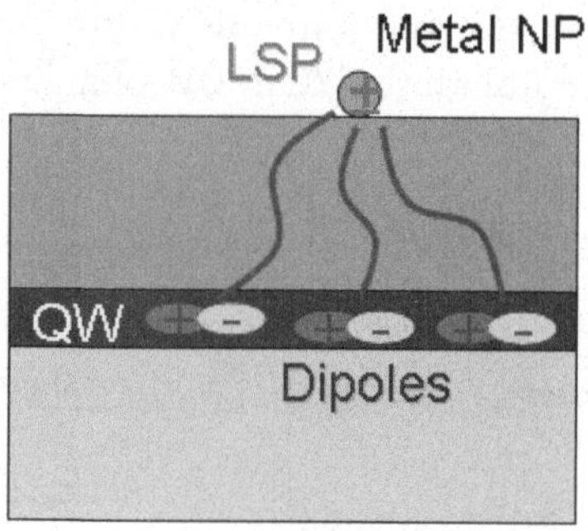

Figure 8.4 Schematic demonstration of the coherent coupling of an LSP mode with more than one radiating dipole.

Several SP-coupled LEDs have been implemented. Three examples are briefly reviewed here. First, it has been demonstrated that the emission intensity of a green single-QW LED at 535 nm of ~7% in IQE was enhanced by up to 120%, when compared to a control sample of the conventional configuration, through SP-QW coupling [13]. In this device, Ag nanoislands of ~100 nm in size were formed on the top of the p-GaN layer by thermally annealing a deposited thin Ag layer. Second, the reduction of the efficiency droop effect through SP coupling has been demonstrated in a green LED of ~515 nm in wavelength [14]. In this implementation, a grid current spreading structure was formed on the top of the LED, which included a 20-nm p-AlGaN layer and a 60-nm p-GaN layer. The current spreading regions consisted of a layer of Ni (5 nm)/Au (5 nm). The square regions between the current spreading grids were deposited with Ag of 10 nm in thickness for inducing SP-QW coupling. Third, partially polarized outputs have been obtained from LEDs with 1D Ag gratings on their tops [15,16]. In such an implementation, periodical grooves of several hundred nm in period, 15 nm in groove depth, and 50% in duty cycle were first formed on the 55-nm p-GaN layer of a 515-nm single-QW LED structure with the techniques of electron-beam lithography and dry etching [15]. Then, the grooves were covered by an Ag layer of 45 nm in thickness. Compared with a reference sample of Ag coverage of the similar thickness on the flat p-GaN layer, the emission efficiency of the metal-grating LED of 500 nm in grating period was enhanced by 87%. Also, its output was partially polarized with the polarization ratio (the intensity of the stronger polarization component over that of the weaker one) at ~1.7.

In this chapter, after we introduced the basic concept of SP coupling with a radiating dipole and reviewed some of the past SP-coupled LED implementations in the current section, we describe the LSP resonance behaviors of surface metal NPs and demonstrate their coupling with the InGaN/GaN QWs beneath for enhancing the emission efficiency in Section 8.2. Then, in Section 8.3, we theoretically study the LSP resonance characteristics of embedded metal NPs in GaN to show that they cannot induce LSP resonance in the blue and green ranges. Next, in Section 8.4, the experimental illustrations of the SP resonance features of metal nanodisks penetrating into GaN are given. Certain LSP resonance features can be used for coupling with the green-emitting QWs to enhance their emission efficiency. A summary is provided in Section 8.5.

8.2 LSP Resonance of Surface Metal Nanoparticles

A simple implementation of SP coupling with the QWs of an LED is to fabricate metal nanostructures on the top surface of the LED. Before LED fabrication, it is important to first understand the QW emission behavior with photoluminescence (PL) measurement when the SP coupling mechanism is applied to the LED structure. To demonstrate such an SP coupling behavior, we fabricate Ag NPs on the top surface of an InGaN/GaN QW structure. Figure 8.5(a) shows the plan-view scanning electron microscopy (SEM) image of the sample showing the regularly arranged Ag NPs on the QW epitaxial structure. The QW epitaxial structure consists of a 2-μm un-doped GaN grown on c-plane sapphire substrate and a three-period yellow-emitting InGaN/GaN QW structure on the top. The well (barrier) layer thickness of the QW is 3 (10 nm). The top barrier layer (the capping layer) has a thickness of ~20 nm. The triangularly patterned Ag NPs are fabricated with nano-imprint lithography. The hole diameter and the center-to-center distance between two nearest neighboring holes (the pitch) on the nano-imprint stamp are 100 and 200 nm, respectively. However, the formed Ag NPs have an average dimension of only ~70 nm. The Ag NPs shown in Fig. 8.5(a) is obtained after a thermal annealing process. Figure 8.5(b) shows the transmittance spectra of the sample. The fast Fabry–Pérot oscillation in the curve of the original data is filtered to give the smooth curve, which shows a minimum around 540 nm. This minimum wavelength corresponds to the LSP resonance of the patterned Ag NPs. This wavelength is shorter than the spectral peak of QW emission around 550 nm.

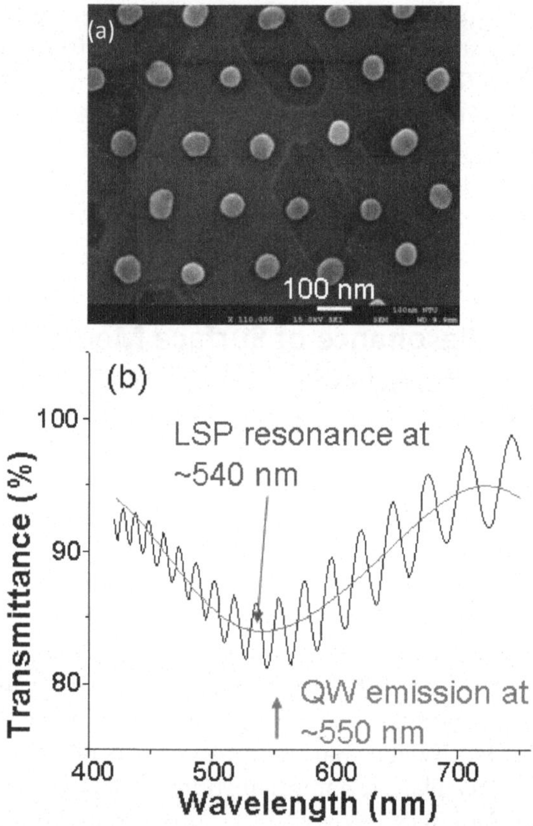

Figure 8.5 (a) Plan-view SEM image of the sample with regularly arranged Ag NPs on a QW epitaxial structure. (b) Transmittance spectra of the sample with the Fabry–Pérot oscillations filtered to give the smooth curve.

Temperature-dependent PL measurement is performed for understanding the QW emission behavior with the Ag NPs on the sample top surface. In the PL measurement, 325-nm HeCd laser is incident from the top surface of the sample for excitation. Because the Ag NPs on the surface may partly block the incident laser, we consider two excitation power levels at 6 and 7 mW, which differ by the surface coverage percentage of the Ag NPs around 17%. Figure 8.6(a) shows the PL spectra of the sample with Ag NPs at 6 (NP1/6 mW) and 7 (NP1/7 mW) mW in excitation power at room temperature. For comparison, the PL spectrum with 6 mW in excitation power of a control sample, which has the same QW epitaxial structure but no Ag NP, at room temperature is also shown in Fig. 8.6(a). Here, one can see that with Ag NPs on the top surface, the PL emission is enhanced. A higher excitation power leads to a larger enhancement. Figure 8.6(b) shows the spectral peak energy variations with temperature of the three cases in Fig. 8.6(a). The spectral peaks are obtained by fitting the spectra with Gaussian curves. In Fig. 8.6(b), one can see that with LSP coupling, the emission spectrum is blue shifted. This is so because the LSP resonance wavelength is shorter than the QW emission spectral peak. With LSP coupling, the short-wavelength side of the emission spectrum is more enhanced so that effectively the emission spectrum is blue shifted. At room temperature, the PL spectral peaks of the cases of control, NP1/6 mW, and NP1/7 mW are 556.1, 554.4, and 553.7 nm, respectively. The S-shaped temperature-dependent variations of the two curves with Ag NPs are similar. However, their variation amplitudes are significantly smaller than that of the control sample. The less significant S-shaped variation in the results of LSP coupling is due to the stronger LSP coupling with the free carriers, when compared with the localized carriers, in the QWs [17]. Because of the formation of indium-rich nano-clusters in an InGaN/GaN QW, the localized carrier states are generated besides the free carrier states [18,19]. Figure 8.6(c) shows the normalized integrated PL intensity variations with temperature of the three cases. Their levels at 300 K can be regarded as the IQEs of QW emissions. The IQEs in the cases of control, NP1/6 mW, and NP1/7 mW are 0.5, 0.97, and 1.48%, respectively. With LSP coupling, the IQE is significantly increased. Also, with a higher PL excitation power, the IQE is further enhanced.

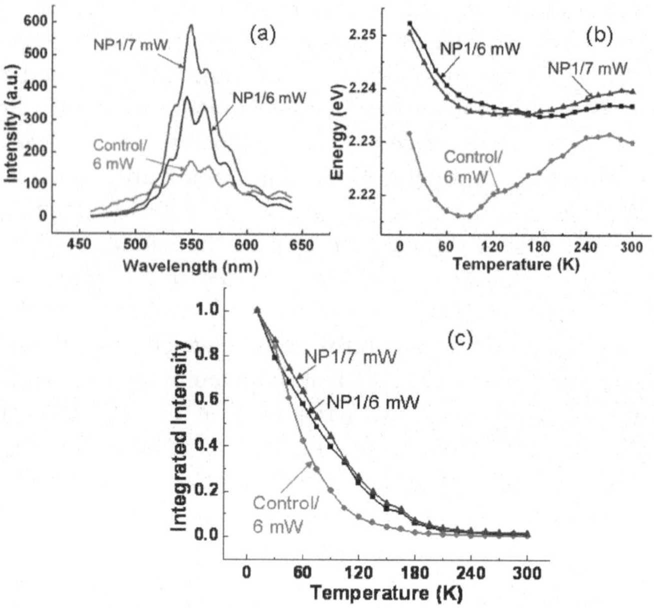

Figure 8.6 (a) PL spectra of the sample with Ag NPs shown in Fig. 8.5 at 1.6 (NP1/6 mW) and 7 (NP1/7 mW) mW in excitation power at room temperature. For comparison, the PL spectrum with 6 mW in excitation power of the control sample at room temperature is also shown. (b) Spectral peak energy variations with temperature of the three cases in part (a). (c) Normalized integrated PL intensity variations with temperature of the three cases.

It is noted that in practical application, the surface Ag NPs need to be covered by a certain dielectric material for protecting the NPs from oxidation and other damages. Also, in the packaging process, usually an epoxy layer is applied to the top surface of the device. However, when the refractive index of the surrounding medium of a metal NP is increased, the LSP resonance wavelength is expected to be red shifted. The results in Fig. 8.7 demonstrate such a shift trend. In Fig. 8.7(a), the plan-view SEM image of another sample with regularly arranged Ag NPs of slightly smaller sizes (average size at about 60 nm) is shown. The Ag NPs are covered with a 100-nm SiO_2 layer. The plan-view SEM image of the sample after SiO_2 coverage is shown in Fig. 8.7(b). The transmittance spectra of the sample before and after SiO_2 coverage are shown in Fig. 8.7(c). Here, again, after filtering the Fabry–Pérot oscillations, the major minima of the two smooth curves show the spectral shift of the LSP resonance. With SiO_2 coverage, the surrounding refractive index is increased from unity to ~1.5 such that the LSP resonance wavelength red shifts from 527 to 548 nm. For practical application, this red shift needs to be considered in designing the NP geometry for matching the QW emission wavelength with the LSP resonance wavelength after the NPs are covered by the designated dielectric material in an LED fabrication.

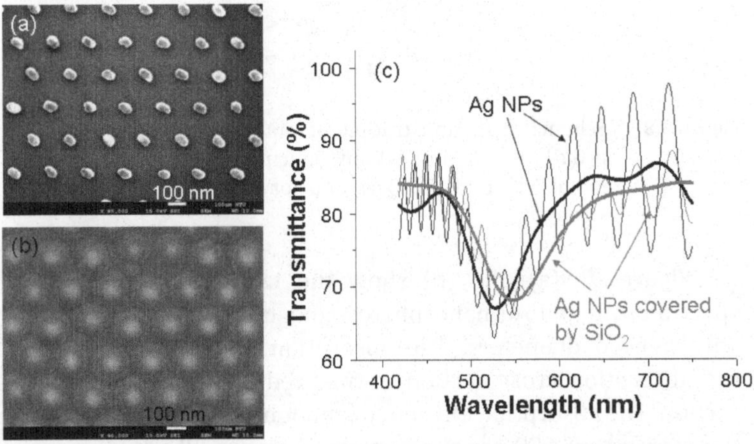

Figure 8.7 (a) Plan-view SEM image of another sample with regularly arranged Ag NPs of slightly smaller sizes. (b) Plan-view SEM image of the sample after the Ag NPs are covered by a 100-nm SiO_2 layer. (c) Transmittance spectra of the samples before and after SiO_2 coverage with the Fabry–Pérot oscillations filtered to show the spectral shift of the LSP resonance.

8.3 LSP Resonance of Embedded Metal Nanoparticles

The SP coupling strength depends on the distance between the light emitter and the metal structure for generating SP resonance. To enhance the coupling strength, it has been proposed to embed metal NPs in the p-GaN or n-GaN layer of an LED structure such that the metal NPs can be close to the QWs of the LED, as schematically shown in Fig. 8.8 [20,21]. Although the enhancements of LED emission efficiency have been observed based on this technique, the enhancements are not necessarily caused by SP coupling. Non-resonant NP scattering can also enhance LED emission efficiency. The key point is the LSP resonance wavelength of a metal NP within GaN, which has a refractive index between 2.4 and 2.5.

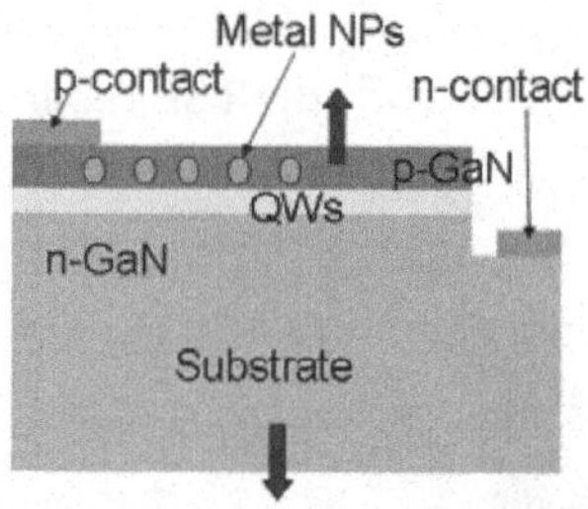

Figure 8.8 Schematic demonstration of embedding metal NPs in the p-GaN layer in an LED for reducing the distance between the metal NPs and the QWs and for enhancing the SP coupling strength.

Figures 8.9(a) and (b) show the LSP absorption efficiency spectra of Ag and Au nanospheres embedded in GaN, respectively, with labeled diameters. The absorption efficiency is defined as the absorption cross section normalized by the geometric cross section. The absorption efficiency cannot be used for understanding the strength of LSP coupling with a radiating dipole. However, it shows the LSP resonance wavelength and the relative resonance strength of the fundamental and higher-order LSP modes. In Figs. 8.9(a) and 8.9(b), the long- (short-) wavelength peak corresponds to the resonance of the fundamental (higher-order) LSP mode if two peaks are observed in a curve. When the NP diameter is small, the higher-order mode can be too weak to be observed. The results in Figs. 8.9(a) and (b) are obtained by using the Mie theory [22] and by assuming 2.4 as the refractive index of GaN. The experimental data of the dielectric constant of Ag are used for numerical calculation [23]. It is noted that the broad humps between 400 and 500 nm in Fig. 8.9(b) are due to the band-to-band absorption of Au and are not caused by LSP resonance. Figure 8.10 shows the spectral peak wavelengths of the fundamental (-F) and higher-order (-HO) LSP modes for Ag and Au NPs as functions of NP diameter, indicating their spectral locations of LSP resonances. Here, one can see that with Au NPs embedded in GaN, the resonance wavelengths of the fundamental and higher-order LSP modes are always longer than 650 nm. Meanwhile, with Ag NPs embedded in GaN, the resonance wavelength of the fundamental LSP mode is always longer than 560 nm. Although the resonance wavelength of the higher-order LSP mode can be extended into the green range, the resonance is usually quite weak, as shown in Fig. 8.11. Here, the absorption efficiency ratios of the higher-order mode over the fundamental mode as functions of NP diameter for Ag and Au NPs are illustrated. One can see that when the NP diameter is smaller than 60 nm, the resonance strength of the higher-order mode is significantly weaker than that of the fundamental mode. Actually, from Fig. 8.9(a) we can see that only when the diameter of Ag NP is around 50 nm, LSP resonance (the higher-order mode) can be observed in the green range. The parameter window for implementing such LSP coupling is quite critical. It is noted that when the shape of the metal NP is different from a sphere, the LSP resonance wavelength becomes longer, further away from the green range.

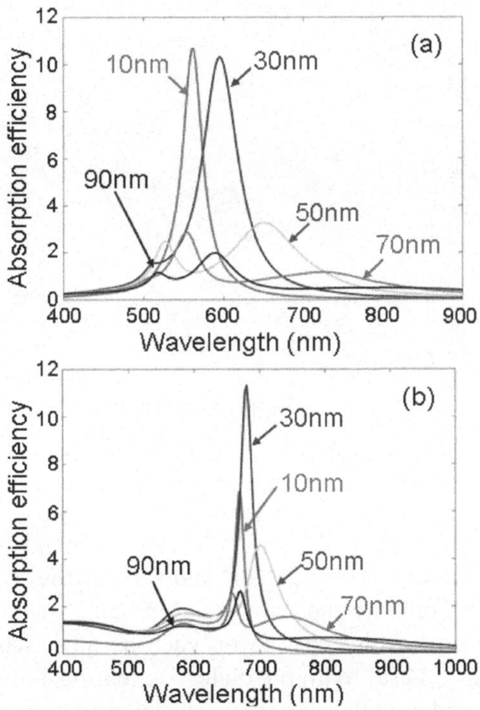

Figure 8.9 LSP absorption efficiency spectra of (a) Ag and (b) Au nanospheres embedded in GaN with labeled diameters.

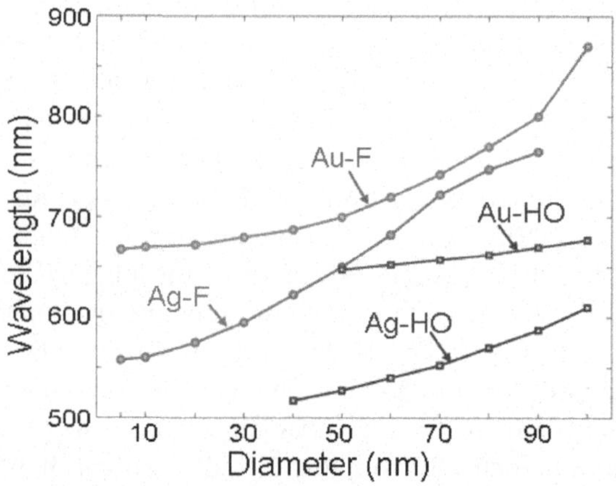

Figure 8.10 Spectral peak wavelengths of the fundamental (-F) and higher-order (-HO) LSP modes for Ag and Au NPs as functions of NP diameter, indicating their spectral locations of LSP resonances.

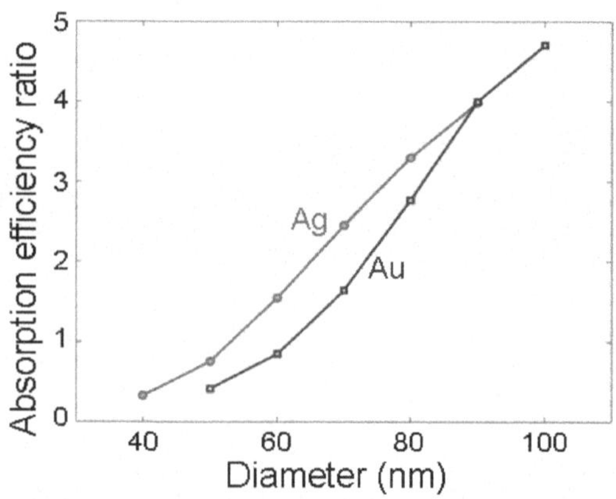

Figure 8.11 Absorption efficiency ratios of the higher-order mode over the fundamental mode as functions of NP diameter for Ag and Au NPs.

One possible approach for shortening the LSP resonance wavelength to fall into the green or even blue range is to cover the metal NPs with a dielectric layer, such as SiO_2, schematically as shown in Fig. 8.12(a). With the dielectric coating, both fundamental and higher-order LSP resonance wavelengths can be blue-shifted into the green and blue ranges. Figure 8.12(b) shows the resonance wavelengths of the fundamental and higher-order LSP modes of Ag

NPs embedded in GaN when they are coated with SiO_2 layers of 5 and 10 nm in thickness. Here, one can see that the resonance wavelength of the fundamental LSP mode can reach 425 nm. The embedment of such coated metal NPs in GaN can be a good approach for generating strong LSP coupling with the QWs in an LED. However, the synthesis of such a coated metal NP can be a challenging task. Also, the LSP resonance strength can be reduced by coating a dielectric layer onto a metal NP. Further study is needed to explore the optimized condition in maximizing the LSP coupling strength.

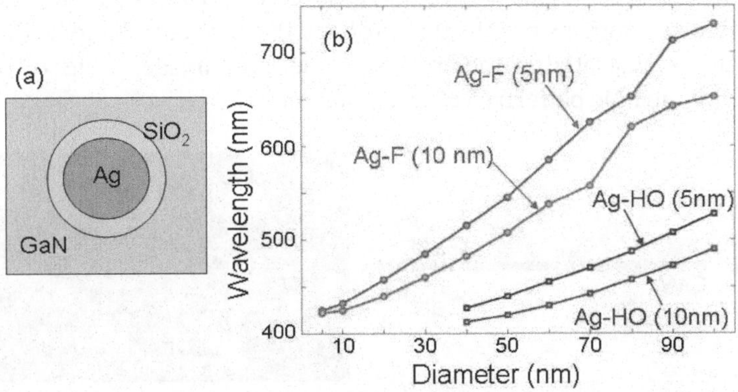

Figure 8.12 (a) Schematic drawing of a SiO_2-coated Ag NP embedded in GaN. (b) LSP resonance wavelengths of the fundamental and higher-order LSP modes of Ag NPs embedded in GaN when they are coated with SiO_2 layers of 5 and 10 nm in thickness.

8.4 SP Resonance of Metal Nanodisks Penetrating into Semiconductor

The other method for implementing effective SP coupling is to form metal nanoposts or nanodisks penetrating into the p-GaN layer for approaching the QWs in an LED structure. Figure 8.13(a) shows a schematic drawing of such a device structure. In such a structure, the bottoms of the metal (Ag is shown) nanoposts, or nanodisks, can be quite close to the QW such that the SP feature induced around the bottom can effectively interact with the QW. The fabrication of such a metal nanostructure starts with the nano-imprint patterning of holes on an LED. For demonstrating only the SP resonance features in this section, we use a semiconductor sample of a thick

GaN layer on c-plane sapphire substrate. Four triangular hole patterns are fabricated, including the patterns of 180 nm/400 nm, 180 nm/500 nm, 250 nm/500 nm, and 250 nm/700 nm. Here, the first and the second number represent the hole diameter and the pitch or the center-to-center distance between two nearest neighboring holes, respectively, of a hole pattern. Figures 8.13(b) and 8.13(c) show the atomic force microscopy (AFM) scanning result and plan-view SEM image of the hole pattern of 250 nm/700 nm when the hole depth in GaN is around 35 nm. After the hole pattern is fabricated, an Ag layer of ~150 nm is deposited onto the top surface. In this process, the holes are filled with Ag, as demonstrated in the cross-sectional SEM image in Fig. 8.13(d) with the hole pattern of 250 nm/500 nm.

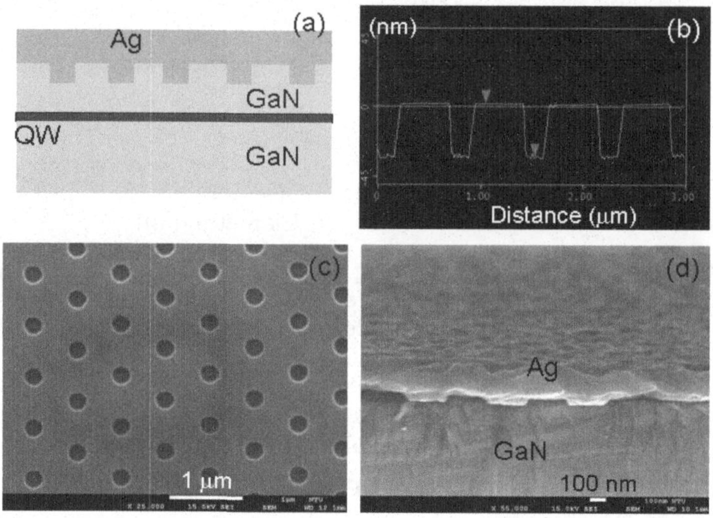

Figure 8.13 (a) Schematic demonstration of metal nanodisks penetrating into the p-GaN layer for approaching the QWs in an LED structure. (b) AFM scanning profile of a hole pattern prepared for fabricating the metal nanodisks. (c) Plan-view SEM image of the hole pattern. (d) Cross-sectional SEM image of the metal nanodisks penetrating into the GaN layer.

The un-polarized reflection spectra of the four hole patterns with the hole depth at 35 nm are shown in Fig. 8.14(a). The reflection measurement is implemented by using a fiber bundle of seven guiding fibers with the central one for light incidence and the surrounding six for collecting the reflecting photons. A Xenon lamp is used as the un-polarized light source. In Fig. 8.14(a), each major dip may correspond to an SP resonance feature. However, only two major dips in each curve can be clearly identified for their SP resonance characteristics, as indicated by the vertical thick arrows. It is noted that among the four hole patterns, the two with the same hole diameter have the two short-wavelength dips at similar wavelengths. Also, the spectral location of the long-wavelength dip strongly depends on the pitch of the hole pattern. The patterns of 180 nm/400 nm and 180 nm/500 nm have the short-wavelength dips between 470 and 480 nm. The patterns of 250 nm/500 nm and 250 nm/700nm have the short-wavelength dips between 530 and 540 nm. The spectral location of the short-wavelength dip is also slightly affected by the pattern pitch. A smaller pitch leads to a slightly longer dip wavelength. For the long-wavelength dips, a larger pitch results in a longer wavelength. The cause for the double-dip feature of the pattern of 180 nm/500 nm between 700 and 750 nm is still unclear. The left dip of this double-dip feature is very close to the dip of the 250 nm/500 nm pattern because the two patterns have the same pitch. Generally speaking, in the whole spectrum, the short-wavelength dip is related to the LSP resonance around the bottom of the nanodisk. The long-wavelength dip is caused by a larger-scale LSP resonance with its field distributed between two neighboring nanodisks or a hybrid SP resonance mixing LSP and SPP characteristics. More detailed studies are needed for clarifying this point.

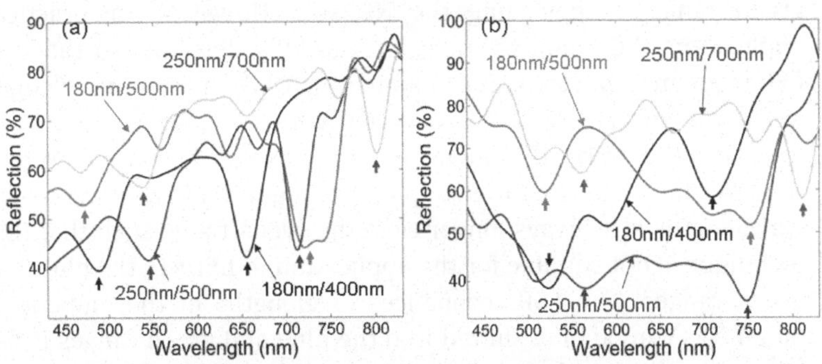

Figure 8.14 (a) Un-polarized reflection spectra of the four hole patterns with the hole depth at 35 nm. (b) Un-polarized reflection spectra of the four hole patterns with the hole depth at 75 nm.

Figure 8.14(b) shows the similar results to those shown in Fig. 8.14(a) except that the hole depth is increased to 75 nm. In this situation, all the concerned dip features are red shifted. For instance, the short-wavelength dips of the patterns of 180 nm/400 nm and 180 nm/500 nm are shifted into the green range (both around 520 nm), which is suitable for green LED application. Also, the short-wavelength dips of the patterns of 250 nm/500 nm and 250 nm/700 nm are shifted to around 565 nm. The red-shifts the SP resonance features can be attributed to the increase of the height of the metal nanodisk. The dimension increase of a metal nanostructure can usually red-shift its SP resonance features. However, the detailed attributions still require further investigation.

8.5 Discussions and Summary

We have first introduced the basic concept of SP coupling with a radiating dipole for enhancing its emission efficiency. After reviewing some of the past SP-coupled LED implementations, the SP resonance behaviors and their potential applications to QW coupling for enhancing LED emission efficiency of three metal nanostructures were discussed, including surface metal NPs, embedded metal NPs in GaN, and metal nanodisks penetrating into GaN. With a surface metal NP arrangement, we demonstrated the QW emission enhancement when the emission wavelength of the QWs beneath roughly coincided with LSP resonance wavelength. The LSP resonance wavelength can be controlled to fall into the green range by designing the NP size. However, for practical application, the metal NP must be carefully designed so that the LSP resonance wavelength can match the QW emission wavelength after the surface NPs are covered with a dielectric material. Although the embedded metal NPs in GaN can become closer to the QWs for enhancing the coupling strength, their LSP resonance wavelengths are always longer than the green range such that this technique is not suitable for the application to LEDs in the blue and green ranges. The LSP resonance wavelengths of the embedded metal NPs can be blue shifted into the blue and green ranges if the NPs are coated with a dielectric layer (such as SiO_2) before they are embedded in GaN. Finally, the experimental results of the SP resonance features of metal nanodisks penetrating into GaN were illustrated. Certain features can be used for coupling with the green-emitting QWs to enhance their emission efficiency. Such a metal nanostructure can be used for producing SP-QW coupling in a vertical LED configuration.

References

1. Maier, S. A. (2007). *Plasmonics: Fundamentals and Applications*, 1st ed (Springer, New York, USA).
2. Barnes, W. L., Dereux, A., and Ebbesen, T. W. (2003). Surface plasmon subwavelength optics, *Nature*, **424**, pp. 824–830.
3. Kelly, K. L., Coronado, E., Zhao, L. L., and Schatz, G. C. (2003). The optical properties of metal nanoparticles: The influence of size, shape, and dielectric environment, *J. Phys. Chem. B*, **107**, pp. 668–677.
4. Neogi, A., Lee, C. W., Everitt, H. O., Kuroda, T., Tackeuchi, A., and Yablonvitch, E. (2002). Enhancement of spontaneous recombination rate in a quantum well by resonant surface plasmon coupling, *Phys. Rev. B*, **66**, p. 153305 (1–4).
5. Okamoto, K., Niki, I., Shvartser, A., Narukawa, Y., Mukai, T., and Scherer, A. (2004). Surface-plasmon-enhanced light emitters based on InGaN quantum wells, *Nat. Mater.*, **3**, pp. 601–605.
6. Sun, G., Khurgin, J. B., and Soref, R. A. (2007). Practicable enhancement of spontaneous emission using surface plasmons, *Appl. Phys. Lett.*, **90**, p. 111107 (1–3).
7. Khurgin, J. B., Sun, G., and Soref, R. A. (2007). Enhancement of luminescence efficiency using surface plasmon polaritons: Figures of merit, *J. Opt. Soc. Am. B*, **24**, pp. 1968–1980.
8. Kwon, M. K., Kim, J. Y., Kim, B. H., Park, I. K., Cho, C. Y., Byeon, C. C., and Park, S. J. (2008). Surface-plasmon-enhanced light-emitting diodes, *Adv. Mater.*, **20**, pp. 1253–1257.
9. Sun, G., Khurgin, J. B., and Yang, C. C. (2009). Impact of high-order surface plasmon modes of metal nanoparticles on enhancement of optical emission, *Appl. Phys. Lett.*, **95**, p. 171103 (1–3).
10. Purcell, E. M. (1946). Spontaneous emission probabilities at radio frequencies, *Phys. Rev.*, **69**, pp. 681–681.
11. Kuo, Y., Ting, S. Y., Liao, C. H., Huang, J. J., Chen, C. Y., Hsieh, C., Lu, Y. C., Chen, C. Y., Shen, K. C., Lu, C. F., Yeh, D. M., Wang, J. Y., Chuang, W. H., Kiang, Y. W., and Yang, C. C. (2011). Surface plasmon coupling with radiating dipole for enhancing the emission efficiency of a light-emitting diode, *Opt. Express*, **19**, A914–A929.

12. Schubert, M. F., Xu, J., Kim, J. K., Schubert, E. F., Kim, M. H., Yoon, S., Lee, S. M., Sone, C., Sakong, T., and Park, Y. (2008). Polarization-matched GaInN/AlGaInN multi-quantum-well light-emitting diodes with reduced efficiency droop, *Appl. Phys. Lett.*, **93**, p. 041102 (1–3).

13. Yeh, D. M., Huang, C. F., Chen, C. Y., Lu, Y. C., and Yang, C. C. (2008). Localized surface plasmon-induced emission enhancement of a green light-emitting diode, *Nanotechnology*, **19**, p. 345201 (1–4).

14. Lu, C. F., Liao, C. H., Chen, C. Y., Hsieh, C., Kiang, Y. W., and Yang, C. C. (2010). Reduction of the efficiency droop effect of a light-emitting diode through surface plasmon coupling, *Appl. Phys. Lett.*, **96**, p. 261104 (1–3).

15. Shen, K. C., Chen, C. Y., Chen, H. L., Huang, C. F., Kiang, Y. W., Yang, C. C., and Yang, Y. J. (2008). Enhanced and partially polarized output of a light-emitting diode with its InGaN/GaN quantum well coupled with surface plasmons on a metal grating, *Appl. Phys. Lett.*, **93**, p. 231111 (1–3).

16. Shen, K. C., Liao, C. H., Yu, Z. Y., Wang, J. Y., Lin, C. H., Kiang, Y. W., and Yang, C. C. (2010). Effects of the intermediate SiO_2 layer on polarized output of a light-emitting diode with surface plasmon coupling, *J. Appl. Phys.*, **108**, p. 113101 (1–8).

17. Huang, C. W., Tseng, H. Y., Chen, C. Y., Liao, C. H., Hsieh, C., Chen, K. Y., Lin, H. Y., Chen, H. S., Jung, Y. L., Kiang, Y. W. and Yang, C. C. (2011). Fabrication of surface metal nanoparticles and their induced surface plasmon coupling with subsurface InGaN/GaN wells, *Nanotechnology*, **22**, p. 475201 (1–8).

18. Lin, Y. S., Ma, K. J., Hsu, C., Feng, S. W., Cheng, Y. C., Liao, C. C., Yang, C. C., Chuo, C. C., Lee, C. M., and Chyi, J. I. (2000). Dependence of composition fluctuation on indium content in InGaN/GaN multiple quantum wells, *Appl. Phys. Lett.*, **77**, pp. 2988–2990.

19. Wang, H. C., Lu, Y. C., Chen, C. Y., and Yang, C. C. (2007). Ultrafast pump-probe spectroscopy in the UV-blue range with an extremely broad probe spectrum for the carrier relaxation study in an InGaN thin film with indium-rich nano-clusters, *Opt. Express*, **15**, pp. 3417–3425.

20. Cho, C. Y., Lee, S. J., Song, J. H., Hong, S. H., Lee, S. M., Cho, Y. H., and Park, S. J. (2011). Enhanced optical output power of green light-emitting diodes by surface plasmon of gold nanoparticles, *Appl. Phys. Lett.*, **98**, p. 051106 (1–3).

21. Cho, C. Y., Kim, K. S., Lee, S. J., Kwon, M. K., Ko, H., Kim, S. T., Jung, G. Y., and Park, S. J. (2011). Surface plasmon-enhanced light-emitting diodes

with silver nanoparticles and SiO_2 nano-disks embedded in p-GaN, *Appl. Phys. Lett.*, **99**, p. 041107 (1–3).

22. Ishimaru, A. (1991) *Electromagnetic Wave Propagation, Radiation, and Scattering*, 1st ed (Prentice-Hall, Englewood Cliffs, USA).
23. Palik, E. D. (1985). *Handbook of Optical Constants of Solids,* 1st ed (Academic Press, Boston, USA).

Chapter 9

Nanosphere-Patterned Gallium Nitride Optical Microcavities

Kwai Hei Li and Anthony H. W. Choi

Department of Electrical and Electronic Engineering, The University of Hong Kong, Pokfulam, Hong Kong

khei@hku.hk, hwchoi@hku.hk

9.1 Introduction

The blooming optoelectronics market is driving ever-increasing demands for blue-ultraviolet lasers for a wide range of domestic/commercial/scientific applications, such as high-density optical storage, high-speed printing, high-resolution laser scanning microscopy, optical spectroscopy, and projection displays among others. Although diode-pumped solid-state (DPSS) lasers are now widely available even for short wavelength lasing, the laser configuration involves crystal-based, non-linear frequency up-conversation, a process that is complex and energy inefficient. Remarkable advancements in group III nitride-based semiconductors [1], such as the $In_xGa_{1-x}N$ and $Al_xGa_{1-x}N$ alloys, have made possible direct-bandgap emission throughout the ultraviolet to visible spectral bands, hence the emergence of an entirely new range of lasers. Undoubtedly, the long-term prospect is the elimination

Handbook of Optical Microcavities
Edited by Anthony H. W. Choi

ISBN 978-981-4463-24-9 (Hardcover), 978-981-4463-25-6 (eBook)
www.panstanford.com

of the need for frequency up-conversion by adopting nitride semiconductors for lasing applications [2,3], which offers many advantages including energy efficiency, low heat generation, simple fabrication, and assembly processes as well as miniature footprints. In spite of all its attractiveness and promises, there still are numerous challenges to be overcome before nitride laser diodes become as popular as their counterparts based on other III–V materials at longer wavelengths, owing to relatively steep pricing and shorter lifespans. This, of course, will change as nitride technology matures over time. Also, nitride laser diodes are predominately of the edge-emitting type at the moment. Surface-emitting laser, on the contrary, offers numerous advantages over edge-emitting lasers but are presently restricted to proof-of-concept small-scale demonstrations in laboratories. Development of processes feasible and suitable for mass production of surface-emitting optical cavities for the nitride material system is definitely a task of high priority.

Fabry–Pérot (F-P) [4–6], whispering-gallery (WG) [7–9], and 2D photonic crystal (PhC) [10–12] defect cavities are the common light-confinement mechanisms adopted for semiconductor lasers [13]. The optical path lengths of such cavities, dependent on the vertical heights of F-P cavities, or lateral dimensions of WG circular cavities and photonic bandgap (PBG) structures, are crucial factors determining the lasing characteristics. To confine propagating photons in controllable and predictable manner, these optical cavities are usually designed to be of the order of wavelength or even at sub-wavelength scales. Unlike conventional AlInGaP red laser diodes that can be processed by standard microlithographic techniques, the blue/UV cavities require further dimensional shrinkage in order to fulfill constraints associated with short-wavelength light confinement, so that traditional optical patterning techniques are no longer able to offer the required resolutions due to the diffraction limit. Although direct-write techniques such as electron-beam lithography (EBL) [14,15] and focused ion beam (FIB) [16,17] milling are capable of producing arbitrary 2D feature accurately down to the nanometer scale, they also each have their own drawbacks. High equipment cost, time-consuming point-by-point processing, and thus low throughput make large-volume manufacturing impractical.

To overcome such limitations, an effective, high-throughput, and cost-effective approach is introduced in this chapter that is particularly suitable for the processing of hard nitride semiconductors: nanosphere lithography (NSL) [18]. Uniform spherical objects have the capability of self-assembling into ordered hexagonal arrays over large areas by proper coating skills and techniques; the coated monolayer of nanospheres serves perfectly as lithographic masks for pattern transfer onto the substrate surface, typically forming closed-packed arrays of pillars or rods. In conjunction with the large range of nanosphere dimensions available, NSL promises an effective solution for generating an extensive range of optical cavities. The resultant sub-wavelength pillar arrays may function as PhCs with the capability of manipulating and trapping the light at specific sites and in specific directions, while micrometer-sized pillar structures serve as high-finesse optical cavities for photon confinement, from visible wavelengths down to the ultraviolet. In this chapter, the main focus is placed on the fabrication and optical properties of nanosphere-patterned optical cavities leading to the demonstration of optically pumped lasing. Following an in-depth introduction of the NSL process including the methods of coating and the subsequent fabrication flow, various potential applications in nitride-based material are highlighted. Specifically, three blue/UV nitride cavities fabricated by NSL combined with micro-fabrication technologies are described. The optical properties of these sphere-patterned cavities will be discussed in detail, based on diverse geometries and resonant mechanisms.

9.2 Nanosphere Lithography

NSL [18] is an inexpensive and ultra-efficient nano-patterning technique with attractive abilities of producing well-ordered periodic arrays over large areas with minimal processing time. With wide ranges of commercially available sphere dimensions from hundreds of micrometer down to tens of nanometers, the spheres can self-assemble into monolayer and multi-layer of periodically ordered array. Nanospheres have previously found uses in technologies such as catalysis, biochemical devices, cell cultures, surface-enhanced Raman spectroscopy, and sensing application. In the field of optoelectronics, ordered periodic structures based on nanospheres may be used as PhCs to manipulate the flow of light, in particular, 3D PhCs [19,20] that can interact with light in both the vertical and lateral directions. In this chapter, we shall mainly focus on lithographic processes for generating 2D monolayers of silica spheres, which serve as high etch selectivity masks for GaN materials during dry etch, with the target of producing various 2D regular nanostructure arrays on the surface of the wafer. We shall firstly discuss the fundamental steps in the formation of self-assembled ordered monolayer in hexagonal close-packed (HCP) arrangement for pattern transfer.

Figure 9.1 FE-SEM images of (a) multi-layers and (b) monolayer of spheres.

9.2.1 Process Development

The formation of an ordered HCP monolayer over large areas on a substrate is commonly achieved via one of three coating strategies, including vertical deposition, dip-coating, and spin-coating, as illustrated in schematic diagrams in Fig. 9.2. The successes of these strategies are dependent on various factors determining quality of the coated monolayer. If the processing conditions are sub-optimal, the spheres will simply aggregate to form undesirable clusters, or spread out loosely without any particular order. Due to the low withdrawal speed in dip-coating and slow evaporation rate in vertical deposition, long processing time (of the order of hours) may be required to establish a monolayer over centimeter-scale areas. Moreover, it is often difficult to maintain precise control of the ambient conditions, such as temperature, humidity, and pressure, which strongly affect the evaporation rate. On the contrary, the spin-coating method, which is mainly governed by rotation speed, concentration, size of spheres, and wettability of substrate, is a more reliable method with distinct advantages of higher throughput and better reproducibility. When these parameters are fine-tuned, monolayer arrays of closed-packed nanospheres over centimeters can be obtained within a matter of minutes.

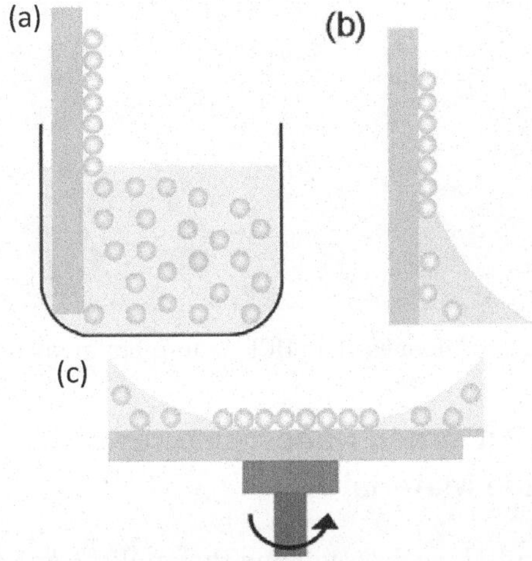

Figure 9.2 Schematic diagrams showing various nanosphere-coating procedures, including (a) vertical deposition, (b) dip-coating, and (c) spin-coating.

Uniform silica nanospheres are initially diluted in de-ionized water to produce the optimal volume concentration of cv nearly 2%. The diluted colloidal suspension is then mixed with sodium dodecyl sulfate at a volume ratio of 10:1. Introduction of a surfactant lowers the surface tension of the colloidal suspension and thus facilitates the spreading of nanosphere to prevent particle agglomeration or aggregation. The well-mixed colloidal suspension is then carefully dispensed onto the sample surface by mechanical micro-pipetting or other means. Optimized rotational speed is necessary to balance the centrifugal force with the solvent capillary force. During spin-coating, the excess suspension is gradually flung off and spheres spread laterally, self-assembling into a monolayered HCP array across the sample. The coated spheres then act as a lithographic mask and the pillar pattern is transferred to the wafer by inductive-coupled plasma (ICP) etching. The etched sample is finally immersed in deionized water under sonication to remove the spheres, leaving behind a nano-pillar array. The resultant HCP nano-pillar array after etching is shown in Fig. 9.3.

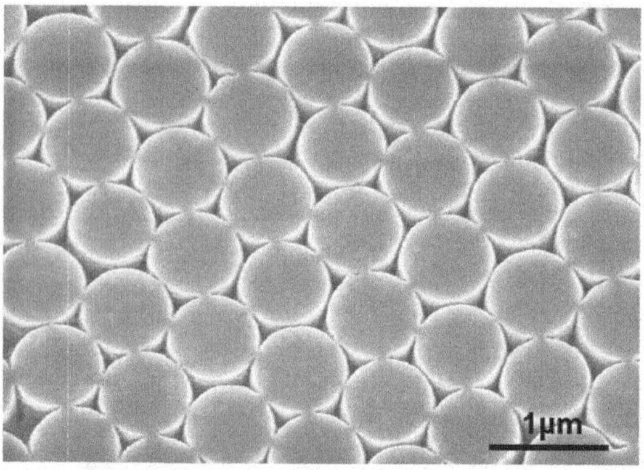

Figure 9.3 FE-SEM images of an HCP nano-pillar array patterned by NSL.

9.2.2 NSL-Patterned Nanostructures on Nitride Material

Some of the initial works involving the application of self-assembly NSL processes to GaN-based material date back to 2008 [21], developed mainly to deal with the light extraction issues of GaN LEDs to suppress light confinement through reduction of total internal reflections at the GaN/air interfaces. Using silica spheres with diameters of 500 nm, Ng et al. successfully fabricated HCP pillars onto an LED surface and measured a two-fold increase in electroluminescence compared with the as-grown. However, the optical effects of the sphere-patterned PhC were not fully characterized at that time, and it is worth noting that such HCP "weak" PhCs can potentially be modified to become "strong" PhCs [22], which can forbid propagation of photons within a certain range of frequencies. This forbidden band, also known as a PBG, is extremely useful in molding the flow of emitted light from an LED [23]. The question is how nanosphere-patterned HCP PhCs can become "strong" in the first place. One of the solutions was offered by Fu et al. [24], who reported the fabrication of hemi-ellipsoidal PBG structure. They make use of a monolayer of 192 nm spheres as the etch template and reduce the etch selectivity of SiO_2 over GaN by using an etch plasma containing a mixture of Cl_2 and CHF_3. The spheres gradually shrink laterally while the pattern is being transferred to GaN, resulting in the formation of hemi-ellipsoid array (Fig. 9.4b). Alternatively, PBG structures can also be realized by non-close packed (NCP) pillar arrays (Fig. 9.4a). Li et al. [25] have proposed to introduce a finite interstitial spacing between adjacent spheres before the pattern is transferred. The authors employed a dimension-adjusting process to shrink the spheres on the plane without altering the sites of spheres; therefore, the packing of the modified array remains largely regular. With different extents of reduction of spheres, the PBG can be shifted to a desired spectral position in a controllable manner. In addition, Li et al. [26] have reported the fabrication of a NCP clover-shaped PBG structure (Fig. 9.4c), which allows optimal overlap between the PGB and emission wavelengths in the blue spectral regions. These modified NSL processes overcame the restrictions of closed-packed patterning and have been applied to GaN LED materials for realizing enhancement of light extraction.

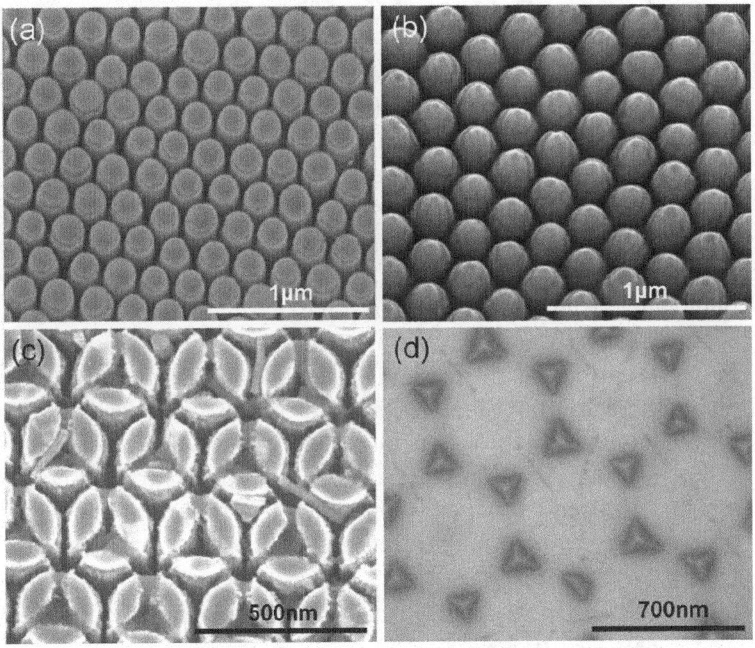

Figure 9.4 FE-SEM images showing (a) air-spaced nano-pillar array, (b) array of hemi-ellipsoid PBG structure, (c) clover-shaped PBG structure, and (d) triangular nano-particles formed by e-beam deposition, followed by lift-off from the monolayer (with permission for reproduction from American Institute of Physics).

Apart from nano-pillar arrays, the inverse pattern of a pillar array can be obtained by depositing a thin layer of desired materials over the nanosphere monolayer. After removal of the spheres, an inverse pattern of pillar array is observed, as shown in Fig. 9.4(d). As the voids formed by three adjacent closed-packed spheres are much smaller than the spheres, the coated triangular nano-particles are also much smaller than the spheres. As reported by G. Y. Mak et al. [27], Ag, Au, and Al triangular-shaped nano-particles have been coated onto GaN-based LEDs via NSL followed by a lift-off process. Pronounced enhancement to the photoluminescence (PL) signal has been observed, which has been attributed to the localized surface plasmon (LSP) effect.

9.3 Photons Confinement in Various Sphere-Pattened Cavities

As discussed in the previous section, versatile nano-patterns can be created through various smart tweaks of the NSL process, such as applying additional deposition and/or dry etch steps on the HCP nanosphere template. As opposed to enhancing light extraction, the sphere-patterned structures described in this section act as optical cavities that support resonant mode in the strong confinement regime, enabling blue/violet lasing at room temperature. The morphologies of the resultant ordered nanostructures and the optical characteristics of three different sphere-patterned nitride cavities based on various lasing mechanisms will be discussed. Firstly, nano-pillar PhC arrays are demonstrated to produce blue lasing via the PhC band-edge effect. This is followed by two WG cavities, namely the nano-ring array and metal-clad nano-pillar arrays, both of which have been fabricated via modified NSL processes and confine prorogating photons along closed circular paths.

9.3.1 Photonic Crystal Band-edge Lasing from Nanopillar Array

Generally, there are two types of lasing mechanisms associated with PhCs, the first of which, and probably the more common mode, is defect-mode lasing that makes use of a localized state defect mode as a laser cavity. The second type is the band-edge micro-laser, whereby, as the name suggests, lasing occurs at the photonic band edge. Specifically, this type of laser operates at the band edge near the point where the group velocity of light approaches zero. Fu et al. [28] developed an ordered nano-pillar structure by a two-step process on the surface of a wafer containing InGaN/GaN MQW structures. The wafers were grown by metal-organic chemical vapor deposition (MOCVD) on a c-plane sapphire substrate. The layer structure consisted of a 25 nm thick GaN buffer layer, a 3 μm thick Si-doped GaN layer, 5 periods of 2/5 nm thick InGaN/GaN MQWs, a 50 nm thick Mg-doped AlGaN layer, and topped with a 250 nm thick Mg-doped GaN layer. A monolayer of 190 nm silica nanosphere was first coated onto a laser-diced 10 × 10 mm sample by vertical deposition. This monolayer of nanosphere, which self-assembled into a HCP array, acts as a hard mask for the subsequent etch step. The sample was then dry-etched by ICP using a gas chemistry comprising 15 sccm of Cl_2 and 8 sccm of Ar for 90 s, at coil and plate powers of 350 W and 135 W, respectively. The pattern was naturally transferred from the hard mask to form ordered HCP nano-pillars. The oxide nanosphere residue was then removed by 10% HF acid wet etching for 10 min followed by sonication in acetone for another 10 min. The resultant nano-pillar array is illustrated in the field-emission scanning electron microscope (FE-SEM) image in Fig. 9.5(b). The height of the nano-pillars is around 150 nm, leaving the active MQW region beneath intact.

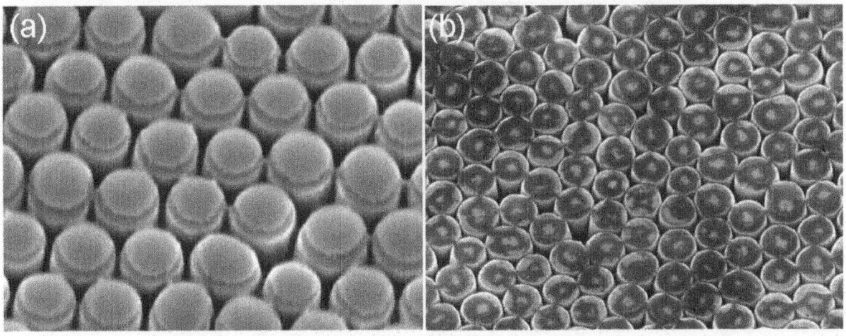

Figure 9.5 FE-SEM images showing HCP nano-pillars (a) before and (b) after removal of spheres (with permission for reproduction from American Institute of Physics).

The authors performed a PL measurement to evaluate the optical properties of the nano-pillars. A frequency-tripled ND: YLF diode-pumped solid state (DPSS) laser with 349 nm emission peak and pulse duration of nearly 4 ns at 1 kHz repetition rate was used as an excitation source. The emitted signal was collected by a fiber bundle and channeled to a 150 mm spectrograph (Acton SP-2150i) and dispersed with a 1200 l/mm grating. A Peltier-cooled CCD camera (Acton Pixis 256) senses the dispersed light; the entire system has an optical resolution of better than 0.3 nm. The emission image can be simultaneously monitored by a high-resolution CCD camera confocal to the optical path. Under varying excitation power density, PL peaks centered at 373 nm and 393 nm were detected as shown in Fig. 9.6, representing the PL signal from the exposed n-GaN and p-AlGaN layers deeper in the structure after partial removal of the p-GaN cap layer. The F-P oscillations appear to have been suppressed as a result of the textured surface. Most significantly, two sharp lasing peaks centered at 415.6 nm and 427.8 nm are observed. Varying excitation power introduces no obvious shift to the lasing peaks, despite the spectral shift experienced by the broader emission centered at 440 nm on which the lasing peaks are superimposed. At room temperature (300 K), the quality factor Q of the 415.6 nm lasing peak is evaluated to be in the range of 600–700 according to the deification:

$$Q = \frac{E}{\Delta E} \tag{9.1}$$

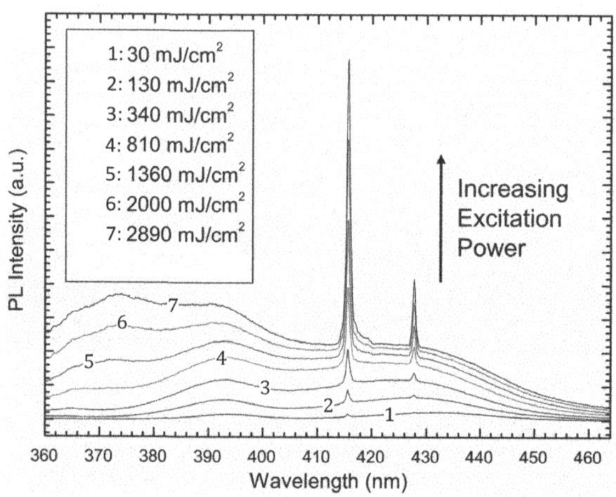

Figure 9.6 Room temperature PL spectrum of nano-pillar array under varying excitation energy density (with permission for reproduction from American Institute of Physics).

Figure 9.7 shows two captured CCD images comparing emission from the as-grown and nano-pillar samples, respectively, under identical excitation energy density of 340 mJ/cm^2. The as-grown sample emits blue light uniformly across the circular region. As for the nano-pillar sample, an intense spot of light is observed in the center, surrounded by blue light around it. The spot appears white due to saturation of the detector. Also, the circle is distorted due to the textured surface.

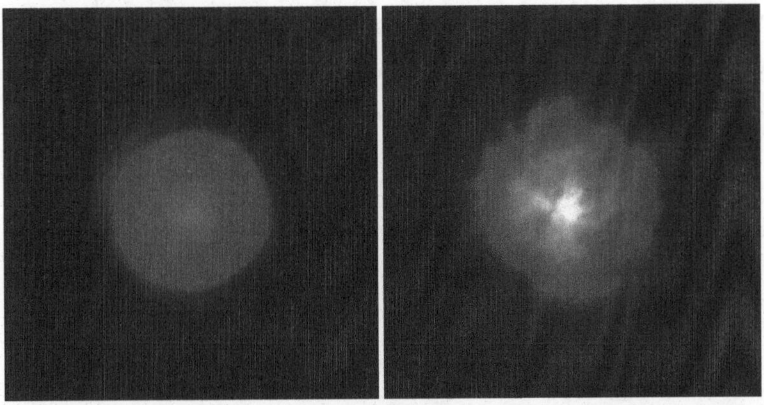

Figure 9.7 CCD images showing PL emission from as-grown (left) and nano-pillar sample (right) (with permission for reproduction from American Institute of Physics).

Figure 9.8(a) shows a plot of integrated PL intensity as a function of excitation power density corresponding to the lasing peak at 415.6 nm. The plot illustrates that the signal intensity grows superlinearly with excitation energy density, with a lasing threshold of nearly 30 mJ/cm^2. Rearranging the experimental data by putting the PL intensity in logarithmic scale as illustrated in Fig. 9.8(b), the spontaneous emission coupling factor β, which is inversely proportional to the difference in PL intensity before and after the lasing threshold on the logarithmic scale, can be estimated to be 2.1×10^{-2}. This value of β can then be used to calculate the Purcell factor F_p with the following equation [29]:

$$F_p = \frac{\beta}{1-\beta} \tag{9.2}$$

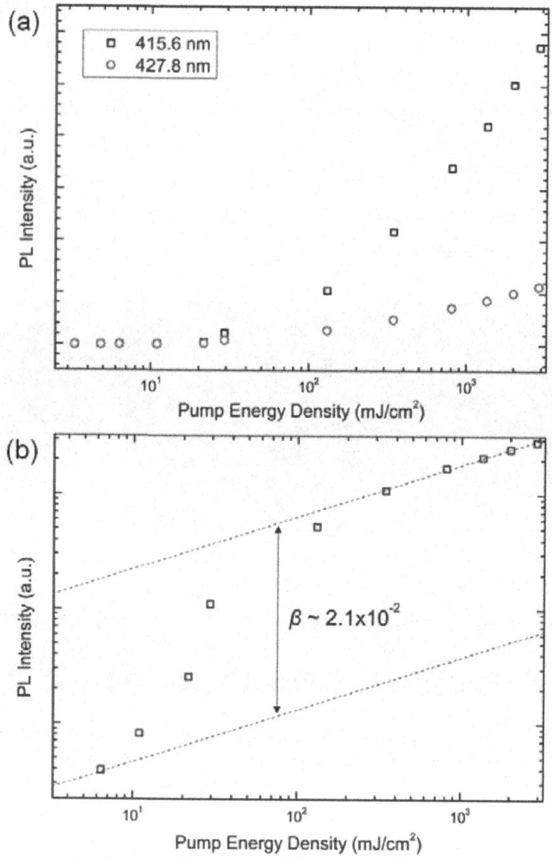

Figure 9.8 (a) Integrated PL intensity of the two lasing peaks of nano-pillar array; (b) integrated PL intensity of the 415.6 nm lasing peak replotted in logarithmic scale (with permission for reproduction from American Institute of Physics).

The Purcell factor F_p is determined to be about 2.1 × 10^{-2}, which is then used to derive the effective cavity volume V_{eff}. An equation derived from the Fermi's golden rule is given by [29,30]

$$F_p = \frac{3Q(\lambda_c/n)^3}{4\pi^2 V_{eff}} \quad (9.3)$$

where Q is the Q factor, λ_c is the lasing peak wavelength, and n is the refractive index of the material. In our calculation, we use the average value of 646, for Q; lasing peak wavelength λ_c of

415.6 nm and refractive index of GaN is 2.55. The estimated value of effective cavity volume Veff is thus 9.8 μm^3. As the Purcell factor is determined to be smaller than unity, the Purcell enhancement effect is thus not present with this photonic crystal structure, which is reasonable, as a physical micro-cavity is absent. Instead, the photonic mode spreads across multiple nano-pillars, contributing to the relatively large estimated effective cavity volume. The observed lasing is therefore attributed solely to the photonic crystal band-edge effect. The experimentally derived value for the spontaneous emission coupling factor, which is of the order of 10^{-2}, is comparable to recent studies of GaN lasing devices with spontaneous emission coupling factor enhancement, for instance, a vertical cavity surface emitting laser [29] or a photonic crystal nanocavity light source [31]. The fabricated nanopillars, unlike other approaches involving deposition, are composed of high-quality single-crystal GaN material, although the dimension of the effective cavity volume may be affected by the loss induced by surface plasma damage. Also, the presence of MQWs within the nanopillars ensured high gain and thus low-threshold lasing.

To understand the observed lasing phenomenon from the nano-structured sample, we consider the conditions for different possible modes of lasing mechanisms. For F-P type lasing, a pair of parallel facet must exist for establishing a resonant standing wave within an optical cavity. However, the fabricated nano-pillars are of dimensions smaller than the wavelength of light, resulting in diffraction and enhanced light extraction, which in turn suppresses the occurrence of F-P modes. This is verified through the disappearance of interference fringes from the PL spectra of the nano-pillar sample. On the contrary, in order to qualify for WG mode lasing, a circular cavity must exist to provide light confinement for resonance, and that the size of the cavity has to be much larger than the wavelength of light. In our circular nano-pillars of subwavelength dimensions, although a WG mode is supported as predicted from finite-difference time-domain (FDTD) simulations, it is however too leaky to acquire sufficient gain. Moreover, as each individual nano-pillar is in close contact with neighboring units due to its hexagonal close-packing, any possible WG mode is likely to be disrupted, as the modes of individual nano-pillars will interfere with each other. With other mechanisms ruled out, the photonic crystal properties associated with the periodic nanostructure pattern are the most likely causes contributing to the lasing phenomena.

Figure 9.9(a) displays the photonic band structure simulated by Rsoft BANDSOLVE. Using a lattice constant of a = 190 nm, in terms of normalized frequency (a/λ), the gain region of the MQWs lies in the range of 0.4 and 0.465. With an apparent absence of a PBG, the lasing phenomenon is expected to be contributed by the photonic band-edge (or dispersion) effect. The three possible lasing points are situated at the band-edge of Γ-point, with normalized frequencies of 0.437, 0.446, and 0.457 as shown in Fig. 9.9(b), in which the light wave will be subjected to a maximum feedback due to zero group velocity ($d\omega/dk = 0$). Compared with the experimental results, we are not able to observe the lasing peak that corresponds to normalized frequency of 0.437. This is caused by blue-shifting of the gain region upon high-intensity optical pumping.

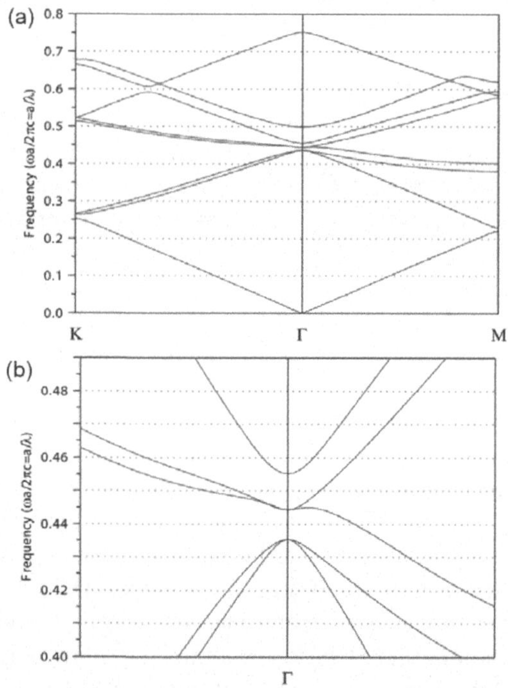

Figure 9.9 (a) Simulated photonic band diagram of nano-pillar array; (b) the zoomed-in region of band diagram (with permission for reproduction from American Institute of Physics).

9.3.2 Whispering Gallery Mode Lasing from Sphere-Patterned Cavities

9.3.2.1 Overview

WG-mode dielectric resonators have received considerable attention, as they offer the potential to pursue high-Q factor and low threshold lasing. The resonator is generally designed as a disk-shaped element to take full advantage of total internal reflections; light is guided by continuous total internal reflection along a curved surface. The pioneering work of WG modes lasing from nitride-based material has been reported by Choi et al. [32]; an array of pivoted GaN microdisks is fabricated on a GaN/Si material. The circular mask pattern with diameters of 20 μm is defined on a GaN-on-Si wafer by photolithography and the microdisk structure is transferred onto the GaN by Cl_2/BCl_3 plasma etching. The sample is then wet-etched in mixture of HNO_3, H_2O, and NH_4F to create an undercut, as this isotropic wet etchant removes the Si surrounding the GaN microdisks. The tiny Si pedestal supports the microdisk and provides a pathway for current injection, as revealed in the SEM images in Fig. 9.10(a) and 9.10(b). The microdisk structure strongly confines WGM in the disk active layer by total internal reflection at semiconductor/air boundaries. Multiple resonant modes, corresponding to WGM, are observed in PL spectra in Fig. 9.11, with mode spacing of nearly 1.1 nm and measured Q factor of 80.

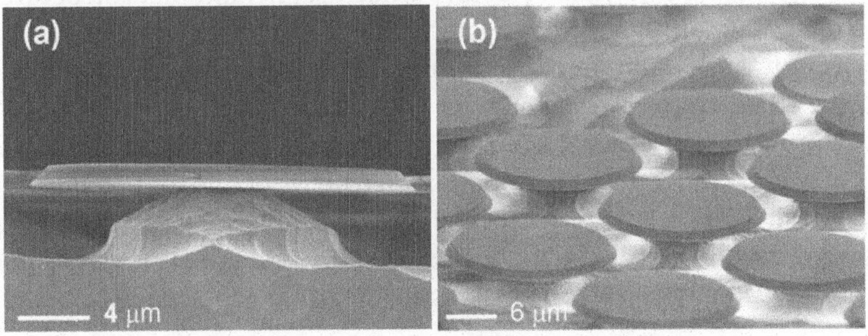

Figure 9.10 FE-SEM images showing (a) the cross-sectional view and (b) tile-view of pivoted microdisk structure (with permission for reproduction from American Institute of Physics).

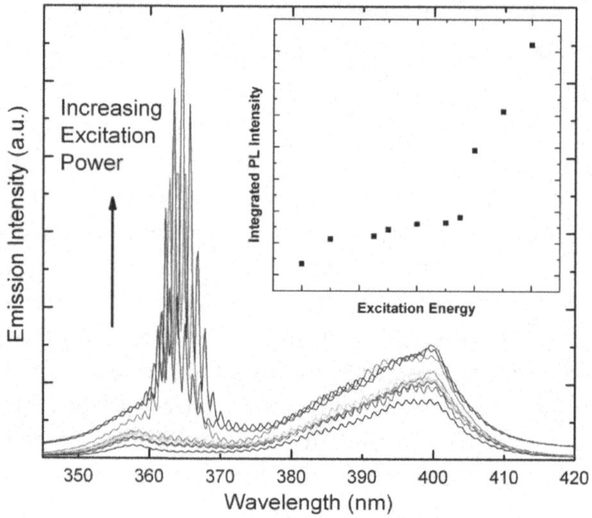

Figure 9.11 PL spectrum of the microdisk at increasing excitation power; the inset shows a plot of integrated PL intensity versus excitation energy (with permission for reproduction from American Institute of Physics).

It is worth to note that the resonant frequency and mode spacing is strongly dependent on the dimension of disk-shape element. Microdisk structures defined by photolithography generally have narrow mode spacing, resulting in dense resonant peaks distributed throughout the spectrum. In order to achieve single WG mode lasing for short visible wavelength and even UV radiation, dimensional downscaling of circular structure is necessary. Nano-pillar structures of submicron dimensions would have larger mode spacing, which is suitable for minimizing the number of modes overlapping the gain spectrum, thus achieving single-mode stimulated emission. During the pattern transfer from the nanosphere mask, the resultant circular-shaped pillars are potentially well-suitable for establishing WGM. As such, sphere-patterned optical cavities may be a promising approach due to their ability to produce structures of very small dimensions, as well as high-density integration on a wafer.

9.3.2.2 UV single-mode lasing from metal-clad pillar

NSL is employed to fabricate a large-area ordered hexagonal nano-pillar array and the process flow is described in detail in an earlier work [21,33]. Silica spheres with mean diameters of 990 nm are spin-coated onto a wafer, whose structure consists of InGaN/GaN MQWs sandwiched between n-GaN and p-GaN layers, grown by MOCVD on c-plane sapphire. The periodical nanosphere pattern is transferred to GaN by inductively coupled plasma (ICP) etching. After sphere mask removal, a metal coating is deposited over the nano-pillar array by e-beam evaporation. The coated sample was dry etched for 180 s giving an etch depth and pillar diameter of approximately 750 nm and 980 nm, respectively, evident from Fig. 9.12(left). To ensure that the air-gaps between neighboring pillars are fully filled with metal, a higher deposition rate of around 3 Å/s was used. The pillars were almost completely covered by the metal clad, as shown in Fig. 9.12(right).

Figure 9.12 FE-SEM images show the cross-sectional views of pillars before (left) and after (right) metal coating (with permission for reproduction from Optical Society of America).

The optical properties of the nano-pillars are evaluated by conducting room-temperature (r.t.) PL measurements. Instead of exciting PL from the p-GaN side, the focused beam with spot diameter of 60 μm was incident perpendicularly through the bottom polished sapphire surface, as the metal layer strongly attenuates the incident UV beam. The PL signal was collected with a fiber probe at a detection angle of nearly 80° to the normal. Figure 9.13 shows the r.t. PL spectra of four nano-pillar cavity samples without and with different metal claddings–namely Au, Ag, and Al–under excitation density of nearly 0.85 MW/cm^2. Two broadband emissions centered at nearly 366.43 nm and nearly 433.07 nm were observed from the nano-pillar sample, corresponding to spontaneous emissions from GaN and InGaN/GaN MQWs. The intensity of MQW emission is much lower than that of GaN emission, as the excitation beam was incident through the substrate side. A significant portion of the incident photons has been absorbed by the GaN layer before propagating to the MQWs region, resulting in domination of GaN UV emission. To promote resonant modes within the nano-pillar, Au, Ag, and Al metal layers are coated onto the surface to encapsulate the cavities. From the PL spectra, a resonant mode was observed only from the Al-coated sample, while the other samples experience mild PL enhancements. The Al coating, with the highest UV reflectivity, is thus demonstrated to be capable of enhancing the confinement effects of a pillar cavity. Figure 9.15 shows excitation-varying PL spectra from the Al-coated sample. Two spectral peaks centered at nearly 365 nm and nearly 373 nm can be resolved. However, when the excitation density exceeds threshold (extracted from the plot to be ~0.42 MW/cm^2), the latter peak becomes dominant. The plot in the inset of Fig. 9.15 verifies the non-linear increase in PL intensity above threshold, characteristics of stimulated emission. Figure 9.14(a) and 9.14(b) show captured CCD images of the PL spot below and above threshold, respectively. Above threshold, speckle-like patterns are clearly observed, attributed to coherent emission.

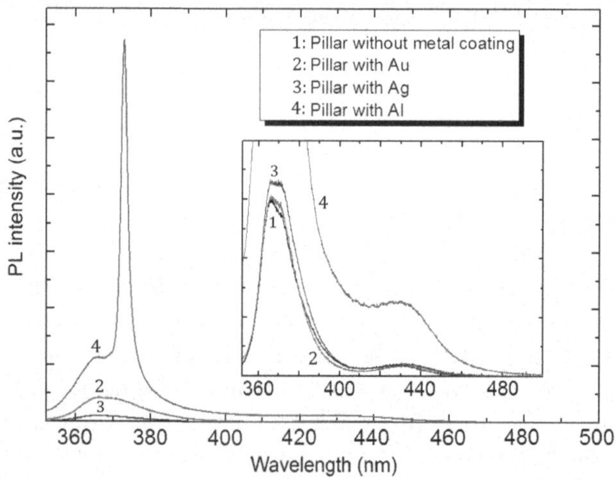

Figure 9.13 R.T. PL spectra of nano-pillars with various metal coatings at high excitation power density of ~0.85 MW/cm^2 (with permission for reproduction from Optical Society of America).

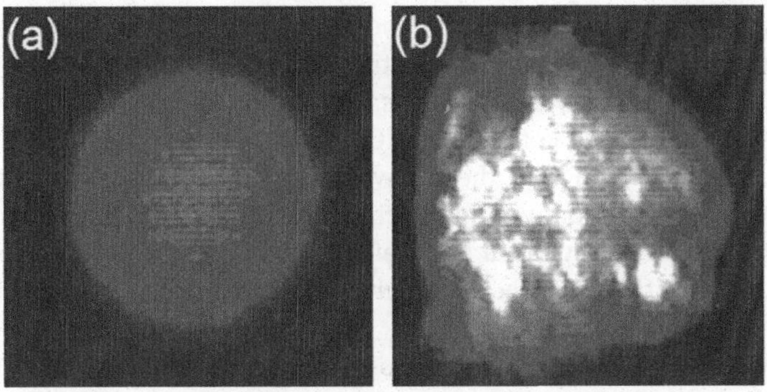

Figure 9.14 CCD-captured images of PL emission from Al-clad pillar below (left) and above threshold (with permission for reproduction from Optical Society of America).

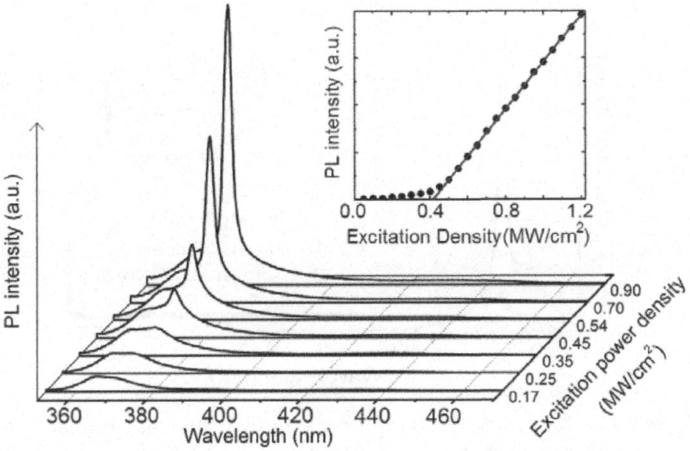

Figure 9.15 PL spectra of Al-coated nano-pillars under varying excitation power density, with lasing mode observed at 373 nm; the inset shows integrated PL intensity versus pump power density (with permission for reproduction from Optical Society of America).

Although the geometries of nano-pillars support WG modes, the presence of a metal coating may also introduce vertical F-P modes. To distinguish between the mechanisms, an Al layer was coated onto an as-grown wafer for comparison. Both the nano-pillar and as-grown samples are then subjected to UV laser excitation at 0.85 MW/cm^2, the PL spectra of which are plotted in Fig. 9.11(c), indicating that no optical modes can be observed from the Al-coated as-grown sample; the likelihood of F-P mode lasing is therefore slim. With a single mirror coating, photons can still escape from the opposite GaN/sapphire or sapphire/air interfaces, resulting in inefficient optical confinement. On the contrary, pillar cavities with metal coating on both the sidewall and top surface provide strong lateral confinement within the circular cavity. This is particularly so for Al, which provides a high reflectivity in the UV spectral region, according to the reflectivity curves for the corresponding metals [34] shown in the inset of Fig. 9.16. For reference, with a lower UV reflectivity, the Ag-clad nano-pillars would also lase at a higher threshold of 0.97 MW/cm^2.

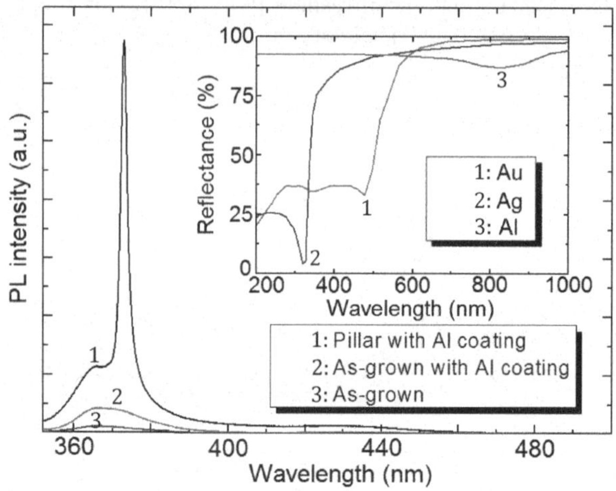

Figure 9.16 Effects of Al-coating on as-grown and nano-pillar samples; the inset shows the reflectance curves for Au, Ag, and Al (with permission for reproduction from Optical Society of America).

The WG resonator frequency and mode spacing of such circular cavities are evaluated by the relations [35]

$$2\pi Rn = m\lambda \tag{9.4}$$

and

$$\Delta\lambda_{WG} = \frac{\lambda^2}{2\pi Rn} \tag{9.5}$$

where R is the pillar radius (490 nm) and n is the effective refractive index (2.3). The lasing peak extracted from the plot is 373 nm, corresponding to m = 19 while the adjacent modes of m =20 and

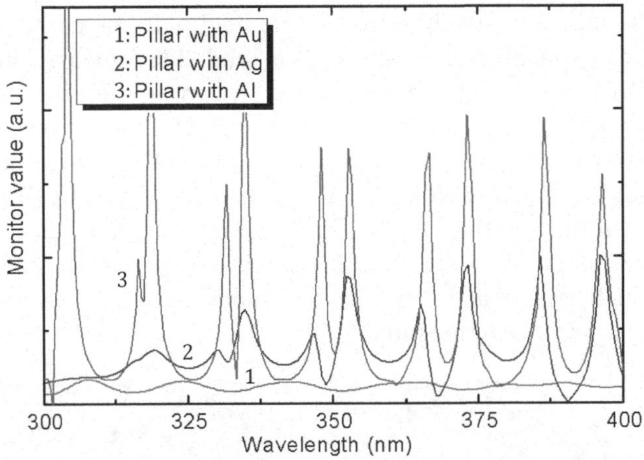

Figure 9.17 FDTD simulated spectra showing resonant characteristics of metal-clad nano-pillar cavities (with permission for reproduction from Optical Society of America).

$m = 18$ are calculated to be 354 nm and 393.4 nm, respectively, with mode spacing of around 19.6 nm. As only the resonant mode at 373 nm overlaps strong GaN emission, single WG mode resonance is observed.

FDTD simulations are performed to verify the resonant characteristics of the nano-pillars with different metal claddings. The computational area and the mesh size (Δx) are fixed at 1.5 × 1.5 μm^2 and 5 nm, respectively, with the time step set to 0.00967 fs satisfy the Courant stability condition

$$\Delta t_{\max} = \frac{\Delta x}{c\sqrt{2}} = 0.0118\,fs \tag{9.6}$$

The simulated transverse electric (TE) mode spectra for the Al, Ag, and Au cladded nano-pillars in the wavelength range of 400–500 nm are plotted in Fig. 9.17. In the spectrum of the Al-clad pillars, a first-order mode at nearly 365 nm and a higher order mode at nearly 373 nm are observed, overlapping with the peak of the GaN emission spectrum. Also note that its modal intensities, thus optical confinement, remain high across the frequency range of interest, while that for Ag-clad pillars decline with reducing wavelengths, whereas no resonant modes are observed for the Au-clad pillars. This is consistent with the reflectivities of the respective metal coatings as plotted in the inset of Fig. 9.16(c). The

FDTD-simulated steady-state electric field patterns corresponding to the resonant modes at 365 nm and 373 nm are shown in Fig. 9.18, neglecting coupling effects. It can be seen that the field corresponding to the first-order mode at 365 nm is concentrated around the periphery of the pillar, and therefore would be prone to scattering losses along the roughened plasma-etched sidewalls [36]. On the contrary, the field pattern for the higher order mode at 373 nm is established within the pillar and will not suffer from optical losses, explaining why this mode becomes the dominant lasing mode above threshold.

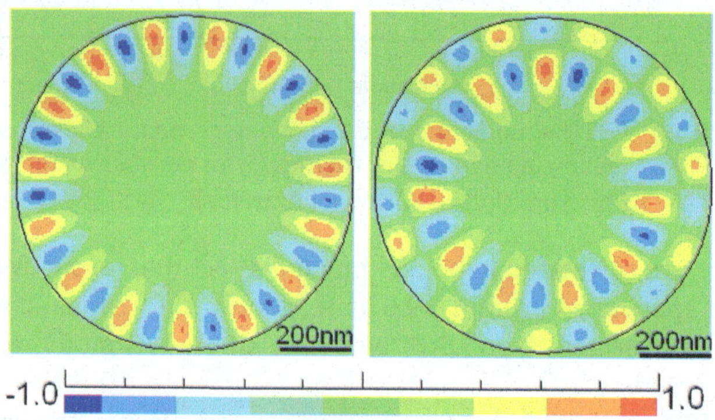

Figure 9.18 FDTD-computed steady-state electric field patterns of the resonant modes at (left) 365 nm and (right) 373 nm (with permission for reproduction from Optical Society of America).

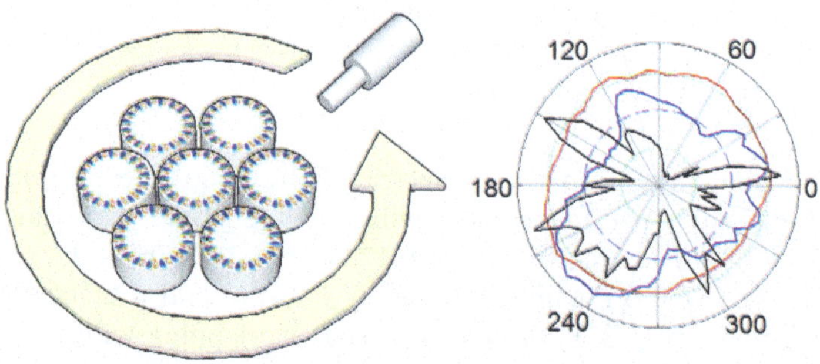

Figure 9.19 (Left) Schematic diagram showing the setup for measuring in-plane angular emission pattern; (Right) Angular emission patterns of Al-clad pillars from the in-plane (black) and vertical (blue) directions. The curve in red corresponds to in-plane measurement from an as-grown sample (with permission for reproduction from Optical Society of America).

With WG emission modes, a highly directional emission pattern can be expected in the planar directions [37]. To verify this, angular-resolved PL in the near-planar direction is measured from a pillar array by rotating a fiber around the central axis of the sample, with the collection angle maintained at nearly 80° to the normal, as illustrated in Fig. 9.19 (left). A high degree of emission directionality is clearly present, with multiple intensity peaks at various discrete angles, deduced from the curve in black from the polar plots of Fig. 9.19 (right). As the emission modes are scattered out of the cavities through roughened (etched) sidewalls, the recorded pattern is asymmetrical; geometrical imperfections of individual pillars and non-uniformities between pillars add on to the asymmetry. Such periodicity is clearly absent from either the angular-resolved PL pattern as measured in the near-normal direction (blue curve) or from the as-grown sample (red curve).

A final note on the cold-cavity quality factor Q has been evaluated as 150 at an excitation power density of 0.4 MW/cm^2. The relatively low magnitude may be attributed to sidewall scattering losses, intracavity coupling, and most importantly, the absence of a gain region. However, the low gain is merely a limitation of the PL measurement geometry. In the electroluminescent device under development, emission from the MQWs will be dominant, providing sufficient gain to overcome the losses. Nevertheless, the present optical-pumped demonstration verifies the feasibility of this approach for forming optical micro-cavities.

9.3.2.3 High-Q WG mode from nanoring array

As mentioned in Section 9.2, modification of the fabrication flow can possibly lead to the formation of novel nano-scale geometries. On the basis of such strategies, we have developed nanoring arrays supporting WG mode, with high Q-factor and low lasing threshold. The fabrication of the nano-ring arrays evolves from standard processes for forming closed-packed nanosphere monolayers, with additional steps for tweaking the dimensions. Figure 9.20 illustrates the entire process flow for the fabrication of closed-packed nano-ring arrays by NSL. The LED wafer used contains InGaN/GaN MQWs grown by MOCVD on c-plane sapphire substrate, with emission center wavelength of nearly 465 nm. Silica (SiO_2) spheres with mean diameters of 1000 nm were self-assembled on the wafer.

The coated monolayer of spheres, which acted as an etch mask, was subsequently transferred to the wafer forming nano-pillars by ICP etching using Cl_2 as etchant gas at a flow rate of 30 sccm. The spheres were subsequently shrunk in diameters though reactive ion etching (RIE) process using CHF_3-based plasmas under low RF power conditions to minimize lateral motion; this non-closed packed sphere array now serves as a lift-off mask. By electron-beam evaporation, a SiO_2 layer (150 nm) was deposited over the entire surface. Following lift-off by sonication in deionized water, a ring of SiO_2 was left behind at the periphery of the nano-pillars, the width of which depends on the extent of the shrinking in the previous step. These oxide rings acted as a mask during the second ICP etch, resulting in the inner-hole region being etched, and the formation of nano-rings. The residue was removed by immersing the wafer into 10% hydrofluoric (HF) acid. Surface morphologies of the nano-ring arrays were imaged using a Seiko Nanopics atomic force microscope (AFM). The AFM images in Fig. 9.21 revealed assembled nano-ring array with internal diameters of 720 nm, wall widths of 140 nm, and heights of 500 nm. The arrays are regularly closed-packed in a hexagonal arrangement.

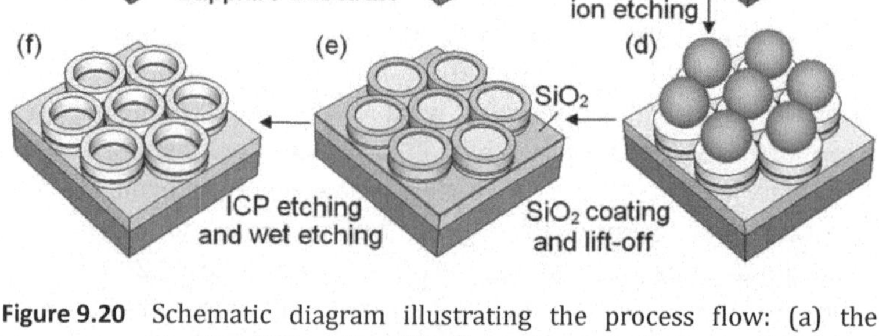

Figure 9.20 Schematic diagram illustrating the process flow: (a) the starting LED wafer; (b) silica spheres coated on p-GaN layer; (c) pattern transfer to GaN by ICP etching; (d) shrinkage of spheres using RIE etching; (e) SiO_2 coating followed by lift-off; (f) removal of SiO_2, giving final structure (with permission for reproduction from American Institute of Physics).

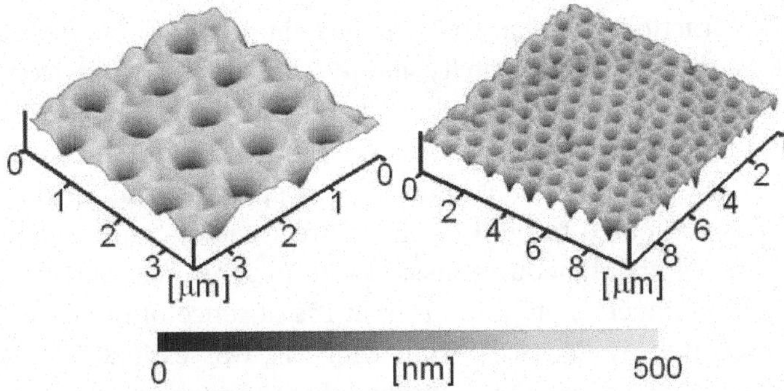

Figure 9.21 AFM images of the HCP nano-ring arrays (with permission for reproduction from American Institute of Physics).

The optical characteristics were evaluated by time-integrated PL at room temperature. A Spectra-Physics DPSS UV laser at 349 nm emission and pulse duration of 4 ns at 1 kHz repetition rate was used as an excitation source. The focused beam was incident nearly perpendicular to the sample surface with a spot diameter of around 150 μm. The PL signal was collected at 30° to the normal with an optical fiber bundle, and the signal coupled to a spectrometer comprising an Acton SP2500A 500 mm spectrograph and a Princeton Instrument PIXIS open-electrode CCD, which offered optical resolutions better than 0.04 nm. PL spectra of the nano-rings at increasing excitation densities are plotted in Fig. 9.22. Under low excitation, the spectrum exhibits weak spontaneous emission from MQWs at λ = 460 nm with FWHM of 34 nm. As the excitation energy density exceeds the threshold of 16.8 mJ/cm^2, a sharp spectral peak emerged at 444 nm; beyond 305.4 mJ/cm^2, a second spectral peak at 472 nm was observed. Zooming in on the spectrum, three satellite spectral peaks at 443.88, 444.08, and 444.26 nm with FWHM as low as 0.09 nm could be resolved as evident in Fig. 9.23 (top); their quality factors, defined by Eq. (9.1), are evaluated as 4455, 5025, and 4660, respectively.

Figure 9.23(bottom) plots intensities of the 444.075 nm lasing peak as a function of excitation energy density in logarithmic scale. Nonlinear increases in PL intensities were observed beyond the extracted threshold value of nearly 10 mJ/cm^2. To characterize the optical cavity, the spontaneous emission coupling factor β, which represent a figure of merit for the cavity, is evaluated. B, defined as the fraction of total spontaneous emission from the source that is coupled into a cavity mode, can be estimated from the ratio of the output intensity below and above the lasing threshold [38]. In practice, it is always less than unity due to the concurrent emission of non-resonant modes. β of spectral peak at 444.075 nm was estimated to be 0.75. The Purcell factor, which quantifies the ratio of spontaneous emission rate in a resonant cavity mode to the spontaneous emission rate in the absence of the cavity, can be calculated by Eq. (9.2), from which F_p was estimated to be 3. Being greater than unity, it is implied that the spontaneous emission rate has indeed been enhanced by the cavity and that excitons can radiate much faster in the cavity than in free space. To identify the position of WGM lasing frequency, the resonant frequency and mode spacing for a ring cavity were calculated from the Eqs. (9.4) and (9.5), where n is the refractive index of GaN, R the ring radius, and m is the mode number. The calculated mode at m = 17 corresponds to the observed lasing peak at 444 nm while the mode spacing is evaluated to be in the range of 26–29 nm, correlating well with the PL spectra.

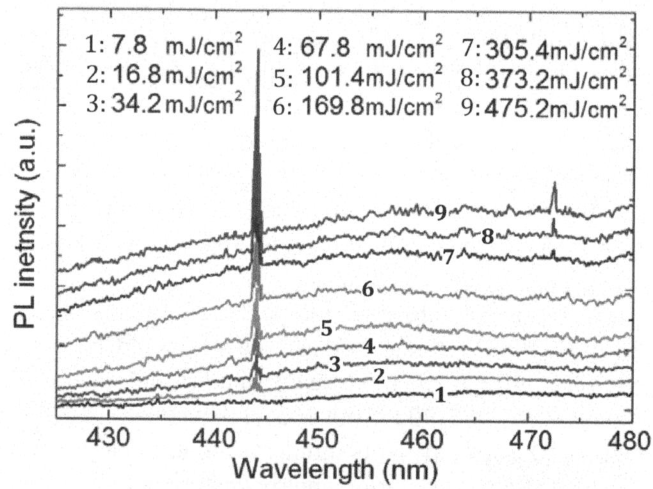

Figure 9.22 Room temperature PL spectra of nano-ring array under varying excitation energy density (with permission for reproduction from American Institute of Physics).

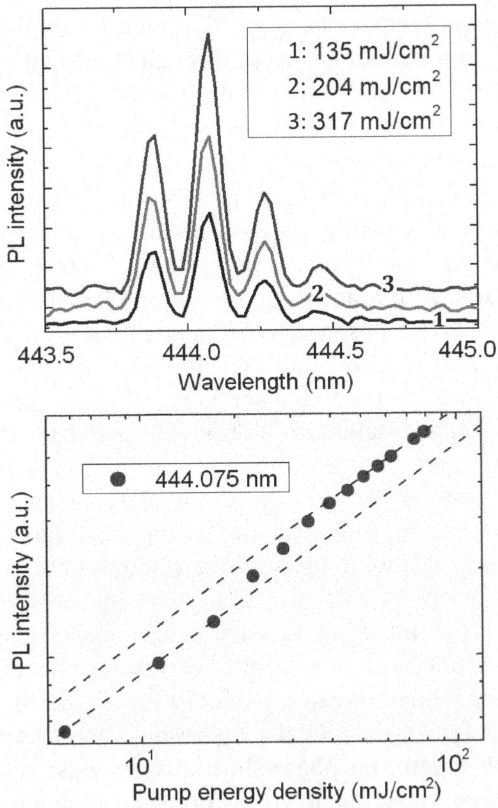

Figure 9.23 (Top) An enlarged view of the PL spectrum near the lasing frequency. (Bottom) PL intensity of the lasing peak at 444.075 nm as a function of pump energy density in logarithmic scale (with permission for reproduction from American Institute of Physics).

The mechanism for stimulated emission from the nano-ring array is investigated. F-P mode lasing requires the existence of a longitudinal cavity for light confinement [39,40]. Without mirror coatings, such a cavity would not exist and F-P type lasing can thus be ruled out. On the contrary, the circular geometry of the nanostructures hints at the possibility of WGM resonance in the planar direction. Indeed, nano-ring cavities are known to support optical confinement modes, which are guided by total internal reflections circling around the peripheries of the rings. By employing effective index method, simulation of a cut-through of the material illustrates that the simulated optical-mode propagation at the wavelength of 444 nm is well confined with the ring with width of 140 nm and height of 500 nm, as shown in Fig. 9.24. This set of data verifies that the mode is well guided and localized in ring wall region.

To verify resonant properties of the nano-ring cavity, FDTD simulations are performed on a ring structure with internal diameters of 720 nm and wall widths of 140 nm. To yield better accuracy, the FDTD mesh size is set to less than 1/20 of wavelength. The computational grid contains 150 × 150 nodes over an area of 3 × 3 μm^2, and the time step set to 0.045 fs, sufficiently small to satisfy the Courant stability condition. The simulated transmittance spectra for three different cell structures are shown in Fig. 9.25, which correlate well with the calculated resonant frequencies and mode spacing, and also the lasing positions from the PL spectra. Compared with a single-ring cell, the optical confinement efficiencies of the double-ring and triple-ring cells are increased, implying that multi-coupled ring arrays have stronger optical confinement effects. Apart from such enhancements, splitting of the resonant modes is also introduced for the multi-coupled ring structure [41]. As individual ring resonators possess the same optical path length, no phase-shift changes exist within a single ring. On the contrary, due to strong optical coupling between rings, there exists round-trip phase shifts when light passes through the couplers, thus resulting in the optical splitting effect [42]. It explains why multiple peaks appear in the vicinity of 444 nm. In addition, sidewall roughness-induced back-reflection into the counter-propagating modes may yet be another mode-splitting factor [43]. The performance of traveling-wave resonators is affected when the forward- and backward-propagation modes are phase-matched. Such contra-directional coupling leads to annihilation of directionality of a traveling-wave resonator and causes the resonant peak to split into multiple modes. The simulated results shown in Fig. 9.26 illustrate the formation of standing waves in single ring structure when a continuous wave at 444 nm was injected at the leftmost position indicated by the red arrow. At the resonant frequencies, photons are effectively confined within the ring structures establishing a WG mode for the single ring cavity.

For the double-ring and triple-ring cells, high-intensity standing waves are set up across the multi-coupled rings structure, indicating that strong light coupling between adjacent rings occurs, and a resonance mode was established due to constructive interference. The ring can thus serve as a resonator and interact further with adjacent rings. This supports the postulation of WG mode lasing observed from PL spectra.

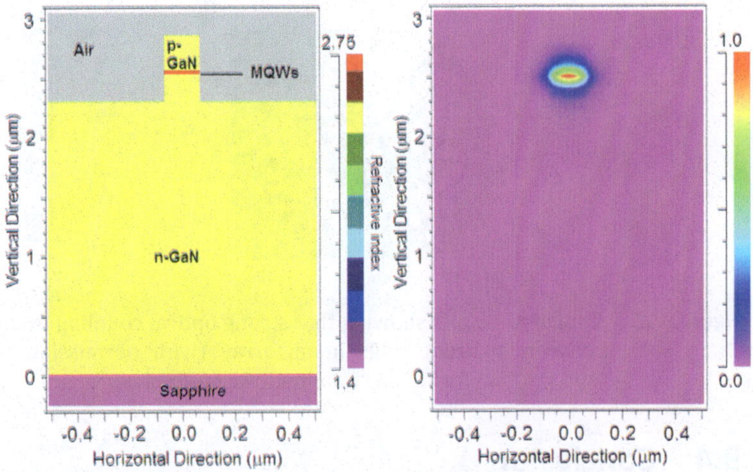

Figure 9.24 Simulation of a cut-through of the structure by the effective index method.

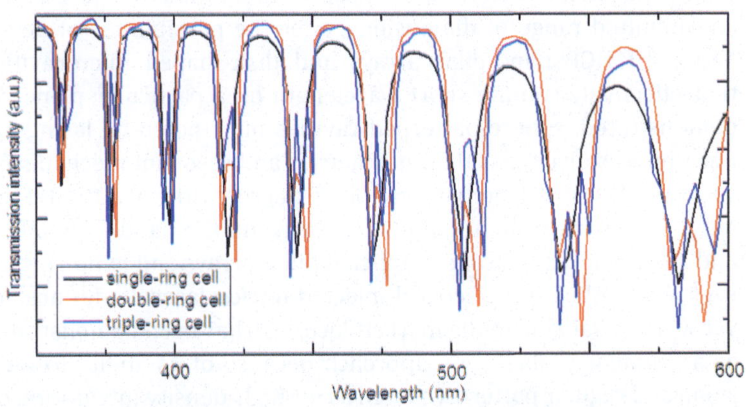

Figure 9.25 FDTD-simulated transmittance plot showing resonant modes for three different cell structures (with permission for reproduction from American Institute of Physics).

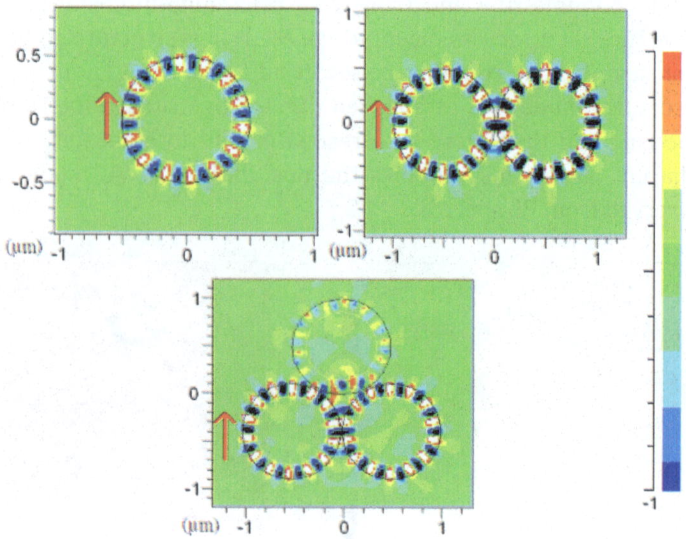

Figure 9.26 Simulated results showing that strong optical coupling occurs between adjacent rings in an array (with permission for reproduction from American Institute of Physics).

9.4 Conclusion

We have described and highlighted several microcavities achieved with nanosphere patterning. Nanospheres and microspheres, with an unlimited range of dimensions, offer the possibilities of developing 2D HCP nano-pilar arrays and disk-shaped microcavities that effectively confine short wavelength on a particular plane. As demonstrated, sphere-patterned cavities offer potential lasing applications as diverse as their geometries and resonant mechanisms, such as UV lasing from vertical pillar microcavities via the F-P effect, blue lasing from a nano-pillar array via the band-edge effect, and WG mode lasing from metal-clad pillar arrays and nano-ring arrays. As pointed out, electrical-injected nitride laser diodes are not yet widely available in the marketplace. NSL has been demonstrated as a promising fabrication approach because of its ability to form small and regular patterns, as well as for high-density integration on a wafer. Further development on electrically injected lasing operation based on NSL will certainly prove to be a great step forward in advancing nitride laser technologies.

References

1. Ponce, F. A., and Bour, D. P. (1997). Nitride-based semiconductors for blue and green light-emitting devices, *Nature*, **386**, pp. 351–359.
2. Nakamura, S., Senoh, M., Nagahama, S., Iwasa, N., Yamada, T., Matsushita, T., Kiyoku, H., and Sugimoto, Y. (1996). InGaN-based multi-quantum-well-structure laser diodes, *Jpn. J. Appl. Phys.*, **35**, pp. L74–L76.
3. Akasaki, I., Sota, S., Sakai, H., Tanaka, T., Koike, M., and Amano, H. (1996). Shortest wavelength semiconductor laser diode, *Electron Lett.*, **32**, pp. 1105–1106.
4. Redwing, J. M., Loeber, D. A. S., Anderson, N. G., Tischler, M. A., and Flynn, J. S. (1996). An optically pumped GaN-AlGaN vertical cavity surface emitting laser, *Appl. Phys. Lett.*, **69**, pp. 1–3.
5. Someya, T., Werner, R., Forchel, A., Catalano, M., Cingolani, R., and Arakawa, Y. (1999). Room temperature lasing at blue wavelengths in gallium nitride microcavities, *Science*, **285**, pp. 1905–1906.
6. Krestnikov, I. L., Lundin, W. V., Sakharov, A. V., Semenov, V. A., Usikov, A. S., Tsatsul'nikov, A. F., Alferov, Z. I., Ledentsov, N. N., Hoffmann, A., and Bimberg, D. (1999). Room-temperature photopumped InGaN/GaN/AlGaN vertical-cavity surface-emitting laser, *Appl. Phys. Lett.*, **75**, pp. 1192–1194.
7. Tamboli, A. C., Haberer, E. D., Sharma, R., Lee, K. H., Nakamura, S., and Hu, E. L. (2007). Room-temperature continuous-wave lasing in GaN/InGaN microdisks, *Nat. Photons*, **1**, pp. 61–64.
8. Chang, S. S., Rex, N. B., Chang, R. K., Chong, G., and Guido, L. J. (1999). Stimulated emission and lasing in whispering-gallery modes of GaN microdisk cavities, *Appl. Phys. Lett.*, **75**, pp. 166–168.
9. Mair, R. A., Zeng, K. C., Lin, J. Y., Jiang, H. X., Zhang, B., Dai, L., Botchkarev, A., Kim, W., Morkoc, H., and Khan, M. A. (1998). Optical modes within III-nitride multiple quantum well microdisk cavities, *Appl. Phys. Lett.*, **72**, pp. 1530–1532.
10. Lu, T. C., Chen, S. W., Lin, L. F., Kao, T. T., Kao, C. C., Yu, P., Kuo, H. C., Wang, S. C., and Fan, S. H. (2008). GaN-based two-dimensional surface-emitting photonic crystal lasers with AlN/GaN distributed Bragg reflector, *Appl. Phys. Lett.*, **92**, p. 011129.
11. Lu, T. C., Chen, S. W., Kao, T. T., and Liu, T. W. (2008). Characteristics of GaN-based photonic crystal surface emitting lasers, *Appl. Phys. Lett.*, **93**, p. 111111.

12. Lai, C. F., Yu, P., Wang, T. C., Kuo, H. C., Lu, T. C., Wang, S. C., and Lee, C. K. (2007). Lasing characteristics of a GaN photonic crystal nanocavity light source, *Appl. Phys. Lett.*, **91**, p. 041101.
13. Vahala, K. J. (2003). Optical microcavities. *Nature*, **424**, pp. 839–846.
14. Oder, T. N., Shakya, J., Lin, J. Y., and Jiang, H. X. (2003). III-nitride photonic crystals, *Appl. Phys. Lett.*, **83**, pp. 1231–1233.
15. Oder, T. N., Kim, K. H., Lin, J. Y., and Jiang, H. X. (2004). III-nitride blue and ultraviolet photonic crystal light emitting diodes, *Appl. Phys. Lett.*, **84**, pp. 466–468.
16. Motayed, A., Davydov, A. V., Vaudin, M. D., Levin, I., Melngailis, J., and Mohammad, S. N. (2006). Fabrication of GaN-based nanoscale device structures utilizing focused ion beam induced Pt deposition, *J. Appl. Phys.*, **100**, p. 024306.
17. Lanyon, Y. H., De Marzi, G., Watson, Y. E., Quinn, A. J., Gleeson, J. P., Redmond, G., and Arrigan, D. W. M. (2007). Fabrication of nanopore array electrodes by focused ion beam milling, *Anal. Chem.*, **79**, pp. 3048–3055.
18. Haynes, C. L., and Van Duyne, R. P. (2001). Nanosphere lithography: A versatile nanofabrication tool for studies of size-dependent nanoparticle optics, *J. Phys. Chem. B*, **105**, pp. 5599–5611.
19. Ozin, G. A., and Yang, S. M. (2001). The race for the photonic chip: Colloidal crystal assembly in silicon wafers, *Adv. Funct. Mater.*, **11**, pp. 95-104.
20. Zhang, Q., Li, K. H., and Choi, H. W. (2012). Polarized Emission from InGaN Light-emitting Diode with Self-assembled Opal Coating, *IEEE Photon. Technol. Lett.*, **24**, pp. 1642–1645.
21. Ng, W. N., Leung, C. H., Lai, P. T., and Choi, H. W. (2008). Photonic crystal light-emitting diodes fabricated by microsphere lithography, *Nanotechnology*, **19**, p. 255302.
22. Wiesmann, C., Bergenek, K., Linder, N., and Schwarz, U. T. (2009). Photonic crystal LEDs designing light extraction, *Laser Photon. Rev.*, **3**, pp. 262–286.
23. Joannopoulos J. D., Johnson S. G., Winn J. N., and Meade R. D. (2008). *Photonic Crystals: Molding the Flow of Light,* 2nd Ed. (Princeton University Press, Princeton).
24. Fu. W. Y., Wong K. Y., and Choi H. W. (2009). Closed-packed hemi-ellipsoid arrays: A photonic bandgap structure patterned by nanosphere lithography, *Appl. Phys. Lett.*, **95**, p. 133125.

25. Li, K. H., Ma, Z. T., and Choi, H. W. (2012). Air-spaced GaN nanopillar photonic bandgap structures patterned by nanosphere lithography, *J. Appl. Phys.*, **109**, p. 023107.

26. Li, K. H., Ma, Z. T., and Choi, H. W. (2012). Tunable clover-shaped GaN photonic bandgap structures patterned by dual-step nanosphere lithography, *Appl. Phys. Lett.*, **100**, p. 141101.

27. Mak, G. Y., Fu, W. Y., Lam, E. Y., and Choi, H. W. (2009). Metallic nanoparticle array on GaN by microsphere lithography, *Physica Status Solidi (c)*, **6**, pp. S654–S657.

28. Fu, W. Y., Wong, K. K. Y., and Choi, H. W. (2010). Room temperature photonic crystal band-edge lasing from nanopillar array on gan patterned by nanosphere lithography, *J. Appl. Phys.*, **107**, p. 063104.

29. Lu, T. C., Kao, C. C., Kuo, H. C., Huang, G. S., and Wang, S. C. (2008). CW lasing of current injection blue GaN-based vertical cavity surface emitting laser, *Appl. Phys. Lett.*, **92**, p. 141102.

30. Gerard, J. M., and Gayral, B. (1999). Strong Purcell effect for InAs quantum boxes in three-dimensional solid-state microcavities, *J. Lightwave Technol.* **17**, pp. 2089–2095.

31. Lai, C. F., Yu, P., Wang, T. C., Kuo, H. C., Lu, T. C., Wang, S. C., and Lee, C. K. (2007). Lasing characteristics of a GaN photonic crystal nanocavity light source, *Appl. Phys. Lett.*, **91**, p. 041101.

32. Choi, H. W., Hui, K. N., Lai, P. T., Chen, P., Zhang, X. H., Tripathy, S., Teng, J. H., and Chua, S. J. (2006). Lasing in GaN microdisks pivoted on Si, *Appl. Phys. Lett.*, **89**, p. 211101.

33. Ng, W. N., Leung, C. H., Lai, P. T., and Choi, H. W. (2008). Nanostructuring GaN using microsphere lithography, *J. Vac. Sci. Technol. B*, **26**, pp. 76–79.

34. Lide. D. R. (2001). *CRC Handbook of Chemistry and Physics*, 82nd (CRC Press).

35. Neumark G. F., Kuskovsky I., and Jiang H. X. (2007). *Wide Bandgap Light-Emitting Materials and Devices*, (Wiley).

36. Tamboli, A. C., Haberer, E. D., Sharma, R., Lee, K. H., Nakamura, S., and Hu, E. L. (2007). Room-temperature continuous-wave lasing in GaN/InGaN microdisks, *Nat. Photon.*, **1**, pp. 61–64.

37. Apalkov, V. M., and Raikh, M. E. (2004). *Phys. Rev. B*, **70**, p. 195317.

38. Takashima, H., Fujiwara, H., Takeuchi, S., Sasaki, K., and Takahashi, M. (2008). Control of spontaneous emission coupling factor β in fiber-coupled microsphere resonators, *Appl. Phys. Lett.*, **92**, p. 071115.

39. Chen, R., Sun, H. D., Wang, T., Hui, K. N., and Choi, H. W. (2010). Optically pumped ultraviolet lasing from nitride nanopillars at room temperature, *Appl. Phys. Lett.*, **96**, p. 241101.

40. Jewell, J. L., Harbison, J. P., Scherer, A., Lee, Y. H., and Florez, L. T. (1991). Surface-emitting lasers: Design, growth, fabrication, characterization, *IEEE J. Quantum Elect.*, **27**, pp. 1332–1346.

41. Guider, R., Daldosso, N., Pitanti, A., Jordana, E., Fedeli, J. M., and Pavesi, L. (2009). NanoSi low loss horizontal slot waveguides coupled to high Q ring resonators, *Opt. Express*, **17**, pp. 20762–20770.

42. Smith, D. D., Chang, H., Fuller, K. A., Rosenberger, A. T., and Boyd, R. W. (2004). Coupled-resonator-induced transparency, *Phys. Rev. A*, **69**, p. 063804.

43. Little, B. E., Laine, J. P., and Chu, S. T. (1997). Surface-roughness-induced contradirectional coupling in ring and disk resonators, *Opt. Lett.*, **22**, pp. 4–6.

Chapter 10

Epitaxial Lift-Off of GaN Films

Keyan Zhang[a] and Soo Jin Chua[a,b]

[a]*Institute of Material Research and Engineering, 3, Research Link, Singapore*
[b]*Department of Electrical and Computer Engineering,*
National University of Singapore, 21 Lower Kent Ridge Rd, Singapore

elecsj@nus.edu.sg

Gallium nitride–based semiconductors continue to attract much attention for the applications in light-emitting diodes, laser diode, and transistors. The high exciton binding and oscillator strength in GaN and its related heterostructures make microcavity structures very promising for room-temperature operation of devices such as edge-emitting lasers, polariton lasers, and optical amplifiers. Heteroepitaxy is an attractive approach to realize the microcavity, but is restricted by the limited materials quality and the narrow high-reflectivity stopbands. A heterogeneous approach is much more attractive, in which the lift-off high-quality GaN thin film is bonded to a metallic or dielectric mirror. Research on lift-off of GaN thin film has been conducted over the last two decades. This chapter reviews the various methods to lift-off GaN film and its related progress in the field.

Handbook of Optical Microcavities
Edited by Anthony H. W. Choi

ISBN 978-981-4463-24-9 (Hardcover), 978-981-4463-25-6 (eBook)
www.panstanford.com

10.1 Introduction

III-nitride based materials system, having a wide and direct band gap, good thermal stability, and conductivity, is a prominent compound semiconductor for applications in optoelectronics, that is, light-emitting diodes (LEDs) and laser diodes (LDs), high power, high speed electronic devices, and solar cells [1]. However, the advantageous properties of GaN-based materials are hindered by the lack of a suitable substrate on which GaN can be grown.

Due to the high melting temperature and the high equilibrium vapor pressure to synthesize III-nitrides, large bulk single crystals are difficult to be realized for use as substrates in homoepitaxy and are also expensive. A common method to produce GaN is to grow it heteroepitaxially on substrates of dissimilar materials, such as the following commonly used substrates: sapphire, silicon carbide, and silicon. For some substrates, the device performance is degraded because of their relatively poor thermal and electrical conductivity [2,3]. These result in inefficient dissipation of heat generated by GaN-based high-current devices, such as LDs and high-power transistors. The poor electrical conductivity of these foreign substrates, especially sapphire, also complicates the contact and packaging schemes, resulting in higher spreading resistance and higher voltages in device operation. Thus such substrates hinder further improvement and limit III-nitrides to be fully explored for its potential applications and functionalities.

Hence, to eliminate the constraints imposed by the substrates, it is highly desirable for the GaN epilayer to be separated from its original substrates and be transferred to other favorable substrates, so that problems such as limited light extraction, poor heat dissipation, poor reliability, and yield can be eliminated to realize high-efficiency and high-power optoelectronic and microelectronic devices. Apart from meeting the needs of these discrete devices, the heterogeneous integration of III–V compound semiconductor-based optical and electrical systems with silicon-based VLSI circuits is a long pursued goal to enhance the functionality of a microsystem that cannot be delivered by Si technology alone. The combination of high piezoelectric coefficient and strong optical emissions of GaN suggests a variety of possibilities for MEMS and microcavity applications. Beyond all these applications, a simple and cost-effective technique is desirable to controllably separate nitride compound semiconductor from the substrates.

Driven by the above-mentioned applications, considerable effort has been put into generating methods to release and transfer GaN epitaxial layers from their dissimilar substrates since 1990s, including laser lift-off (LLO), photoelectrochemical lift-off, chemical lift-off, and mechanical lift-off. Good progress has been made in heterogeneous integration resulting in improvements in performance of optoelectronics and electronic devices, devices on flexible substrates, MEMS, photonic devices, as well as complicated system in chip level. Compared with well established semiconductor materials system GaAs/AlGaAs, in which an epitaxial lift-off (ELO) [4] method was well established since 1990s due to the availability of the wet etching selectivity between GaAs and AlAs, GaN-based material system does not offer such a sacrificial layer that is compatible with conventional growth method for high material quality and that can be wet etched with high selectivity. Here, we will review the different lift-off techniques, ranging from optical method (LLO) to optical-assisted electrochemical, electrochemical, chemical, and mechanical methods and present their state-of-the-art performance.

10.2 Lift-Off Techniques and Their Applications

10.2.1 Laser Lift-Off

LLO method is developed for the rapid separation of GaN epilayers from their UV transparent sapphire substrates.

In this process, a sequence of high-power laser pulses is incident on the GaN epilayer through the sapphire substrate on which it is grown, as illustrated in Fig. 10.1. The sapphire substrate is transparent to the incident photons but are strongly absorbed by GaN at its interface with sapphire, resulting in localized heating at the GaN/sapphire interface and raising the temperature to above 800°C. This results in thermal decomposition of interfacial GaN into gaseous nitrogen and metallic gallium. The GaN film is subsequently removed by melting Ga above 39°C [5].

In the case of GaN on sapphire, the widely explored laser systems are the third harmonic of a Nd:YAG laser (λ = 355 nm) [6] or excimer lasers (λ = 248 nm) [5,7]. Using Nd:YAG laser as an example, the melted zone [8] caused by a single pulse was found to increase monotonically in thickness from 0 nm at a threshold intensity of 250 mJ/cm^2 to a few hundred nm at 600 mJ/cm^2. At a higher laser intensity, the etch depth saturates due to the accumulation of metallic Ga at the interface. The lateral homogeneities of the laser beam have a drastic effect on the macroscopic quality of the process. Nowadays, with the development of the excimer lasers with homogeneous beam with fluence variation typically less than 1 mJ/cm^2 over an illuminated area of 5 mm^2, the lift-off time can be shorter than 1 min for a 2 inch diameter wafer [9].

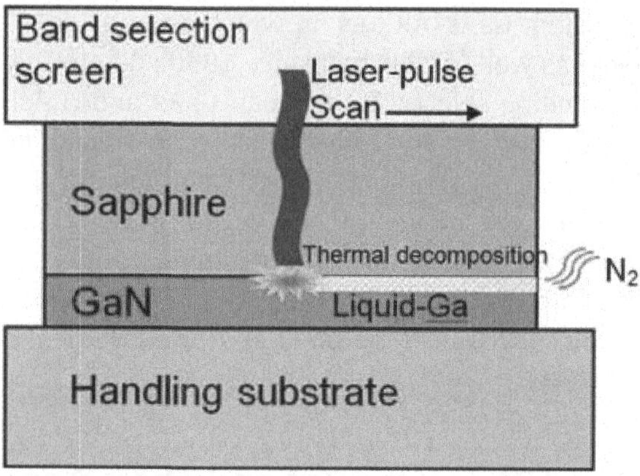

Figure 10.1 Schematic diagram of LLO.

Having a fast etching rate and being a simple process, LLO is a widely developed process available for creating thick free-standing GaN [8], GaN-based vertical thin film LEDs, and photonic crystal LEDs with improved external efficiency [10,11] and for LDs [12].

It should be noted that during the laser irradiation process, due to intense localized heating, a high degree of disorder is created, resulting in cracks and fractures, a result of uneven thermal expansion, making the production of uniform and crack-free films difficult and resulting in an increase in reverse bias leakage current [13]. Thus, the strain state in the GaN film plays an important role during the lift-off process and is the main cause for cracks to form. To reduce the wafer bowing-induced cracks, due to the difference in the thermal expansion coefficient for freestanding GaN (thickness >100 μm) and the substrate, Miskys et al. [8] performed LLO process above 600°C to reduce the bowing. To prevent cracks being formed in the thin film GaN (a few μm thick), which is mechanically fragile, usually the GaN film is wafer bonded onto a substrate acting as a mechanical support with favorable thermal and electrical properties. GaN-based devices could be integrated into receptor substrates with better properties or for increased functionalities such as on Si with logic circuits. Various wafer-bonding processes, such as using Pd-In metal [14], direct wafer fusion [15], AuGe-mediated wafer bonding [16], SiO_2 interlayer bonding [17], electroplating [18], were developed to bond GaN onto Si, GaAs, metal, and other substrates. Individual devices on a GaN wafer are normally isolated prior to wafer bonding to prevent the propagation of the cracks and reduce the damage caused by the non-uniformity at the edge of the laser beam.

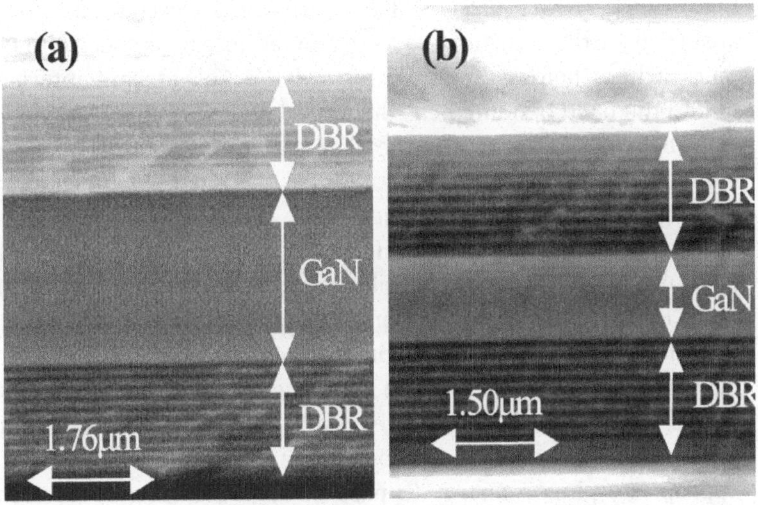

Figure 10.2 Cross-sectional SEM images of GaN microcavities formed using SiO_2/ZrO_2 DBR mirrors and LLO (a) without ICP etch-back and (b) after ICP etch-back, reducing the GaN thickness from 2 to 0.8 μm.

An example is illustrated in Fig. 10.2 for the LLO application in GaN microcavity [19], in which GaN thin film is sandwiched between two oxide-multilayer distributed Bragg reflectors. First silica/zirconia mirror is deposited on the as-grown GaN before bonding onto a silicon substrate for LLO process to separate the GaN from the sapphire. After the removal of the sapphire substrate, inductive coupled plasma etching was then used to thin down the lift-off GaN film for the control of the cavity length and modification of the exposed surface. The second silica/zirconia multilayer mirrors were then deposited by using e-beam evaporation.

10.2.2 Photo-Assisted Electrochemical Lift-Off

An alternative method to lift off GaN-based epilayers is photoelectrochemical (PEC) selective lateral etching. The method was first reported by Minsky et al. [20] and later further explored by Youtsey et al. [21] and Haberer et al. [22], among others. PEC etching uses a high-power UV light to selectively excite a sacrificial layer designed to be incorporated in GaN/sapphire structure during the MOCVD growth process. The energy of the excitation is then absorbed in the sacrificial layer to generate excess electron-hole pairs, where holes are drawn toward the GaN-etchant interface due to the band bending near the material surface, followed by the electrochemical reaction at the GaN/electrolyte interface for the lateral etching process.

Figure 10.3 shows a typical PEC setup, using a UV lamp (1000 W Xe/Hg) and maintaining an electrical contact between the platinum cathode and GaN sample anode, immersed into dilute KOH or HCl as an electrolyte. Similar to LLO, PEC process is band gap selective in etching, depending on the photon energy of the incident light, as compared with the band gap energy of the sacrificial GaN-based layer. Hence, by using backside illumination through the sapphire substrate, the transparent sapphire allows light with appropriate photon energy to pass through resulting in electron-hole pairs generation in the sacrificial layer causing the etching of GaN near the interface. The photo-generated holes assist the oxidation on the surface and subsequent dissolution of the GaN into aqueous solution. The excess electrons are conducted away from the GaN to the Pt cathode in solution, in which they concurrently participate in a reduction reaction. The electrochemical reaction is shown in the following equations:

$$2\mathrm{GaN} + 6\mathrm{h}^{+} \rightarrow 2\mathrm{Ga}^{3+} + \mathrm{N}_2$$

$$\mathrm{Ga}^{3+} + 6\mathrm{OH}^{-} \rightarrow \mathrm{GaO}_3^{3-} + 3\mathrm{H}_2\mathrm{O}$$

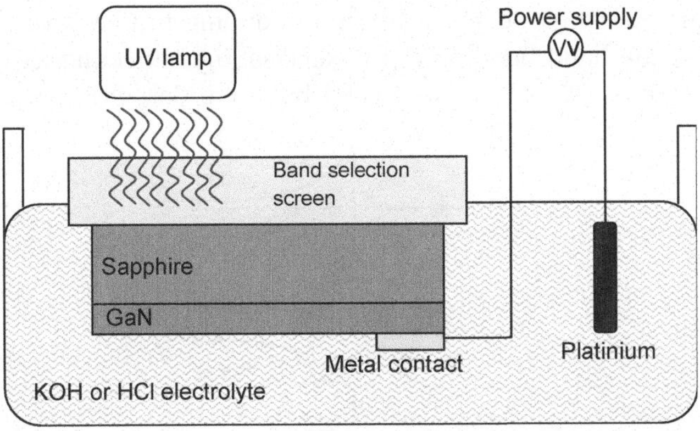

Figure 10.3 Schematic diagram of photoelectrochemical lift off [23].

Early result demonstrated by Stonas et al. [24] shows the lateral etching of unintentional n-doped GaN, leaving a lifted-off p-GaN layer, with a rough etching morphology. Various sacrificial layers are developed later, including thick InGaN and InGaN superlattice [22] for successful lift-off of GaN and improving the selectivity and smoothness of the etching surface.

A key factor of the PEC etching is the initiation of etching through the formation of photogenerated holes in the sacrificial layer. Hence, selection of band gap of the sacrificial layer and the photon energy of the incident light needs to be carefully considered. For example, Stones et al. used InGaN as a sacrificial layer, and the photon energy of the incident light is carefully adjusted to fall above the band gap energy of the InGaN layer, but below that of the GaN.

Another important factor that needs to be considered to achieve uniform etching is the polarization-induced piezoelectric field or the band bending and carrier confinement in the InGaN layer (Haberer et al. [26]). The photo-generated carriers need be confined to the interface with the electrolyte so that they can react with the electrolyte before they diffuse and recombine. In case of insufficient confinement, the carriers have more chances to recombine before participating in the reaction, causing the low etching rate or incomplete etching along the dislocation lines. In addition, the doping of the epilayer also affects the etching selectivity; as discussed earlier, the holes that have been generated are drawn to the electrolyte interface for the reaction by the band-bending [25].

At low electrolyte concentration, the doping in GaN layer largely affects the band bending at the surface, thus affecting its etching rate. Hence, n-GaN could be etched, but not p-GaN or i-GaN [27].

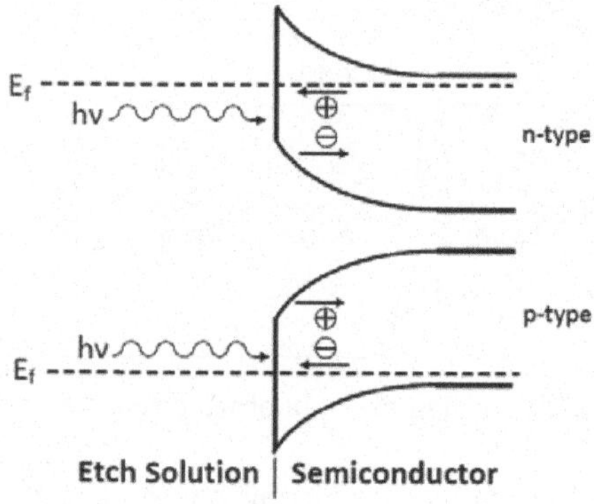

Figure 10.4 Band diagram of the electrolyte-GaN interface showing band bending effects on photo-generated electron hole pairs. Holes are forced toward the surface in an n-type semiconductor and toward the bulk in a p-type semiconductor [25].

Apart from the points mentioned above, there are other factors affecting the etching rate and the surface morphology, including (1) selection of sacrificial layer, (2) intensity of incident light, (3) selection of electrolytes and its concentration, (4) external bias, and (5) the doping of the epilayer. The sacrificial layer in the PEC process is generally a conventional GaN-based material available in the epitaxial growth process, including InGaN and InAlN, where the materials quality is not seriously degraded. In the PEC etching reported, generally no external bias (0 V) is applied. But very low external bias was reported to enhance the surface smoothness [28]. The electrolyte and its concentration largely affect the surface morphology of the etching surface. For example, Youtsey et al. [29] studied the KOH concentration and its etching morphology, and they found that KOH concentration below 0.04 M produced a smoothly etched GaN surface, but it is strongly diffusion-limited etching with a low etching rate. At higher KOH concentration, the etching rate increases and changes to reaction-limited process, where anisotropic crystallographic etching happens [30].

By using PEC wet etching process, various structures have been successfully demonstrated, including microdisk resonator [31] and cantilever beam for MEMS application [32].

Harber et al. reported an optically pumped GaN/InGaN microdisk laser fabricated by PEC etching. The epitaxial structure for the microdisk laser is shown in Fig. 10.5, including disk layers for the gain medium and the sacrificial postlayers for undercut. The disk layers consist of a 34 nm GaN upper cladding layer, an InGaN multiple quantum well active region, and a bilayer lower cladding layer of 14 nm GaN and 20 nm $Al_{0.2}GaN$. The sacrificial poststructure consisted of a 5 period 20 nm $In_{0.04}$ GaN and 20 nm $In_{0.07}GaN$ super lattice. Pt was deposited around the disk area as a cathode after the dry etching process to define the disk area. By doing so, the group was able to selectively etch the lower composition InGaN superlattice postregion without damaging the higher composition InGaN MQWs active layer. An unintentionally doped GaN filter was used to preferentially filter out light with photon energy greater than the bandgap of GaN, allowing only the lower energy light to pass. A dilute 0.004 M HCl solution was used as the electrolyte during the PEC etching. The undercut etch rate was nearly 25–30 nm/min. Subsequently, by using electron-beam lithography and PEC etching, GaN microdisks with smaller diameter, smooth side walls, good circularity, and uniformity across the sample were reported by Tamboli, demonstrating the first room-temperature continuous-wave lasing in GaN/InGaN microdisks. The 1.2 μm mushroom disk showed a very low threshold power density of 300 W/cm^2.

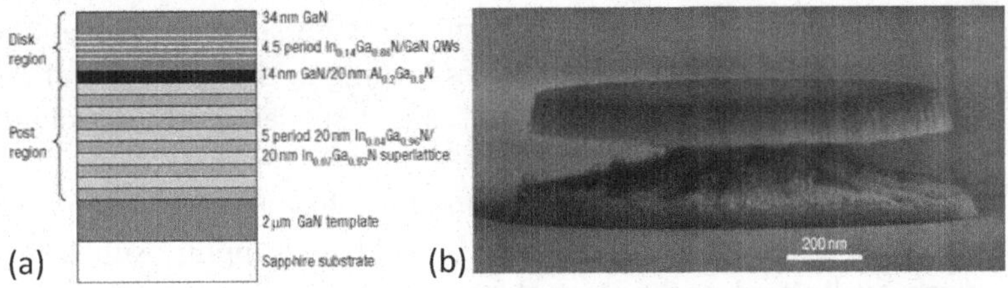

Figure 10.5 (a) Epitaxial structure of material used to fabricate microdisks. Light is confined to the disk region, and the postregion is selectively etched to form an undercut structure. (b) SEM image of a 1.2 μm microdisk fabricated by using PEC lateral etching [31,33] (Reprinted with permission. Copyright 2006 Nature Publishing Group).

Being simple, low cost, the PEC etching is proven to be an effective method to achieve selective lateral etching of GaN structures.

However, being diffusion limited, PEC etching rate is low, generally less than 100 nm/min, and it is difficult to achieve a smooth etching surface in the reaction-limited regime, which hinders its large-scale application. Nevertheless, the method demonstrates its potential applications in novel nano-devices.

10.2.3 Electrochemical Lift-Off

In recent years, a few electrochemical etching methods were developed to simplify the etching process without using photo-excitation, but using bias voltage. The technique is to use either "doping selective" or "defect selective" mechanism to enhance the electrochemical reaction without using photo excitation. The methods were developed with simple design and with improved controllability and better uniformity.

By applying an external bias to the sample, the electrochemical reaction is activated due to a change in the electrical potential, leading to a more than proportionate change in reaction rate. At a higher positive potential, a stronger band bending will occur, driving charged carriers to a partial spatial separation [34]. If electrons are made available, the potential at the surface becomes more negative, as excess electrons accumulate at the interface waiting for the reduction-oxidation (redox) reaction to occur through carrier exchange across the semiconductor-electrolyte (S-E) interface. Redox induces a change in oxidation number thus breaking one or more bonds. The applied external voltage source removes the bonding electrons by injecting holes leading to the oxidation of the semiconductor with the resulting oxides subsequently dissolving into the electrolyte [30].

J. Park [35] reported the doping-selective electrochemical etching method to release large area GaN from sapphire substrates, using oxalic acid as an electrolyte. A highly doped n-typed ($n > 1 \times 10^{18}$ cm^{-3}) GaN layer is used as a current conductive layer to be anodized and removed during the EC etching process. The procedure is similar to the anodic formation of porous aluminum oxide. The electron current flows through the highly doped n-GaN layer from an Indium contact. Nanopores are initially formed (Fig. 10.6) in the n-GaN layer and nanopores then merged together to form larger pores, and finally the n-GaN is laterally etched away. A lateral etch rate of 12 μm/min was achieved under 60 V bias for n-type doping concentration of 8×10^{18} cm^{-3}. A GaN epitaxial layer patterned into 300×300 μm^2 was demonstrated to be released from the sapphire substrate.

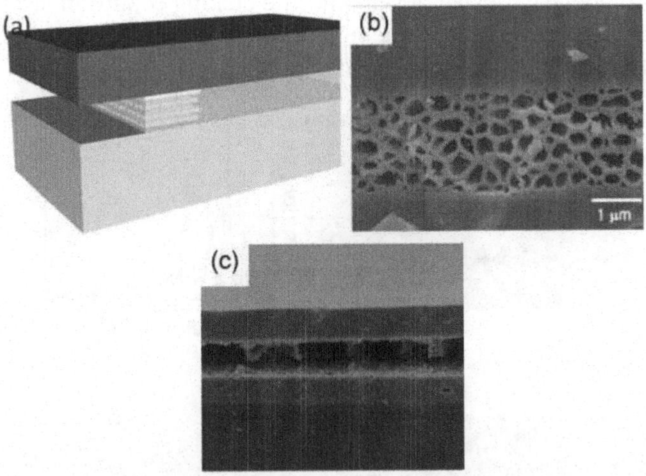

Figure 10.6 (a) Schematic diagram of doping-selective lateral electro-chemical (EC) etching of GaN for lift-off, (b) cross-sectional SEM image of nanopores created during the EC etching, and (c) cross-sectional SEM image of lift-off region of undoped GaN with the etching away of n-doped GaN [35] (Reprinted with permission. Copyright 2009 American Institute of Physics).

Y. Zhang [36] further studied the electrochemical etching conditions, that is, doping concentration and applied voltage and their effects on the etching characteristics. Three regimes are identified: (i) no etching, (ii) formation of nanopores, and (iii) electropolishing depending on the etching conditions. The group also demonstrated the possibility of tuning the etching pathways at microscale through designing the conductivity and changing the electrical voltage. Various structures are reported: microdisk, distributed Bragg reflectors, cantilever, and large area nanostructured free-standing GaN film [37]. Using n-GaN as a sacrificial layer is completely compatible with the conventional GaN system with no degradation in material quality during the MOCVD growth and has flexibility in the design of the structures. It offers a flexible choice in the structure design, for example, GaN/air and GaN nanoporous DBR. On the contrary, as the doping layer plays an important role for the conduction of the electron current and subsequently being etched away, there is a limitation in the selection of device structure, where both p-type and n-type doping are needed for the device. Device design and the etching condition are needed to be considered simultaneously to realize the applications of this method in lift-off of GaN-based optoelectronic devices.

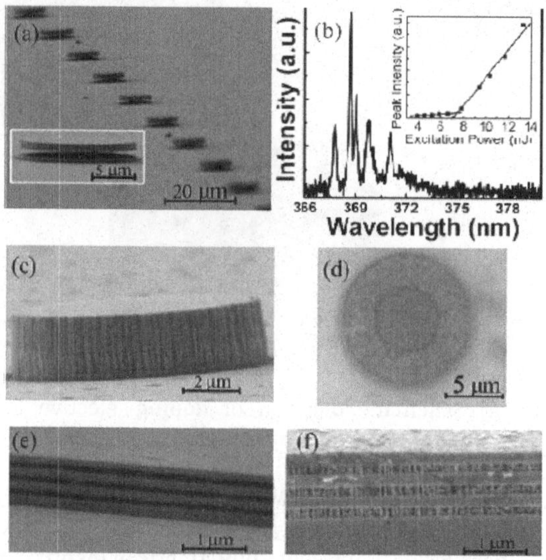

Figure 10.7 (a) GaN microdisk array prepared by selective lateral undercut etching, (b) light emission spectrum from an optically pumped GaN disk above threshhold. The inset shows the emission intensity versus excitation power (c) SEM and (d) optical microscope image of the GaN microdisk with a nanoposous layer. (e) and (f) SEM images of GaN/air and GaN/porous GaN DBRs [36]. (Reprinted with permission. Copyright 2010 John Wiley and Sons.)

The methods discussed above all use conventional GaN materials where generally the dislocation density is on the order of 10^8–10^{10} cm^{-2}. Our group developed a lift-off method via "defect selective" electrochemical etching that has the advantage that in the preparation of the sacrificial layer, a better quality GaN film results, which is later lifted off from the substrate.

A nanoporous GaN template was first fabricated by ICP etching using anodic aluminum oxide (AAO) films as etch mask. The average pore diameter is 65 nm, interpore distance is 110 nm, and the depth of the pores is 70 nm. Next, GaN growth is carried out over the nanoporous GaN by MOCVD. Nano-airbridges are formed over the pores and merge to form a continuous film. The quality of the nano-airbridge GaN has its threading dislocation density reduced by at least one order of magnitude compared with that of the conventional GaN grown on sapphire [38]. The epitaxial film formed over the nanopores results in nano-voids being embedded in the overgrown GaN layer. The continuous layer of air-bridge mediated lateral overgrowth is shown in Fig. 10.8. The nanoporous GaN interlayer can serve as an effective buffer to reduce the strain and defect density in the overgrown GaN improving its surface morphology and optical quality [39].

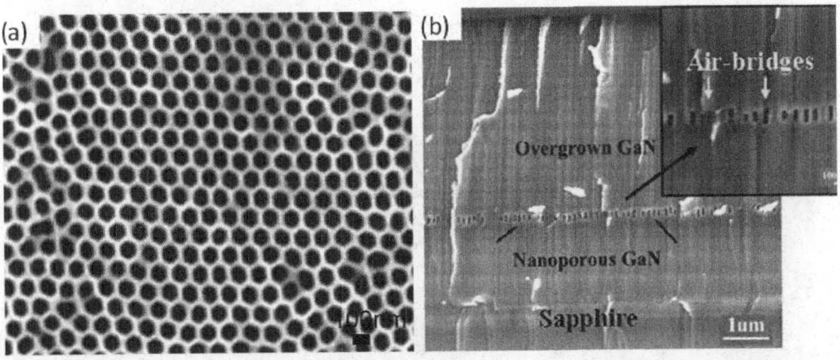

Figure 10.8 (a) SEM images of the nanoporous GaN film. The average diameter of the pores in GaN is about 65 nm and the pores are uniformly distributed over the GaN surface. (b) Cross-sectional SEM image of the overgrown GaN on the nano-porous GaN template. The inset shows the formation of nanoscale air bridges. (Reprinted with permission. Copyright 2005 American Institute of Physics.)

The nanoporous "defective" GaN was then used as the sacrificial layer for EC etching. The nanopores act as channels for the transport of the reactant and offer a large area exposed to etching. The epitaxial GaN, referred to as nano-airbridge GaN (NAB GaN), is separated from the layer below by the pores. It is thus "weakly" bonded to the layer below through nanowires and nanowalls bounding the pores. The electrochemical etching process proceeds by etching the nanowires and nanowalls exposed to the KOH electrolyte. An Au contact is deposited on the n-GaN while a Pt electrode acts as the anode. Via the electric field induced across the sample and with the transport of the reactant through the pores, the porous layer is etched away with a current of about 100 μA. The GaN layer is lifted off across the nanoporous interface (Fig. 10.9). Cross-sectional SEM shows the separation of the NAB GaN at the nanoporous layer leaving a smooth surface. In some areas, inverted cone structures are also observed, as seen on surfaces of previously PEC etching samples. It is suggested that their formation is related to the reaction of GaN with KOH, where the etch rate is different for different GaN facets.

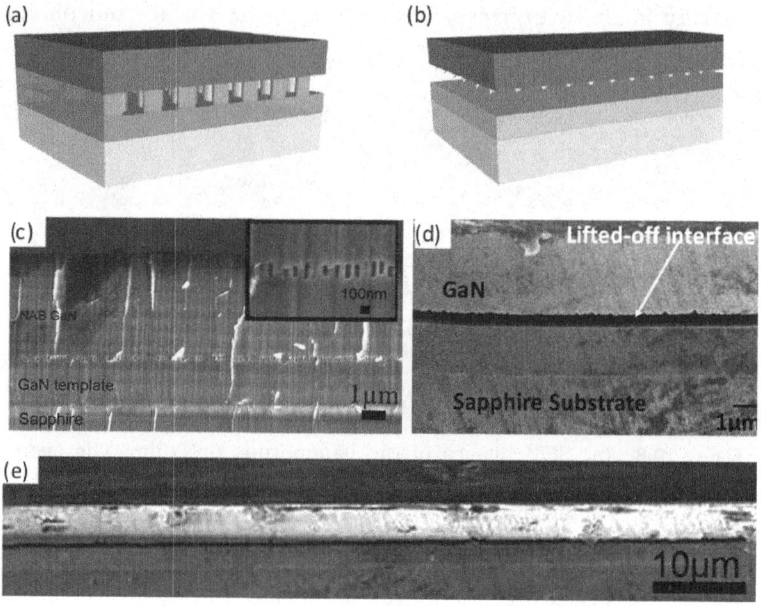

Figure 10.9 (a) Schematic diagram showing the side view of NAB GaN. NAB GaN is grown by MOCVD over the nanopore-GaN template. (b) Electrochemical etching releases the NAB GaN epilayer. The separation happens at the NAB interface. (c) Cross-sectional SEM images of NAB GaN before the release and (d) after the release; Insets are enlarged images of NABs. (e) A cross-sectional SEM image of a lift-off structure showing the uniformity in the etching.

Electrochemical etching, to lift-off the NAB GaN layer from substrate, is optimized with the use of 10 M KOH concentration. Localized etching with an etch rate of 0.1 μm/min is successfully carried out.

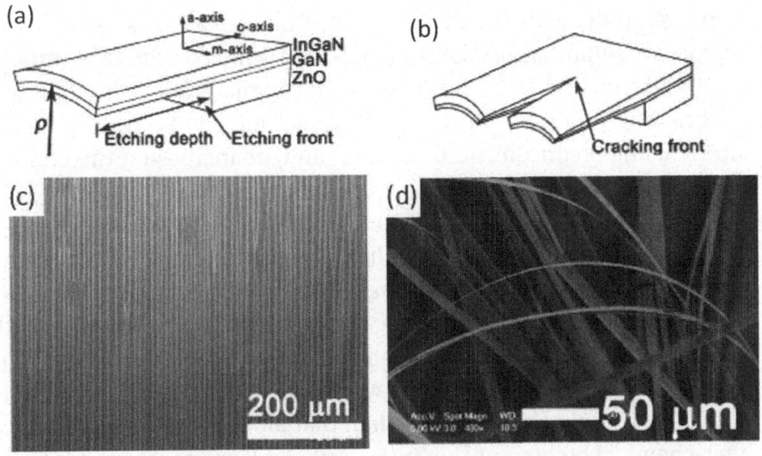

Figure 10.10 Schematic diagram of the (a) curling and (b) cracking of the InGaN/GaN bilayer film when the sacrificial ZnO template is laterally etched in solution. (c) Microscopy image recorded from sapphire substrate side showing the nanobelt-flaking. (d) SEM image of the released nanobelts [44].

We have introduced above the application of the EC lift-off technique on nano-air-bridge GaN to obtain a GaN film with high material quality. The method comprises of a nano-epitaxial lateral overgrowth on a porous GaN template followed by EC etching. The high-quality GaN thin film is formed, as the nanopores act as a "filter" in annihilating some of the threading dislocations through the bending of the dislocations during the overgrowth. This method provides a technique for yielding a high-quality GaN thin film layer, where GaN devices could be build upon. However, the etching rate needs to be improved for large-scale application.

10.2.4 Wet Chemical Lift-Off Using Sacrificial Layer

An alternative and simple way to release GaN epilayer from a foreign substrate is by chemical lift off. CrN [40], InAlN [41], and ZnO [42] were reported to be grown separating the foreign substrate and the GaN epilayer, and then be laterally, selectively chemically etched away in order to lift-off the GaN epilayer grown over it.

Our group [43] reported the chemical lift-off of MOCVD epitaxially grown InGaN/GaN heterostructures by using a sputtered ZnO layer acting as a buffer and sacrificial layer on c-sapphire and r-sapphire substrates. By using a black-wax technique and a vapor wax elimination method, a 10 × 10 mm^2 crack-free InGaN/GaN epilayer was lifted off from the c-sapphire substrate followed by bonding to a glass substrate. This method can be readily scaled up by using a concentrated solution and/or increasing the etching time.

By using the chemical lift-off method and strain engineering, Liu et al. [44] reported high-quality GaN-based nanobelts arrays via strain-controlled cracking of thin solid films. By epitaxially growing InGaN/GaN bilayer film on an r-plane sapphire substrate with ZnO as a sacrificial layer on sapphire, the in-plane elastic stress is asymmetrically stored in the bilayer film with the component along [1-100] axis being much larger than that along [0001] axis. A lateral wet-chemical etching of the ZnO sacrificial layer causes the InGaN/GaN bilayer film to lift off and crack into strips parallel to [11–20] direction while perpendicular to [1-100] and [0001]. The crack fronts proceed along the [0001] axis with the increase in etching time, forming paralleled alligned nanobelt arrays. The structure and doping of these belts can vary, allowing greater control for potential applications in piezotronic and optoelectronic devices.

One thing to be noted is that for the lift-off, strain in the GaN layer plays a very important role in the final lifted off geometry, morphology, and is readily linked to the performance of the devices built on it.

The purely chemical lift-off approach is rather simple and inexpensive compared with the other lift-off method. However, the challenge is that the material quality is far less optimized up at the present stage of research, because a nonoptimized buffer layer is used during the MOCVD growth, compared with the conventional growth of GaN using low-temperature GaN or AlN buffer. Hence, for the commercial application of this technique, an optimized buffer layer and improved materials quality need to be realized. From another point of view, this technique enables flexible design in nanostructure for the applications in piezotronics, optoelectronic, and PV devices.

10.2.5 Nanoepitaxial Lift-Off

As discussed above, a purely chemical lift-off process is preferred for simplicity, low cost, and batch processing to achieve separation and release of GaN thin film. However, the material quality achieved for the structure that can be chemically lifted-off is not optimized compared with the conventional GaN. Hence, a structure that has high material quality and meanwhile can be chemically lifted-off would be an advantage for producing high-quality, free-standing GaN thin films for the applications in heterogeneous integration of silicon logic devices, III–V power devices, and other optoelectronic devices.

A technique [45] has been developed comprising a simple wet chemical etching method to release high-quality epitaxial III-nitride films from their substrates. This method builds on a nanoepitaxial lateral overgrowth (NELO) structure that provides III-Nitride films with low dislocation densities. NELO is accomplished using a nanoporous SiO_2 mask layer patterned on GaN substrates. We [46] reported the improved material quality with low dislocation density ($\sim 10^7$ cm^{-2}) in the NELO structure and the LEDs built on it demonstrated a large improvement in light emission (Fig. 10.11). Chemical removal of the SiO_2 layer after growth of III-Nitride overlayer causes fracture at the interface between the GaN film and the original GaN substrate, resulting in a free-standing GaN film with a nanostructured surface on one side (Fig. 10.12). These layers can be transferred to other substrates, and the nano-structured surface can be used in photonic devices, or power devices. This method offers the advantages of a low-cost simple process with infinite etching selectivity and it produces GaN of high crystal quality.

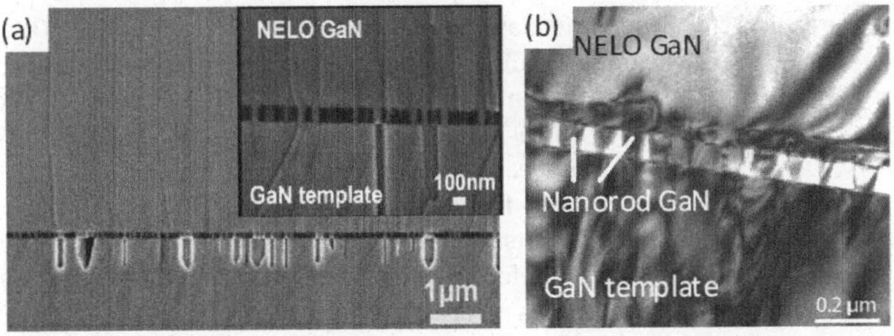

Figure 10.11 (a) Cross-sectional SEM image of nanoscale lateral epitaxially overgrown GaN. Inset is a higher magnification SEM image near the interface of SiO_2 mask. (b) Bright field TEM image of the NLEO GaN, which shows the reduction of threading dislocations by the blockage imposed by the SiO_2 nanomask and subsequent bending and annihilation with one another near the interface.

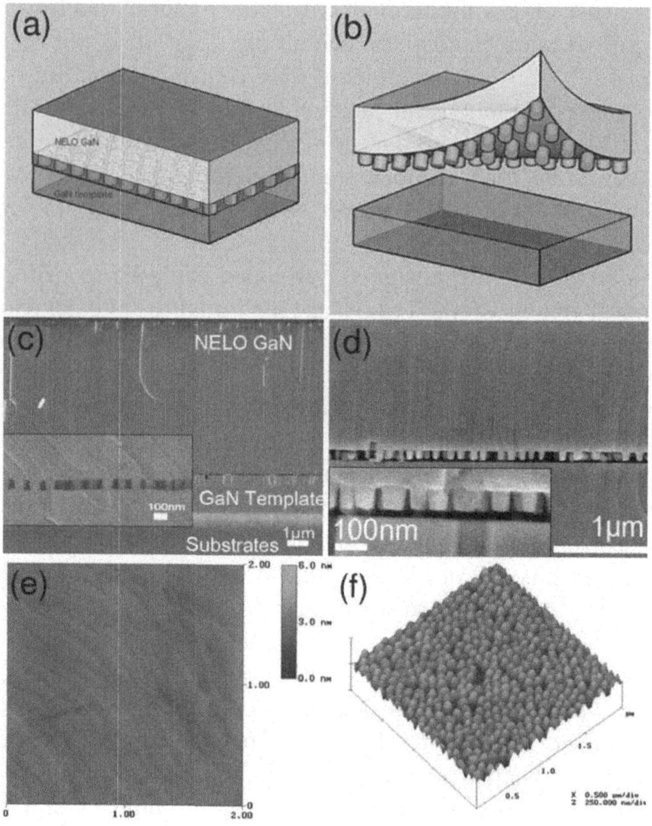

Figure 10.12 (a) Schematic diagram of structures of NELO GaN. NELO GaN is grown by MOCVD through the nanopores of SiO_2 mask that are fabricated on GaN substrate. (b) Nanoepitaxial chemical release of NELO GaN epilayers by using HF solution. The chemical removal of SiO_2 mask leads to the NELO GaN being separated from the GaN template with GaN nanorods remaining on its surface. (c) Cross-sectional SEM images of NELO GaN before release and (d) after release; Insets are enlarged images near the interface of nanostructured SiO_2 mask and GaN nanorods. (e) AFM images of the top surface of released NELO GaN epilayer showing low defect density. (f) 2D view of bottom surface of released NELO GaN showing the dense array of nanorods. The diameter of the rod is ~60 nm, and the height of the nanorod is ~100 nm. It indicates that the separation happens at the interface of NELO GaN and the GaN template.

A lateral etch rate in HF solution at a concentration of nearly 26.5 wt% at room is estimated at nearly 15 μm/min for a 100 nm thick SiO_2 mask with diameter of the nanopores of nearly 60 nm and interpore distance of nearly 100 nm. A free-standing 300 μm^2 GaN film was lifted off and transferred to PDMS (Fig. 10.13). Further experiments are being conducted in order to fully understand the lift-off mechanism. We believe that if combined with an improved transfer and bonding technology, this lift-off method could provide a simple and yet effective way to integrate GaN-based devices with Si circuits or other favorable substrates for application in power devices and high-efficiency optoelectronic devices.

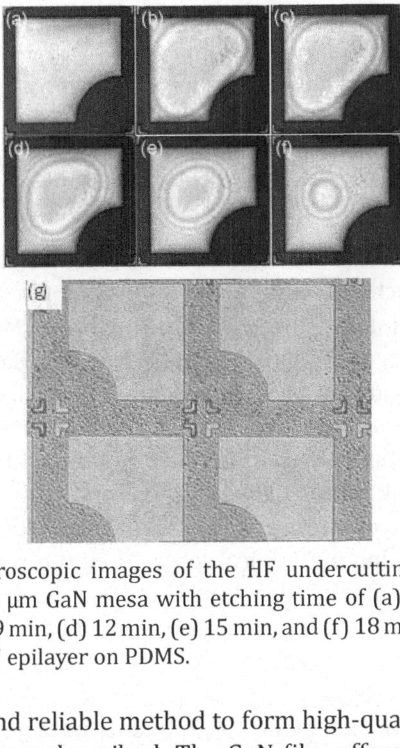

Figure 10.13 Microscopic images of the HF undercutting of a 300 μm × 300 μm GaN mesa with etching time of (a) 3 min, (b) 6 min, (c) 9 min, (d) 12 min, (e) 15 min, and (f) 18 min. (g) Transferred GaN epilayer on PDMS.

A simple and reliable method to form high-quality freestanding GaN film has been described. The GaN film offers a better crystal quality, beneficial for high efficiency and high power applications. The short nanorods left on the GaN surface after the lift-off can be used for specific applications such as for improvement of the light extraction efficiency for LED or to create hierarchical structures for nanophotonic applications. The method is potentially of interest for applications of heterogeneous integration of III-nitride based optical and power devices on Si or flexible substrates, as well as various other applications, for example, MEMs system, microcavities and microoptics systems, and nanophotonics.

This method offers the advantages of simple design, low cost, and can be scaled up to a larger wafer size, and at the same time offers an improved crystal quality and device performance. The key factors in this lift-off method include (i) mask material enabling the selective removal by HF acid; (ii) nanostructured mask layer enabling the self-debonding of nano-ELO GaN at the interface after removal of silicon diode mask. (iii) The GaN layer is self lifted off from the substrate near the interface mask by wet chemical etching. (iv) NELO GaN has improved crystal quality, low defect density, and can be used at a large scale with high yield. (v) The aperiodic array of nanorods left on the surface after lift-off can be used for light extraction or absorption.

10.2.6 Mechanical Lift-Off

Apart from the methods discussed above, mechanical lift-off is generating much interest owing to its simplicity and potential applications in large scale and throughput. Mechanical lift-off was first reported by Chung et al. [47], where GaN was grown on ZnO/graphene and was mechanically released at the graphene layer. Graphene attracts extensive research interest recently. It has a layered structure of strong covalently bonded hexagonal carbon atoms and weak bonds between the atomic layers. These weakly bonded inter-layers act as an excellent media that can be simply separated by using mechanical force. They reported the growth of GaN-based LED device structure using ZnO as a buffer layer and the graphene layer as a mechanical release layer. The LED devices were released and transferred to various substrates, metal, glass, and plastic substrates.

Similar to the structure of graphene, hexagonal boron nitride (h-BN) also has a layered structure. Kobayashi et al. [48] reported using hexagonal boron nitride (h-BN) layer to mechanically release GaN-based device structures (LEDs and HEMTs) to be placed onto foreign substrates. The h-BN layer serves two purposes: as a buffer layer for the growth of high quality of GaN, and as a shear plane to mechanically release the GaN epilayer. They demonstrated a 2 ×

2 cm^2 of GaN layer that was mechanically lifted off and transferred to a 1 × 1 cm^2 Indium sheet without showing any cracks (Fig. 10.14).

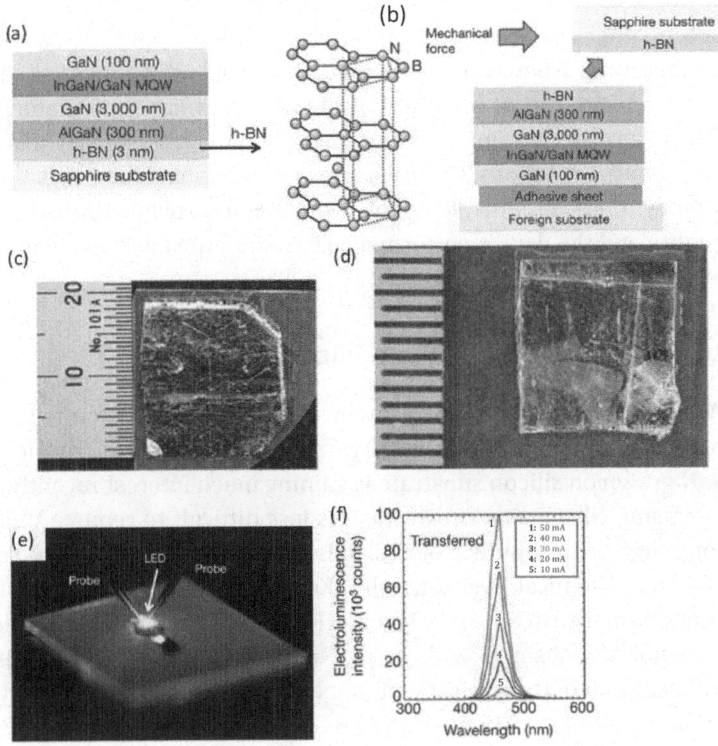

Figure 10.14 (a) Schematic diagram of GaN-based structure using h-BN as an intermediate layer for its release and transfer; on the right, the crystal structure of h-BN is shown; (b) the MQW structure is flipped upside down and placed on a foreign substrate via an Indium sheet. After heat-sealing the Indium, the MQW structure is released from the host sapphire by mechanical force and transferred to a foreign substrate. (c) Photograph of the transferred AlGaN/GaN structure attached to the Indium sheet on the foreign sapphire substrate; the smallest division in the scale at left is 500 μm. (d) Photograph of the transferred MQW structure attached to the Indium sheet on the foreign sapphire substrate. One division in the scale at left is 500 μm. (e) Optical image of the blue-light electroluminescence from the transferred vertical LED. (f) Electroluminescence spectra from the transferred LED with currents ranging from 10 to 50 mA [48]. (Reprinted with permission. Copyright 2012 Nature Publishing Group)

The mechanical lift-off method, compared with conventional lift-off methods, is a promising lift-off technique that has the advantages of very fast release rate, simple design, low cost, and potentially large area application. The process is operated at room temperature, where no heating and chemicals are being applied. It opens up the use of a variety of bonding and handling materials, and meanwhile minimizes the damages that may result from heating or chemical reactions. As the use of graphene and/or BN is at its infancy, there is still plenty of room for improving the material quality and the device performance based on the mechanical lift-off process.

10.2.7 Lift-Off of GaN-on-Silicon

With the development of growth techniques for GaN on large-scale silicon substrate, driven by cost, the release and transfer of GaN grown on silicon substrate is gaining much interest recently.

Using silicon as a substrate, it is less difficult to remove silicon compared with sapphire or SiC. Si is easy to be dry etched by XeF_2 [49–51], chemical wet etch by KOH [52], TMAH (tetramethyl ammonium hydroxide) [53] or HF/HNO_3 [54], or by chemical mechanical polishing with a high removal rate. GaN micro-disk was fabricated and optically pumped and lasing has been reported in these disks [51]. Micron thin film GaN-based transistors and LEDs released and transferred from silicon substrate and transferred to flexible and stretchable substrates have been successfully demonstrated [52,53]. A big concern for GaN on silicon is the material quality and the bonding techniques compared with the separation techniques. GaN is relatively more difficult to be grown on silicon substrate than on sapphire and SiC substrates due to the large mismatch in lattice constants and thermal expansion coefficients. A large dislocation density and cracks are usually generated in GaN film. Hence, GaN grown on silicon substrate usually consists of a thick AlGaN-based buffer layer, including AlN and graded AlGaN as a strain engineering layer to prevent cracks and to reduce the threading dislocation density. For better device performance, the thick buffer layer nearly 1–2 μm is preferably removed, in the way of back polishing or by other dry etching method.

10.2.8 Comparison of the Merits of Different Etching Techniques

We reviewed the state-of-art technology for release of GaN epilayers from its original substrates. As they are at very different levels of maturity, a comparison of their present merits is not very straightforward. Among different reported methods, LLO is widely studied and it is now available in the industry. It is compatible with conventional GaN growth on sapphire or SiC, offering good materials quality. However, it needs the use of a high power laser and hence high cost. The technique generates a lot of heat and hence there are restrictions on the selection of the bonding materials. Considering this, mechanical lift-off has the advantages of flexibility in materials selection, where no heating and chemicals are used. However, the handling of the brittle GaN materials is a challenge. Chemical lift-off is simple, but has a slow release rate, and there is also a restriction on the selection of the handling materials due to the use of chemicals. In many respects, these methods have common issues and problems, including stress in the film, bonding, and transfer materials.

GaN is commonly known to have a high value of built-in residual stress and/or micro-cracks. There is a change in stress before and after release that may cause the bowing of the film and finally resulting in cracks after the release from the substrate. An example of stress-induced crack during the release stage is discussed in Section 10.2.4: cracking of InGaN/GaN into ribbons due to a large difference in in-plane strain when grown on r-plane sapphire substrates. In this case, the stress has been used for a useful purpose to produce nanobelts. Other examples are the formation of the tubes, spirals, and wrinkle structures during the release process [55,56]. Nanoepitaxial lift-off utilize the strain interaction of the nano-GaN with the mask [44]; Tomita et al. [56] reported the self-separation of HVPE GaN grown on striped GaN seeding layers on sapphire substrate. The separation of the thick GaN layer takes place during the growth sequence at the interface of GaN/sapphire induced by stress. The stress in the thin film is also of general concern, which might lead to device performance degradation and long-term reliability problems.

GaN thin films are mechanically fragile; hence, the method to handle them during release process is a key parameter for the lift-off process. As discussed earlier, wafer bonding to a support is a commonly adopted method. Various wafer-bonding processes such as using Pd-In metal [14], direct wafer fusion [15], AuGe-mediated wafer bonding [16], SiO_2 interlayer bonding [17], and electroplating were developed to bond GaN onto Si, GaAs, metal, and other substrates. An alternative choice is to rely on the Van der Waals forces of the newly formed interface to assure adhesion of the thin film. Rogers et al. [53] used soft stamp to pick-up the microstructures and place them on to other substrates by changing the release rate of the handling material. Full automation in the pick-and-place technique is available in laboratory scale. For wafer bonding, a high bonding strength at low temperature with high yield and reliability is a requirement.

10.3 Summary and Outlook

GaN, a wide band gap compound semiconductor, offers excellent optoelectronic properties, as well as high-frequency/power performance. The combination of GaN with highly developed Si circuits has been pursued over the past decades. Applications include integrating optoelectronic devices on Si, high electron mobility transistor (HEMT) on Si, sensors, actuators, photonic devices, and integrated circuits to achieve more functionality. Although the monolithic approach of growing GaN heteroepitaxially on silicon substrate has stimulated much interest for its simplicity, reliability, and economic potential, there is still a big challenge to acquire high material quality. The growth of GaN structures on Si (100) or Si(110) substrates by molecular beam epitaxy (MBE) and metalorganic chemical vapor deposition (MOCVD) has been reported [58,59] for the integration of GaN devices with Si circuits. Compared with heteroepitaxy, thin-film lift-off and transfer technology offers flexibility in design and high material quality at current stage. On-wafer integration of Si (100) MOSFETs and GaN HEMTs [60,61], GaN optical interconnects [62], GaN photonic circuits [63], and thin film transistors on flexible substrate [52] has been demonstrated by using release and transfer techniques, such as LLO or chemical removal of substrates, layer transfer, and wafer-bonding technique. The ability to separate dissimilar substrates from GaN semiconductor thin film enables better device performance, where the substrate properties can be optimized separately, for example, for thermal conductivity for use in high power applications, reflectivity for use as integrated mirror layers, and electrical conductivity for LED applications. Moreover, lift-off and transfer permit the integration of GaN with Si circuits and promote advances in circuits and systems, for example, high power integrated wireless transmitters, power gating and power conversion circuits, and optical interconnections on Si CMOS.

References

1. S. Nakamura, S. Pearton, G. Fasol, *The Blue Laser Diode*, Springer, Berlin (2000).
2. P. Kozodov, J. P. Ibbetson, H. P. Marchand, T. Fini, S. Keller, J. S. Speck, S. P. DenBaars, and U. K. Mishra, Electrical characterization of GaN p-n junctions with and without threading dislocations, *Appl. Phys. Lett.*, **73**, 975 (1998).
3. J. E. Northrup, L. T. Romano, J. Neugebauer, Surface energetics, pit formation, and chemical ordering in InGaN alloys, *Appl. Phys. Lett.*, **74**, 2319 (1999); S. D. Lester, F. A. Ponce, M. G. Craford, and D. A. Steigerwald, High dislocation densities in high efficiency GaN-based light-emitting diodes, *Appl. Phys. Lett.*, **66**, 1249 (1996).
4. E. Yablonovitch, T. Gmitter, J. P. Harbison, and R. Bhat, Extreme selectivity in the lift-off of epitaxial GaAs films, *Appl. Phys. Lett.*, **51**, 2222 (1987).
5. W. S. Wong, T. Sands, and N. W. Cheung, Damage-free separation of GaN thin films from sapphire substrates, *Appl. Phys. Lett.*, **72**, 599–601 (1998).
6. M. K. Kelly, R. P. Vaudo, V. M. Phanse, L. Gorgens, O. Ambacher, and M. Stutzmann, Large free-standing GaN substrates by hydride vapor phase epitaxy and laser-induced liftoff, *Jpn. J. Appl. Phys.*, **38**, 217 (1999).
7. W. S. Wong, T. Sands, N. W. Cheung, M. Kneissl, D. P. Bour, P. Mei, L. T. Romano, and N. M. Johnson, Fabrication of thin film InGaN light-emitting diode membranes by laser liftoff, *Appl. Phys. Lett.*, **75**, 1360–1362 (1999).
8. C. R. Miskys, M. K. Kelly, O. Ambacher, and M. Stutzmann, Freestanding GaN-substrates and devices, *Phys. Stat. Sol* (*c*), **0(6)**, 1627–1650 (2003).

9. B. Fechneer, J. Brune, and R. Delmdahl, Thin film processing with UV Excimer lasers, Proceedings of LPM 2010–the 11th International symposium on laser precision microfabrication, **1** (2010).
10. J. J. Wierer, A. David, and M. M. Megens, III-nitride photonic-crystal light-emitting diodes with high extraction efficiency, *Nat. Photonics*, **3**, 163 (2009).
11. A. David, T. Fujii, B. Moran, S. Nakamura, S. P. DenBaars, and C. Weisbuch, Photonic crystal laser lift-off GaN light-emitting diodes, *Appl. Phys. Lett.*, **88**, 133514 (2006).
12. M. Kneissl, W. S. Wong, D. W. Treat, M. Teepe, N. Miyashita, and N. M. Johnson, Continous-wave operation of InGaN multiple-quantum-well laser diodes on copper substrates obtained by laser lift-off, *IEEE J. Sel. Top. Quantum Electron.*, **7**, 188 (2001).
13. E. A. Stach, M. Kelsch, and E. C. Nelson, Structural and chemical characterization of free-standing GaN films separated from sapphire substrates by laser lift-off, *Phys. Lett.*, **77**(12), 1819 (2000); M. K. Kelly, Optical patterning of GaN films, *Appl. Phys. Lett.*, **69**, 1749 (1999).
14. W. Wong, T. Sands, N. Cheung, M. Kneissl, D. Bour, P. Mei, L. Romano, and N. Johnson, Ubiquitous blue LEDs: the integration of GaN thin films with dissimilar substrate materials by wafer bonding and laser lift-off, *Compound Semiconductor*, Nov/Dec, 54–56 (1999); W. S. Wong, Y. Cho, N. J. Quitoriano, T. Sands, A. B. Wengrow and N. W. Cheung, Integration of GaN thin films with dissimilar substrate materials by Pd-In metal bonding and laser liftoff, *J. Electronic Mater.*, **28**, 1409–1413 (1999).
15. J. Jasinski, Z. Liliental-Weber, S. Estrada, and E. Hu, Microstructure of GaAs/GaN interfaces produced by direct wafer fusion, *Appl. Phys. Lett.*, **81**, 3152 (2002).
16. M. Funato, S. Fujita, and S. Fujita, Integration of GaN with Si using a AuGe-mediated wafer bonding technique, *Appl. Phys. Lett.*, **77**, 3959 (2000).
17. J. W. Chung, E. L. Piner, and T. Palacios, N-face GaN/AlGaN HEMTs fabricated through layer transfer technology, *IEEE Electron Device Lett.*, **30**, 113 (2009).
18. H. K. Cho, S. K. Kim, D. K. Bae, B-C Kang, J. S. Lee, and Y-H Lee, Laser liftoff GaN thin film photonic crystal GaN-based light-emitting diodes, *IEEE Photon. Technol. Lett.*, **20**(24) 2096 (2008).
19. R. W. Martin, H. S. Kim, Y. Cho, P. R. Edwards, I. M. Watson, T. Sands, N. W. Cheung, and M. D. Dawson, GaN microcavities formed by laser lift-off and plasma etching, *Mater. Sci. Eng.*, **B93**, 98 (2002).

20. M. S. Minsky, A. M. White, and E. L. Hu, Room-temperature photoenhanced wet etching of GaN, *Appl. Phys. Lett.*, **68**, 1531 (1996).
21. C. Youtsey, I. Adesida, and G. Bulman, Broad-area photoelectrochemical etching of GaN, *Electron. Lett.*, **33**, 245 (1997); C. Youtsey, L. T. Romano, and I. Adesida, *Appl. Phys. Lett.*, **73**, 797 (1998); C. Youtsey, G. Bulman, and I. Adesida, Dopant-selective photoenhanced wet etching of GaN, *J. Electron. Mater.*, **27**, 282 (1998).
22. E. D. Haberer, R. Sharma, A. R. Stonas, S. Nakamura, S. P. DenBaars, and E. L. Liu, Removal of thick (100 nm) InGaN layers for optical devices using band-gap-slective photoelectrochemical etching, *Appl. Phys. Lett.*, **85**, 762 (2004); E. D. Haberer, R. Sharma, C. Meier, A. R. Stonas, S. Nakamura, S. P. DenBaars, E. L. Hu, Development of selective lateral photoelectrochemical etching of InGaN/GaN for lift-off applications, *Appl. Phys. Lett.*, **85**, 5179 (2004).
23. E. L. Hu, A. R. Stonas, Photoelectrochemical undercut etching of semiconductor materials, US6884740 (2005).
24. A. R. Stonas, T. Margalith, S. P. DenBaars, L. A. Coldren, and E. L. Hu, Development of selective lateral photoelectrochemical etching of InGaN/GaN for lift-off applications, *Appl. Phys. Lett.*, **78**, 1945 (2001).
25. J. W. Graff, E. F. Shubert, A. Osinsky, On the reduction of base resistance in GaN-based heterojection bipolar transistors, high performance devices, 2000. Proceedings of 2000 IEEE/Cornell Conference, 28–32 (2000).
26. E. D. Haberer, R. Sharma, A. R. Stonas, S. Nakamura, S. P. DenBaars, and E. L. Liu, Removal of thick (100 nm) InGaN layers for optical devices using band-gap-slective photoelectrochemical etching, *Appl. Phys. Lett.*, **85**, 762 (2004).
27. C. Youtsey, G. Bulman, and I. Adesida, Dopant-selective photoenhanced wet etching of GaN, *J. Electron. Mater.*, **27**, 282 (1998).
28. P. Visconti, M. A. Reshchikov, K. M. Jones, F. Yun, D. F. Wang, R. Cingolani, H. Morkoç, C. W. Litton, and R. J. Molnar (2000). Highly selective photoenhanced wet etching of GaN for defect investigation and device fabrication. *MRS Proceedings*, **639**, G3.14 doi:10.1557/PROC-639-G3.14; P. Visconti, M. A. Reshchikov, K. M. Jones, D. F. Wang, R. Cingolani, H. Morkoc, R. J. Molnar, and D. J. Smith, Highly selective photo electro-chemical etching of nitride materials for defect nvestigation and device fabrication, *J. Vac. Sci. Technol. B*, **19**, 1328 (2001).
29. C. Youtsey, L. T. Romano, and I. Adesida, Gallium nitride whiskers formed by selective phtoenhanced wet etching of dislocations, *Appl. Phys. Lett.*, **73**, 797 (1998); C. Youtsey, I. Adesida, L. T. Romano and

G. Bulman, Smooth n-type GaN surfaces by photoenhanced wet etching, *Appl. Phys. Lett.*, **72**, 560 (1998).

30. D. Zhuang, and J. H. Edgar, Wet etching of GaN, AlN, and SiC: a review, *Mater. Sci. Eng. R*, **48**(1), 1 (2005).
31. A. C. Tamboli, E. D. Haberer, R. Sharma, K. H. Lee, S. Nakamura, and E. L. Liu, room-temperature continuous-wave lasing in GaN/InGaN microdisks, *Nat. Photonics*, **1**, 61 (2007).
32. A. R. Stonas, N. C. MacDonald, K. L. Turner, S. P. DenBaars, and E. L. Hu, Photoelectrochemical undercut etching for fabrication of GaN microelectromechanical systems, *J. Vac. Sci. Tech. B*, **19**, 2838 (2001).
33. E. D. Haberer, R. Sharma, C. Meier, A. R. Stonas, S. nakamura S. P. DenBaars, and E. L. Liu, Free-standing, optically pumped, GaN/InGaN microdisk lasers fabricated by photoelectrochemical etching, *Appl. Phys. Lett.*, **85**, 5179 (2004).
34. L. Macht, *"An electrochemical study of photoetching of heteroepitaxial GaN: kinetics and morphology". J. Cryst. Growth*, **273**, 347–356 (2005).
35. J. Park, K. M. Song, S-R Jeon, J. H. Baek, and S-W Ryu, Doping selective lateral electrochemical etching of GaN for chemical lift-off, *Appl. Phys. Lett.*, **94**, 221907 (2009).
36. Y. Zhang, S.-W. Ryu, C. Yerino, B. Leung, Q. Sun, Q. Song, H. Cao, and J. Han, A conductivity-based selective etching for next generation GaN devices, *Phys. Status Solidi B*, **247**, 1713 (2010); Y. Zhang, Q. Sun, B. Leung, J. Simon, M. L. Lee and J. Han, The fabrication of large-area, free-standing GaN by a novel nanoetching process, *Nanotechnology*, **22**, 045603 (2011).
37. Y. Zhang, Q. Sun, B. Leung, J. Simon, M. L. Lee, and J. Han, The fabrication of large-area, free-standing GaN by a novel nanoetching process, *Nanotechnology*, **22**, 045603 (2011).
38. Y. D. Wang, K. Y. Zang, S. J. Chua, S. Tripathy, P. Chen, and C. G. Fonstad, Nanoair-bridged lateral overgrowth of GaN on ordered nanoporous GaN template, *Appl. Phys. Lett.*, **87**, 251915 (2005).
39. K. Y. Zang, Y. D. Wang, H. F. Liu, and S. J. Chua, Structural and optical properties of InGaN/GaN multiple quantum wells grown on anno-air-bridged GaN template, *App. Phys. Lett.*, **89**, 171921 (2006).
40. H. Goto, S. W. Lee, H. J. Lee, H-J Lee, J. S. Ha, M. W. Cho, and T. Yao, Chemical lift-off of GaN epitaxial films grown on c-sapphire substrates with CrN buffer layers, *Phys. Stat. Sol* (*c*), **5**, 1659–1661 (2008); J.-S. Ha, S. W. Lee, H.-J. Lee, H.-J. Lee, S. H. Lee, H. Goto, T. Kato, K. Fujii, M. W. Cho, and T. Yao, The fabrication of vertical light-emitting diodes using chemical lift-off process, *IEEE Photon. Technol. Lett.*, **20**,

175 (2008); S. W. Lee, J.-S. Ha, H.-J. Lee, and H. Goto, Lattice strain in bulk GaN epilayers grown on CrN/sapphire template, *Appl. Phys. Lett.*, **94**, 082105 (2009).

41. J. Dorsaz, H.-J. Bühlmann, J.-F. Carlin, N. Grandjean, and M. Ilegems, Selective oxidation of AlInN layers for current confinement in III-nitride devices, *Appl. Phys. Lett.*, **87**, 072102 (2005).
42. D. J. Rogers, F. Hosseini Teherani, A. Ougazzaden, S. Gautier, and L. Divay et al., Use of ZnO thin films as sacrificial templates for metal organic vapor phase epitaxy and chemical lift-off of GaN, *Appl. Phys. Lett.*, **91**, 071120 (2007).
43. H. F. Liu, W. Liu, and S. J. Chua, Epitaxial growth and chemical lift-off of GaInN/GaN heterostructures on c- and r-sapphire substrates employing ZnO sacrificial templates, *J. Vac. Sci. Technol. A*, **28**(4), 590 (2010).
44. H. F. Liu, W. Liu, S. J. Chua, and D. Z. Chi, Fabricating high-quality GaN-based nanobelts by strain-controlled cracking of thinfilms for application in piezotronics, *Nano Energy*, **1**, 316 (2012).
45. K. Y. Zang, D. W. C. Cheong, H. F. Liu, H. Liu, J. H. Teng, and S. J. Chua, A new method for lift-off of III-nitride semiconductors for hetrogeneous integration, *Nanoscale Res. Lett.*, **5**, 1051 (2010).
46. K. Y. Zang, Y. D. Wang, S. J. Chua, and L. S. Wang, Nanoepitaxy to improve the efficiency of InGaN light-emitting diodes, *Appl. Phys. Lett.*, **92**, 243126 (2008); K. Y. Zang, and S. J. Chua, Orders of magnitude reduction in dislocation density in GaN grown on Si (111) by nano lateral epitaxial overgrowth, *Phys. Stat. Sol. (c)*, **5**, 1585 (2008); Y. D. Wang, K. Y. Zang, S. J. Chua, S. Tripathy, H. L. Zhou, and C. G. Fonstad, Improvement of microstructural and optical properties of GaN layer on sapphire by nanoscale lateral epitaxial overgrowth, *Appl. Phys. Lett.*, **88**, 211908 (2006).
47. K. Chung, C-H Lee, and G-C Yi, Transferable GaN layers grown on ZnO-coated graphene layers for optoelectronic devices, *Science*, **330**, 655 (2010).
48. Y. Kobayashi, K. Kumakura, T. Akasaka, and T. Makimoto, Layered boron nitride as a release layer for mechanical transfer of GaN-based devices, *Nature*, **484**, 223 (2012).
49. S. Vicknesh, S. Tripathy, V. K. X. Lin, L. S. Wang, and S. J. Chua, Fabrication of deeply undercut GaN-based microdisk structure on silicon platforms, *Appl. Phys. Lett.*, **90**, 071906 (2007).
50. D. Sam-Giao, D. Néel, S. Sergent, B. Gayral, M. J. Rashid et al, High quality factor AlN nanocavities embedded in a photonic crystal waveguide, *Appl. Phys. Lett.*, **100**, 191104 (2012).

51. H. W. Choi, K. N. Hui, P. T. Lai, P. Chen, X. H. Zhang, S. Tripathy, J. H. Teng, S. J. Chua, Lasing in GaN microdisks pivoted on Si, *Appl. Phys. Lett.*, **89**, 211101 (2006).
52. H-S Kim, E. Brueckner, J. Song, Y. Li, S. Kim, C. Liu, J. Sulkin, K. Choquette, Y. Huang, R. G. Ruzzo, and J. A. Rogers, Unusual strategies for using indium gallium nitride grown on silicon (111) for solid-state lighting, *Proc. Natl. Acad. Sci. USA* (*PNAS*), **108**, 10072 (2011).
53. J-H Ahn, H-S Kim, K. J. Lee, S. Jeon, S. J. Kang, Y. Sun, R. G. Nuzzo, and J. A. Rogers, Heterogeneous three-dimensional electronics by use of printed semiconductor nanomaterials, *Science*, **314**, 1754 (2006).
54. K. J. Lee, J. Lee, H. Hwang, Z. J. Reitmeier, R. F. Davis, J. A. Rogers, and R. G. Nuzzo, A printable form of single-crystalline gallium nitride for flexible optoelectronic systems, *Small*, **12**, 1164 (2005).
55. Y. Mei, D. J. Thurmer, C. Deneke, S. Kiravittaya, Y-F. Chen, A. Dadgar, F. Bertram, B. Bastek, A. Krost, J. Christen, T. Reindl, M. Stoffel, E. Coric, and O. G. Schmidt, Fabrication, self-assembly, and properties of ultrathin AlN/GaN porous crystalline nanomembranes: Tubes, spirals, and curved sheets, *ACS Nano*, **3**(7), 1663 (2009).
56. P. Cendula, S. Kiravittaya, Y. F. Mei, Ch. Deneke, and O. G. Schmidt, Bending and wrinkling as competing relaxation pathways for strain free-hanging films, *Phys. Rev. B*, **79**, 085429 (2009).
57. K. Tomita, T. Kachi, S. Nagai, A. Kojima, S. Yamasaki, and M. Koike, Self-separation of freestanding GaN from sapphire substrates by hydride vapor phase epitaxy, *Phys. Stat. Sol.* (*a*), **194**, 563 (2002).
58. S. Joblot, F. Semond, Y. Cordier, and P. Lorenzimi, J. Massies, High-electron-mobility AlGaN/GaN heterostructures grown on Si(001) by molecular-beam epitaxy, *Appl. Phys. Lett.*, **87**, 133505 (2005).
59. F. Schulze, O. Kisel, A. Dadgar, A. Krtschil, J. Blasing, M. Kunze, I. Daumiller, T. Hempel, A. Diez, R. Clos, and A. Krost, Crystallographic and electric properties of MOVPE grown AlGaN/GaN-based. FET's on Si(001) substrates, *J. Cryst. Growth*, **299**, 399 (2007).
60. J. W. Chung, J. K. Lee, and E. L. Piner, Seamless on-wafer integration of Si(100) MOSFETs and GaN HEMTs, *IEEE Electron Device Lett.*, **30**, 1015 (2009).
61. 60. N. M. Jokerst, M. A. Brooke, S.-Y. Cho, S. Wilkinson, M. Vrazel, S. Fike, J. Tabler, Y. J. Joo, S.-W. Seo, and D. S. Wills, A. Brown, The heterogeneous integration of optical interconnections into integrated microsystems, *IEEE J. Sel. Top. Quantum Electron.*, **9**, 350 (2003).

62. N. W. Cheung, Proceedings of 7th International Conference on solid-state and integrated circuits technology, 1–3, 2167 (2004).
63. C. Xiong, W. Pernice, K. K. Ryu, C. Schuck, K. Y. Fong, T. Palacios, and H. X. Tang, Integrated GaN photonic circuits on silicon (100) for second harmonic generation, *Opt. Express*, **19**, 10462 (2011).

Chapter 11

Gallium Arsenide Disk Optomechanical Resonators

Lu Ding,[a] Christophe Baker,[a] Alessio Andronico,[a] David Parrain,[a] Pascale Senellart,[b] Aristide Lemaitre,[b] Sara Ducci,[a] Giuseppe Leo,[a] and Ivan Favero[a]

[a]*Matériaux et Phénomènes Quantiques, Université Paris Diderot, Sorbonne Paris Cité, CNRS UMR 7162, 10 rue Alice Domon et Léonie Duquet, Paris, 75013 France*
[b]*Laboratoire de Photonique et Nanostructures, CNRS, Route de Nozay, Marcoussis, 91460 France*

dingl@imre.a-star.edu.sg, ivan.favero@univ-paris-diderot.fr

11.1 Introduction

The field of optomechanics explores the coupling between light and the motion of mechanical systems [1–3]. Research in optomechanics finds many similarities with concepts developed in atomic physics. For example, the principles of optical cooling of atoms are today applied to optical cooling of mechanical oscillators [4]. Conversely, optical tweezers used to trap atoms can be considered as optomechanical realizations, the atom being the moving mechanical element [5,6]. Optomechanical effects like optical control, optical tuning, or optical actuation of mechanical motion generate prospects for innovative experiments in sensing, metrology, or mesoscopic quantum physics. For example, advances in the optical cooling of

Handbook of Optical Microcavities
Edited by Anthony H. W. Choi

ISBN 978-981-4463-24-9 (Hardcover), 978-981-4463-25-6 (eBook)
www.panstanford.com

a mechanical oscillator into its quantum regime offer the exciting perspective to reveal and study the "quantumness" of macroscopic solid state mechanical systems [7–9]. In sensing experiments, nanoscale optomechanical systems allow to combine the extreme sensitivity of nanomechanical sensors to optical read-out techniques [10]. For a few years now, the principles of optomechanics find regularly novel usance.

High frequency mechanical elements are commonly used in optomechanics applications to ease the access to the quantum regime, or to develop high-speed sensing systems, but they must be combined with a large optical–mechanical coupling. The optomechanical coupling indeed quantifies how efficiently one can turn mechanical into optical energy and vice versa. Optomechanical coupling is generally measured by one of the two following parameters: g_{om}, the frequency-pull parameter (in Hz/nm) giving the shift of the optical cavity resonance (in Hz) per unit of displacement of the mechanical system (in nm) or g_0 the vacuum optomechanical coupling giving the cavity resonance shift for a displacement equal to the zero-point fluctuation of the mechanical resonator. For example, optomechanics experiments inserting a vibrating nanomechanical element in a radio frequency (RF) strip-line cavity reach g_{om} value of order 10 to 100 kHz/nm [11], a rather limited value due to the fact that the mechanical element is small compared to the RF photons wavelength. More recent experiments using mechanical membranes in the RF cavities have boosted this value to a few 10 MHz/nm [8], approaching values obtained with optical photons in millimeter-long cavities, in the 100 MHz/nm range [12,13]. Silica-based whispering gallery structures reach a few GHz/nm with a typical 50–100 μm extension of the resonator [14,15]. In general, the optomechanical coupling is enhanced when both the mechanical and optical mode are confined in a smaller interaction volume where they adopt a comparable wavelength. In the context of optomechanics applications, it is then highly desirable to shrink the size of the mechanical and optical systems below the millimeter–micrometer scale employed in most current experiments, and venture into nanoscale optomechanics. This has been the focus of the recently emerging field of nano-optomechanics [16–20].

In this chapter, we will present nanoscale gallium arsenide (GaAs) disk optomechanical resonators (Fig. 11.1) that consist of a GaAs disk cap (diameter of a few micrometers, typical thickness of 200 nm) suspended over a nearly cylindrical AlGaAs pedestal (typical height 1.5 μm). These resonators possess high-Q optical whispering gallery modes (WGMs), combined with high-frequency (GHz) mechanical modes. With their ability to confine both the optical and mechanical modes in a sub-wavelength optomechanical interaction volume, these resonators exhibit extremely strong optomechanical coupling g_{om}, in principles up to the THz/nm. The optomechanical coupling in a disk resonator can be understood as follows: the photons circulating in the disk gallery mode exert a centrifugal radiation pressure force that excites breathing mechanical modes of the structure, conversely, when the disk "breathes" mechanically, the photons in the disk are modulated. This mutual action and back-action, of the optics on the mechanics and of the mechanics on the optics, is at the basis of optomechanical phenomena. The GaAs optomechanical resonators presented here are compatible with the insertion of optically active material, and could for example allow exploring novel situations where a (quantum) mechanical oscillator would be coupled to a (quantum) photon emitter embedded in the host material [21].

Figure 11.1 A GaAs disk optomechanical resonator.

This chapter reviews the fundamental of optical properties of dielectric disk cavities; first by introducing the terminology (Section 11.2.1) and later presenting analytical and numerical methods to compute the WGMs of a disk (Section 11.2.2). The evanescent coupling between a disk WGM and a straight optical waveguide via their evanescent field is discussed, notably within the coupled mode theory (CMT) context (Section 11.2.3). Standing-wave WGMs, resulting from the coupling between clockwise (cw) and counter-clockwise (ccw) WGMs are introduced. Sections 11.3 to 11.5 are devoted to experimental investigations. In Section 11.3, we present the nanofabrication process used for GaAs disks resonators and optical fiber tapers used in our experiments. The evanescent optical spectroscopy of GaAs disks using a fiber taper as near-field optical probe is discussed in Section 11.4. Measured optical properties are discussed and analyzed statistically. In Section 11.5, we present optomechanics experiments on GaAs disk resonators. Mechanical modes of the disks are measured optically with ultra-high motional sensitivity. Extreme optomechanical coupling factors are measured and modeled. Finally, we discuss future applications of GaAs optomechanical resonators.

11.2 Fundamentals of Whispering Gallery Disk Resonators

11.2.1 Terminology of Whispering Gallery Modes in Disk Resonators

High refractive index dielectric disks can store light in optical whispering gallery modes (WGMs). These cavities are typically used as optical filters [22], modulators [23], and delay lines [24], thanks to their fine frequency selectivity. Moreover, their ability to confine optical energy in a small mode volume greatly enhances light–matter interaction and results in strong nonlinear phenomena, such as lasing.

WGMs are optical modes of rotationally symmetric cavities that arise from total internal reflection at the cavity bent sidewalls (see Fig. 11.2(a)). There are two characteristic parameters frequently used to represent the degree of temporal and spatial confinement of a WGM: (1) the quality factor (Q) and (2) the effective mode volume (V_{eff}).

The Q factor is proportional to the photon lifetime inside the cavity τ_{ph}, so that a large Q indicates a strong temporal confinement. On the other hand, the effective mode volume represents the volume physically occupied by the WGM: a smaller volume corresponds to a stronger spatial confinement. Here, we summarize useful expressions involving these parameters [25–27].

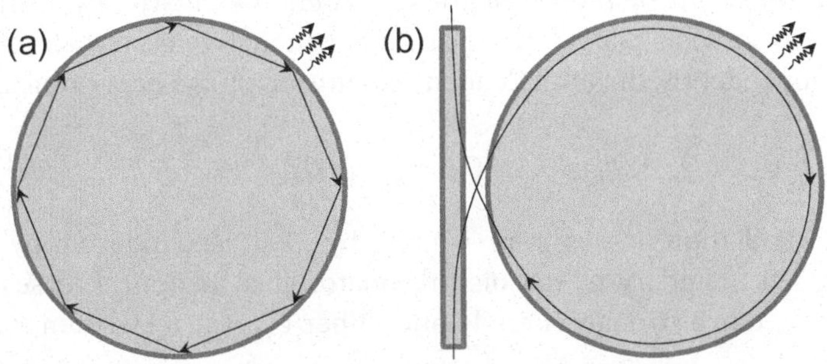

Figure 11.2 Schematics presentations of (a) a WGM in a disk seen in a geometric optics viewpoint and (b) the evanescent coupling between a single mode waveguide and a disk.

11.2.1.1 Quality factor, photon lifetime, and optical losses

The quality factor, Q is a function of the total energy stored in the cavity (U) and the energy dissipated per cycle:

$$Q \equiv 2\pi \times \frac{\text{Energy Stored}}{\text{Energy Dissipated per Cycle}} = \omega_0 \times \frac{\text{Energy Stored}}{\text{Power Loss}} \tag{11.1}$$

For a WGM of frequency ω_0 (or vacuum wavelength λ_0), the time dependence of the energy stored in the cavity is:

$$U(t) = U(t_0)e^{-(\omega_0/Q)/t} = U(t_0)e^{-t/\tau_{\text{ph}}} \tag{11.2}$$

where $\tau_{\text{ph}} \equiv Q/\omega_0 = 1/\delta\omega$ is the cavity photon lifetime defined as the time after which the energy is $1/e$ of its original value, and where $\delta\omega$ is the full width at half maximum (FWHM) of the WGM resonance in angular frequency. The quality factor Q can then be spectrally measured (either in frequency or wavelength)

$$Q = \frac{\omega_0}{\delta\omega} = \frac{\lambda_0}{\delta\lambda} \tag{11.3}$$

For a WGM, there are two main classes of loss mechanisms: (1) intrinsic losses (represented by Q_{in}) and (2) extrinsic losses (represented by Q_{ex}). Intrinsic losses relate to the physical properties of the optical resonator itself, whereas extrinsic losses are produced by the resonator's surrounding environment. Under the hypothesis that all losses are small enough to treat their effects independently, the total (loaded) quality factor can be written as

$$Q_{tot}^{-1} \equiv Q_{in}^{-1} + Q_{ex}^{-1} = Q_{rad}^{-1} + Q_{abs}^{-1} + Q_{scatt}^{-1} + Q_{ext}^{-1} \tag{11.4}$$

Radiation losses represented by Q_{rad} are induced by the curved boundary of the disk resonator, just as optical losses are induced in a strongly curved optical fiber. Q_{rad} can be written as:

$$Q_{rad} = \frac{\mathrm{Re}(\tilde{\omega})}{2\mathrm{Im}(\tilde{\omega})} \tag{11.5}$$

where $\tilde{\omega}$ is the complex optical eigenfrequency of the resonator obtained when solving Maxwell's equations for the dielectric disk resonator with circular boundary and perfectly lossless dielectric response.

Absorption losses represented by Q_{abs} include not only the linear optical absorption of the dielectric material (the lossy part of the linear dielectric response), but also nonlinear loss mechanisms such as two-photon absorption.

In addition, surface roughness or departure from a perfectly circular geometry can cause scattering of the photons trapped in a disk cavity. In such case photons leave the cavity; from the point of view of the cavity, it is a loss mechanism but it is not absorption since the optical energy does not transform into heat in this process. This scattering loss mechanism can be described by an effective quality factor Q_{scatt} [26]. The scattering of photons by surface roughness cannot be neglected in experiments and has important consequences, such as lifting the degeneracy of cw and ccw WGMs of a disk, as will be discussed later.

All optical losses due to a coupling of the cavity mode to the optical modes of nearby structures, such as bus optical waveguide or other resonators, are included in the term Q_{ext}. Figure 11.2(b) plots the case of evanescent coupling between a disk and a single-mode straight waveguide. The coupling strength is determined by the overlap integral between the waveguide mode and the disk WGM. Experimentally, the coupling strength can be adjusted by simply modifying the gap distance between the disk and waveguide.

11.2.1.2 Free spectral range, group velocity, round-trip time, and finesse

The free spectral range (FSR) of a cavity is defined as the frequency (or equivalently wavelength) spacing $\Delta\omega_{FSR}$ ($\Delta\lambda_{FSR}$) between two spectrally adjacent resonant modes belonging to the same "family". For example in a disk cavity, the FSR refers to the spacing between successive azimuthal order (m) WGMs with the same radial (p) order (see Section 11.2.2 for the definition of the p and m numbers). Practically, the choice of the optimal FSR for a cavity is application dependent. A large FSR can be useful when developing single-mode lasers, while small FSR is good for multi-mode purposes [28].

From an experimental point of view, the FSR provides an indirect measure of the group velocity $v_g \equiv d\omega/d\beta$, where β is the phase propagation constant. Since $\beta = m/r$ for the azimuthal phase propagation in a WGM of azimuthal number m, the group velocity relates to the FSR.

$$v_g \equiv \frac{\partial\omega}{\partial\beta_\phi} = \frac{\Delta\omega_{FSR}}{\frac{m+1}{r}-\frac{m}{r}} = r\Delta\omega_{FSR} \tag{11.6}$$

In a ray optics picture, WGMs with low p orders are confined close to the periphery of the disk. For these modes, we can use the approximation $r \approx R$ and the group velocity is $v_g \approx R\Delta\omega_{FSR}$. The group velocity is important for light packets propagating in a dispersive medium. The propagation constant in a dispersive material is $\beta = n(\omega)\omega/c$, where $n(\omega)$ is the refractive index of the material. The group velocity becomes $v_g = d\omega/d\beta = c/\left|n+\omega(dn/d\omega)\right|$. However, for GaAs, the dispersion relation is negligible around the telecommunication wavelengths ($\lambda \sim 1.5$ μm), which gives equal phase and group velocity, $v_g \approx v_\phi = c/n$. In a disk cavity, the mode effective index n_ξ should better be used instead of the bulk refractive index. The FSR is then expressed as

$$\Delta\omega_{FSR} = \frac{c}{n_\xi R} \text{ (in angular frequency)}$$

or

$$\Delta\lambda_{FSR} = \frac{\lambda_0^2}{2\pi n_\xi R} \text{ (in wavelength)} \tag{11.7}$$

WGMs with higher p orders have their optical energy density displaced from the periphery slightly toward the disk center, so that we can replace r in Eqs. (11.6) and (11.7) by a smaller "effective radius" R_{eff}. The round trip time τ_{rt} for a propagating light packet in the cavity is given as

$$\tau_{\text{rt}} = L_{\text{rt}}/v_{\text{g}} = 2\pi/\Delta\omega_{\text{FSR}}$$

The finesse F of an optical cavity is a measure of the ability of the cavity, when used as spectral etalon, to spectrally resolve fine structures. It is defined as

$$F \equiv \frac{\Delta\omega_{\text{FSR}}}{\delta\omega} \tag{11.8}$$

For a WGM with quality factor Q, it can be expressed as $F = Q/m$, where m is the WGM azimuthal number. In the time domain, the finesse is linked to the number of cavity round trips a photon is able to complete before leaving the cavity.

$$\tau_{\text{ph}}/\tau_{\text{rt}} = F/2\pi$$

11.1.2.3 Effective mode volume

The effective mode volume V_{eff} describes the volume that a given WGM physically occupies in space.

$$V_{\text{eff}} = \frac{\int_V \varepsilon_0 \varepsilon_r |\vec{E}(\vec{r})|^2 dV}{\max(\varepsilon_0 \varepsilon_r |\vec{E}(\vec{r})|^2)} \tag{11.9}$$

Where $\int_V \ldots dV$ denotes integration over the infinite volume V. Confining a large optical energy density in a small volume is another key factor to enhance light–matter interaction (discussed in the context of the optomechanics section).

11.2.2 Eigenvalue Equations for WGMs

A description of the Eigenmodes of a disk optical resonator starts from Maxwell's equations in the absence of free charges and currents in the material:

$$\begin{cases} \nabla \times \vec{E} = -\dfrac{\partial \vec{B}}{\partial t} & \nabla \cdot \vec{D} = 0 \\ \nabla \times \vec{H} = \dfrac{\partial \vec{D}}{\partial t} & \nabla \cdot \vec{B} = 0 \end{cases} \quad \text{with} \quad \begin{cases} \vec{D} = \varepsilon_0 \varepsilon_r \vec{E} \\ \vec{B} = \mu_0 \mu_r \vec{H} \end{cases} \tag{11.10}$$

The above equations lead to an eigenvalue equation in a symmetric format for both the electric and magnetic fields, which can be used to find the eigenfrequencies and spatial field distributions of the optical modes

$$\nabla^2 \vec{F} + n^2 \frac{\omega^2}{\omega^2} \vec{F} = 0 \tag{11.11}$$

where $\vec{F}$ is a field representing either $\vec{E}$ or $\vec{H}$, $c = 1/\sqrt{\varepsilon_0 \mu_0}$ is the speed of light, $n = 1/\sqrt{\varepsilon_r \mu_r}$ is the bulk refractive index. The relative permittivity, ε_r and the relative permeability, μ_r are constants for isotropic materials or tensors for anisotropic materials. We limit our discussion to dielectric materials that have isotropic and homogeneous ε_r and μ_r with little dispersion in the frequency range of interest. In the chapter, we adopt the convention that the electromagnetic field has a sinusoidal time dependence $e^{-i\omega t}$, where ω is the angular frequency.

Figure 11.3 plots the schematics of a disk resonator with radius R and thickness h in cylindrical coordinates (r, ϕ, z). The fully vectorial wave equation, Eq. (11.11) in three dimensions (3D) cannot be analytically solved in general for such disk. However, by making reasonable assumptions and by taking advantage of the axial symmetry of the disk cavity, we can reduce this complex 3D problem into a (2 + 1)D.

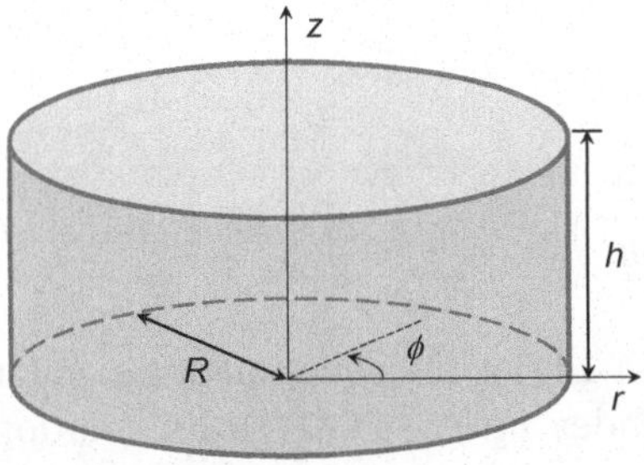

Figure 11.3 Schematics of a disk resonator with radius R and thickness h in cylindrical coordinates.

11.2.2.1 The semi-analytical effective index approach

To analytically solve Eq. (11.11), we will make several assumptions on the field F and reduce the problem to solvable differential equations. Writing the Laplace operator of Eq. (11.11) in cylindrical coordinates (r, φ, z), we find

$$\left[\frac{1}{r}\frac{\partial}{\partial r}\left(r\frac{\partial}{\partial r}\right)+\frac{1}{r^2}\frac{\partial^2}{\partial \phi^2}+\frac{\partial^2}{\partial z^2}+n^2\frac{\omega^2}{c^2}\right]\vec{F}(r,\phi,z)=0 \tag{11.12}$$

We first assume that there are two sets of uncoupled optical modes with orthogonal polarizations, labeled as TM and TE modes. For a TM mode, the electric field is polarized perpendicularly to the disk plane and the magnetic field lies in the plane of the cavity $(H^r, H^\phi, E^z) \neq 0$, whereas for a TE mode, the non-vanishing mode components are (E^r, E^ϕ, H^z). This assumption is reminiscent of the usual effective index approach for treating optical slab waveguides: it holds reasonably well as long as the disk thickness is greater than the wavelength in the dielectric $(h > \lambda/n)$. For thinner disks, TE and TM modes start to hybridize and the distinction between TE and TM no longer holds.

The field component F^z, which is the electric field E^z (magnetic field H^z) for TM (TE) modes, can be computed by making a second assumption on the separation of variables in the field's spatial dependence, $F^z(r, \phi, z) = \psi(r)\Phi(\phi)Z(z)$. The other field components are related to F^z by Maxwell's constitutive relations. Using this assumption, we find the following three equations

$$\begin{cases}\dfrac{\partial^2 Z}{\partial z^2}+\dfrac{\omega^2}{c^2}(n-n_\xi^2)Z=0 & \text{(a)}\\[2ex] \dfrac{\partial^2 \Phi}{\partial \phi^2}+m^2\Phi=0 & \text{(b)}\\[2ex] \dfrac{\partial^2 \Psi}{\partial r^2}+\dfrac{1}{r}\dfrac{\partial \Psi}{\partial r}+\left(n_\xi^2\dfrac{\omega^2}{c^2}-\dfrac{m^2}{r^2}\right)\Psi=0 & \text{(c)}\end{cases} \tag{11.13}$$

Equation (11.13a) is the equation of a standard slab waveguide with effective index $n_\xi(\xi$ = TE/TM). Its solutions can be labeled with the integer number l (axial number) (l = 1, 2...). The solution to Eq. (11.13b) are simply $e^{-im\phi}$, m being the azimuthal number (m = 1, 2...). The radial eigenfunction $\psi(r)$ of azimuthal order m is determined solving Eq. (11.13c), and it can be expressed in terms of first-kind Bessel functions (inside the disk) and first-kind Hankel functions (outside the disk):

$$\Psi(r)=\begin{cases} NJ_{\mathrm{m}}\left(\tilde{k}n_{\xi}r\right) & r\leq R \\ N\dfrac{J_{\mathrm{m}}\left(\tilde{k}n_{\xi}R\right)}{H_{\mathrm{m}}^{(1)}\left(\tilde{k}R\right)}H_{\mathrm{m}}^{(1)}(\tilde{k}r) & r>R \end{cases} \tag{11.14}$$

where N is a normalization constant and $\tilde{k}$ the WGM wavenumber. Note that second-kind Hankel functions must replace first-kind functions if using the opposite convention $e^{i\omega t}$ for the field time evolution [29].

The resonance eigenfrequencies can be found by enforcing the continuity of the field's tangential components E^z and H^ϕ (E^ϕ and H^z) for TM (TE) mode at the dielectric–air boundary ($r = R$):

$$\begin{aligned} &n_{\xi}\frac{\dot{J}_{\mathrm{m}}(\tilde{k}n_{\xi}R)}{J_{\mathrm{m}}(\tilde{k}n_{\xi}R)}-\frac{\dot{H}_{\mathrm{m}}^{(1)}(\tilde{k}R)}{H_{\mathrm{m}}^{(1)}(\tilde{k}R)}=0 \qquad \text{TM modes} \\ &\frac{\dot{J}_{\mathrm{m}}(\tilde{k}n_{\xi}R)}{J_{\mathrm{m}}(\tilde{k}n_{\xi}R)}-n_{\xi}\frac{\dot{H}_{\mathrm{m}}^{(1)}(\tilde{k}R)}{H_{\mathrm{m}}^{(1)}(\tilde{k}R)}=0 \qquad \text{TE modes} \end{aligned} \tag{11.15}$$

By solving these equations for the wave vector $\tilde{k}$, we obtain the complex WGM eigenfrequency $\tilde{\omega}=c\tilde{k}$. It is worth reminding that the imaginary part of the resonance frequency stems from the curved geometry of the disk and causes the WGM to have a finite radiative quality factor Q_{rad}, which can be expressed as in Eq. (11.5). Also notice that, if m is fixed, multiple solutions of Eq. (11.15) do exist. These solutions can be labeled with the radial number p (p = 1, 2...). Once F^z is known, the other mode components can be obtained by using Eq. (11.10):

$$\begin{cases} H^r = -\dfrac{m}{\mu_0 r\tilde{\omega}} E^z \\ H^\phi = \dfrac{i}{\mu_0\tilde{\omega}} \dfrac{\partial E^z}{\partial r} \end{cases} \text{(TM modes) and } \begin{cases} E^r = \dfrac{m}{\varepsilon_0 \varepsilon_r r\tilde{\omega}} H^z \\ E^\phi = \dfrac{-i}{\varepsilon_0\varepsilon_r\tilde{\omega}} \dfrac{\partial H^z}{\partial r} \end{cases} \text{(TE modes)} \quad (11.16)$$

Assuming that $Z(z)$ is normalized according to $\int_{-\infty}^{+\infty} Z^2(Z)dz = 1$, the normalization constant N can be determined by normalizing the mode to the azimuthal power flow.

$$P^{\Phi} = \begin{cases} +\dfrac{1}{2}\mathrm{Re}(\int_0^\infty E^z H^{r*}\, dr) \text{ TM modes} \\ -\dfrac{1}{2}\mathrm{Re}(\int_0^\infty H^z E^{r*}\, dr) \text{ TE modes} \end{cases} \quad (11.17)$$

To summarize, WGMs of a disk optical cavity can be labeled by using three indices: the axial order 1 accounting for the vertical confinement along the z-axis; the azimuthal order m expressing the mode's rotational symmetry; and the radial order p, which accounts for the mode's radial profile. Figure 11.4 gives an example of WGMs sustained by a disk cavity: a WGM with $m = 7$, $p = 1$ (left) and a WGM with $m = 7$, $p = 2$ (right).

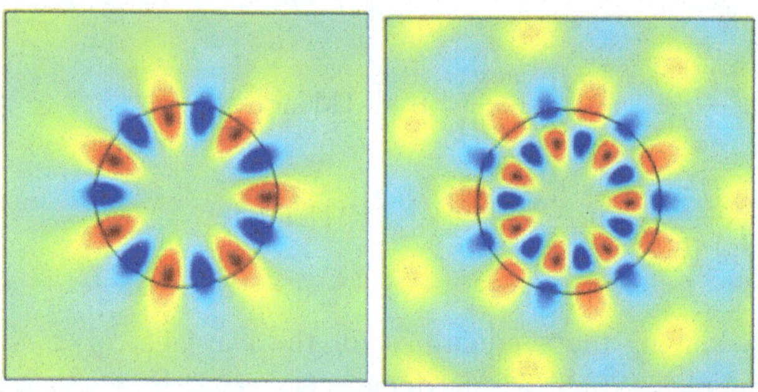

Figure 11.4 Example of WGMs sustained by a dielectric disk: a WGM with $m = 7, p = 1$ (left) and a WGM with $m = 7, p = 2$ (right). The black line represents the disk boundary.

The semi-analytical approach presented here is an excellent way of computing rapidly optical eigenfrequencies for a disk resonator, with pretty good accuracy. Its efficiency relies on the fact that a single 1D implicit equation of the kind Eq. (11.15) needs to be solved numerically at the end of the computation, the rest being fully analytical. Of course, the validity of the approach relies on the fact that the TE/TM separation is reasonable, and disks with thickness below λ/n should better be treated with more advanced computational techniques. The semi-analytical approach tends also to over-estimate the radiative Q factor: indeed there is no energy flow along the z-direction for the WGMs described by the semi-analytical approach, and only the radial outward Poynting vector gives rise to a radiative loss. This does not correctly depict the reality and other more appropriate tools are needed to numerically estimate Q_{rad}.

11.2.2.2 The finite-element approach

Compared to the above semi-analytical approach, a fully vectorial finite element method (FEM) approach provides a more accurate tool to determine the resonance frequencies of WGMs, the related spatial mode profiles and associated radiative quality factors Q_{rad} [30].

Due to the rotational invariance of the disk with respect to its central z-axis, we can factorize the azimuthal dependence of the WGM fields and only treat 2D equations in the (r, z) plane. This saves computer resources and allows shorter computation times for a given meshing of the problem. We combine the code originally developed by M. Oxborrow for 2D FEM simulation of WGMs axisymmetric resonators with the improvement brought by M. Cheema and coworkers [31,32], which includes perfectly matched layer (PMLs) and allows an accurate determination of the WGMs radiative quality factor. We chose to use the commercial software COMSOL because of its ability to solve multiphysics problems, for example, the optomechanical problem that we will treat later.

The FEM equation in "weak form" is written as

$$\int_{\mathrm{V}}\left[(\nabla\times\vec{\tilde{H}}^{*})\cdot\frac{1}{\varepsilon_{\mathrm{r}}}(\nabla\times\vec{H})+\zeta(\nabla\cdot\vec{\tilde{H}}^{*})(\nabla\cdot\vec{H})+c^{-2}\vec{\tilde{H}}^{*}\cdot\left(\mu_{r}\frac{\partial^{2}\vec{H}}{\partial t^{2}}\right)\right]dV=0 \tag{11.18}$$

where $\vec{H}$ is the magnetic field vector and the integral is over the whole volume considered for the simulation (encompassing the disk and its surroundings). ε_r is the relative permittivity of the material, which we assume optically isotropic here. The essence of the axis-symmetric reduction is to factorize the azimuthal dependence of the magnetic field of a given WGM of frequency $f = \omega/2\pi$ and azimuthal order m, in the form $H(r, f, z) = H(r, z)e^{-imf}e^{-i2\pi ft}$ (the azimuthal convention used in Ref. [32] is different due to another orientation of the coordinates system). The 3D nature of Eq. (11.18) reduces then to a 2D problem. $\tilde{\vec{H}}$ is the test magnetic field that appears in the weak form formulation of a partial differential equation (PDE) problem [30]. The second term in the integrand of Eq. (11.18) is a "penalty term" introduced to avoid spurious mode solutions, that is here modes that would violate the equation div(H) = 0. This mathematical formula is applied to all spatial subdomains of the treated computational domain (in the example shown in Fig. 11.5); this includes the semiconductor disk with its pedestal, the air surroundings, and the PMLs. A PML is an artificial absorber placed at the computational domain boundaries in order to avoid reflections, thus emulating an infinite space surrounding the disk resonator. The diagonal relative permittivity and permeability tensors of the absorber must be taken to be $\bar{\varepsilon}_{\mathrm{r}} \equiv \varepsilon_{\mathrm{r}}\bar{\Lambda}$, $\bar{\mu}_{\mathrm{r}} \equiv \mu_{\mathrm{r}}\bar{\Lambda}$, where $\bar{\Lambda}$ reads in cylindrical coordinates [33]:

$$\bar{\Lambda} = \left(\frac{\tilde{r}}{r}\right)\left(\frac{s_{\mathrm{z}}}{s_{\mathrm{r}}}\right)\hat{r} + \left(\frac{r}{\tilde{r}}\right)(s_{\mathrm{z}}s_{\mathrm{r}})\hat{\phi} + \left(\frac{\tilde{r}}{r}\right)\left(\frac{s_{\mathrm{r}}}{s_{\mathrm{z}}}\right)\hat{z} \tag{11.19}$$

with

$$\tilde{r} = \begin{cases} r & 0 \le r \le r_{\mathrm{m}} \\ r - iG\dfrac{(r - r_{\mathrm{m}})^3}{3t_{\mathrm{m}}^2} & r > r_{\mathrm{m}} \end{cases},$$

$$s_{\mathrm{r}} = \begin{cases} 1 & 0 \le r \le r_{\mathrm{m}} \\ 1 - iG\dfrac{(r - r_{\mathrm{m}})^2}{t_{\mathrm{m}}^2} & r > r_{\mathrm{m}} \end{cases},$$

$$s_{\mathrm{z}} = \begin{cases} 1 - iG\left(\dfrac{z_{\mathrm{ml}} - z}{t_{\mathrm{m}}}\right)^2 & z < \\ 1 & z_{\mathrm{ml}} \le z \le z_{\mathrm{mu}} \\ 1 - iG\left(\dfrac{z - z_{\mathrm{mu}}}{t_{\mathrm{m}}}\right)^2 & z > \end{cases}$$

where t_m is the PML thickness, and r_{m}, z_{mu}, and z_{ml} are coordinates (in the radial r and vertical $+z$ and $-z$ directions, respectively) giving the location of the segment representing the inner boundary of the PML. G is a real number roughly equal to 5~6 for a PML thickness of $\lambda/4$ and a FEM mesh thickness of $\lambda/20$. By introducing a PML in the computation, the resonance frequency f of the WGM becomes a complex number and the quality factor Q_{rad} due to radiation loss can be obtained using Eq. (11.5). Figure 11.5 shows the process of 2D FEM simulation of a WGM (p = 3, m = 33) with $Q_{\mathrm{rad}} \sim 10^7$.

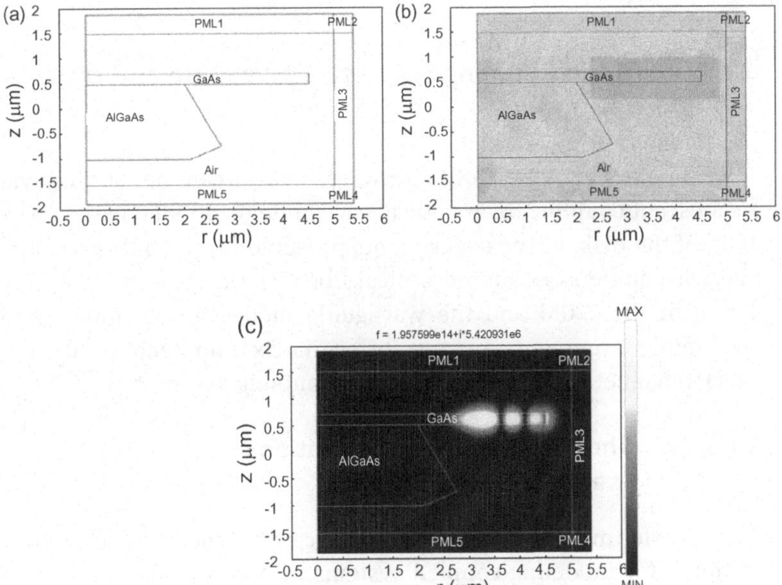

Figure 11.5 (a) The computational domain for an example of a GaAs disk positioned over an AlGaAs pedestal, with different spatial domains of the computational box, (b) the meshing of the problem, and (c) the intensity of the radial component of the electric field for a WGM (p = 3, m = 33) with $Q_{\mathrm{rad}} \sim 10^7$. Note that above the computation box in (c) the complex frequency of the computed WGM.

This FEM computational approach provides us with more accurate values of the eigenfrequency and radiative Q factor of WGMs. However, to obtain a full set of WGMs on a given disk, the computation is more time-consuming than the semi-analytical approach: indeed an azimuthal number m must first be chosen before solving the problem to extract all corresponding WGMs. To obtain a complete mode labeling, each WGM must then be identified as quasi-TE or quasi-TM (when this distinction is granted by construction in the semi-analytical approach), and be associated to a radial number p (which was unambiguous in the semi-analytical approach). Once done, the whole process must be repeated for the next azimuthal $m + 1$ number. Following this procedure, the numerical calculation of a disk's optical spectrum typically takes a full working day.

11.2.3 Mode Coupling between Waveguide and Disk Resonator

Experimentally, WGMs of a disk resonator can be excited via evanescent coupling to an optical waveguide positioned in the near-field of the disk. In the experiments presented later in this chapter, the waveguide is a tapered optical fiber. The evanescent coupling between the WGM and the waveguide mode can be understood by simple arguments in a transfer-matrix approach or by the electromagnetic CMT under the weak coupling assumption.

11.2.3.1 The transfer-matrix description of disk–guide evanescent coupling

The transfer matrix method (TMM) is a convenient approach widely adopted for modeling-coupled systems in optics [34]. It considers the coupling region as a black box with known transmission and reflection coefficients and neglects the details of the coupling mechanism.

Figure 11.6 shows a situation of unidirectional coupling between one WGM of the disk and one single-guided mode of the waveguide. If the coupling is lossless, the interaction transfer matrix can be written as [35]

$$\left(\frac{B_2}{A_2}\right) = \begin{pmatrix} t & i\kappa \\ i\kappa & t \end{pmatrix} \left(\frac{B_1}{A_1}\right) \tag{11.20}$$

where A_i and B_i are the complex field amplitude of the coupled traveling waves normalized such that $|A_i|^2$ and $|B_i|^2$ correspond to the optical power flowing through the relevant cross-section (of the guide or of the disk). The coupling matrix follows the condition $t^2 + \kappa^2 = 1$ (lossless), where t and κ are the amplitude transmission and coupling coefficient, respectively. The reference phase planes have been chosen here in order for t and κ to be real, but the coupling can also be treated with t and κ being arbitrary c-numbers [35].

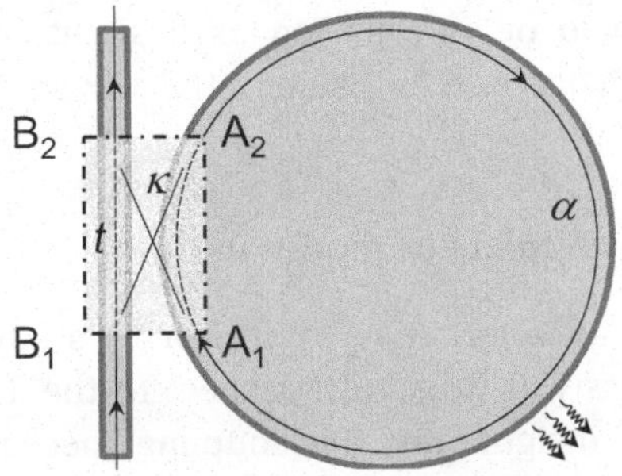

Figure 11.6 The relevant parameters for the TMM.

Using the cavity round-trip relation $A_1 = A_2\alpha e^{-i\phi}$, where α represents the cavity round-trip amplitude loss and ϕ is the cavity round-trip phase shift with the azimuthal phase convention chosen in Section 11.2.2.1, the amplitudes of the output field B_2 and intracavity field A_1 are expressed as a function of B_1.

$$B_2 = \frac{t - \alpha e^{-i\phi}}{1 - t\alpha e^{-i\phi}} B_1, \quad A_1 = \frac{i\sqrt{1-t^2}\,\alpha e^{-i\phi}}{1 - t\alpha e^{-i\phi}} B_1 \tag{11.21}$$

The normalized transmission and intracavity circulating power are

$$\frac{|B_2|^2}{|B_1|^2} = \frac{(t - \alpha\cos\phi)^2 + (\alpha\sin\phi)^2}{(1 - t\alpha\cos\phi)^2 + (t\alpha\sin\phi)^2},$$

$$\frac{|A_1|^2}{|B_1|^2} = \frac{(\sqrt{1-t^2}\,\alpha)^2}{(1 - t\alpha\cos\phi)^2 + (t\alpha\sin\phi)^2} \tag{11.22}$$

At resonance $\phi = 0$ (modulo 2π)

$$\frac{|B_2|^2}{|B_1|^2} = \frac{(t-\alpha)^2}{(1-t\alpha)^2}, \quad \frac{|A_1|^2}{|B_1|^2} = \frac{(\sqrt{1-t^2}\alpha)^2}{(1-t\alpha)^2} \tag{11.23}$$

Notice that the transmitted power vanishes, that is, $|B_2/B_1|^2 = 0$, when $t = \alpha$. This condition, known as the critical coupling, is due to the perfect destructive interference at the guide output between the directly transmitted field tB_1 and the field exiting the disk cavity κA_1. Given the lossless condition relating t and κ, the critical coupling can also be interpreted as a situation where intrinsic losses in the cavity exactly balance the waveguide–disk coupling losses.

11.2.3.2 The coupling of modes in time

CMT provides an alternative approach to study the evanescent coupling process [36–38]. With respect to the TMM, this method provides more insight into the coupling mechanism and allows determining an analytical expression for the coupling coefficient. Using heuristic arguments, Ref. [38] gives the time evolution equation for the amplitude $a(t)$ of a WGM as:

$$\frac{da(t)}{dt} = -i\omega_0 a(t) - \left(\frac{\gamma_{\text{in}}}{2} + \frac{\gamma_{\text{ex}}}{2}\right)a(t) + \kappa_{\text{T}} s_+ \tag{11.24}$$

$$s_- = s_+ - \kappa_{\text{T}}^* a(t)$$

where the amplitude $a(t)$ is normalized so that $|a(t)|^2$ is the total energy stored in the disk cavity (see Fig. 11.7). The signs appearing in Eq. (11.24) correspond to our convention choice for the field phase (and phase planes have been chosen again in order for κ_T to be real). The waveguide input field of amplitude s_+ feeds energy to the WGM. The waveguide output field is denoted by s_-. γ_{in} and γ_{ex} are the intrinsic and extrinsic decay rate of the energy stored in the WGM. In absence of sourcing ($s_+ = 0$) and intrinsic losses ($\gamma_{\text{in}} = 0$), the energy stored in the cavity decreases as $|a(t)|^2 = |a_0|^2 e^{-\gamma_{\text{ex}} t}$, where $|a_0|^2$ is the initial energy. Assuming that the coupling is lossless, the change in energy due to the waveguide coupling $-d|a(t)|^2/dt = \gamma_{\text{ex}}|a(t)|^2$ is identical to the output power $|s_-|^2 = |\kappa_{\text{T}}|^2$

$|a(t)|^2$. Hence, κ_T is related to the extrinsic decay rate γ_{ex} via $|\kappa_T|^2 = \gamma_{ex}$. Assuming time-reversal symmetry, a scattering matrix analysis leads $\kappa_T = i\sqrt{\gamma_{ex}}$.

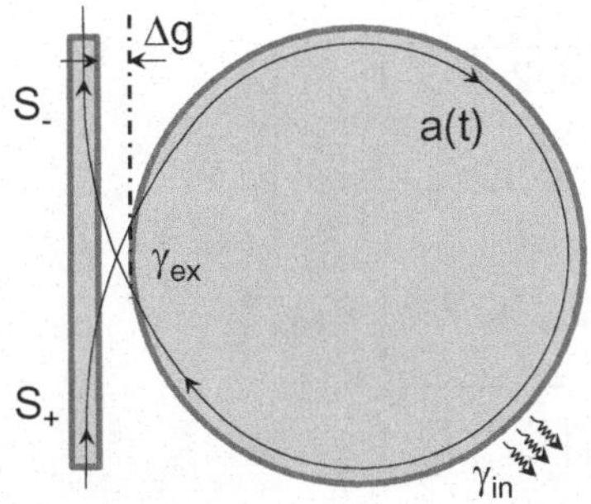

Figure 11.7 Parameters used in the CMT approach for disk–waveguide evanescent coupling.

The coupling term κ_T can be shown to be proportional to the overlap integral between the evanescent field of the WGM and the waveguide mode. Below, to interpret experiments, we will use an approximation where γ_{ex} has exponential dependence on the disk–waveguide gap distance Δg, that is scales like $e^{-\eta\Delta g}$ with η an adjustable parameter (see Fig. 11.7).

The above heuristic CMT approach has been recently put on firmer mathematical grounds [36] that remarkably conclude that the time evolution Eq. (11.24) is essentially correct under a few assumptions [39]. The mathematical treatment starts from Maxwell's equations and uses reciprocity relations. Equation 11.24 is a good approximation when the coupling coefficient κ_T has no frequency dependence and when the shift of the WGM resonance (induced by the coupling to the guide) is neglected.

With a driving source $s_+ \sim e^{-i\omega t}$, the steady state WGM amplitude and the output field amplitude are:

$$a = \frac{i\sqrt{\gamma_{ex}}}{-i\Delta\omega + \left(\frac{\gamma_{in}}{2} + \frac{\gamma_{ex}}{2}\right)} s_+, \quad s_- = \frac{-i\Delta\omega + \left(\frac{\gamma_{in}}{2} - \frac{\gamma_{ex}}{2}\right)}{-i\Delta\omega + \left(\frac{\gamma_{in}}{2} + \frac{\gamma_{ex}}{2}\right)} s_+ \qquad (11.25)$$

where $\Delta\omega$ is the detuning from resonance, $\Delta\omega = \omega - \omega_0$. The intra-cavity-stored energy normalized by the input power reads

$$\frac{|a|^2}{|s_+|^2} = \frac{\gamma_{ex}}{\Delta\omega^2 + \left(\frac{\gamma_{in}}{2} + \frac{\gamma_{ex}}{2}\right)^2} \tag{11.26}$$

and the normalized transmission is

$$\frac{|s_-|^2}{|s_+|^2} = \frac{\Delta\omega^2 + \left(\frac{\gamma_{in}}{2} - \frac{\gamma_{ex}}{2}\right)^2}{\Delta\omega^2 + \left(\frac{\gamma_{in}}{2} + \frac{\gamma_{ex}}{2}\right)^2} \tag{11.27}$$

Equation (11.27) provides us with two important parameters, the FWHM $\delta\omega$ of the transmission resonance and the on-resonance transmission T_{on}

$$\begin{cases} \partial\omega = \gamma_{in} + \gamma_{ex} = \gamma_{in}(1+K) \\ T_{on} = \left(\frac{1-K}{1+K}\right)^2 \end{cases} \tag{11.28}$$

Where $K = \gamma_{ex}/\gamma_{in} = Q_{in}/Q_{ex}$. After measuring the transmission resonance FWHM for varying evanescent coupling conditions (i.e., for varying γ_{ex}), we can extract the intrinsic loss γ_{in} using Eq. (11.28). This will be used later in experimental realizations to measure the intrinsic Q_{in} of a WGM.

Another useful formula is the intracavity circulating power at resonance, normalized to the input power, that can be is derived from Eq. (11.26) as

$$P_{on} = \frac{4K}{\tau_{rt}\gamma_{in}(1+K)^2} = \frac{1-T_{on}}{\tau_{rt}\gamma_{in}} = \frac{F_{in}}{2\pi}(1-T_{on}) = \frac{Q_{in}}{2\pi m}(1-T_{on}) \tag{11.29}$$

where F_{in} is the intrinsic WGM cavity finesse and m the related WGM azimuthal number.

Figure 11.8 shows the dimensionless parameters T_{on} and τ_{rt} $\gamma_{in}P_{on}$ versus K as obtained from Eqs. (11.28) and (11.29).

Three coupling regimes appear:

(1) Under-coupling $\gamma_{ex} < \gamma_{in}$ ($K < 1$): the external coupling to the waveguide is weak and the power injection into the disk cavity is not efficient. When the disk and waveguide are far from one another, the coupling is so weak that the measured resonance FWHM is approximately given by the amount of intrinsic losses. Reducing the disk–waveguide gap distance increases γ_{ex}, progressively enlarges the FWHM $\delta\omega$ and decreases T_{on}. The circulating power P_{on} is also increased, before reaching critical coupling.

(2) Critical-coupling $\gamma_{ex} = \gamma_{in}$ ($K = 1$): external and intrinsic losses are equal. The FWHM is twice the intrinsic one. Theoretically, all the input power is injected into the disk and T_{on} is zero. The normalized circulating power P_{on} reaches its maximum value $P_{on} = 1/\tau_{rt}\gamma_{in}$.

(3) Over-coupling $\gamma_{ex} > \gamma_{in}$ ($K > 1$): the external losses due to the coupling to the waveguide are dominant. Reducing the disk–waveguide gap distance Δg increases K. It further broadens the resonance but T_{on} increases anew. The circulating power decreases.

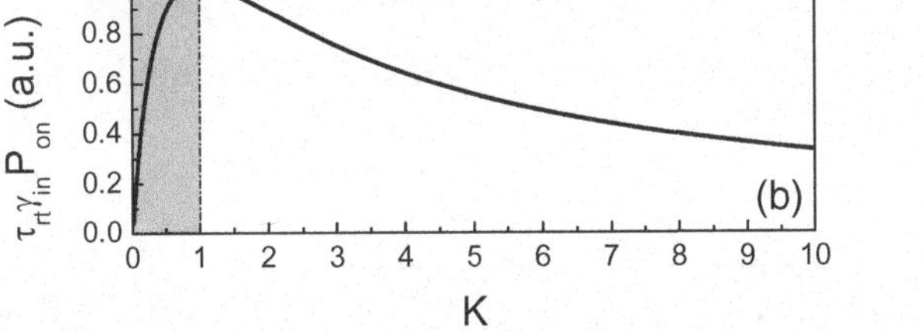

Figure 11.8 (a) The on-resonance transmission and the FWHM of the transmission resonance, as a function of K and (b) the on-resonance intracavity circulating power as a function of K. A dash-dotted line indicates the critical coupling $K = 1$ and separates the under-coupling $K < 1$ (grey) and over-coupling $K > 1$ (white) regimes.

11.2.4 Backscattering and Doublet Splitting

For each chosen radial and azimuthal numbers (p,m), a perfectly symmetric rotationally invariant disk supports two frequency degenerate (p,m) WGMs: the clock-wise (cw) and counter clock-wise (ccw) mode. In reality, small fabrication imperfections, or even the non-isotropic nature of the underlying crystal lattice, can break the symmetry and lift this degeneracy [26,40]. Considering the case in Fig. 11.9, the input optical power is injected into the cw mode, so that:

$$\begin{aligned} \frac{da_{\text{ccw}}}{dt} &= -i\omega_0 a_{\text{ccw}} - \left(\frac{\gamma_{\text{in}}}{2} + \frac{\gamma_{\text{ex}}}{2}\right) a_{\text{ccw}} + i\frac{\gamma_\beta}{2} a_{\text{cw}} \\ \frac{da_{\text{cw}}}{dt} &= -i\omega_0 a_{\text{cw}} - \left(\frac{\gamma_{\text{in}}}{2} + \frac{\gamma_{\text{ex}}}{2}\right) a_{\text{cw}} + i\frac{\gamma_\beta}{2} a_{\text{ccw}} + i\sqrt{\gamma_{\text{ex}}} s_+ \\ s_- &= s_+ + i\sqrt{\gamma_{\text{ex}}} a_{\text{cw}} \end{aligned} \tag{11.30}$$

where γ_β is the energy exchange rate due to photons scattering between the cw and ccw modes (so γ_β is a coupling parameter between the two modes). Introducing the standing-wave modes basis

$$\begin{aligned} a_1 &= \frac{1}{\sqrt{2}}(a_{\text{ccw}} + a_{\text{cw}}) \leftrightarrow a_{\text{ccw}} = \frac{1}{\sqrt{2}}(a_1 + a_2) \\ a_2 &= \frac{1}{\sqrt{2}}(a_{\text{ccw}} - a_{\text{cw}}) \leftrightarrow a_{\text{cw}} = \frac{1}{\sqrt{2}}(a_1 - a_2) \end{aligned} \tag{11.31}$$

the steady-state solution for the standing wave mode amplitudes is given by:

$$a_{1,2} = \frac{i\sqrt{\gamma_{\text{ex}}/2}}{\mp\left[i\Delta\omega - \left(\frac{\gamma_{\text{in}}}{2} + \frac{\gamma_{\text{ex}}}{2}\right)\right] - i\frac{\gamma_\beta}{2}} s_+ \tag{11.32}$$

The total stored optical energy is $|a|^2 = |a_1|^2 + |a_2|^2$. Compared to the traveling-wave modes case, for example Eq. (11.26), each standing wave WGM now receives approximately half the stored energy. The field amplitude of the output wave is $s_- = s_+ + i\sqrt{\gamma_{\text{ex}}/2}\,(a_1 - a_2)$ and the transmission is

$$\frac{|s_-|^2}{|s_+|^2} = \left|1 + i\sqrt{\frac{\gamma_{ex}}{2}}\left(\frac{a_1}{s_+} - \frac{a_2}{s_+}\right)\right|^2 \tag{11.33}$$

Two transmission minima occur at $\omega = \omega_0 \pm \gamma_\beta/2$. In the measurement, the backscattering effect can hence be quantified by the spectral distance γ_β between the two minima, and translated into the quantity $Q_\beta \equiv \omega_0/\gamma_\beta$. If $Q_\beta >> Q_{in}$, the degeneracy of the ccw and cw WGMs still holds (it means that the coupling between the cw and ccw modes is weaker than the coupling of each of them to its environment). When $Q_\beta << Q_{in}$, the converse situation occurs: the cw–ccw degeneracy is lifted and a doublet is resolved in the measured transmission spectrum. The two standing-wave modes are well-resolved spectrally. Figure 11.10 plots the doublet resonance spectrum in case of $\gamma_{in} = \gamma_{ex} = 1$ and $\gamma_\beta = 0, 1,$ and 5.

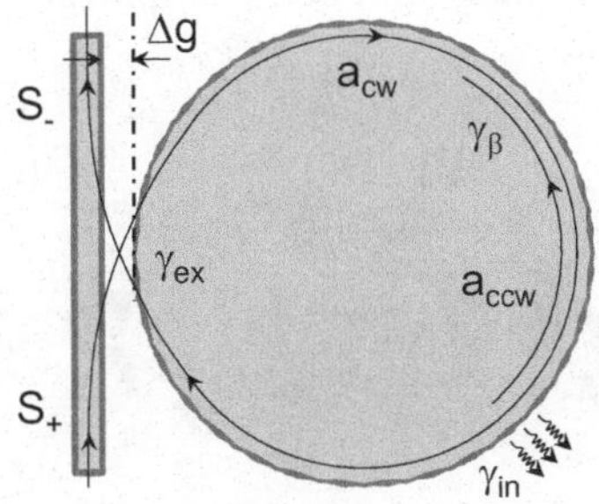

Figure 11.9 CMT parameters in a situation where photons traveling in the cw mode are backscattered into the ccw mode.

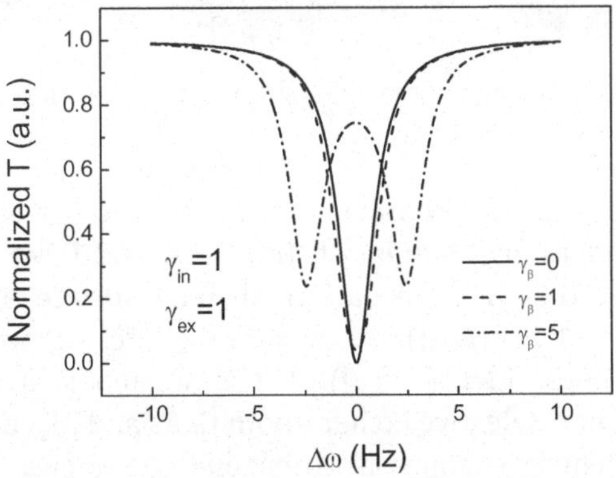

Figure 11.10 The resonance doublet in the transmission spectrum in the case $\gamma_{in} = \gamma_{ex} = 1$ (critical coupling) and $\gamma_\beta = 0, 1,$ and 5.

11.3 Nanofabrication of GaAs Disks and Optical Fiber Tapers

11.3.1 Nanofabrication of a GaAs Disk

The GaAs disk resonators are fabricated out of extra-pure material grown by molecular beam epitaxy (MBE). The nanofabrication process involves e-beam lithography and subsequent wet etching techniques [41]. Epitaxial layers of $Al_{0.8}Ga_{0.2}As$ (thickness 1.8 μm) and GaAs (thickness 200 nm) are subsequently grown on a commercial semi-insulating GaAs wafer (the 200 nm GaAs layer ending up "above" the AlGaAs). The $Al_{0.8}Ga_{0.2}As$ layer is used as sacrificial layer in the fabrication process and serves as constituting material for the pedestal, the 200 nm GaAs layer being used for the disk itself. Typical radii of the employed disks range from 1 to 20 μm.

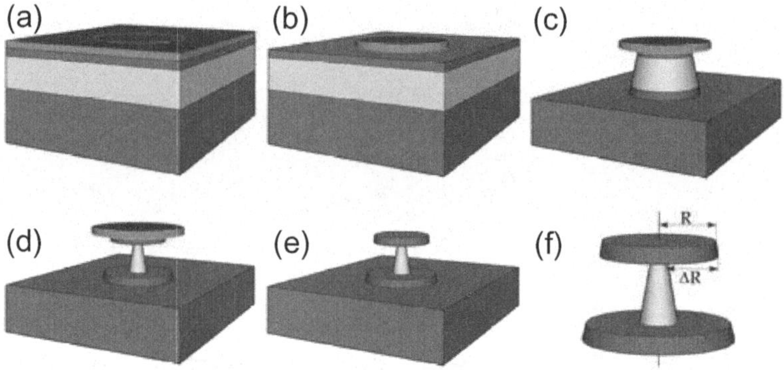

Figure 11.11 (a–e) Steps of the nanofabrication process and (f) 3D cartoon view of resulting disk resonator.

The nanofabrication process is summarized in Fig. 11.11. Using e-beam lithography (Nanomask JEOL JBX 5D2U) with a negative resist (Man 2403), a disk is patterned on the surface of the epitaxial wafer (Figs. 11.11(a),(b)). A two-step wet etching process is then employed (Figs. 11.11(c)–(e)). The first step is a hydrobromic acid–based non-selective etch for both GaAs and $Al_{0.8}Ga_{0.2}As$ layers. This first etching is almost isotropic and etches away the material down to the GaAs substrate, giving rise to a pillar structure capped by the disk-shaped resist (Fig. 11.11(c)). Notice that the lateral etch depth, which is typically two-third of the vertical etch depth for this etching step, must be taken into account in the mask design in order to obtain the desired diameter for the disk. In a second etch step, a hydrofluoric acid solution selectively under-etches the $Al_{0.8}Ga_{0.2}As$ layer while letting the GaAs disk itself barely affected (Fig. 11.11(d)). The resist is then removed by a standard acetone dip (Fig. 11.11(e)). The side view of the finally obtained GaAs disk resonator is shown in Fig. 11.11(f). Figure 11.12 shows scanning electron microscope (SEM) images obtained after the main fabrication steps.

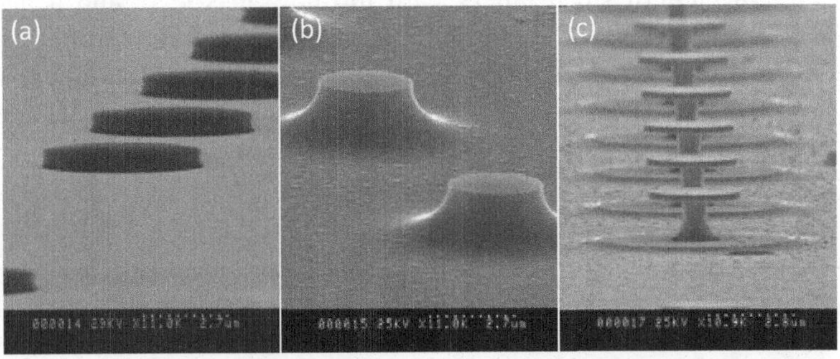

Figure 11.12 SEM images obtained after the main fabrication steps: (a) after mask development, (b) after first wet etching step, and (c) after second etching step [42].

The underetching of the AlGaAs sacrificial layer releases the outer part of the disk plate on a radial distance ΔR (see Fig. 11.11(f)) and enables the disk to vibrate. It also greatly improves the vertical optical confinement of the WGM, thanks to the large refractive index mismatch between GaAs and air. Three groups of GaAs disk resonators fabricated respectively with an under-etch distance ΔR = 1, 2, and 4 μm are discussed in later part of the chapter. A statistical study of the disk's optical WGMs properties as a function of the disk's dimensions is presented in Section 11.4.5.

11.3.2 Fabrication of Tapered Nanofibers

Optical fiber tapers with sub-wavelength cross-section have become popular tools to inject light into small mode-volume cavities by means of evanescent coupling [43–47]. A low loss level of the taper is desirable in this application, which requires an excellent geometric regularity and a low surface roughness. To this aim, we decided to avoid the conventional burner flame technique generally used to produce fiber tapers, because this technique comes with technical difficulties (turbulence of the flame, bursts in the gas flow, and contamination by impure gas in the burner). Since a moderate temperature at 1,100–1,200°C is already enough to reach softening of a silica fiber and start to stretch it [48,49], we adopted a simple ceramic thermoelectric heater to this purpose, which has the advantage of providing a more stable heat source. Using such ceramic heater, we obtain highly reproducible fabrication of sub-wavelength silica tapers with ultralow loss level and a high level of control over their shape and single-mode operation. The fabrication method is low cost (no gas consumption) and easy to implement, a beneficial feature for a broad range of fiber-taper applications.

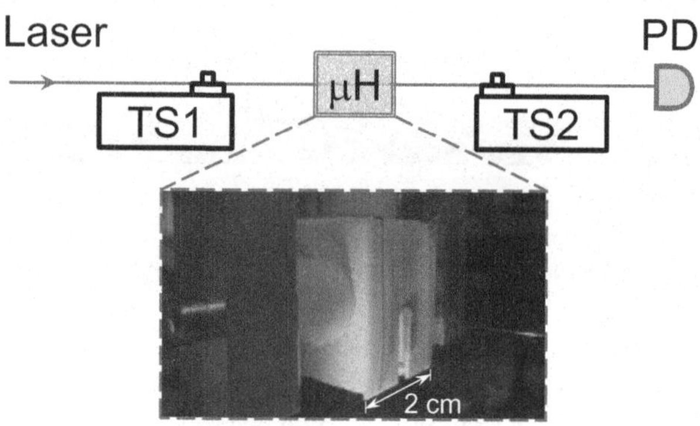

Figure 11.13 Schematic view of the set-up for fiber-taper fabrication. μH, the thermoelectric microheater providing a stable hot zone; TS1 and TS2 are motorized linear translation stages; PD, photo detector. The zoom-in picture is the microheater (white cube), hold by three ceramic disks for thermal isolation, with an optical fiber positioned in the heater's cavity.

The fabrication of tapered fibers consists in softening and then elongating an optical fiber, while monitoring its optical transmission (see Fig. 11.13) [49]. For this purpose, a telecom diode laser beam is launched into a single-mode telecom fiber (Corning SMF-28). The fiber coating is striped over 3 cm of length and the fiber subsequently clamped on two linear translation stages (Newport UTS 100CC). The striped part of the fiber is cleaned with solvents and positioned in the microheater's cavity. This ceramic heater is a centimeter-sized thermoelectric oven originally developed at NTT Pte Ltd. for fiber-coupler fabrication [50] (see Fig. 11.13). The heater's operation temperature is 1,160°C for our purposes. After positioning of the fiber in the heater's cavity, we wait 120 s for its thermal stabilization. The two stages are then set at a constant pulling velocity of 40 μm/s. After about 40 mm of elongation, the stages are stopped and the taper is slowly removed from the cavity. Finally, the tapered fiber is preserved in a dry box to slow down the degradation of its optical properties due to ambient humidity.

The taper profile, that is radius r versus fiber axis coordinate z, is measured with a high magnification ratio optical microscope (500×) and is compared to the constant hot-zone model of Birks and Li [51] (see Fig. 11.14(b)), which predicts:

$$r(z)=\begin{cases} r_w = r_0 e^{-z_0/L_0} & |z| \le L_0/2 \\ r_0 e^{(|z|-z_0-L_0/2)/L_0} & L_0/2 < |z| \le z_0 + L_0/2 \end{cases} \tag{11.34}$$

where L_0 is the hot-zone length (i.e., the length over which the piece of silica fiber can be considered as plastically soft and stretchable), $2z_0$ the elongation length, and $z = 0$ is the center of the taper. Equation (11.34) describes well our results: the fit of the measured profile is excellent when using the measured values $z_0 = 19.75$ mm and $r_0 = 66$ μm, and setting the only adjustable parameter L_0 to 3.96 mm. Here, the waist diameter $2r_w$ is 0.9 ± 0.3 μm. For a few tapers, we double-checked that this estimation is correct using a SEM. The typical taper waist diameter used in the evanescent coupling experiments described later is 800 nm. In the taper fabrication method, we can also vary L_0 by changing the operation temperature of the heater (and consequently the pulling velocity), in order to fabricate tapers of targeted profiles for other applications.

The obtained tapered nanofibers are expected to be highly adiabatic for an optical wavelength of $\lambda = 1.55$ μm. A statistics over 80 tapers, all having about 40 mm of elongation length, shows an average transmission level of 94%. We stress that half of the tapers display a transmission between 95% and 100%, while the best tapers rise to transmission values superior to 99%. If excluding taper transition loss contributions, this corresponds for the uniform waist region to a linear loss level in the 10^{-3} dB/mm range. For what concerns the total taper transmission (including taper transitions), our fabrication method gives reproducibly excellent results above the level reported with other methods [52,53].

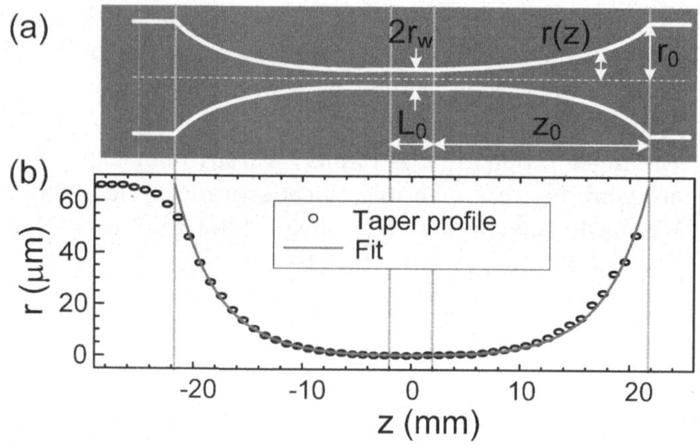

Figure 11.14 Taper profile. (a) Schematic profile of a symmetric fiber taper formed in a constant hot zone, where the total taper length is $2z_0 + L_0$ (z_0, length of the taper transition and L_0, length of the taper waist; hence also, length of the hot-zone). (b) Measured profile of a typical fiber taper fabricated using the ceramic microheater. The fit in red is given by Eq. (11.34).

Straight optical fiber tapers can be employed for evanescent coupling to microscale (typical 100 μm) or millimeter-scale optical cavities, such as silica microspheres or microtoroids [43,44]. But in the case of sub-micrometer mode-volume cavities integrated on-chip, like GaAs disks suspended 1 to 2 μm above the substrate, parasitic coupling to the substrate becomes limiting if using a straight taper for coupling. Building on an earlier method [47], we have implemented a simple technique for fabricating micro-looped fiber tapers that allow spatially selective evanescent coupling to GaAs disks cavities. At the end of our fabrication process, we slowly remove the taper from the heater and the translation stages are both moved 1 mm forward in order to release tension in the taper. With the help of a rotating fiber clamp, we twist one taper end and form a multi-twisted looped structure as shown in Fig. 11.15(a). The taper is then tensioned back and glued on a holder possessing a setscrew. This screw is turned gently to adjust the tension while observing the loop untwist until it reaches the single-loop structure of Fig. 11.15(b). The looped tapers obtained in this manner have a typical diameter of 70 μm (Fig. 11.15(b), inset), and a typical normalized transmission spectrum as shown in Fig. 11.15(c). Periodic oscillations in this spectrum correspond to resonances of the micro ring cavity formed by the loop [54].

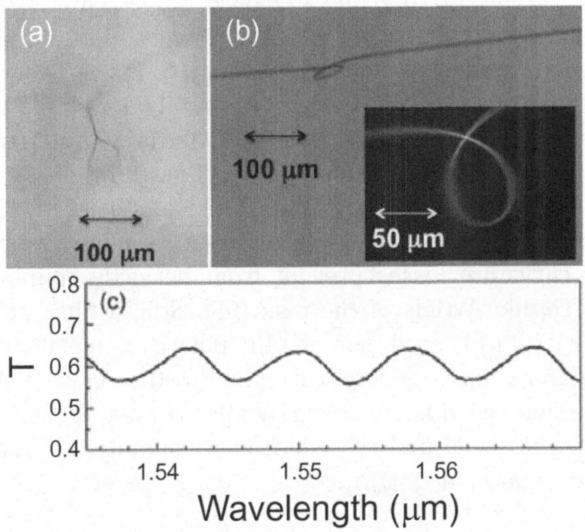

Figure 11.15 Micro-looped taper: (a) Multi-twisted looped taper structure; (b) Single loop fiber under tension; (c) Normalized transmission spectrum of a single-loop fiber taper.

11.4 Optical Spectroscopy of GaAs Disks Resonators

Miniature GaAs disk resonators, typically a few micrometers in diameter and 200 nm thick in our work, combine many assets for optomechanics experiments: high optical Q, high mechanical frequency, nanoscale pg motional mass, and sub-micrometer mode volume to strongly confine both optical and mechanical energy and boost the optomechanical coupling.

First, the optical properties of these resonators are characterized by means of fiber taper transmission experiments. WGMs with optical Q factor up to a few 10^5 are observed. Modes identification, critical coupling, and optical resonance doublet splitting are discussed.

11.4.1 Experimental Set-Up

The set-up for optical characterization of GaAs disk resonators is shown in Fig. 11.16. Using XYZ piezo-stages with high stability, the looped fiber taper is positioned in the disk optical near-field to allow evanescent coupling (see Fig. 11.16, inset). Measurements of the taper transmission spectrum as a function of the taper-disk gap distance are performed using an external cavity tunable laser source (λ = 1,500–1,600 nm) with a linewidth ≤1 MHz. In the looped fiber taper, the polarization state is not perfectly controlled: the birefringence induced by the curvature of the loop and the related curvature losses prevent from perfectly coupling to the TE- or TM-like WGMs of the disk [55]. Still, a fiber polarization controller (FPC) is used to adjust the polarization state in the loop and maximize the evanescent coupling to the selected WGM. The transmission signal is measured by a P-I-N photodiode, monitored on an oscilloscope and recorded on a computer through a data acquisition card (NI-DAQ).

Figure 11.16 Schematics of disk-taper spectroscopy experiments utilizing the fiber taper coupling technique (FPC, fiber polarization controller; PD, photo detector). The inset is a close-up of the fiber-disk system.

11.4.2 Identification of WGMs

When the disk and tapered fiber are evanescently coupled, the disk WGMs appear as a series of transmission dips in the fiber transmission spectrum. Figure 11.17(a) shows a broadband transmission spectrum, from 1,530 to 1,580 nm, of a GaAs disk resonator of large diameter (radius 17.5 μm) probed by a looped fiber taper. The transmission was normalized to the transmission of the fiber taper positioned far away from the disk. In a disk, the electric field of $p = 1$ WGMs is strongly confined at the periphery of the disk while high order radial modes distribute their optical energy density toward the central part of the disk. A smaller pedestal (larger underetch distance ΔR) results then in a better vertical (along z-axis) optical confinement for high order $p > 1$ modes. In Fig. 11.17(a), the underetch distance $\Delta R = 4.7$ μm is relatively small compared to the disk radius ($\Delta R << R$) and only two sets of radial WGMs ($p = 1$ and 2) are observed in the spectrum, while higher radial order WGMs are quenched due to large optical losses induced by the presence of the pedestal. Taking the approximation $2\pi n_\xi R = m\lambda_0$ to estimate the FSR with $\lambda_0 = 1{,}550$ nm and $n_\xi = 2.6$, the effective index of the 200 nm GaAs slab TE mode, we find FSR = 8.4 nm, which agrees correctly with the experimental value of 7.9 nm (Fig. 11.17(a)).

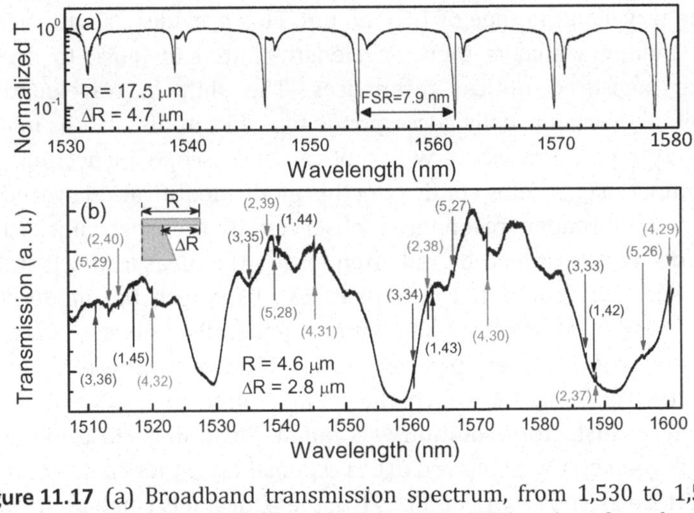

Figure 11.17 (a) Broadband transmission spectrum, from 1,530 to 1,580 nm, of a GaAs disk resonator (R = 17.5 μm) with underetch distance ΔR of 4.7 μm. (b) Broad-band transmission spectrum, from 1,500 to 1,600 nm, of a GaAs disk resonator (R = 4.6 μm). WGMs are classified via the indices (p, m). The underetch distance ΔR = 2.8 μm, defined as the radius difference between the disk and the pedestal, is larger than half the disk radius.

Figure 11.17(b) shows the broadband transmission spectrum of another GaAs disk resonator (R = 4.6 μm), with identified WGM (p, m) indices. Fine TE-like WGM resonances are protruding (highest $Q > 1 \times 10^5$) out of an oscillating background with period ~8 nm attributed to the ring resonances of the looped fiber taper [49]. In the telecom range where we perform these experiments, only the first-order slab-mode l = 1 (vertical confinement) is supported for the in-plane polarization TE. The higher l-orders will not be considered. This holds true for all the disks investigated here (they all have the same thickness 200 nm) and will no longer be mentioned in the text. To identify unambiguously the (p, m) indices of observed WGMs, we use FEM electromagnetic simulations, which is the method of choice for thin disks as explained in Section 11.2.2.2 [31]. Although we are lacking the exact knowledge of each given GaAs disk's geometry, we can estimate the disk's radius with a precision of 300 nm by optical microscope inspection. At this level of accuracy, the 200 nm thickness is guaranteed within a few percent by in situ control during the MBE growth. Using FEM simulations, we compute the eigenfrequency and radiation Q factor Q_{rad} of WGMs (p,m) in the wavelength range of 1.5~1.6 μm. This provides us with a comb of modes, which is then wavelength-shifted in block to overlap the measured optical resonances. The shift in wavelength to obtain overlap usually corresponds to a shift in disk radius inferior to 100 nm, consistent with optical microscope inspection. The computed Q_{rad} helps confirming this mode identification procedure: for the broader resonances observed in the spectrum, whose width is determined by radiation losses, the measured Q is indeed close to the computed one. In cases of interest, the mode comb sequence obtained by simulations generally imposes a unique solution to the identification problem. However, the existence of a class of disks for which the identification procedure does not lead a satisfactory solution was noted. Small imperfections in the disk geometry are believed to be responsible for these discrepancies. In the disk shown in Fig. 11.17(b), the pedestal is small in diameter (large underetch distance $\Delta R > 1/2R$), and we are able to identify WGMs with radial number p ranging from 1 to 5. The very broad structures appearing with a period of about 30 nm are believed to relate to even-higher order p-modes of TE polarization, or to TM modes.

11.4.3 Adjustable Evanescent Coupling Experiments

The external evanescent coupling to a WGM resonance is studied as a function of the gap distance between the disk and the coupling waveguide (the tapered fiber here). Figure 11.18(a) shows the optical resonance spectrum of a single WGM (disk radius, $R = 5.9$ μm) centered at 1,546.8 nm when the taper is positioned laterally ~1.8 μm away from the edge of the disk. The gap distance is kept large in order to minimize WGM optical losses induced by the coupling to the taper (taper loading effects [45,56]). Fitting the mode resonance to a Lorentzian gives a linewidth (FWHM) $\delta\lambda$ of 95 pm, indicating a Q factor of 1.6×10^4. In Fig. 11.18(b), the spectrum is acquired when the taper is positioned at the critical coupling distance ($\Delta g \sim 0.42$ μm). The increasing taper-disk coupling has modified the total loss rate of the disk WGM, yielding a loaded Q of 0.8×10^4. The resonance contrast $\Delta T = 1 - T_{\text{on}}$ has also significantly increased from 2% to 97%. Finally, Fig. 11.18(c) shows the spectrum in the over-coupling regime where the loaded Q reduces to 0.3×10^4 and the resonance contrast lowers to 60%.

For this WGM resonance, the complete evolution of T_{on} and $\delta\lambda$ as a function of the gap distance is shown in Fig. 11.18(d). $T_{\text{on}}(\Delta g)$ is fitted by Eq. (11.28), where we employ the exponential approximation $\gamma_{\text{ex}}(\Delta g) = \gamma_{\text{ex}}(0)e^{-\eta\Delta g}$ with $\gamma_{\text{ex}}(0)$, the "zero-gap" external disk–taper coupling rate, and η the spatial decay rate of the evanescent coupling. The best fit to the data enables extracting the ratio of $\gamma_{\text{ex}}(0)/\gamma_{\text{in}} = 5.2$ (γ_{in} being the intrinsic loss rate of the WGM) and $\eta = 1/390$ nm^{-1} for the evanescent coupling experiments. Using these values extracted from the fit and an extrapolated value $Q_{\text{in}} \approx 1.6 \times 10^4$ inferred from measurements at large gap distance, Eq. (11.28) enables us obtain a fitting curve for the gap dependence of $\delta\lambda = (\lambda_0/Q_{\text{in}})(1 + K)$ (The fitting curve is the solid black curve in Fig. 11.18.). This fitting curve agrees well with experimental data.

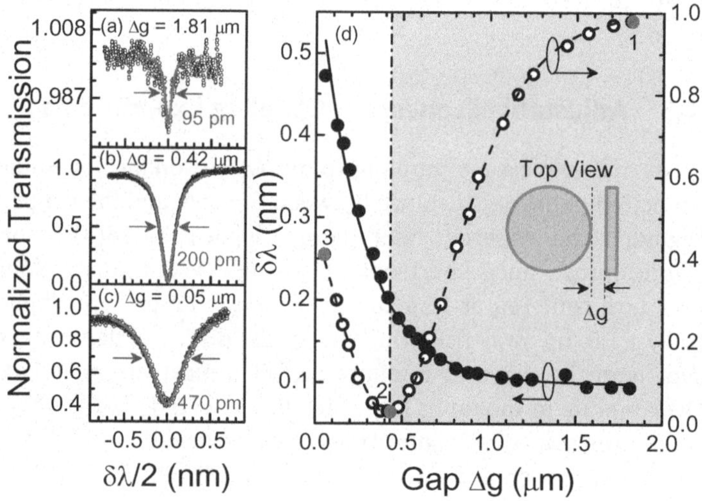

Figure 11.18 Normalized transmission spectrum in the (a) under-coupling, (b) critical coupling, and (c) over-coupling regime for a WGM resonance centered at 1,546.8 nm, measured on a GaAs disk of radius R = 5.9 μm. (d) The on-resonance transmission T_{on} (black circles) and the (FWHM) linewidth $\delta\lambda$ (black dots) as a function of gap distance, together with fitting curves. A dash-dotted line indicates the critical coupling and separates the under-coupling (grey) from the over-coupling (white) regime. Inset shows schematics of the disk-taper positioning parameters. The transmission spectra (a), (b), and (c) are taken at three gap distances indicated by red dots 1, 2, and 3 on the T_{on} curve, respectively.

With the high spectral resolution provided by the fine laser linewidth, we are able to resolve high Q WGMs (up to 3.5×10^5) of GaAs disk resonators. In many cases of interest, the high-Q optical mode resonance splits into a doublet (Fig. 11.19). In other WGM structures (silica microspheres and silicon microdisks), the doublet splitting has been in the past attributed to the lifting of degeneracy of cw and ccw WGMs, as a consequence of Rayleigh scattering on the resonator surface roughness [40,57–59]. The rate at which photons are scattered from the cw to ccw mode is quantified by the doublet splitting $\Delta\lambda$, while the rate at which photons are lost from the resonator is quantified by the intrinsic linewidth $\delta\lambda$ of each individual component of the doublet. Fitting the transmission spectrum with a double Lorentzian, we find in Fig. 11.19(a) $\Delta\lambda$ = 66 pm and $\delta\lambda$ = 10 pm for a GaAs disk of radius R = 4.9 μm. The individual mode loaded quality factor is $Q_{load} = \lambda_0/\delta\lambda = 1.5 \times 10^5$, placing this measurement in the regime $\Delta\lambda >> \delta\lambda$ where the splitting is clearly resolved. In contrast, we obtain in Fig. 11.19(b) $\Delta\lambda$ = 35 pm and $\delta\lambda$ = 29 pm for a disk of radius R = 3.5 μm, which gives an example of $\Delta\lambda \approx \delta\lambda$. The loaded Q factor of this resonance is $Q_{load} = 5.3 \times 10^4$. We typically do not resolve the doublet splitting when Q is inferior to this value.

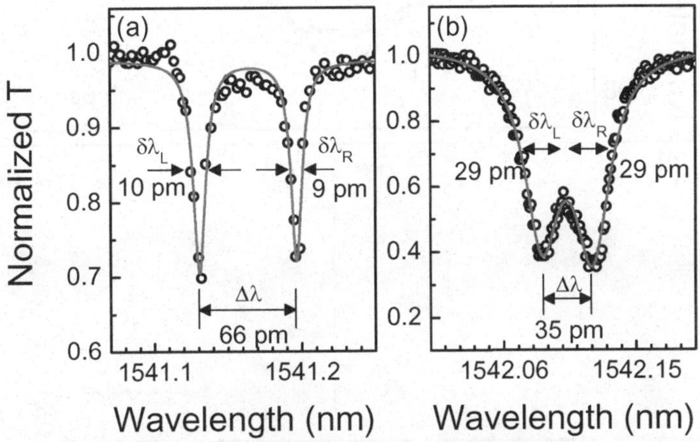

Figure 11.19 High-resolution taper–disk transmission spectra showing a WGM resonance doublet in the case of (a) $\Delta\lambda >> \delta\lambda$ for a disk of radius R = 4.9 μm and (b) $\Delta\lambda \approx \delta\lambda$ for a disk of radius R = 3.5 μm. $\Delta\lambda$ and $\delta\lambda$ ($\delta\lambda_L$ and $\delta\lambda_R$) indicate the splitting of the two standing modes discussed in Section 11.2.4 and the associated individual mode linewidth (left and right), respectively. The condition $\delta\lambda_L = \delta\lambda_R$ generally holds reasonably.

11.4.4 Optical Gradient Force in Evanescent Coupling Experiments

If the fiber taper is mechanically soft enough, the evanescent coupling experiments enable to observe an optomechanical phenomenon of optical force attraction of the fiber taper to the GaAs disk. Theoretical predictions of an optical force acting on evanescently coupled optical waveguides or resonant optical structures was followed by its recent observation in a disk-taper on a silicon nitride (SiN) photonic platform [60–62]. In this work, the authors showed that a milliwatt optical power produces a micrometer-scale displacement of a movable optical fiber taper evanescently coupled to a high-Q SiN disk resonator with diameter D = 44.8 μm and thickness t = 253 nm.

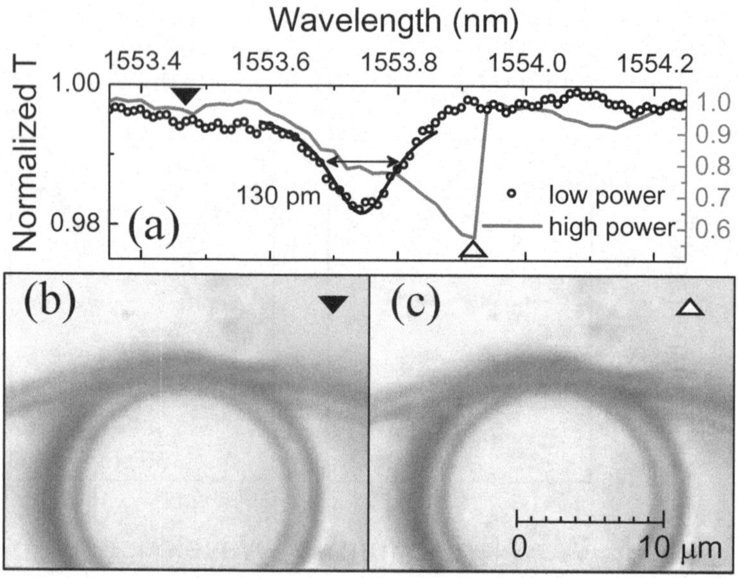

Figure 11.20 (a) Low optical power (black circles) transmission spectrum of a fiber taper coupled to a WGM ($Q \sim 1.2 \times 10^4$) of a GaAs disk (R = 5.7 μm, t = 200 nm). High optical power (solid curve) transmission spectrum of the same WGM showing thermo-optic distortion and a spectral shift in the on-resonance transmission minimum. The black and white triangles indicate the taper transmission level in (b) and (c). (b) An optical micrograph of the taper-disk coupling region taken off WGM resonance at high power. (c) An optical micrograph taken on resonance at high power, showing a significant displacement of the fiber taper waveguide toward the disk.

The GaAs disk resonator shown in Fig. 11.20 has a radius R = 5.7 μm. Optical power is injected into the disk through evanescent coupling from the looped fiber taper as previously. Figure 11.20(a) shows the transmission spectra of a WGM ($Q \sim 1.2 \times 10^4$) at low and high optical power, 310 and 550 μW impinging onto the transmission photodetector respectively. At low power, the symmetric transmission resonance indicates that the disk-taper system behaves linearly. At higher power, thermo-optic effects within the GaAs result in an asymmetric resonance in the scanned transmission spectrum [62–64]. In addition, the on-resonance transmission significantly decreases with increase in the input power. An optical micrograph of the fiber-taper waveguide (taper waist diameter ~1 μm) in the near-field of the GaAs disk is shown in Fig. 11.20(b). When scanning up the laser wavelength toward the transmission resonance minimum, the fiber taper is observed to move toward the disk. This effect is directly visible when comparing Figs. 11.20(b) and (c). The displacement is as large as 500 nm, easily observable with the optical microscope integrated in our experiment. As estimated, an optical force inferior to 1 pN is responsible for this displacement (following the model of Ref. [62]). This value is not necessarily surprising, indeed the looped fiber taper is mechanically very soft in the experiment, making it sensitive to extremely small forces.

11.4.5 Statistical Analysis of Nanofabricated GaAs Disks

The quality of our GaAs disks nanofabrication process is studied statistically. In one unit cell of the electronic lithography mask that we have employed for the results presented here, there are in total 25 distinct disks of varying radius, between 1 and 50 μm. In Fig. 11.21, a pie diagram shows the associated disk size distribution. Let us remind here that in our fabrication process, the final disk radius is determined by the mask design and by the first (nonselective) etch depth step, whereas the underetch distance is set by the duration of the second (selective) underetch. Disks with the final radius R smaller than the underetch distance ΔR simply collapse on the substrate and are not further available for experiments. In our case, we have fabricated three series of GaAs disks with underetch ΔR = 1, 2, and 4 μm, respectively. For each series, we attempt here to give quantitative statistics to estimate the quality and reproducibility of the nanofabrication process. The quality estimation is based on a first coarse optical microscope investigation and on a more accurate optical spectroscopy characterization.

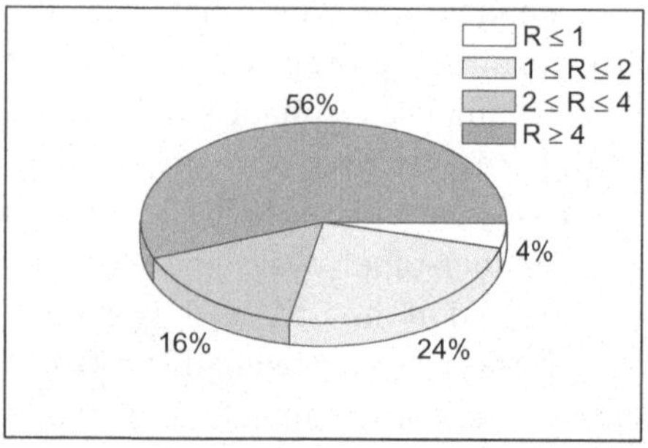

Figure 11.21 Pie diagram for the disk radius distribution, in one unit cell of the employed lithography mask.

We group the disks into five "quality" classes: (1) the collapsed disks, (2) the damaged disks, (3) the disks of poor quality, (4) the disks of average quality, and (5) the excellent disks. The first two classes can be identified immediately by optical microscope inspection (they have simply collapsed, or are typically partly broken or corrupted by an important geometric defect). Poor, average, and excellent disks look "good" under the optical microscope: they have nearly circular geometry and no major visible defect. To further distinguish them, we use optical fiber taper spectroscopy. For "poor quality" disks, no WGMs are observed at all in the transmission spectrum. The "average quality" disks possess reasonably high Q WGMs ($Q_{\text{rad}} \sim 10^4$–10^5) but systematically give rise to small transmission resonance contrast in the evanescent coupling experiments. "Excellent" disks display the best WGM optical resonances, both for what concerns the Q factor and the contrast of the transmission resonance dip. Note that this distinction into five "quality classes" is somewhat arbitrary and is directly motivated by the classes of results that we obtain experimentally. Even if not perfect, this distinction is relevant to qualify the reproducibility of GaAs disks measured optical properties, and may be useful to other groups performing similar experiments in their laboratory.

Figure 11.22 summarizes the statistics obtained on the three series of distinct underetch distance ΔR = 1, 2, and 4 μm. Series 1 (ΔR = 1 μm) contains a total of 132 disks. The distribution of the five "quality classes" is shown as a pie diagram in Fig. 11.22(a). With small underetch distance ΔR = 1 μm, about 99% of the disks do "survive" to the etching process (i.e., do not collapse), 43% out of which do "look good" under the optical microscope. However, only 6% display optical WGM resonances in the experiments, and only 2.27% with $Q_{load} > 10^4$ and transmission contrast over 0.5. To gain better understanding, we plot bar diagrams in Fig. 11.22(d). The *x*-axis in these bar diagrams represents the quantity $1 - \frac{\Delta R}{R}$. This quantity is the ratio of the pedestal radius over the disk radius. Negative values (in light grey background) correspond to disks which have collapsed $\Delta R > R$. Three small symbol disks are placed above Fig. 11.22(d) as indicators. The *y*-axis is the percentage of disks having a given value of $\left(1 - \frac{\Delta R}{R}\right)$.

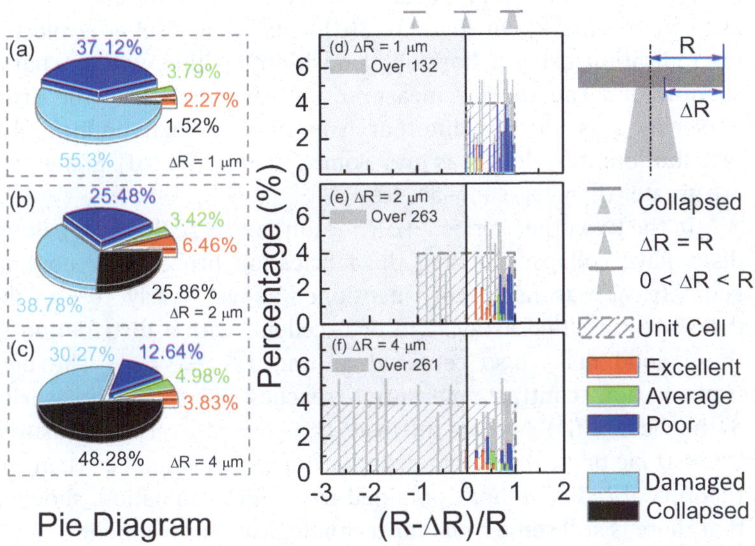

Figure 11.22 (a–f) Statistical analysis of the "quality" of GaAs disks, for three series ΔR = 1, 2, and 4 μm. On the right, histogram of the "quality" as a function of the disk geometry (details discussed in text).

In the first series (ΔR = 1 μm), "poor quality" disks are mainly localized in the region at $\frac{1}{2} < 1 - \frac{\Delta R}{R} < 1$ (the pedestal radius is greater than a half of the disk radius), which corresponds to $R > 2$ μm. From Fig. 11.21, 72% disks per unit cell are $R > 2$ μm. For these disks, 1 μm underetch is too small for the disk to possess a good vertical optical confinement of WGMs. Most of the optical energy of WGMs hence leaks into the disk pedestal. The remaining 28% of disks (for which in the first series the pedestal radius is less than a half of the disk radius) have a radius between 1 μm and 2 μm. This implies a circular boundary with a large curvature, which lead to radiation loss for the WGMs. In our experiments, we find that 6% of disks in the first series on which we can measure the optical resonances. This is not only due to large radiation losses but also to incidents in the fabrication (broken or corrupted disks). We note that the 6% of optically "valid" disks are centered around $1 - \frac{\Delta R}{R} = \frac{1}{3}$, where the pedestal radius is one-third of the disk radius. We obtain similar results on the two other series. Let us stress also that in the mask, the percentage of occurrence of each disk size is 4% (1/25 since there are 25 different nominal radii in the mask). In Fig. 11.22(d), this value of 4% is plotted for indication using a horizontal dash-dotted line, with a sparse shadow underneath. The measured size distribution of the disks in series 1 is slightly different from the nominal uniform 4% distribution: it is plotted as grey combs in Fig. 11.22(d). Some bars are greater than 4%, some are less.

In the two other series 2 ($\Delta R = 2$ μm) and 3 ($\Delta R = 4$ μm), many disks have collapsed during the fabrication process, for obvious geometrical reasons already mentioned above. Nearly 10% of the disks are classified as average or excellent, with a distribution of these good disks also centered around $1 - \frac{\Delta R}{R} = \frac{1}{3}$. The best transmission contrast measured reaches 97% and in series 3 the best WGM has an intrinsic $Q \sim 5 \times 10^5$. This measured WGM Q factor is about two orders of magnitude smaller than the theoretical radiative limit obtained from FEM simulation, showing that there is still some room for technological improvement.

As a conclusion, we would like to stress that the attempt of statistical study shown here is representative of our control on GaAs disks fabrication at a given date (year 2010 for these data). By more careful manipulation, we can rather easily reduce the number of broken disks and increase the yield of average and excellent disks on one sample. Last, it is not perfectly clear yet what role the geometry of a given disk plays on the observation of WGMs resonances in fiber taper spectroscopy. Phase-matching conditions in the evanescent coupling can for example explain why certain classes of geometry do not give rise to WGMs resonances in the spectrum (especially here for large diameter disks of diameter above 15–20 μm): in that case these disks will be classified as "poor quality" in our analysis, when their intrinsic quality is in reality maybe comparable to other. This certainly leads in our present case to an under-estimation of the statistical quality of our GaAs disks. Thus, the analysis shown here must be considered with some care. Its only aim is to provide some experimental information that could help to qualify other fabrication techniques of GaAs disks.

11.5 Optomechanics of GaAs Disk Resonators

In this section we discuss how mechanical modes of GaAs disk resonators couple to their optical WGMs. Indeed, a disk structure is both an optical cavity and a mechanical oscillator, the two being coupled to form an optomechanical resonator. Photons circulating in a gallery mode of a disk exert a centrifugal radiation pressure on the physical boundaries of the disk, giving rise to a radial expansion of the gallery. Conversely, the mechanical radial breathing of the disk structure couples evidently to photons stored in a WGM supported by the disk, by modulating the effective optical length of the gallery cavity. In optomechanics, the degree of optomechanical coupling is generally quantified by one of the two following parameters: $g_{\mathrm{om}} = (d\omega_0/d\alpha)$(in Hz/nm) giving the differential shift of the optical cavity resonance ω_0 (in Hz) per unit of displacement $d\alpha$ of the mechanical system (in nm) or g_0 the vacuum optomechanical coupling giving the cavity resonance shift for a displacement equal to the zero-point fluctuation of the mechanical resonator. These coupling parameters naturally enter the quantum description of any optomechanical system coupled through the radiation pressure coupling Hamiltonian $H_c = \hbar g_{\mathrm{om}}(a^+a)\alpha$, where $a(a^+)$ is the optical annihilation (creation) operator and α the mechanical displacement degree of freedom. A large value of g_{om} is required for efficiently turning mechanical energy into optical energy and vice versa, and is hence beneficial to most experiments relying on optomechanical phenomena. As we will explain below, a miniature GaAs disk resonator can sustain a very large value of g_{om}, up to the THz/nm, placing it at the forefront of optomechanical systems in terms of optomechanical coupling.

Indeed, coming back to the problem of finding the WGMs eigenfrequencies of a dielectric disk by the approximate analytical approach presented in Section 11.2.2.1 [65], let us take the implicit Eq. 11.15 for TM WGMs of azimuthal number, m:

$$n_{xi}\frac{\dot{J}_{\mathrm{m}}(\tilde{k}n_{xi}R)}{J_{\mathrm{m}}(\tilde{k}n_{xi}R)}-\frac{\dot{H}_{\mathrm{m}}^{(1)}(\tilde{k}R)}{H_{\mathrm{m}}^{(1)}(\tilde{k}R)}=0 \tag{11.35}$$

where $\mathrm{Re}(\tilde{k}) = 2\pi/\lambda$. For a given thickness h, the left term of this equation only depends on kR. If k_0 is a solution to this equation for a disk radius R_0, the solution k for radius R is given by $k = k_0R_0/R$ (neglecting the dispersion). As a consequence, in the limit where the approximate analytical approach holds correct, $d\omega_0/dR = -\omega_0/R$ is an exact expression of the optomechanical coupling factor g_{om} for a pure radial displacement of the disk $d\alpha = dR$. The same reasoning also holds for TE WGMS. Thanks to a large refractive index, GaAs disks with radius as small as 1 μm can sustain high quality factor (Q) optical WGMs in the near infrared [41,66,67]. According to the formula $g_{\mathrm{om}} = -\omega_0/R$, g_{om} rises up to the THz/nm range on such disks.

11.5.1 Experimental Optomechanics of GaAs Disk Resonators

To measure mechanical properties of GaAs disk resonators, near-field experiments are performed employing the looped optical fiber taper evanescent coupling technique presented above [49,68]. Tuning the laser wavelength to a flank of the disk optical resonance (inset of Fig. 11.24(a)), we obtain a situation where the disk mechanical vibration modulates the optical transmission T of the fiber. By measuring the noise of this optical transmission, we obtain spectral signatures of the mechanical motion and perform optical vibrational spectroscopy of the disk. The mechanical vibration of our GaAs disks is typically in the RF range. We use a balanced photodetection technique to remove the contribution of laser excess noise in the transmission noise measurement (Fig. 11.23).

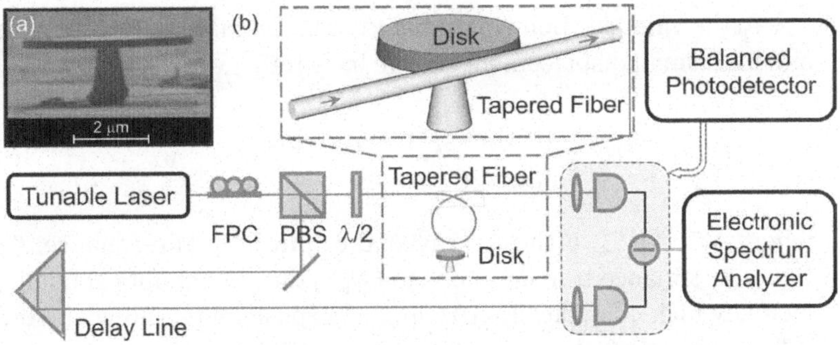

Figure 11.23 Optomechanical measurement on a GaAs disk. (a) SEM view of a GaAs disk (radius 2.25 μm and thickness 200 nm) suspended on an AlGaAs pedestal. (b) Schematics of the near-field optomechanical experiment (FPC, fiber polarization control, PBS, polarization beam splitter). The balanced photodetector operates in the high frequency range, and its AC RF output signal is fed into an electronic spectrum analyzer.

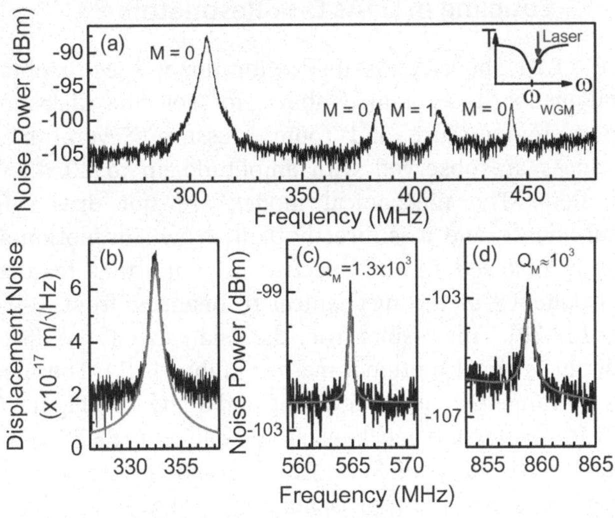

Figure 11.24 Selected vibrational spectra of GaAs optomechanical disk resonators. (a) Optically measured motional noise spectrum of a disk of radius 5.6 μm and thickness 200 nm. (b) Calibrated displacement noise resonance of a disk of radius 4.5 μm, showing a sensitivity of 2×10^{-17} m/$\sqrt{\text{Hz}}$. (c), (d) High-frequency mechanical resonances of a disk of radius 3.6 μm in the 500 MHz – 1 GHz band.

At the inflection point of the flank of a Lorentzian WGM resonance and for a small displacement $\Delta\alpha$, we have [69]

$$T(\Delta\alpha) - T(0) = \frac{dT}{d\omega_0}\frac{d\omega_0}{d\alpha}\Delta\alpha = \frac{3\sqrt{3}}{4}\Delta T\frac{Q_{\text{load}}}{\omega_0}g_{\text{om}}\Delta\alpha \qquad (11.36)$$

where $\Delta T = 1 - T_{\text{on}}$ is the contrast of the optical WGM resonance, T_{on} the on-resonance transmission, and Q_{load} the loaded optical quality factor, which is typically 10^5 in the experiments shown below. ΔT, ω_0, and Q_{load} being measured quantities, and the amplitude of motion $\Delta\alpha$ being known by the equipartition theorem when measuring the Brownian motion, the optomechanical coupling factor g_{om} can be directly inferred from the calibration of the vibrational optical measurement.

11.5.2 Mechanical Modes and Optomechanical Coupling in GaAs Disk Resonators

Figure 11.24 shows GaAs disks vibrational spectra obtained by optical means as explained above, for selected disks measured at room temperature and room pressure. Several mechanical resonances are observed with amplitude up to 20 dB over the noise floor. The mechanical modes are not driven in these measurements, and it is directly their Brownian motion at 300 K which is resolved in the spectra. An important feature is the high frequency of the mechanical resonances, from 100 MHz to 1 GHz [69,70]. This results from the small size of the disks, which also induces small motional masses (Table 11.1). The mechanical Q factors range between a few tens and a few 10^3. Figure 11.24(d) shows for example a mechanical resonance at 858.9 MHz with a mechanical Q factor of 862. This corresponds to a frequency-Q product of 0.7×10^{12}, approaching best values reported in micro-electromechanical systems (MEMS) and nanoelectromechanical systems (NEMS), in the 10^{13} range [71]. A noteworthy point is that this high frequency-Q product value is obtained here in ambient conditions. Mechanical dissipation will not be discussed in detail here but we note that a high purity crystalline GaAs mechanical system should be free of bulk mechanical losses affecting amorphous glasses at low temperature [72]. GaAs material, contrarily to silicon, was not used extensively in micro- and nano-mechanics, but the epitaxial quality of the material could lead in principles to excellent mechanical properties, provided that nanofabrication does not induce mechanical degradation. As a proof of the GaAs potential for micro-/nano-mechanics, a micromechanical oscillator with Q_M factor above 10^5 was recently reported at low temperature [73].

Table 11.1 Optical–mechanical resonances of three GaAs disks, experimental (superscript *e*) and simulated (superscript *s*)

R_{disk} (μm)	λ_0 (nm) (p,m)	Optical *Q*	f_M^e (MHz)	f_M^s (MHz)	m_{eff}^s (pg)	g_{om}^e (GHz/nm)	g_{om}^s (GHz/nm)	Mechanical *Q*
4.5 ± 0.1	1551.7	1.0×10^5	343	339	21.2	71	61	66
	(3,33)		397	380	21.2	65	76	38
4.6 ± 0.1	1535.2	9.5×10^4	336	337	22.1	95	65	120
	(3,34)		395	380	20.9	60	72	66
5.6 ± 0.1	1513.9	5.9×10^4	308	326	36	132	78	73
	(3,45)		383.5	373	22.4	41	35	114

Source: Reprinted with permission from Ref. [69]. Copyright (2010) by the American Physical Society.

Figure 11.24(b) displays a measurement of a disk Brownian motion at 300 K, showing a sensitivity of 2×10^{-17} m/$\sqrt{\text{Hz}}$. The calibration is obtained by using equipartition of energy for the Brownian motion amplitude [69] but it can also be obtained from an independent estimation of g_{om}, as we will see now. Indeed, the formula $g_{om} = -\omega_0/R$ is convenient for simple arguments in whispering gallery optomechanics, and was used extensively in the past, however this formula is expected to be no longer valid when the semi-analytical approach for WGMs does not hold anymore. This happens if the disk mechanical mode has a nonradial component (out-of-plane motion), when the disk is too thin ($<\lambda/n$), or when it is not a perfect cylinder (hence the rotation symmetry of the disk is lost). With a 200 nm thickness, we expect our GaAs disk resonators to deviate from $g_{om} = -\omega_0/R$. To understand the optomechanical coupling problem, we must adopt a more complete description and address the general problem of how an optical resonator Eigenfrequency ω_0 depends on the deformation of the resonator. This problem can be solved by a perturbative treatment of Maxwell's equations [74]. The most convenient approach is to represent the resonator deformation by the displacement α of a chosen point of the resonator having maximum displacement amplitude. If the normalized displacement profile is additionally known, α is sufficient to represent the complete resonator deformation field. In this approach g_{om} is expressed as the integral of an optomechanical function over the rigid resonator boundaries:

$$g_{om} = \frac{\omega_0}{4} \iint (\vec{q} \cdot \vec{n}) \left[\Delta\varepsilon \left| \vec{e}_{\parallel} \right|^2 - \Delta\varepsilon^{-1} \left| \vec{d}_{\perp} \right|^2 \right] dA \tag{11.37}$$

In this expression $\vec{q}$ is the normalized displacement profile vector such that $\max|\vec{q}(r)| = 1$, $\vec{n}$ the normal unit vector on the boundary, $\Delta\varepsilon = \varepsilon - 1$ with ε the dielectric constant of the resonator material, $\Delta(\varepsilon^{-1}) = \varepsilon^{-1} - 1$, $\vec{e}$ is the electric complex field vector normalized such that $\frac{1}{2}\iiint \varepsilon |\vec{e}|^2 dV = 1$, and $\vec{d} = \varepsilon \vec{e}$ [18]. It is important to understand that Eq. (11.37) implies that g_{om} depends both on the optical and mechanical mode under consideration. This also shows that the formula $g_{om} = -\omega_0/R$, which is independent on the details of the optical and mechanical modes, is too simple to fully describe the optomechanical coupling in whispering gallery structures.

In the experiments, we identify both optical and mechanical modes in the measurement, in order to fully model the optomechanical coupling. The WGM identification is carried out as explained in Section 11.4.2 [68].

The identification of mechanical modes measured in our experiments is more difficult. Indeed, their frequencies depend on the exact geometry and composition of the disk and of its pedestal. We identify each mechanical mode by studying its frequency variation as a function of the disk dimensions, using for that purpose disks of different radius but on the same sample, hence made out of the same material and resulting from the same fabrication process. The disk thickness and pedestal height are controlled with a high precision during the epitaxial growth. The other geometrical parameters are measured with a precision of ±50 nm in a SEM and serve as input for numerical simulations, where we simplify our modeling by treating the GaAs material as mechanically isotropic. We use standard elastic parameters at 300 K for GaAs: Young modulus E = 85.9 GPa, Poisson ratio = 0.31, and density 5.316 g/cm^3, and for $Al_{0.8}Ga_{0.2}As$ E = 83.86 GPa, Poisson ratio = 0.39, and density

4.072 g/cm^3. The numerical simulations are performed with the COMSOL Multphyiscs software. They provide us for each disk with a series of mechanical modes that can be classified using radial and azimuthal numbers P and M. The displacement profile of a given mechanical mode can be factorized in the form $q(\rho, \theta) = \cos(M\theta) \times f_P(\rho, z)$ as a result of the rotational symmetry of the disk.

Figure 11.25 shows mechanical modes frequencies measured on GaAs disks of radius varying between 3 and 6 μm. The shadowed guides to the eyes are obtained from simulations and displayed with a thickness of ±30 MHz corresponding to the uncertainty concerning the exact disk geometry. Figure 11.25 allows identifying the prominent mechanical resonances in the spectrum as having $M = 0$ and $M = 1$ azimuthal numbers (modes A, C, and D of Fig. 11.25 correspond to $M = 0$ and mode B to $M = 1$). This identification is further confirmed by looking at the related optomechanical coupling. Indeed, as a consequence of the disk rotational symmetry, the integral of Eq. (11.37) contains an azimuthal prefactor of the form

$$\int_{\theta=0}^{2\pi} \cos(M\theta)\cos^2(m\theta)d\theta = \frac{1}{2}\delta_{M,0} + \frac{1}{4}\left[\delta_{M+2m,0} + \delta_{M-2m,0}\right] \quad (11.38)$$

where $m(M)$ is the azimuthal number of the optical (mechanical) mode under consideration. This prefactor shows that optomechanical coupling g_{om} is maximal for mechanical modes of azimuthal number $M = 0$. This is confirmed by our measurement where prominent resonances in the spectra always correspond to $M = 0$ modes.

For the GaAs disks inspected here, the WGMs resonances in our wavelength range correspond to m between 30 and 50 and consequently, $\delta_{M+2m,0} = 0$. Apart from $M = 0$, the only other case of finite optomechanical coupling is hence expected for $M = 2m$. Mechanical modes with $M = 2m$ are of very high order and practically not visible in the spectra with our current sensitivity. This said, we observe in our spectra mechanical modes with small azimuthal numbers ($M = 1, 2...$). Even if their weight in the mechanical spectrum remains small in comparison to $M = 0$ modes, their presence shows that the rotational symmetry of the disk and the above selection rules are only approximate. The rotation invariance can be broken by small imperfections in the pedestal or in the disk geometry. Also the GaAs crystal itself is not rotation invariant hence some degree of asymmetry is expected even if the geometrical envelope of the disk is perfectly rotation-invariant.

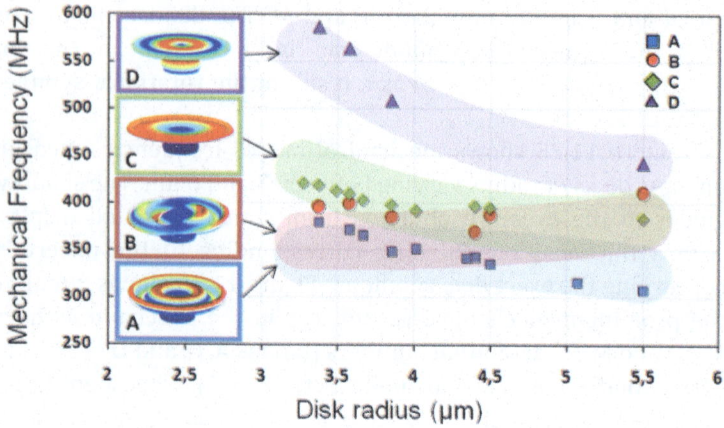

Figure 11.25 Frequency dispersion of GaAs disks mechanical modes measured as a function of the disk radius. The disks thickness is 200 nm. Data points were acquired on different disks of the same sample. Reprinted with permission from Ref. [69]. Copyright (2010) by the American Physical Society.

11.5.3 Quantitative Estimation of the Optomechanical Coupling

Once both optical and mechanical modes are identified in the measurement, we export the computed normal displacement of the mechanical mode from the FEM simulation and import it into the optical FEM simulation to evaluate numerically the surface integral using Eq. (11.37). The integral is performed numerically on each surface of the disk via the COMSOL postprocessing tools and the different contributions added at the end. This sum gives the "simulated" optomechanical coupling g_{om}^{s} for each optical–mechanical combination. The mechanical resonance shown in Fig. 11.24(b) corresponds, for example, to the $M = 0$ mode labeled A in Fig. 11.25. The optical resonance used in Fig. 11.24(b) has been identified to be the WGM ($p = 3$, $m = 33$). A numerically "simulated" value g_{om}^{s} of the coupling between these two modes is calculated to be $g_{\mathrm{om}}^{s} = 61$ GHz/nm. But the effective motional mass of the mechanical mode also leads the Brownian motion amplitude through the Equipartition theorem $m_{\text{eff}}\Omega_{\text{M}}^2\langle\Delta\alpha\rangle^2 = k_{\text{B}}T$. Our measurements are performed at 300 K and the magnitude of the thermo-optic shift of the WGM resonance indicates that heating induces negligible modification of the Brownian motion amplitude [64]. The Brownian motion can then also be used as a motional reference to calibrate the measurement and extract in a second independent manner an "experimental" value g_{om}^e of the optomechanical coupling. g_{om}^e and g_{om}^s values are generally in reasonable agreement.

The parameters relating to optical and mechanical resonances observed on three selected GaAs disks are summarized in Table 11.1. The disk effective motional masses are typically around 10 pg, with optomechanical coupling spanning from 30 to 150 GHz/nm. These high levels of optomechanical coupling, combined with optical Q factors in the 10^5 range, lead to a typical displacement sensitivity of 10^{-17} m/$\sqrt{\text{Hz}}$. For comparison, displacement sensitivity on MHz nanomechanical motion at the level of 10^{-16} m/$\sqrt{\text{Hz}}$ was obtained using a single electron transistor at mK temperature [75], 10^{-15} m/$\sqrt{\text{Hz}}$ using a rf strip line cavity at mK temperature [11] and 10^{-15} m/$\sqrt{\text{Hz}}$ reachable using a high finesse fiber cavity or a silica toroid as a mechanical–optical transducer [17,20]. A sensitivity in the 10^{-17} m/$\sqrt{\text{Hz}}$ range was recently reported in optomechanical crystals [18]. The extreme motional sensitivity obtained in GaAs optomechanical resonators places them at the forefront of optical detection of high frequency nanomechanical motion. As we will see below, this 10^{-17} m/$\sqrt{\text{Hz}}$ motional sensitivity also holds for GHz motion of the disk, and is remarkably obtained at room temperature and room pressure.

The merits of GaAs disk resonators for optomechanics applications are further illustrated by shrinking down their size to the smallest dimension possible. This lower limit, around the photon wavelength in radius extension, corresponds to a transition point where WGMs start to experience too much radiation loss as a consequence of the strong disk curvature. Ultra-strong optomechanical coupling is observed on wavelength-sized GaAs optomechanical disk resonators [76]. We will present here exemplary results on disks having a typical radius R = 1 μm and thickness 200 nm, and standing on Christmas-tree-like $\text{Al}_{0.8}\text{Ga}_{0.2}\text{As}$ pedestals as seen in Fig. 11.26(a). The precise shape of the pedestal depends on steering of the sample in the selective etchant solution, and does not have any wanted functionality here. For a disk as shown in Fig. 11.26(a), FEM simulations predict one unique optical WGM resonance in the 1,500–1,600 nm range, at a wavelength $\lambda_0 \approx 1{,}550$ nm. The corresponding WGM is a TE ($p = 1$, $m = 7$) mode, with p and m the radial and azimuthal numbers, respectively. The electric field intensity is localized at the periphery of the disk (Fig. 11.24(b)), as always for $p = 1$ WGMs. The FSR between two consecutive m numbers of this radial WGM is ~140 nm, hence one unique resonance can be measured in our experimental set-up where the tunability range of our laser is ~80 nm. The breathing mechanical modes of the same disk are simulated by FEM simulation. The axis-symmetry breathing mode with azimuthal number $M = 0$, which couples maximally to WGMs photons [69], is numerically found at a frequency around 1.5 GHz. Its 3D deformation profile is shown in Fig. 11.24(c).

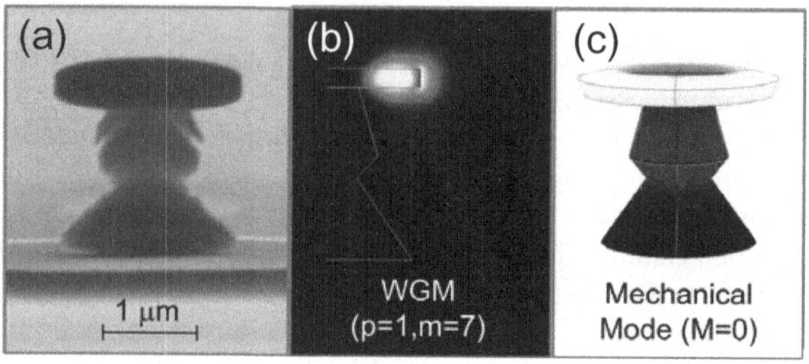

Figure 11.26 (a) SEM view of a wavelength-sized GaAs disk (1 μm radius and 200 nm thickness) suspended on an AlGaAs pedestal. (b) Electric field intensity $|E(r)|^2$ of the optical mode ($p = 1$, $m = 7$). (c) Deformation profile of an $M = 0$ mechanical breathing mode observed in our measurements.

The optical and mechanical responses of the GaAs disk are shown in Fig. 11.27. By measuring the symmetric optical resonance at low optical power, the optical Q factor of the GaAs disk is measured at a level of 1×10^4, mainly limited at this wavelength by radiation losses. At high optical power, the optical resonance experiences an important thermo-optic distortion. The strong confinement of optical energy in the disk combines with residual absorption and a poor thermal anchoring to produce significant heating (seen in Fig. 11.27(a)).

Figure 11.27(b) shows the calibrated RF mechanical spectrum obtained by optical means as previously. The mechanical frequency of the breathing mode is found at 1.4 GHz, in reasonable agreement with numerical simulation. At high optical power, the optically induced heating barely modifies the material's elasticity so that the associated mechanical frequency shift, about 1%, is negligible.

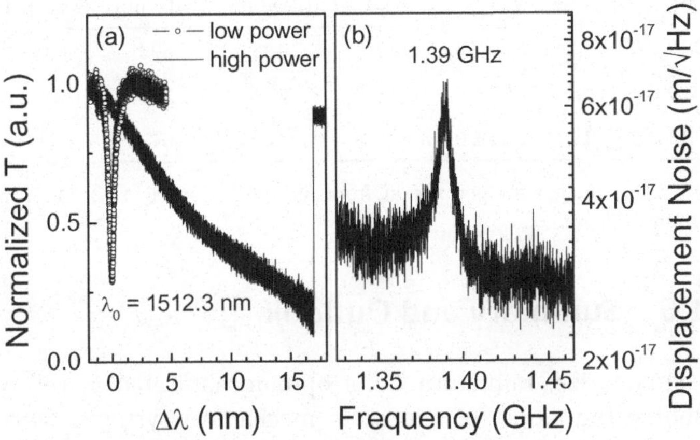

Figure 11.27 (a) Optical and (b) mechanical response of a wavelength-sized GaAs disk.

The optomechanical coupling between the identified optical and mechanical modes is computed numerically as above, yielding a simulated optomechanical coupling of g^s_{om} = 690 GHz/nm, with a related vacuum coupling of g^s_0 = 1.1 MHz. The second "experimental" estimation, inferred from the Brownian motion amplitude, yields an experimental optomechanical coupling of g^e_0 = 0.8 MHz (g^e_{om} = 485 GHz/nm), a factor 1.4 times smaller than the FEM-estimated value. Similar discrepancies were observed in photonic crystals structures [77] and probably stem from an imperfect control of the geometry obtained by nanofabrication. The calibrated Brownian motion gives a sensitivity of a few 10^{-17} m/$\sqrt{\mathrm{Hz}}$, comparable to that obtained above on slightly larger disks. Table 11.2 lists optical, mechanical and optomechanical parameters for three distinct wavelength-sized GaAs disks. The second disk in the table possesses a larger pedestal than the disk seen in Fig. 11.24(a). This results in a less confined optical WGM and a reduced effective motional mass. The optomechanical coupling is reduced with respect to the two other disks.

Table 11.2 Experimental (superscript *e*) and simulated (superscript *s*) mechanical and optomechanical parameters of three miniature GaAs disks

R_{disk} (μm)	f_M^e (GHz)	f_M^s (GHz)	m_{eff}^s(pg)	g_{om}^e (GHz/nm)	g_{om}^s (GHz/nm)	g_0^e (MHz)	g_0^s (MHz)
0.98 ± 0.05	1.39	1.39 ± 0.05	2.2	485	693	0.8	1.14
1.04 ± 0.05	1.36	1.37 ± 0.05	1.4	277	396	0.58	0.84
1.03 ± 0.05	1.34	1.34 ± 0.05	2.4	429	613	0.7	1.0

Source: Reprinted with permission from Ref. [76]. Copyright [2011], American Institute of Physics.

11.6 Summary and Outlook

To summarize, miniature GaAs optomechanical disk resonators combine the assets of nanoscale mechanical systems, with high frequency (GHz) and low motional mass in the picogram (pg) range, and the merits of semiconductor optical microcavities, with sub-micrometer mode volumes and optical quality factors above 10^5. Thanks to the miniature optical–mechanical interaction volume, the optomechanical coupling reaches several hundreds of GHz/nm, enabling optical detection of the disk motion with an ultra-high sensitivity at room temperature, approaching the ultimate quantum limit of detection. Besides offering high-frequency mechanical oscillators with an ultrasensitive optical readout, miniature GaAs optomechanical resonators are naturally suited for integration in geometry of arrays on a chip and for interfacing with single emitters in form of InGaAs quantum dots [78], and promise to open a vast research playground at the interface of quantum optomechanics and active semiconductor photonics.

References

1. Favero, I. and Karrai, K. (2009). Optomechanics of deformable optical cavities, *Nat. Photonics*, **3**, pp. 201–205.

2. Marquardt, F. and Girvin, S. M. (2009). Optomechanics, *Physics*, **2**, p. 40.
3. Aspelmeyer, M., Gröblacher, S., Hammerer, K. and Kiesel, N. (2010). Quantum optomechanics: Throwing a glance, *J. Opt. Soc. Am. B*, **27**, pp. A189–A197.
4. Vuletic, V. and Chu, S. (2000). Laser cooling of atoms, ions, or molecules by coherent scattering, *Phys. Rev. Lett.*, **84**, pp. 3787–3790.
5. Ashkin, A., Dziedzic, J. M., Bjorkholm, J. E. and Chu, S. (1986). Observation of a Single-beam gradient force optical trap for dielectric particles, *Opt. Lett.*, **11**, pp. 288–290.
6. Block, S. M., Goldstein, L. S. B. and Schnapp, B. J. (1990). Bead movement by single kinesin molecules studied with optical tweezers, *Nature*, **348**, pp. 348–352.
7. O'Connell, A. D., Hofheinz, M., Ansmann, M., Bialczak, R. C., Lenander, M., Lucero, E., Neeley, M., Sank, D., Wang, H., Weides, M., Wenner, J., Martinis, J. M. and Cleland, A. N. (2010). Quantum ground state and single-phonon control of a mechanical resonator, *Nature*, **464**, pp. 697–703.
8. Teufel, J. D., Donner, T., Li, D., Harlow, J. W., Allman, M. S., Cicak, K., Sirois, A. J., Whittaker, J. D., Lehnert, K. W. and Simmonds, R. W. (2011). Sideband cooling of micromechanical motion to the quantum ground state, *Nature*, **475**, pp. 359–363.
9. Chan, J., Alegre, T. P. M., Safavi-Naeini, A. H., Hill, J. T., Krause, A., Groblacher, S., Aspelmeyer, M. and Painter, O. (2011). Laser cooling of a nanomechanical oscillator into its quantum ground state, *Nature*, **478**, pp. 89–92.
10. Li, M., Pernice, W. H. P. and Tang, H. X. (2009). Broadband all-photonic transduction of nanocantilevers, *Nat. Nanotechnol.*, **4**, pp. 377–382.
11. Teufel, J. D., Donner, T., Castellanos-Beltran, M. A., Harlow, J. W. and Lehnert, K. W. (2009). Nanomechanical motion measured with an imprecision below that at the standard quantum limit, *Nat. Nanotechnol.*, **4**, pp. 820–823.
12. Arcizet, O., Cohadon, P. F., Briant, T., Pinard, M. and Heidmann, A. (2006). Radiation-pressure cooling and optomechanical instability of a micromirror, *Nature*, **444**, pp. 71–74.
13. Gigan, S., Bohm, H. R., Paternostro, M., Blaser, F., Langer, G., Hertzberg, J. B., Schwab, K. C., Bauerle, D., Aspelmeyer, M. and Zeilinger, A. (2006). Self-cooling of a micromirror by radiation pressure, *Nature*, **444**, pp. 67–70.

14. Schliesser, A., Del'Haye, P., Nooshi, N., Vahala, K. J. and Kippenberg, T. J. (2006). Radiation pressure cooling of a micromechanical oscillator using dynamical backaction, *Phys. Rev. Lett.*, **97**, p. 243905.
15. Park, Y. S. and Wang, H. (2009). Resolved-sideband and cryogenic cooling of an optomechanical resonator, *Nat. Phys.*, **5**, pp. 489–493.
16. Favero, I. and Karrai, K. (2008). Cavity cooling of a nanomechanical resonator by light scattering, *New J. Phys.*, **10**, p. 095006.
17. Favero, I., Stapfner, S., Hunger, D., Paulitschke, P., Reichel, J., Lorenz, H., Weig, E. M. and Karrai, K. (2009). Fluctuating nanomechanical system in a high finesse optical microcavity, *Opt. Express*, **17**, pp. 12813–12820.
18. Eichenfield, M., Chan, J., Camacho, R. M., Vahala, K. J. and Painter, O. (2009). Optomechanical crystals, *Nature*, **462**, pp. 78–82.
19. Li, M., Pernice, W. H. P. and Tang, H. X. (2009). Reactive cavity optical force on microdisk-coupled nanomechanical beam waveguides, *Phys. Rev. Lett.*, **103**, p. 223901.
20. Anetsberger, G., Arcizet, O., Unterreithmeier, Q. P., Rivière, R., Schliesser, A., Weig, E. M., Kotthaus, J. P. and Kippenberg, T. J. (2009). Near-field cavity optomechanics with nanomechanical oscillators, *Nat. Phys.*, **5**, pp. 909–914.
21. Wilson-Rae, I., Zoller, P. and Imamoglu, A. (2004). Laser cooling of a nanomechanical resonator mode to its quantum ground state, *Phys. Rev. Lett.*, **92**, p. 075507.
22. Djordjev, K., Choi, S. J. and Dapkus, R. D. (2002). microdisk tunable resonant filters and switches, *IEEE Photonic. Tech. L.*, **14**, pp. 828–830.
23. Zhou, L. and Poon, A. W. (2006). Silicon electro-optic modulators using pin diodes embedded 10-micron-diameter microdisk resonators, *Opt. Express*, **14**, pp. 6851–6857.
24. Luo, X. and Poon, A. W. (2009). Many-Element Coupled-Resonator Optical Waveguides Using Gapless-Coupled Microdisk Resonators, *Opt. Express*, **17**, pp. 23617–23628.
25. Srinivasan, K. (2006). *Semiconductor Optical Microcavities for Chip-Based Cavity Qed* (dissertation). Pasadena, CA: California Institute of Technology.
26. Borselli, M. (2006). *High-Q Microresonators as Lasing Elements for Silicon Photonics* (dissertation). Pasadena, CA: California Institute of Technology.

27. Heebner, J. E. (2003). *Nonlinear Optical Whispering Gallery Micro-resonators for Photonics* (dissertation). Rochester, NY: University of Rochester.

28. McCall, S. L., Levi, A. F. J., Slusher, R. E., Pearton, S. J. and Logan, R. A. (1992). Whispering gallery mode microdisk lasers, *Appl. Phys. Lett.*, **60**, pp. 289–291.

29. Andronico, A. (2008). *Electromagnetic Study of AlGaAs Integrated Emitters of Infrared and THz Light* (dissertation). Paris, France: Université Paris Diderot.

30. Monk, P. (2003). *Finite Element Methods for Maxwell's Equations.* USA: Oxford University Press.

31. Oxborrow, M. (2007). Traceable 2-D finite-element simulation of the whispering-gallery modes of axisymmetric electromagnetic resonators. *IEEE T. Microw. Theory*, **55**, pp. 1209–1218.

32. Cheema, M. I. and Kirk, A. G. (2010). *Implementation of the Perfectly Matched Layer to Determine the Quality Factor of Axisymmetric Resonators in Comsol.* Proceedings of the COMSOL Conference. October 10, 2010, Boston.

33. Greenwood, A. D. and Jin, J. M. (1999). A novel efficient algorithm for scattering from a complex bor using mixed finite elements and cylindrical Pml, *IEEE T. Antenn. Propag.*, **47**, pp. 620–629.

34. Yariv, A. and Yeh, P. (2007). *Photonics Optical Electronics in Modern Communications*, 6th Ed. USA: Oxford University Press.

35. Yariv, A. (2000). Universal relations for coupling of optical power between microresonators and dielectric waveguides, *Electron. Lett.*, **36**, pp. 321–322.

36. Haus, H. A. and Huang, W. (1991). Coupled-mode theory, *P. IEEE*, **79**, pp. 1505–1518.

37. Little, B. E., Chu, S. T., Haus, H. A., Foresi, J. and Laine, J. P. (1997). Microring resonator channel dropping filters, *J. Lightwave Technol.*, **15**, pp. 998–1005.

38. Manolatou, C., Khan, M. J., Fan, S., Villeneuve, P. R., Haus, H. A. and Joannopoulos, J. D. (1999). Coupling of modes analysis of resonant channel add-drop filters, *IEEE J. Quantum. Elect.*, **35**, pp. 1322–1331.

39. Kamalakis, T. and Sphicopoulos, T. (2006). Frequency dependence of the coupling coefficients and resonant frequency detuning in a nanophotonic waveguide-cavity system, *IEEE J. Quantum. Elect.*, **42**, pp. 827–837.

40. Weiss, D. S., Sandoghdar, V., Hare, J., Lefevre-Seguin, V., Raimond, J. M. and Haroche, S. (1995). Splitting of high-Q mie modes induced by light backscattering in silica microspheres, *Opt. Lett.*, **20**, pp. 1835–1837.
41. Peter, E., Dousse, A., Voisin, P., Lemaitre, A., Martrou, D., Cavanna, A., Bloch, J. and Senellart, P. (2007). Highly directional radiation pattern of microdisk cavities, *Appl. Phys. Lett.*, **91**, p. 151103.
42. Peter, E. (2006). *Couplage Fort Exciton-Photon Pour Une Boîte Quantique De GaAs En Microdisque* (Disertation). Orsay, Franch: Université Paris Sud (in French).
43. Knight, J. C., Cheung, G., Jacques, F. and Birks, T. A. (1997). Phase-matched excitation of whispering-gallery-mode resonances by a fiber taper, *Opt. Lett.*, **22**, pp. 1129–1131.
44. Cai, M. and Vahala, K. (2000). Highly efficient optical power transfer to whispering-gallery modes by use of a symmetrical dual-coupling configuration, *Opt. Lett.*, **25**, pp. 260–262.
45. Srinivasan, K., Barclay, P. E., Borselli, M. and Painter, O. (2004). Optical-fiber-based measurement of an ultrasmall volume high-Q photonic crystal microcavity, *Phys. Rev. B*, **70**, p. 081306.
46. Hwang, I. K., Kim, S. K., Yang, J. K., Kim, S. H., Lee, S. H. and Lee, Y. H. (2005). Curved-microfiber photon coupling for photonic crystal light emitter, *Appl. Phys. Lett.*, **87**, p. 131107.
47. Grillet, C., Monat, C., Smith, C. L., Eggleton, B. J., Moss, D. J., Frédérick, S., Dalacu, D., Poole, P. J., Lapointe, J., Aers, G. and Williams, R. L. (2007). Nanowire coupling to photonic crystal nanocavities for single photon sources, *Opt. Express*, **15**, pp. 1267–1276.
48. Orucevic, F., Lefèvre-Seguin, V. and Hare, J. (2007). Transmittance and near-field characterization of sub-wavelength tapered optical fibers, *Opt. Express*, **15**, pp. 13624–13629.
49. Ding, L., Belacel, C., Ducci, S., Leo, G. and Favero, I. (2010). Ultralow loss single-mode silica tapers manufactured by a microheater, *Appl. Opt.*, **49**, pp. 2441–2445.
50. Takeuchi, Y. and Noda, J. (1992). Novel fiber coupler tapering process using a microheater, *IEEE Photonic. Tech. L.*, **4**, pp. 465–467.
51. Birks, T. A. and Li, Y. W. (1992). The shape of fiber tapers, *J. Lightwave Technol.*, **10**, pp. 432–438.
52. Leon-Saval, S., Birks, T., Wadsworth, W., St. J. Russell, P. and Mason, M. (2004). Supercontinuum generation in submicron fibre waveguides, *Opt. Express*, **12**, pp. 2864–2869.

53. Brambilla, G., Xu, F. and Feng, X. (2006). Fabrication of optical fibre nanowires and their optical and mechanical characterisation, *Electron. Lett.*, **42**, pp. 517–519.

54. Sumetsky, M., Dulashko, Y. and Hale, A. (2004). Fabrication and study of bent and coiled free silica nanowires: Self-coupling microloop optical interferometer, *Opt. Express*, **12**, pp. 3521–3531.

55. Ulrich, R., Rashleigh, S. and Eickhoff, W. (1980). Bending-induced birefringence in single-mode fibers, *Opt. Lett.*, **5**, pp. 273–275.

56. Borselli, M., Srinivasan, K., Barclay, P. E. and Painter, O. (2004). Rayleigh scattering, mode coupling, and optical loss in silicon microdisks, *Appl. Phys. Lett.*, **85**, p. 3693.

57. Little, B. E., Laine, J. P. and Chu, S. T. (1997). Surface-roughness-induced contradirectional coupling in ring and disk resonators, *Opt. Lett.*, **22**, pp. 4–6.

58. Gorodetsky, M. L., Pryamikov, A. D. and Ilchenko, V. S. (2000). Rayleigh scattering in high-Q microspheres, *J. Opt. Soc. Am. B*, **17**, pp. 1051–1057.

59. Borselli, M., Johnson, T. J. and Painter, O. (2006). Measuring the role of surface chemistry in silicon microphotonics, *Appl. Phys. Lett.*, **88**, p. 131114.

60. Povinelli, M. L., Loncar, M., Ibanescu, M., Smythe, E. J., Johnson, S. G., Capasso, F. and Joannopoulos, J. D. (2005). Evanescent-wave bonding between optical waveguides, *Opt. Lett.*, **30**, pp. 3042–3044.

61. Povinelli, M. L., Johnson, S. G., Loncar, M., Ibanescu, M., Smythe, E. J., Capasso, F. and Joannopoulos, J. D. (2005). High-Q enhancement of attractive and repulsive optical forces between coupled whispering-gallery-mode resonators, *Opt. Express*, **13**, pp. 8286–8295.

62. Eichenfield, M., Michael, C. P., Perahia, R. and Painter, O. (2007). Actuation of micro-optomechanical systems via cavity-enhanced optical dipole forces, *Nat Photon*, **1**, pp. 416–422.

63. Vogel, M., Mooser, C., Karrai, K. and Warburton, R. J. (2003). Optically tunable mechanics of microlevers, *Appl. Phys. Lett.*, **83**, p. 1337.

64. Carmon, T., Yang, L. and Vahala, K. (2004). Dynamical thermal behavior and thermal self-stability of microcavities, *Opt. Express*, **12**, pp. 4742–4750.

65. Andronico, A., Caillet, X., Favero, I., Ducci, S., Berger, V. and Leo, G. (2008). Semiconductor microcavities for enhanced nonlinear optics interactions, *J. Europ. Opt. Soc. Rap. Public.*, **3**, p. 08030.

66. Gayral, B., Gerard, J. M., Lemaitre, A., Dupuis, C., Manin, L. and Pelouard, J. L. (1999). High-Q wet-etched GaAs microdisks containing InAs quantum boxes, *Appl. Phys. Lett.*, **75**, p. 1908.

67. Michael, C. P., Srinivasan, K., Johnson, T. J., Painter, O., Lee, K. H., Hennessy, K., Kim, H. and Hu, E. (2007). Wavelength-and material-dependent absorption in GaAs and AlGaAs microcavities, *Appl. Phys. Lett.*, **90**, p. 051108.

68. Ding, L., Senellart, P., Lemaitre, A., Ducci, S., Leo, G. and Favero, I. (2010). GaAs micro-nanodisks probed by a looped fiber taper for optomechanics applications, *Proc. SPIE*, **7712**, p. 771211.

69. Ding, L., Baker, C., Senellart, P., Lemaitre, A., Ducci, S., Leo, G. and Favero, I. (2010). High frequency GaAs nano-optomechanical disk resonator, *Phys. Rev. Lett.*, **105**, p. 263903.

70. Carmon, T. and Vahala, K. J. (2007). Modal spectroscopy of optoexcited vibrations of a micron-scale on-chip resonator at greater than 1 Ghz frequency, *Phys. Rev. Lett.*, **98**, p. 123901.

71. Wang, J., Butler, J. E., Feygelson, T. and Nguyen, C. T. C. (January 25–29, 2004). *1.51-Ghz Nanocrystalline diamond micromechanical disk resonator with material-mismatched isolating support, Proceedings 17th International IEEE Conference on Micro Electro Mechanical Systems*, pp. 641–644.

72. Pohl, R. O., Liu, X. and Thompson, E. J. (2002). Low-temperature thermal conductivity and acoustic attenuation in amorphous solids, *Rev. Mod. Phys.*, **74**, p. 991.

73. Mahboob, I. and Yamaguchi, H. (2008). Bit storage and bit flip operations in an electromechanical oscillator, *Nat. Nanotechnol.*, **3**, pp. 275–279.

74. Johnson, S. G., Ibanescu, M., Skorobogatiy, M. A., Weisberg, O., Joannopoulos, J. D. and Fink, Y. (2002). Perturbation theory for Maxwell's equations with shifting material boundaries, *Phys. Rev. E*, **65**, p. 066611.

75. Naik, A., Buu, O., LaHaye, M. D., Armour, A. D., Clerk, A. A., Blencowe, M. P. and Schwab, K. C. (2006). Cooling a nanomechanical resonator with quantum back-action, *Nature*, **443**, pp. 193–196.

76. Ding, L., Baker, C., Senellart, P., Lemaitre, A., Ducci, S., Leo, G. and Favero I. (2011). Wavelength-sized GaAs optomechanical resonators with gigahertz frequency, *Appl. Phys. Lett.*, **98**, p. 113108.

77. Safavi-Naeini, A. H., Mayer Alegre, T. P., Winger, M. and Painter, O. (2010). Optomechanics in an ultrahigh-Q two-dimensional photonic crystal cavity, *Appl. Phys. Lett.*, **97**, p. 181106.

78. Peter, E., Senellart, P., Martrou, D., Lemaître, A., Hours, J., Gérard, J. and Bloch, J. (2005). Exciton-photon strong-coupling regime for a single quantum dot embedded in a microcavity, *Phys. Rev. Lett.*, **95**, p. 67401.

Chapter 12

Whispering Gallery Mode in ZnO with Hexagonal Cross Section

Rui Chen, Van Duong Ta, and Handong Sun
Division of Physics and Applied Physics, School of Physical and Mathematical Sciences, Nanyang Technological University, Singapore
hdsun@ntu.edu.sg

In recent years, enormous attention has been paid to optical microresonators, which hold a great promise for microlasers as well as fundamental studies in cavity quantum electrodynamics. Among various kinds of optical microresonators, whispering gallery mode (WGM) cavity is one of the most-interesting configurations because of its intrinsically high quality factor, low lasing threshold, and relatively simple fabrication. In this chapter, we will describe the WGM resonators in ZnO nano/microsturcutres with hexagonal cross section. The content of this chapter are arranged as following. Firstly, a general introduction of the basic properties of ZnO and WGM will be presented. Secondly, the experimental techniques about different methods to synthesize ZnO and knowledge about optical characterizations will be introduced. In the third section, some recent key research findings about the modification of spontaneous emission, different kinds of WGM lasing as well as exciton polariton in hexagonal ZnO will be summarized and

Handbook of Optical Microcavities
Edited by Anthony H. W. Choi

ISBN 978-981-4463-24-9 (Hardcover), 978-981-4463-25-6 (eBook)
www.panstanford.com

discussed. Finally, the potential device application and future directions of ZnO WGM will be drawn. At the end of the chapter, a selection of references will be given for further reading that penetrate deeper into the topic or give more detail description.

12.1 Introduction

The development of optical microresonators have attracted extensive research interest [1–4], not only because of their small mode volume and high quality (Q) factor for fundamental physics investigations [5,6], but also for potential applications in such as laser sources, active filters, and photonic sensing devices for label-free detection [7,8]. With the rapid development of nanoscience and nanotechnology, different kinds of structures and fabrication techniques have been adopted to construct optical microresonators. Laser emissions have been observed from resonators with various geometries, such as Fabry–Pérot (F-P) cavity [9,10], photonic crystal [11,12], random cavity [13,14], plasmonic cavity [15,16], distributed feedback resonator [17], as well as whispering gallery modes (WGM) [7,18]. In the case of WGM resonator, the light is trapped inside the cavity by multiple total internal reflections at the interface near the structure and the surrounding medium [19]. Compared to other resonators, WGM exhibits much lower loss and thus is expected to lead to low lasing threshold and high Q lasers.

Till date, fabrication of solid-state microresonators is primarily based on either top-down or bottom-up technique [20]. In the top-down approach, a microresonator can be obtained by etching an epitaxially grown planar cavity [9]. This approach critically relies on epitaxial growth, thus requires costly apparatus and sophisticated grow techniques. Meanwhile, the strict requirement on lattice matching between the epitaxial layers and the substrate severely limits material choices. Moreover, this method introduces optical losses at the boundary due to the imperfection of the etching process, leading to lower Q factor of the microresonators. In comparison, for the bottom-up approach, the microcavities are directly formed by self-assembled crystallisation of nanostructures, which shows inherent advantages such as high material quality and high throughput assembly. By controlling the experimental parameters, different shapes and sizes of the microresonators can be obtained [18,21–23]. Meanwhile, the approach takes advantage of both the radiative recombination in the material as internal light source and the resonator formed by the structure itself.

During the past decade, low-dimensional semiconductor structures have attracted extensive research interest and various notable achievements have been achieved [24–30]. Among various kinds of semiconductors, ZnO is recognized as potential building blocks for nanometer-scale electronic and optoelectronic devices operated in the blue and ultraviolet (UV) spectral regions due to its wide band gap (~3.37 eV) and large exciton binding energy (~60 meV) at room temperature [25,31]. Moreover, the richness of nanostructures of ZnO facilitates the realization of various interesting optical confinement and laser cavities. The resonant process in ZnO nano/microstructures can be classified into three different mechanisms:

(1) Random lasing from randomly distributed ZnO powders or nanowires. In the case of random lasing, the light is amplified along the closed-loop feedback paths by the recurrent scattering at the boundaries. Generally speaking, random lasing presents indefinite lasing modes with a very high threshold due to high loss during the random scattering path [14,32].
(2) F-P lasing from a single ZnO structure. In the case of F-P lasing, the mirrors are formed by two end facets. For ZnO, the F-P modes suffer strongly from the mirror losses at the interface between ZnO and air due to a reflectance of about 20%. This is why it is difficult to obtain lasing from F-P cavity in ZnO [33].
(3) WGM lasing from hexagonal ZnO structures. In the case of WGM, light propagates circularly in the cavity due to multiple total internal reflections at the inner walls of ZnO hexagon. Due to a lower loss and better confinement, ZnO WGM cavity is expected to demonstrate a high Q factor as well as a low threshold compared to random cavity and F-P cavity [18,34].

The WGM resonate phenomena have been investigated in ZnO nanodisks [21], nanonails [35,36], nano/microwires [37,38], and microdisks [18]. However, it is only until recently that distinct WGM modes have been identified from individual ZnO microwire laser at low temperature [37]. After that, various kinds of WGM lasing and potential device application using hexagonal ZnO have been demonstrated. Since this is a rapidly expanding field in which some unique properties have been discovered as well as some potential applications have been achieved, this chapter intends to provide the state-of-the-art research activities about the WGM in ZnO with hexagonal cross section.

12.2 Experimental and Instrumental Methodology

In this section, the main experimental means used for ZnO nano/microstructures synthesis and optical characterization will be briefly described. The principle of the experimental techniques will be introduced. The purpose of this section is to provide a fundamental description of the experimental and instrumental methodology that is necessary for understanding the experimental results to be presented later.

12.2.1 Material Growth

The most popular approach for ZnO nano/microstructure growth is vapor phase transport (VPT) method [39–42]. Generally speaking, the mechanism of the VPT approach is the introduction of a catalytic liquid alloy phase that can rapidly adsorb a vapor to super saturation levels, and from which crystal growth can subsequently occur from nucleated seeds at the liquid–solid interface. In the ZnO growth process, substrates such as silicon wafer or *c*-plane sapphire was pre-coated with a ~20 nm Au film. The source materials of the mixture of high purity ZnO (99.99%) and graphite powder (99.99%) with a 1:1 weight ratio were placed at the end of a slender one-end sealed quartz tube. The source was placed in the center of a horizontal tube furnace maintained at around 950–1000°C and the substrate temperature at about 600°C. After growth for 40 min, the ZnO nano/microstructures can be obtained. More detailed description of the growth process can be found in [39,43] and the references therein.

FESEM images of ZnO microstructures were shown in Fig. 12.1. The wires or the disks can be scratched from the substrate and isolated on other substrates. This preparation technique enables one to investigate the optical properties from a single structure. It is interesting to note that most of the nanowires and microdisks have a perfect hexagonal shape and the trace with boundaries are regularly arrayed with rotation angles of 60°, which confirms the high quality of the ZnO structures.

Figure 12.1 (a) Field emission scanning electron microscopy (FESEM) image of ZnO nanowires. The inset shows a nanowire with higher magnification. Reprinted with permission from Ref. [22]. Copyright 2010, WILEY-VCH Verlag GmbH & Co. KGaA, Weinheim. (b) FESEM image of ZnO microdisks. Reprinted with permission from Ref. [39]. Copyright 2004, AIP Publishing LLC.

12.2.2 Optical Characterization

Optical property of individual ZnO structure with hexagonal cross section can be investigated with the so-called micro-photoluminescence (μ-PL) method, and the experimental setup is shown schematically in Fig. 12.2. The common excitation source for spontaneous emission from ZnO is continuous wave (CW) 325 nm laser line from a He-Cd laser, while the high-density excitation is pulsed neodymium doped yttrium aluminum garnet (Nd:YAG) laser operating at 266 nm or 355 nm with repetition rate of 60 Hz and pulse width of 1 ns. The excitation laser was focused to a spot with a grazing angle normal to the ZnO sample. And, the sample was put on a sample holder placed on an X–Y–Z movable stage. The signal emitted from individual disk was collected with a UV objective and delivered to a camera for imaging, or focused into an optical fiber and directed into monochromator combined with a charged coupled device (CCD) for spectral recording. An attenuator for high power lasers was used to set the excitation intensity for power dependent PL measurement. To achieve lasing from ZnO WGM cavity, the excitation density always needs to exceed the order of kilowatt per square centimeter, which also depends on pumping and experimental temperature. For determining the polarization-dependent measurement of the emitted light, a polarizer should be inserted in the detection path. It is important to mention that suitable long-pass filters are necessary at the entrance slit of the monochromator to block the laser beam. Compared to the steady-state PL, time-resolved PL (TRPL) spectroscopy can provide information about the complex dynamics in the semiconductor. It has been used for investigation of WGM lasing in ZnO. To obtain the time domain information, the detector can be replaced by either a time-correlated single photon counting unit or a streak camera [44].

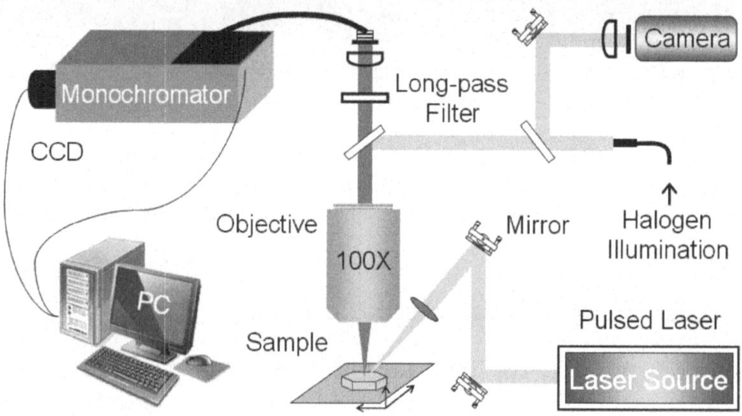

Figure 12.2 Schematic diagram of the setup for micro-photoluminescence (μ-PL) experiments.

12.3 Key Research Findings

At ambient conditions, the crystal structures of ZnO can be hexagonal wurtzite (WZ), cubic zinc blende, or cubic. The WZ structure has a hexagonal unit cell, belonging to the space group $P6_3mc$ with lattice parameters $a = 0.3296$ and $c = 0.52065$ nm. The structure of ZnO can be simply described as a number of alternating planes composed of tetrahedrally coordinated O^{2-} and Zn^{2+} ions, stacked alternately along the *c*-axis. Therefore, ZnO will demonstrate a hexagonal pillar or disk when grown along the *c*-axis, which will form WGM cavity if the crystalline quality of the sample is high [45]. As mentioned above, the WGM cavity investigation in ZnO structures was just initiated in the past several years. In fact, there are not many papers discussing about this issue, and some of the mechanism is still controversial. This section will summarize the cavity effects in terms of the modification of spontaneous emission, different kinds of WGM lasing, as well as exciton polariton in low-dimensional ZnO structures.

12.3.1 Spontaneous Emission

The optical properties of ZnO have been studied by a variety of experimental techniques. It is well known that at room temperature the PL emission from ZnO typically consists of two emission bands. One is the UV emission band located at ~380 nm, which is related to the near band edge (NBE) transition of ZnO, namely, the recombination of the free excitons or its phonon replica, while the broad emission band is found between 420 and 700 nm regardless of growth conditions, which is related to the deep level emission (DLE) [46]. Until now, various DLE has been reported, which is due to recombination of intrinsic or extrinsic defects such as zinc vacancies (V_{Zn}) [47], interstitial zinc (Zn_i) [48], oxygen vacancies (V_O) [49,50], and interstitial oxygen (O_i) [51–53]. For ZnO semiconductor with hexagonal cross section, due to the WGM cavity effect, the resonance phenomenon and the modification of the two emission bands have been observed and investigated. It is found that the NBE is polarized preferable perpendicular to the wire axis, while the DLE is weakly polarized [54]. The observation is due to the fact that in ZnO, the excitons that are allowed to couple to light perpendicular to the optical axis (*c*-axis, which is parallel to the wire axis) have a lower energy ground state, leading to a preferred emission perpendicular to the wire axis [55].

From the journal papers, several investigations about the modification of NBE from ZnO by WGM cavity have been reported [35,42, 54,56]. For example, Fig. 12.3(a) shows the cathodoluminescence (CL) spectra from the vertically grown isolated hexagonal nanodisk, which shows intense NBE while no DLE can be observed. As the monochromatic CL image at 378 nm shown in Fig. 12.3(b), the intensity distributed of the vertically oriented ZnO hexagonal nanodisk is not uniform. The spatial distribution of intensity across the nanodisk is shown in Fig. 12.3(c), and the stronger intensity near the boundary can be clearly seen, which is ascribed to WGM-like-enhanced emission. The darkened area is one-quarter of the total area under the intensity profile. To verify the WGM cavity effect, the authors also carried out multi-Gaussians function fitting based on the cavity mode position according to the size of the nanodisk. They found that the estimated portion of the WGM-like-enhanced emission in the total intensity at 378 nm is approximately 25 ± 10%, which is in good agreement with the result obtained from the spatial distribution of luminescence intensity.

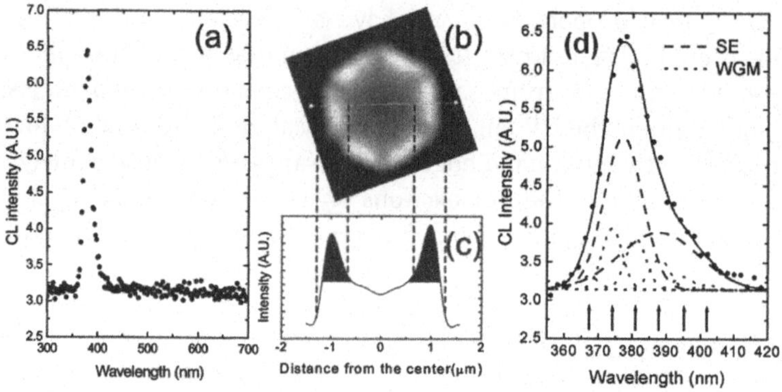

Figure 12.3 (a) CL spectra of a ZnO nanodisk measured at room temperature. (b) Monochromatic CL image (at 378 nm) of a vertically oriented ZnO nanodisk. The side length of the ZnO hexagonal nanodisk is approximately 1.6 μm. (c) The intensity profile across the nanodisk. (d) The contribution from WGM-like-enhanced luminescence. Reprinted with permission from Ref. [54]. Copyright 2006, AIP Publishing LLC.

It is also possible to observe the resonance effect of ZnO emission in DLE [57]. For example, Fig. 12.4(a) presents such characteristic resonance spectra of typical polarization-resolved μ-PL spectra of ZnO microcavities. It can be seen clearly that the emission is modulated with resonance lines of certain width and spacing. Due to the refractive index difference between transverse electric (TE) (electric field perpendicular to wire axis) and transverse magnetic (TM) (electric field parallel to the wire axis) mode, the resonance peaks position is different. Considering a simple plane-wave model, the resonant in WGM cavity from ZnO with a hexagonal cross section can be written as,

$$N = \frac{3\sqrt{3}nD}{2\lambda} - \frac{\pi}{6}\tan^{-1}\left(\beta\sqrt{3n^2 - 4}\right), \tag{12.1}$$

where D is the diameters of the outer circle of the resonator, n is the effective refractive index for the optical mode, and λ is the resonant wavelength. The integer N defines the interference order of the resonator. β characterizes the polarization and $\beta = n$ for TE polarization and $\beta = 1/n$ for TM polarization [58]. Figure 12.4(b) shows the dependence of the WGM spectra on the diameters of the ZnO wires. It can be seen that the positions of the TM polarized resonant peaks correspond well with the predicted theoretical values for all the wire diameters. The symbols represent the experimentally observed peak positions with the interference orders while the solid curves are the calculation from the equation above. It can be seen that a larger diameter will lead to smaller interference peak spacing and higher interference orders observed in the same spectral range [38].

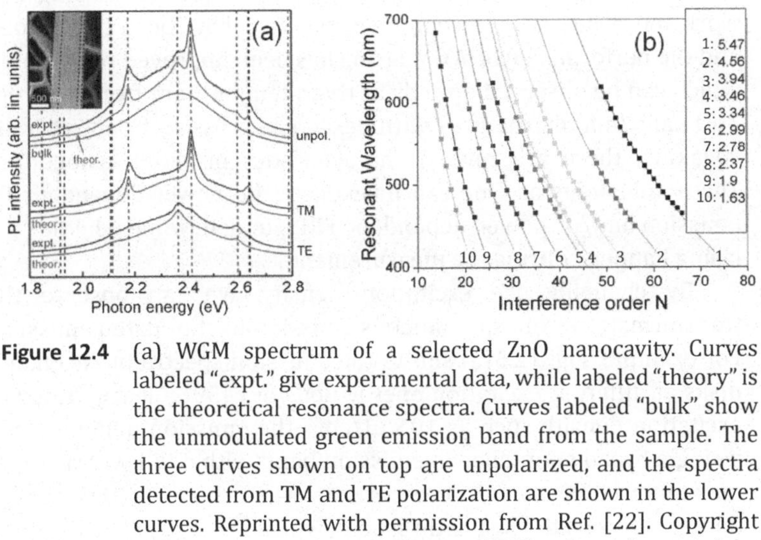

Figure 12.4 (a) WGM spectrum of a selected ZnO nanocavity. Curves labeled "expt." give experimental data, while labeled "theory" is the theoretical resonance spectra. Curves labeled "bulk" show the unmodulated green emission band from the sample. The three curves shown on top are unpolarized, and the spectra detected from TM and TE polarization are shown in the lower curves. Reprinted with permission from Ref. [22]. Copyright 2005, American Physical Society. (b) The dependences of the resonant wavelength of TM polarization on the interference order N for different wire diameters. Reprinted with permission from Ref. [38]. Copyright 2011, AIP Publishing LLC.

The spectrum modification of both the NBE and DLE bands in ZnO verifies the existence of interaction between ZnO and WGM cavity. However, due to the CW excitation, only enhanced phenomenon or periodic peaks can be observed. If the CW laser is replaced by high-density pulse excitation, we may be able to observed stimulated emission. However, it is needed to mention that only NBE can demonstrate the WGM lasing, because the DLE is related to defects emission with much lower gain. In the following section, WGM lasing from ZnO will be introduced and discussed.

12.3.2 WGM Lasing

The realization of semiconductor lasers emitting short wavelength UV light is of considerable interest for a wide range of applications including optical computing, high-density data storage, and material processing, and chemical/biochemical microanalysis. However, due to the difficulties in achieving high-quality ZnO samples, it was until recently that the distinct WGM lasing modes was observed and identified from single-photon excited individual ZnO microwires at low temperature (10 K) [37]. Following this work, more and more relevant investigations have been carried out, and the performance of WGM lasing has been improved remarkably, which can be observed directly at room temperature from different ZnO nano/microstructures with hexagonal cross section. Generally speaking, there are several characteristics of WGM lasing from ZnO semiconductor. One can investigate the power-dependent PL emission and/or power-dependent TRPL to determine whether the lasing happens during the measurement.

By changing the excitation density, we can observe the transformation from spontaneous emission to stimulated emission. For example, Fig. 12.5(a) shows the emission spectra of ZnO nano-disks at different excitation energy densities [59]. Clearly, at lower excitation density such as 105 μJ/cm^2, the emission demonstrates a single broad spontaneous emission band with a full width at half maximum of 80 meV. In comparison, when the excitation energy density exceeds the lasing threshold, which is ~205 μJ/cm^2, several narrow peaks emerge in the emission spectrum. The line widths of these peaks are <15 meV, which are significantly smaller than the line width of the spontaneous emission peak below the threshold. Above the threshold, the integrated PL emission increases rapidly with the pumping energy density, as shown in Fig. 12.5(b). The narrow line width and rapid increase of the emission intensity indicate that the stimulated emission takes place in these ZnO sample. From the non-linear increase of the new emission peaks, the threshold of the stimulated emission can be determined.

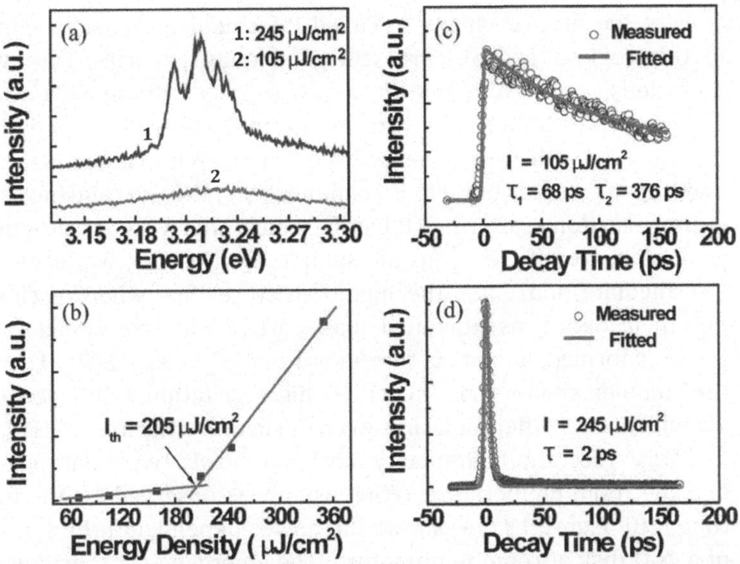

Figure 12.5 (a) PL spectra from the ZnO nanodisks below and above threshold. (b) The integrated PL intensity from nanodisks as a function of the excitation energy density. TRPL spectra from the ZnO nanodisks below (c) and above (d) threshold under femtosecond pulse laser excitation. Reprinted with permission from Ref. [59]. Copyright 2007, AIP Publishing LLC.

Another approach to determine whether the lasing takes place is the TRPL measurement, which is presented in Fig. 12.5(c and d) where the ZnO sample is excited by femtosecond pulse laser at 324 nm (~3.83 eV). For excitation density below the lasing threshold, the PL decay can be well fitted by a bi-exponential function with 68 and 376 ps [33]. In comparison, when the excitation density is above the threshold, the transient PL is dominated by a very fast decay process of ~2 ps, which is shown in Fig. 12.5(d). The narrow emission with an ultrafast lifetime of 2 ps implies the opening up of additional relaxation pathway, which verifies the existence of stimulated emission or lasing.

12.3.2.1 Lasing mechanism

An exciton is a quasi-particle consisting of a bound state of an electron and a hole in semiconductor crystals by the electrostatic Coulomb force. One reason that intensive research interest paid to ZnO is its exciton phenomenon. When increasing the excitation density, we first reach the so-called intermediate density regime. In this regime, excitons are still good quasi-particles. However, their density is so high that they start to interact with each other. This kind of interaction is very important because new emission mechanisms with even higher efficiency than exciton recombination may appear, such as biexciton recombination (M-band) and inelastic exciton–exciton scattering (P-band), which offer very large optical gain. Moreover, if we pump the sample even harder, we leave the intermediate and enter the high-density regime, where excitons lose their identity as individual quasi-particles, and a new collective phase is formed, known as the electron-hole plasma (EHP). Due to the complex situation of ZnO under high excitation, in the section, we will discuss different lasing mechanism of ZnO lasing [60–62].

Very recently, high-quality ZnO microdisks were fabricated, and the room temperature WGM lasing was observed for the first time [18]. Figure 12.6(a) shows the power-dependent μ-PL spectra of a ZnO disk at room temperature. The emergence of sharp peaks and their non-linear increase of the emission intensity indicate lasing phenomenon in the ZnO disk. A μ-PL spectrum under low excitation density is enlarged and plotted in Fig. 12.6(b). At a low pumping intensity, a spontaneous emission peak located at ~380 nm (3.2632 eV) is observed, which can be attributed to the recombination of free exciton (FX) in ZnO. At higher excitation density, the FX becomes saturated and the emission at ~390 nm (3.1795 eV) is more pronounced. For inelastic exciton–exciton scattering, two different cases may happen. One of the excitons may be scattered into an excited state (P_n-band emission) or into a continuum state (P-band emission). Under this situation, the energy of an emitted photon can be calculated by [61]

$$E_n = E_{ex} - E_{ex}^b\left(1 - \frac{1}{n^2}\right) - \frac{3}{2}k_B T \quad (n = 2, 3, 4, \ldots, \infty), \tag{12.2}$$

where E_{ex} is the energy of FX, E_{ex}^b is the exciton binding energy, n is the quantum number of the envelope function, and $k_B T$ is the thermal energy. The energy difference between the two bands is deduced to be 83.7 meV, which is close to the calculated value (85.8 meV). Therefore, the emission ~390 nm can be ascribed to the P-band emission of ZnO, which indicates excitonic lasing existing in the disk.

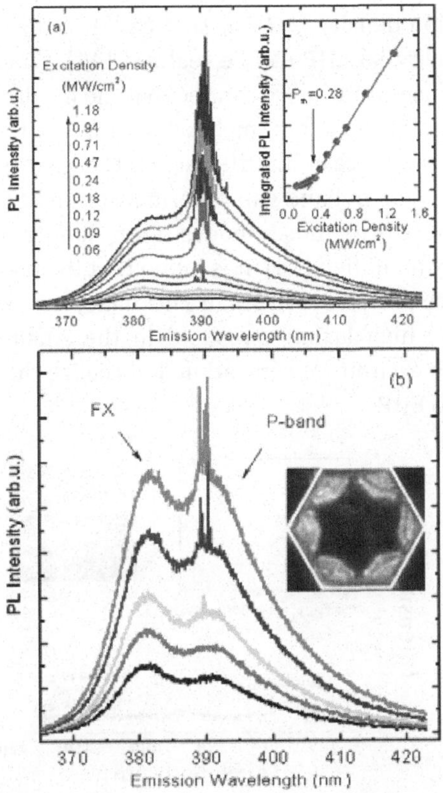

Figure 12.6 (a) The μ-PL spectrum of an individual ZnO disk under different excitation density. The inset shows the dependence of integrated PL intensity on the excitation density. (b) The enlarged part of the μ-PL spectrum under low excitation density. Inset shows the PL image of ZnO disk with diameter of 12.85 μm. Reprinted with permission from Ref. [18]. Copyright 2011, WILEY-VCH Verlag GmbH & Co. KGaA, Weinheim.

Besides excitonic lasing, EHP lasing has also been observed in ZnO microstructures [63]. It is needed to mention that this also depends on the quality of the sample and the excitation density. Fig. 12.7(a) shows the lasing emission spectra for a ZnO microrod with diameter of 8 μm. At low excitation density, only a typical spontaneous emission bands centered at around 380 and 390 nm were detected. When the excitation intensity increased to 330 kW/cm^2, sharp peaks emerged in the longer wavelength region, and the excitation intensity yielded the carrier density higher than

Mott transition density (~3.7×10^{19} cm^{-3}) in the hexagonal WGM ZnO microcavity, so EHP can be realized [45]. When the excitation density increases further, we can see that the maximum lasing peaks emerge at around 400 nm, meanwhile, the exciton emission was suppressed. In the case of EHP lasing, the gain is shifted to low energy region due to band gap renormalization at higher excitation densities, therefore, those modes with longer wavelength are preferentially amplified [64]. It is found that the observation of EHP lasing is more preferable in small ZnO nano/microstructures. That is due to the high lasing threshold in the small cavity helps to generate high exciton concentration and realize the transition from exciton gas to EHP.

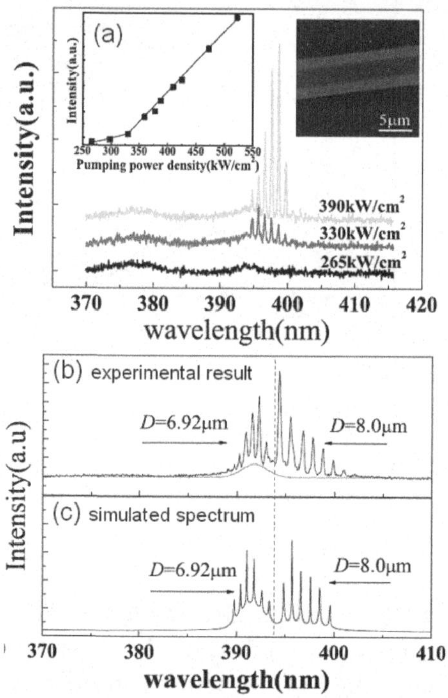

Figure 12.7 (a) The emission spectra under different excitation power densities for an individual ZnO microrod with diameter of 8 μm. Inset shows the relation between peak intensity with excitation power density. (b) and (c) The experimental and simulated spectrum of lasing spectrum from two ZnO microrods. Reprinted with permission from Ref. [63]. Copyright 2009, AIP Publishing LLC.

Lasing phenomena could not happen without appropriate cavity feedback. Various feedback mechanisms may account for the lasing. As discussed above, WGM is expected to have high Q factor and low threshold for hexagonal ZnO. In fact, the hexagonal-shaped dielectric cavity can support quasi-WGM (Fig. 12.8(a)) and WGM (Fig. 12.8(b)), where the light is totally reflected by the three (six) lateral sides of the hexagonal ZnO with a 30° (60°) incident angle because the critical angle of total internal reflection is about 25.8° at the ZnO/air boundary [62]. To determine the cavity feedback, we can consider the Q factor from $Q = \lambda/\Delta\lambda$, where λ is the peak wavelength and $\Delta\lambda$ is the line width of the peak, respectively. Then compare the Q factor calculated from the following equation for an m-facet polygonal cavity [65],

$$Q = \frac{\pi D m n R^{m/4}}{2\lambda(1 - R^{m/2})} \sin\left(\frac{2\pi}{m}\right), \tag{12.3}$$

where D is the diameter of the circle that circumscribes the polygon, m is the number of facets, and R is the reflectivity of the facets. If the laser was quasi-WGM or WGM, it can be deduced that the effective reflectivity is about 89.35% for quasi-WGM cavity and 93.71% for WGM cavity. In fact, for WGM cavity, the boundary does not have to be hit in the center between two corners, as depicted by the dashed line in Fig. 12.8(c). This is why an intense optical field distributed at the corners of the hexagonal disk but with very low intensity in the disk center, as shown in the inset of Fig. 12.6(b). Therefore, the WGM is more preferable due to a better optical confinement. Nevertheless, quasi-WGM lasing from ZnO has been also found and investigated [66].

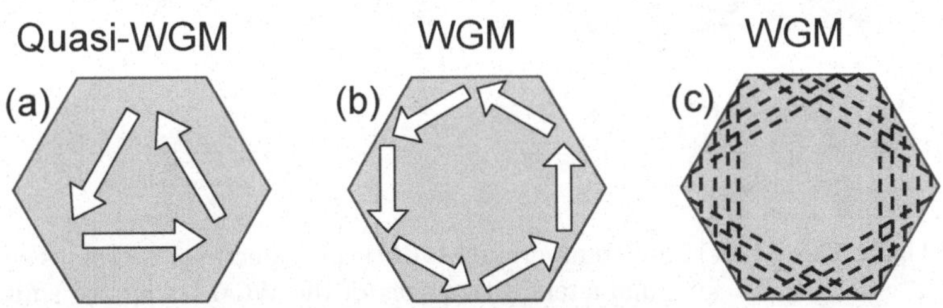

Figure 12.8 Different kinds of cavity feedback in hexagonal ZnO structure. Reprinted with permission from Ref. [18]. Copyright 2011, WILEY-VCH Verlag GmbH & Co. KGaA, Weinheim.

12.3.2.2 Lasing characteristics

As mentioned above, we can obtain different sizes of the hexagonal ZnO structure by different growth parameters, which enable us to investigate the size dependent WGM lasing. Figure 12.9 presents the lasing spectrum of ZnO microdisks with various diameters, while the insets show the dependence of integrated PL intensity on pumping density, and the corresponding PL images of WGM lasing. It is found that the lasing threshold decreases with the increase in the diameter of ZnO microdisks, which is summarized in Fig. 12.9(d). Previous studies show that the WGM lasing threshold strongly depends on Q and confinement factor (Γ). And, a $1/D^2$ relationship is expected for disk diameter and lasing threshold, because lasing threshold is inversely proportional to Q and Γ, respectively [18,21]. The increase of lasing threshold can be attributed to the smaller disk cavity volume, which provides less spatial overlap between the WGM and the ZnO gain medium, resulting in more light leakage

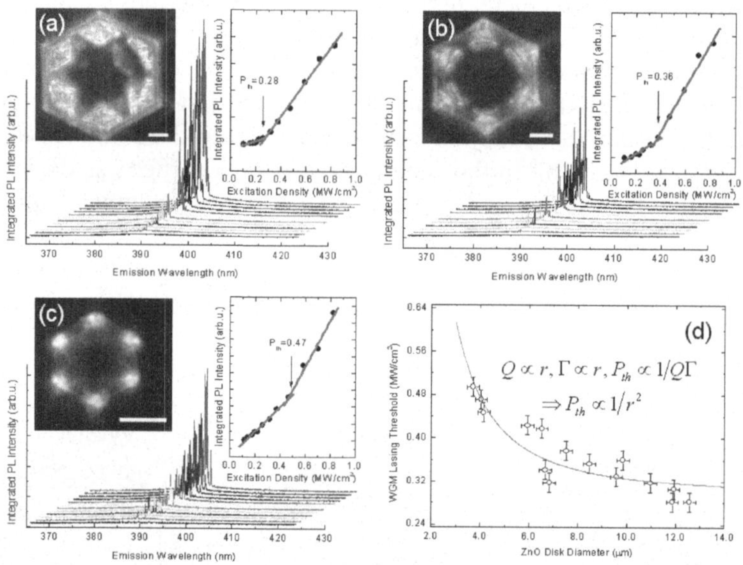

Figure 12.9 (a–c) Room temperature WGM lasing spectra of ZnO disks with various diameters. PL images of the WGM lasing are shown with a scale bar of 2 μm. (d) The room temperate WGM lasing threshold changes with the diameter of the ZnO disk. Reprinted with permission from Ref. [18]. Copyright 2011, WILEY-VCH Verlag GmbH & Co. KGaA, Weinheim.

(optical loss) from the WGM cavity. The interpretation is also supported by the differences of PL image as shown in the inset of Fig. 12.9. It can be seen that with the decrease of the diameter of the disk, optical field distribution of the WGM lasing decreases, and finally, intense optical field only distributes at the corners for small ZnO disk.

Moreover, numeric method can also be used to calculate the near field intensity patterns $|E^2|$ of the ZnO WGM resonances, such as the wavefront-matching condition based on ray optics and finite element method supported from COMSOL Multiphysics [67,68]. For example, the normalized cavity field distribution of TE modes WGM in hexagonal ZnO with diameter of 2 μm is carried out, and the results are plotted in Fig. 12.10(a–d). The patterns depict the meaning of m as an azimuthal mode number, which is equal to the number of field maxima in angular direction along the hexagonal edges. As

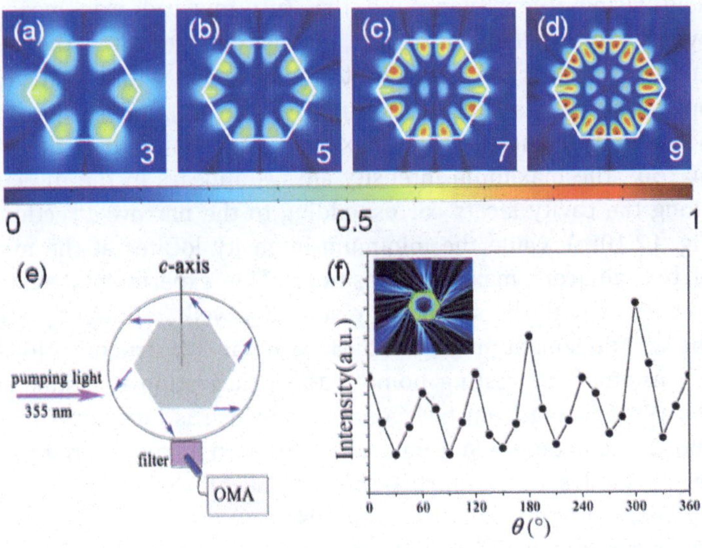

Figure 12.10 (a)–(d) Normalized cavity mode field distribution of WGM in hexagonal ZnO with diameter of 2 μm (boundaries are presented in the white lines) as a function of mode number m (bottom right corners). All patterns are indicated in the same linear color scale (below the figures). (e) The schematic diagram for the angle-dependent lasing measurement. (f) The far-field lasing intensity distribution around the c-axis of the hexagonal ZnO microrod. Reprinted with permission from Ref. [34]. Copyright 2009, AIP Publishing LLC.

shown in the figures, the electric field distribution increases with increasing of mode numbers m. For the smaller m, the electric field is highly lost due to strong scattering at the boundaries of microresonator as Fig. 12.10(a and b). However, for a larger m (or equal to the same wavelength, but with larger cavity size), better confinement of light can be achieved. As a result, the large microresonator has higher Q factor, therefore, is more appropriate for realization of WGM lasing at room temperature. Moreover, the WGM character vanishes for increasing mode number m [22].

Based on the above simulation, it is known that the ZnO WGM lasing is predominately emitted out at the corners of hexagon, and the lasing intensity is periodically distributed around the cavity with a period angle of $\pi/3$. Recently, angle dependent WGM lasing experiment has been carried out, and the schematic diagram is shown in Fig. 12.10(e) [34]. The spatial distribution of the far field lasing intensity around the ZnO micorod was measured by moving the fiber on a circle. At each detected sites, the collect facet of the fiber was oriented to the direction with the strongest optical signal. As can be seen in Fig. 12.10(f), where displays a periodic distribution of the maximum and minimum in the range of 360°, the maximum intensity always appears in the directions along the cavity facets corresponding to the narrow directions in Fig. 12.10(e), while the minimum intensity locates at the middle of two adjacent maximum positions. This experimental result is consistent with the simulated result very well and directly proves the WGM resonant process in the hexagonal ZnO structures [69].

Another interesting point is the polarization dependence of the WGM lasing from hexagonal ZnO, which is different from the spontaneous emission. Under low excitation at room temperature, the TM modes were found to be much weaker than the corresponding TE modes. From detailed comparison, two physical reasons for this behavior were found: (1) the low temperature leads to a more dominant polarization of the emitted light perpendicular to the ZnO c-axis due to a narrowing of the Fermi function with decreasing temperature, and (2) according to the plane wave model, the difference between the TM and TE line-width is smaller at photon energies above the isotropy point, where the lasing peaks are observed, compared to the spectral region below the isotropy point, were the low excitation WGM were observed [22].

12.3.2.3 Multi-photon absorption induced lasing

ZnO is a non-linear semiconductor material, which is useful in 3D optical data storage, frequency upconversion, optical limiting, and photomedical therapy. For multi-photon-induced lasing experiments, ultrafast femtosecond laser from optical parametric amplifier should be used to reach the regime of GW/cm^2~TW/cm^2 for high excitation. Very recently, multi-photon-induced WGM lasing from hexagonal ZnO structures have been reported, such as nanodisks [70], microrods [71], micronails [72,73], and twin-rods microstructure [74]. Similar to the phenomenon discussed above, the multi-photon-induced WGM lasing from ZnO shows the threshold behavior with distinct lasing peaks that can be determined by the size of the cavity. It is also found that the intensity of two-photon absorption and three-photon absorption induced emission is quadratic or cubic depending on the excitation power density. In fact, the multi-photon-induced WGM lasing is generally reported from hexagonal ZnO in micrometer range, because the large-sized cavity is expected to have sufficiently long optical gain path and low optical loss for the lasing generation. At the same time, it is needed to mention that ZnO has no center of inversion, which shows phenomena that come with even powers of the electric file such as $\chi^{(2)}$ effects in addition to odd powers, for example in $\chi^{(3)}$ phenomena. Therefore, accompanying with the multi-photon-induced WGM lasing, it is also possible to observe the double frequency signal of the excitation light, which is attributed to the second harmonic generation from the hexagonal ZnO structures [72,75].

12.3.3 Exciton Polaritons

An exciton has a dynamical electric dipole, which can couple with electromagnetic wave, or light. Therefore, inside a suitable optical cavity or photonic crystal structure, excitons will strongly couple with the cavity mode and form cavity polaritons. This phenomenon commonly refers to the strong coupling regime. In principle, this coupling between particle (exciton) and field (electromagnetic cavity mode) can be used to achieve zero-threshold laser diodes. Till date, all optical techniques have applied to investigate the properties of exciton polaritons and various kinds of materials from low-dimensional semiconductors to organic thin films, and even hybrid structures have been used for discussion.

Historically, the concept of polariton was first proposed and described in early 1960s by Hopfield and Thomas [76]. It is a quasi-particle resulting from strong interaction between polarization wave and photon. The new quasi-particle is represented in the energy versus wave-vector dispersion by the lower and upper polariton branch, respectively. And, the energy gap between the two levels is also known as the normal-mode splitting or Rabi splitting, which is used to measure the strength of the exciton-photon interaction. Generally speaking, the dispersion curve of exciton-polaritons in microcavities can directly obtain by measuring the reflectivity or photoluminescence. It is needed to mention that the exciton polariton is differed from other kind of polaritons such as the surface polariton that only occurs at the interface between metal and dielectric. Moreover, strong confined optical mode in microcavities leads to unique properties of exciton-polaritons compared with in a structure without confinement. However, the properties of exciton polariton are very difficult to study in bulk structure. It is known that the light and matter interaction in nanostructures is enhanced, which is more preferable to study the exciton polariton. Moreover, it is found that the Rabi splitting energy is proportional to coupling constant of the exciton, which makes ZnO to be one of the most potential candidates for realization of exciton polariton due to its large exciton oscillator strength [77].

As discussed above, the ZnO nano/microstructures have a perfect hexagonal cross section, which is appropriate to obtain WGM due to the total internal reflection at ZnO–air interface. In fact, the strong coupling regime and a few evidences of polariton lasing have been observed in various microcavities including the WGM microcavity [78–81]. For example, a ZnO tetrapod grown by VPT method was used to investigate the strong coupling regime in WGM cavity, as shown in Fig. 12.11. Interestingly, a tapered arm of the ZnO tetrapod has continuous varying radius, which leads to a tunability of WGM cavity by excited to different position on the tapered. Therefore, by using a μ-PL spectrometer to scan excitations along the tapered, a transition from the weak coupling with optical WGM to strong coupling regime with a splitting of lower and upper exciton-polariton branch was observed, as plotted in Fig. 12.11(c).

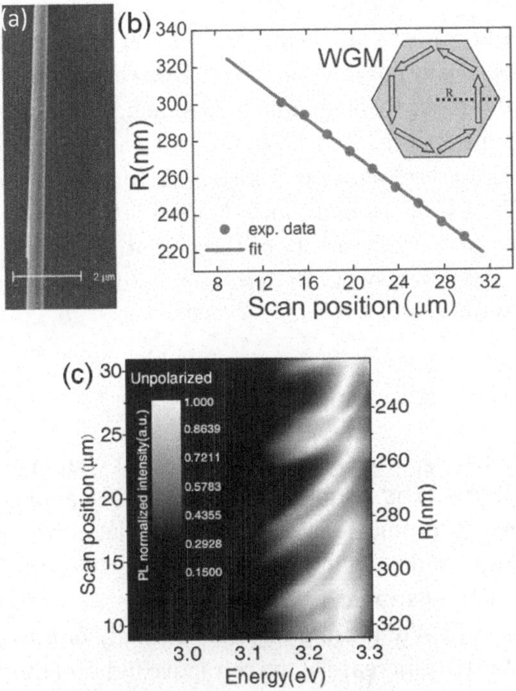

Figure 12.11 (a) Magnified FESEM image of the ZnO tetrapod resonator shows perfect surface quality. (b) Corresponding radius of the tetraped as a function of scanning position. (c) Spatially resolved PL mapping along the tapered arm with the excitation laser polarization $E \parallel c$-axis and detection unpolarized. Reprinted with permission from Ref. [79]. Copyright 2008, American Physical Society.

Very recently, applying the same approach, strong coupling regime from WGM ZnO microrod cavity has been reported [81]. Therefore, crystalline ZnO nano/microstructures with hexagonal cross section, which is easily grown by conventional techniques, is promising candidate for investigation of strong coupling regime at room temperature. Research down this line will point to the realization of polariton lasing from hexagonal ZnO with WGM microcavity, which calls for more in-depth investigations.

12.4 Potential Applications

ZnO is an interesting semiconductor for application in optoelectronic devices, and various applications such as light-emitting diodes (LEDs), electrically pumped lasers, electron collection in photovoltaic cells, and as transparent conducting oxide have been demonstrated [82]. As mentioned above, the hexagonal ZnO structure provides a natural WGM microcavity with high Q factor, low lasing threshold, distinct WGM lasing modes, and definite lasing output directions. In this section, we will introduce the potential applications such as refractive index sensor and microlaser diode based on hexagonal ZnO with WGM cavity.

12.4.1 Refractive Index Sensor

Due to the high Q factor and easy evanescent coupling to a tapered fiber, WGM cavity is widely used for sensing application. Based on this idea, photonic sensing devices for label-free detection or even single virus detection from the reactive shift of a WGM cavity have been demonstrated [8]. As discussion above, it is note that the threshold of WGM lasing depends strongly on the surrounding medium. With the increase of the refractive index of the surrounding medium, reflectivity at the boundary between ZnO and the medium will decrease, which leads to smaller Q and Γ factor, resulting in a higher WGM lasing threshold.

The proof-of-concept sensing application based on ZnO WGM lasing has been demonstrated recently [18]. The comparison of the lasing spectrum from ZnO microdisk in air and in alcohol is carried out. Figure 12.12 plots the integrated PL intensity with the change of excitation density for two measurements, with the schematic diagrams showing the difference between two measurements. As expected, the WGM lasing threshold of ZnO in alcohol increase from 0.35 to 0.47 MW/cm^2, while the integrated PL intensity decreases more than four times. The demonstration shows that the WGM lasing from hexagonal ZnO structure is very sensitive to the surrounding medium, therefore, can be used as an effective refractive index sensor.

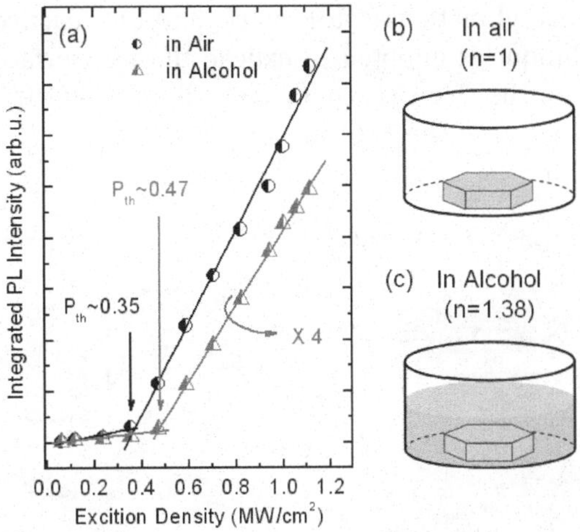

Figure 12.12 The integrated PL intensity of a ZnO disk in air and alcohol under different excitation densities. Reprinted with permission from Ref. [18]. Copyright 2011, WILEY-VCH Verlag GmbH & Co. KGaA, Weinheim.

12.4.2 Microlaser Diode

Regarding the electrically driven light-emitting devices and lasers, the problem of the lack of p-type ZnO has been circumvented in the past by n-type ZnO/p-type homojunctions. Up till now, the p-type conductivity in ZnO is still under intense debate and is not yet technologically mature. Nevertheless, the heterostructures combined with other p-type semiconductor materials, such as p-Si and p-GaN, can effectively realize the electroluminescence (EL) alternatively. Up till now, there are some ZnO random lasing device structures were designed and demonstrated [83,84].

Until very recently, the electrically pumped WGM lasing from hexagonal ZnO is reported [85]. Figure 12.13 shows the schematic procedure for fabricating a ZnO microrod/GaN microlaser diode. For this approach, a 20 nm Zn thin film was sputtered between the p-GaN substrate and the ZnO microrod. After thermal annealed at 500°C in oxygen ambient for 1 h, the Zn thin film into a ZnO buffer layer, which plays an important role in the contact between the ZnO microrod and the GaN substrate. The structure is buried in Poly (methyl methacrylate) matrix and the reactive ion etching (RIE) technique is adopted to expose the top surface of the ZnO microrod for the electrical contact. Finally, Au and Ni/Au electrodes were deposited on the ZnO microrod and p-GaN by electron beam evaporation.

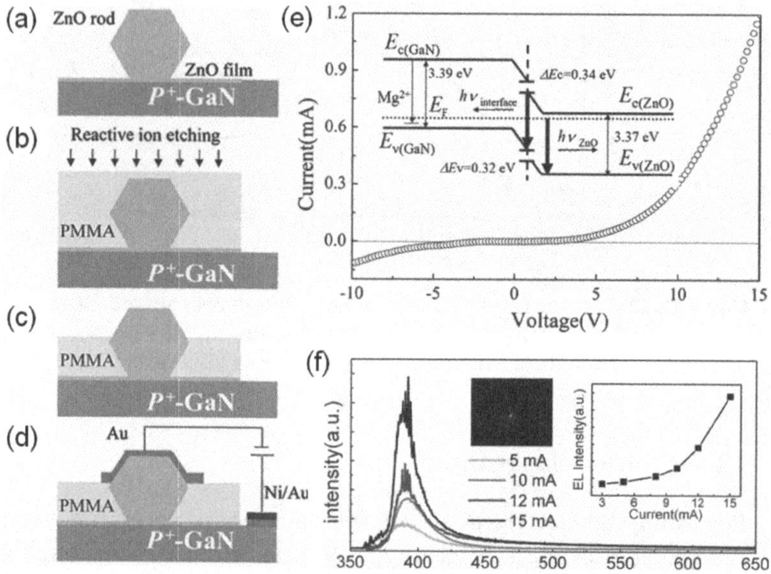

Figure 12.13 (a–d) The fabrication process of a ZnO microrod/GaN heterojunction diodes. (e) I–V curve for the ZnO microrod/GaN heterojunction diodes. The inset shows a schematic of the band diagram. (f) The EL spectra from the ZnO microrod/GaN heterojunction diodes under different applied current. The left inset shows a photograph of the EL emission from sample and the right inset shows the relationship between the EL intensity with the applied current. Reprinted with permission from Ref. [85]. Copyright 2011, WILEY-VCH Verlag GmbH & Co. KGaA, Weinheim.

Figure 12.13(e) shows a current–voltage (I–V) curve of the device, where the I–V curve presents a typical rectifying curve of the heterostructured diode device. The forward turn-on voltage was determined to be about 4 V. The inset in shows the energy bands diagram of the n-ZnO/p-GaN heterojunction. Figure 12.13(f) shows the room temperature EL emission spectra of the device at different applied currents, inset with an image of the EL spot at the current of 15 mA and the relationship between the EL emission intensity and the current. From the investigation, it is found that under low applied current, the EL spectrum is consisted by the emissions from the ZnO, p-GaN, and the interface between the ZnO/GaN heterostructure. In comparison, it is found that under higher applied current such as 12 mA, beside the broad spontaneous emission, four discrete sharp peaks emerge on the EL emission band. When the current is increased to 15 mA, more resonant peaks appear on the spectrum, meanwhile the emission intensities of the peaks enhance rapidly. From the calculation, the authors confirm that the sharp emission peaks come from WGM lasing action through electrically driven ZnO microrod. In addition, those lasing peaks have approximately the same wavelengths as those in the optically pumped lasing spectrum, which indicates that the hexagonal ZnO microstructure has the potential to be applied in low threshold electrically pumped microlaser diodes.

12.5 Conclusions and Future Perspective

In summary, recent developments of the WGM cavity in ZnO structures with hexagonal cross section have particularly been highlighted in this chapter. First of all, a general introduction of ZnO and WGM cavity are presented. Secondly, the experimental techniques about the synthesis of ZnO nano/microstructures and basic knowledge about optical characterizations are introduced. Then, recent key research findings about modification of spontaneous emission, different kinds of WGM lasing, as well as exciton polaritons due to WGM cavity from hexagonal ZnO structure are discussed. Finally, potential applications based on hexagonal ZnO were discussed. The main conclusions drawn from the literatures are summarized as following.

(1) Due to WGM cavity effect, PL spectrum of UV emission from NBE and the visible DLE will be modified and changed to periodic peaks of which the positions can be calculated from the size of the cavity. The intensive field will be distributed at the corner of the hexagonal due to the cavity resonance.

(2) UV WGM lasing have been observed from various hexagonal ZnO. The lasing gain mechanism can be classified to be inelastic excition–excition scattering (P-band) as well as the

EHP. The lasing intensity is periodically distributed around the cavity with a period angle of $\pi/3$. Besides single-photon, multi-photon-induced WGM lasing has been also observed in various hexagonal ZnO structures. Finally, strong coupling regime between exciton and WGM cavity in hexagonal ZnO is discussed.

(3) The WGM cavity is very sensitive to the surrounding medium, which shows potential application in sensing such as refractive index sensing. Moreover, electrically pumped WGM lasing from ZnO microrod is reported.

There is considerable interest in further study of ZnO semiconductor with hexagonal cross-section. For materials fabrication, new fabrication technologies are still being developed to achieve high crystalline quality and size-tunable ZnO structure on various substrates (such as flexible substrates) with large scale. The high-quality WGM cavity should be able to provide a good platform for studying the light and matter interaction, especially the exciton polariton. Moreover, using the optical fiber to couple the WGM lasing from the cavity is needed for practical application. Research down this line may help to advance ultra-sensitive sensor or UV photonic devices based on hexagonal ZnO structure, and will receive more and more attentions due to its unique properties. There is still a lot of room for future development in this area.

References

1. Vahala KJ (2003). Optical microcavities, *Nature*, **424**, 839–846.
2. Barrelet CJ, Bao J, Lončar M, Park H-G, Capasso F, Lieber CM (2005). Hybrid single-nanowire photonic crystal and microresonator structures, *Nano Lett*, **6**, 11–15.
3. Park Y-S, Cook AK, Wang H (2006). Cavity QED with diamond nanocrystals and silica microspheres, *Nano Lett*, **6**, 2075–2079.
4. Humar M, Musevic I (2010). 3D microlasers from self-assembled cholesteric liquid-crystal microdroplets, *Opt Express*, **18**, 26995–27003.
5. Gibbs HM, Khitrova G, Koch SW (2011). Exciton-polariton light-semiconductor coupling effects, *Nat Photonics*, **5**, 273–273.
6. Reithmaier JP, Sek G, Loffler A, et al. (2004). Strong coupling in a single quantum dot-semiconductor microcavity system, *Nature*, **432**, 197–200.

7. Humar M, Ravnik M, Pajk S, Musevic I (2009). Electrically tunable liquid crystal optical microresonators, *Nat Photonics*, **3**, 595–600.
8. Vollmer F, Arnold S (2008). Whispering-gallery-mode biosensing: Label-free detection down to single molecules, *Nat Methods*, **5**, 591–596.
9. Chen R, Sun HD, Wang T, Hui KN, Choi HW (2010). Optically pumped ultraviolet lasing from nitride nanopillars at room temperature, *Appl Phys Lett*, **96**, 241101.
10. Someya T, Werner R, Forchel A, Catalano M, Cingolani R, Arakawa Y (1999). Room temperature lasing at blue wavelengths in gallium nitride microcavities, *Science*, **285**, 1905–1906.
11. Strauf S, Hennessy K, Rakher MT, et al. (2006). Self-tuned quantum dot gain in photonic crystal lasers, *Phys Rev Lett*, **96**, 127404.
12. Painter O, Lee RK, Scherer A, et al. (1999). Two-dimensional photonic band-gap defect mode laser, *Science*, **284**, 1819–1821.
13. Chen R, Bakti Utama MI, Peng Z, Peng B, Xiong Q, Sun H (2011). Excitonic properties and near-infrared coherent random lasing in vertically aligned CdSe nanowires, *Adv Mater*, **23**, 1404–1408.
14. Cao H, Zhao YG, Ho ST, Seelig EW, Wang QH, Chang RPH (1999). Random laser action in semiconductor powder, *Phys Rev Lett*, **82**, 2278–2281.
15. Ma RM, Oulton RF, Sorger VJ, Bartal G, Zhang X (2011). Room-temperature sub-diffraction-limited plasmon laser by total internal reflection, *Nat Mater*, **10**, 110–113.
16. Nezhad MP, Simic A, Bondarenko O, et al. (2010). Room-temperature subwavelength metallo-dielectric lasers, *Nat Photonics*, **4**, 395–399.
17. Samuel IDW, Turnbull GA (2007). Organic semiconductor lasers, *Chem Rev*, **107**, 1272–1295.
18. Chen R, Ling B, Sun XW, Sun HD (2011). Room temperature excitonic whispering gallery mode lasing from high-quality hexagonal ZnO microdisks, *Adv Mater*, **23**, 2199–2204.
19. Snee PT, Chan Y, Nocera DG, Bawendi MG (2005). Whispering-gallery-mode lasing from a semiconductor nanocrystal/microsphere resonator composite, *Adv Mater*, **17**, 1131–1136.
20. Ning CZ (2010). Semiconductor nanolasers, *Phys Status Solidi B*, **247**, 774–788.
21. Gargas DJ, Moore MC, Ni A, et al. (2010). Whispering gallery mode lasing from zinc oxide hexagonal nanodisks, *ACS Nano*, **4**, 3270–3276.

22. Czekalla C, Nobis T, Rahm A, et al. (2010). Whispering gallery modes in zinc oxide micro- and nanowires, *Phys Status Solidi B*, **247**, 1282–1293.

23. Tamboli AC, Haberer ED, Sharma R, Lee KH, Nakamura S, Hu EL (2007). Room-temperature continuous-wave lasing in GaN/InGaN microdisks, *Nat Photonics*, **1**, 61–64.

24. Chen R, Xing GZ, Gao J, Zhang Z, Wu T, Sun HD (2009). Characteristics of ultraviolet photoluminescence from high quality tin oxide nanowires, *Appl Phys Lett*, **95**, 061908.

25. Tang ZK, Wong GKL, Yu P, et al. (1998). Room-temperature ultraviolet laser emission from self-assembled ZnO microcrystallite thin films, *Appl Phys Lett*, **72**, 3270–3272.

26. Chen R, Li D, Liu B, et al. (2010). Optical and excitonic properties of crystalline ZnS nanowires: Toward efficient ultraviolet emission at room temperature, *Nano Lett*, **10**, 4956–4961.

27. Zhou J, Gu Y, Fei P, et al. (2008). Flexible piezotronic strain sensor, *Nano Lett*, **8**, 3035–3040.

28. Duan X, Huang Y, Agarwal R, Lieber CM (2003). Single-nanowire electrically driven lasers, *Nature*, **421**, 241–245.

29. Xiang J, Lu W, Hu Y, Wu Y, Yan H, Lieber CM (2006). Ge/Si nanowire heterostructures as high-performance field-effect transistors, *Nature*, **441**, 489–493.

30. Law M, Greene LE, Johnson JC, Saykally R, Yang P (2005). Nanowire dye-sensitized solar cells, *Nat Mater*, **4**, 455–459.

31. Klingshirn C, Fallert J, Zhou H, et al. (2010). 65 years of ZnO research–old and very recent results, *Phys Status Solidi B*, **247**, 1424–1447.

32. Cao H (2003). Lasing in random media, *Waves Random Media*, **13**, R1–R39.

33. Huang MH, Mao S, Feick H, et al. (2001). Room-temperature ultraviolet nanowire nanolasers, *Science*, **292**, 1897–1899.

34. Dai J, Xu CX, Zheng K, Lv CG, Cui YP (2009). Whispering gallery-mode lasing in ZnO microrods at room temperature, *Appl Phys Lett*, **95**, 241110.

35. Liu J, Ngo QM, Park KH, et al. (2009). Optical waveguide and cavity effects on whispering-gallery mode resonances in a ZnO nanonail, *Appl Phys Lett*, **95**, 221105

36. Wang D, Seo HW, Tin C, Bozack MJ, Williams JR, Park M, Tzeng Y (2006). Lasing in whispering gallery mode in ZnO nanonails, *J Appl Phys*, **99**, 093112.

37. Czekalla C, Sturm C, Schmidt-Grund R, Cao BQ, Lorenz M, Grundmann M (2008). Whispering gallery mode lasing in zinc oxide microwires, *Appl Phys Lett*, **92**, 241102.

38. Dai G, Zhang Y, Liu R, Wan Q, Zhang Q, Pan A, Zou B (2011). Visible whispering-gallery modes in ZnO microwires with varied cross sections, *J Appl Phys*, **110**, 033101.

39. Xu CX, Sun XW, Dong ZL, Yu MB (2004). Zinc oxide nanodisk, *Appl Phys Lett*, **85**, 3878–3880.

40. Yang P, Yan H, Mao S, et al. (2002). Controlled growth of ZnO nanowires and their optical properties, *Adv Funct Mater*, **12**, 323–331.

41. Willander M, Nur O, Zhao QX, et al. (2009). Zinc oxide nanorod based photonic devices: Recent progress in growth, light emitting diodes and lasers, *Nanotechnology*, **20**, 332001.

42. Yang YH, Zhang Y, Wang NW, Wang CX, Li BJ, Yang GW (2011). ZnO nanocone: Application in fabrication of the smallest whispering gallery optical resonator, *Nanoscale*, **3**, 592–597.

43. Wang ZL (2003). Nanobelts, nanowires, and nanodiskettes of semiconducting oxides-from materials to nanodevices, *Adv Mater*, **15**, 432–436.

44. O'Connor DV, Phillips D (1984). *Time correlated single photon counting*, Academic Press, London.

45. Ozgur U, Alivov YI, Liu C, et al. (2005). A comprehensive review of ZnO materials and devices, *J Appl Phys*, **98**, 041301.

46. Djurisic AB, Leung YH (2006). Optical properties of ZnO nanostructures, *Small*, **2**, 944–961.

47. Zeng H, Li Z, Cai W, Liu P (2007). Strong localization effect in temperature dependence of violet-blue emission from ZnO nanoshells, *J Appl Phys*, **102**, 104307–104304.

48. Djurisic AB, Leung YH, Tam KH, et al. (2006). Green, yellow, and orange defect emission from ZnO nanostructures: Influence of excitation wavelength, *Appl Phys Lett*, **88**, 103107–103103.

49. Jeong SH, Kim BS, Lee BT (2003). Photoluminescence dependence of ZnO films grown on Si(100). by radio-frequency magnetron sputtering on the growth ambient, *Appl Phys Lett*, **82**, 2625–2627.

50. Tam KH, Cheung CK, Leung YH, et al. (2006). Defects in ZnO nanorods prepared by a hydrothermal method, *J Phys Chem B*, **110**, 20865–20871.

51. Ahn CH, Kim YY, Kim DC, Mohanta SK, Cho HK (2009). A comparative analysis of deep level emission in ZnO layers deposited by various methods, *J Appl Phys*, **105**, 013502–013505.

52. Djurisic AB, Leung YH, Tam KH, et al. (2007). Defect emissions in ZnO nanostructures, *Nanotechnology*, **18**, 095702.

53. McCluskey MD, Jokela SJ (2009). Defects in ZnO, *J Appl Phys*, **106**, 071101–071113.

54. Kim C, Kim Y, Jang E, Yi G, Kim HH (2006). Whispering-gallery-modelike-enhanced emission from ZnO nanodisk, *Appl Phys Lett*, **88**, 093104.

55. Wang NW, Yang YH, Yang GW (2010). Fabry-Pérot and whispering gallery modes enhanced luminescence from an individual hexagonal ZnO nanocolumn, *Appl Phys Lett*, **97**, 041917.

56. Kim SS, Kim YJ, Yi GC, Cheong H (2009). Whispering-gallery-modelike resonance of luminescence from a single hexagonal ZnO microdisk, *J Appl Phys*, **106**, 094310.

57. Dong H, Chen Z, Sun L, et al. (2010). Single-crystalline hexagonal ZnO microtube optical resonators, *J Mater Chem*, **20**, 5510–5515.

58. Nobis T, Kaidashev EM, Rahm A, Lorenz M, Grundmann M (2004). Whispering gallery modes in nanosized dielectric resonators with hexagonal cross section, *Phys Rev Lett*, **93**, 103903.

59. Yu D, Chen Y, Li B, Chen X, Zhang M, Zhao F, Ren S (2007). Structural and lasing characteristics of ultrathin hexagonal ZnO nanodisks grown vertically on silicon-on-insulator substrates, *Appl Phys Lett*, **91**, 091116.

60. Klingshirn C (2007). ZnO: From basics towards applications, *Phys Status Solidi B*, **244**, 3027–3073.

61. Klingshirn CF (2007). *Semiconductor Optics,* 3rd ed., Springer-Verlag, Germany.

62. Ursaki VV, Burlacu A, Rusu EV, Postolake V, Tiginyanu IM (2009). Whispering gallery modes and random lasing in ZnO microstructures, *J Opt A Pure Appl Opt*, **11**, 075001.

63. Dai J, Xu CX, Ding R, et al. (2009). Combined whispering gallery mode laser from hexagonal ZnO microcavities, *Appl Phys Lett*, **95**, 191117.

64. Dai J, Xu CX, Wu P, Guo JY, Li ZH, Shi ZL (2010). Exciton and electron-hole plasma lasing in ZnO dodecagonal whispering-gallery-mode microcavities at room temperature, *Appl Phys Lett*, **97**, 011101.

65. Bhowmik AK (2000). Polygonal optical cavities, *Appl Opt*, **39**, 3071–3075.

66. Sun LX, Dong HX, Xie W, An ZH, Shen XC, Chen ZH (2010). Quasi-whispering gallery modes of exciton-polaritons in a ZnO microrod, *Opt Express*, **18**, 15371–15376.

67. Ning M, Tung FKL, Lui SF, Poon AW (2003). Hexagonal micropillar cavities: multimode resonances and open-loop resonance linewidth broadening, in *Physics and Simulation of Optoelectronic Devices Xi*, vol. 4986, (Osinski M, Amano H, Blood P, eds.), pp. 153–160.

68. Tung FKL, Ma N, Poon AW (2003). FDTD simulation of hexagonal micropillar cavities, in *Smart Sensors, Actuators, and Mems, Pts 1 and 2*, vol. 5116, (Chiao JC, Varadan VK, Cane C, eds.), pp. 488–494.

69. Wiersig J (2003). Hexagonal dielectric resonators and microcrystal lasers, *Phys Rev A*, **67**, 023807.

70. Zhang CF, Zhang F, Sun XW, Yang Y, Wang J, Xu J (2009). Frequency-upconverted whispering-gallery-mode lasing in ZnO hexagonal nanodisks, *Opt Lett*, **34**, 3349–3351.

71. Dai J, Xu CX, Sun LX, Chen ZH, Guo JY, Li ZH (2011). Multiphoton absorption-induced optical whispering-gallery modes in ZnO microcavities at room temperature, *J Phys D Appl Phys*, **44**, 025404.

72. Dai J, Xu CX, Sun XW (2011). Single-photon and three-photon absorption induced whispering-gallery mode lasing in ZnO micronails, *Opt Communi*, **284**, 4018–4021.

73. Zhu GP, Xu CX, Zhu J, Lv CG, Cui YP (2009). Two-photon excited whispering-gallery mode ultraviolet laser from an individual ZnO microneedle, *Appl Phys Lett*, **94**, 051106.

74. Dai J, Xu CX, Shi ZL, et al. (2011). Three-photon absorption induced whispering gallery mode lasing in ZnO twin-rods microstructure, *Opt Mater*, **33**, 288–291.

75. Zhang Y, Zhou HJ, Liu SW, Tian ZR, Xiao M (2009). Second-harmonic whispering-gallery modes in ZnO nanotetrapod, *Nano Lett*, **9**, 2109–2112.

76. Hopfield JJ, Thomas DG (1963). Theoretical and experimental effects of spatial dispersion on the optical properties of crystals, *Phys Rev*, **132**, 563–572.

77. Zamfirescu M, Kavokin A, Gil B, Malpuech G, Kaliteevski M (2002). ZnO as a material mostly adapted for the realization of room-temperature polariton lasers, *Phys Rev B*, **65**, 161205.

78. van Vugt LK, Ruhle S, Ravindran P, Gerritsen HC, Kuipers L, Vanmaekelbergh D (2006). Exciton polaritons confined in a ZnO nanowire cavity, *Phys Rev Lett*, **97**, 147401.

79. Sun LX, Chen ZH, Ren QJ, et al. (2008). Direct observation of whispering gallery mode polaritons and their dispersion in a ZnO tapered microcavity, *Phys Rev Lett*, **100**, 156403.

80. Sun LX, Chen ZH, Ren QJ, et al. (2009). Polarized photoluminescence study of whispering gallery mode polaritons in ZnO microcavity, *Phys Status Solidi C*, **6**, 133–136.

81. Dai J, Xu CX, Sun XW, Zhang XH (2011). Exciton-polariton microphotoluminescence and lasing from ZnO whispering-gallery mode microcavities, *Appl Phys Lett*, **98**, 161110.

82. Yang P, Yan R, Fardy M (2010). Semiconductor nanowire: What's next? *Nano Lett*, **10**, 1529–1536.

83. Chu S, Wang GP, Zhou WH, et al. (2011). Electrically pumped waveguide lasing from ZnO nanowires, *Nat Nanotechnol*, **6**, 506–510.

84. Zhu H, Shan C-X, Zhang J-Y, et al. (2010). Low-threshold electrically pumped random lasers, *Adv Mater*, **22**, 1877–1881.

85. Dai J, Xu CX, Sun XW (2011). ZnO-microrod/p-GaN heterostructured whispering-gallery-mode microlaser diodes, *Adv Mater*, **23**, 4115–4119.

Chapter 13

Nanoscale Semiconductor Ring Lasers

Min W. Kim and P.-C. Ku

Department of Electrical Engineering and Computer Science, University of Michigan, 1301 Beal Ave, Ann Arbor, MI 48109, USA

minwkim@umich.edu, peicheng@umich.edu

13.1 Introduction

Chip-scale integration of photonic components can decrease the system footprint, eliminate reliability issues due to optical misalignment, and lower the overall system cost. Microring resonators have been widely used as one of the fundamental building blocks for photonics integration [1]. A ring cavity is advantageous for integration because it is a traveling-wave cavity with no etched or cleaved facets, and hence can be densely fabricated using state-of-the-art semiconductor patterning technologies. Furthermore, a ring cavity can be easily coupled to an on-chip optical waveguide via evanescent coupling as illustrated in Fig. 13.1. Semiconductor lasers employing ring cavities were first demonstrated nearly 40 years ago [2,3] and have since undergone continuous downsizing, from over 100 μm in diameter initially to a sub-wavelength dimension recently [4]. In ring lasers, light is confined in the circular waveguide

Handbook of Optical Microcavities
Edited by Anthony H. W. Choi

ISBN 978-981-4463-24-9 (Hardcover), 978-981-4463-25-6 (eBook)
www.panstanford.com

and experiences losses due to waveguide bending. As the ring size decreases, the confined optical mode suffers from more bending loss. As a result, the laser threshold increases and makes reaching the lasing action more and more difficult. When the optical modal volume approaches the diffraction limit, the radiation loss in the cavity can become so large that further scaling of the gain media can no longer provide adequate optical gain to reach the lasing condition. To curb the radiation loss when downscaling, one must suppress it from being generated in the first place as much as possible. To this end, metallic coating of the ring cavity can enhance optical confinement in the gain media and reduce the overall optical loss despite of the added loss from metal absorption [5–7]. In this Chapter, we will summarize the developments behind the first semiconductor nanoring laser employing metal-clad ring cavities, with a specific emphasis on the role of metal in enabling the scaling of a semiconductor ring laser to a sub-wavelength dimension.

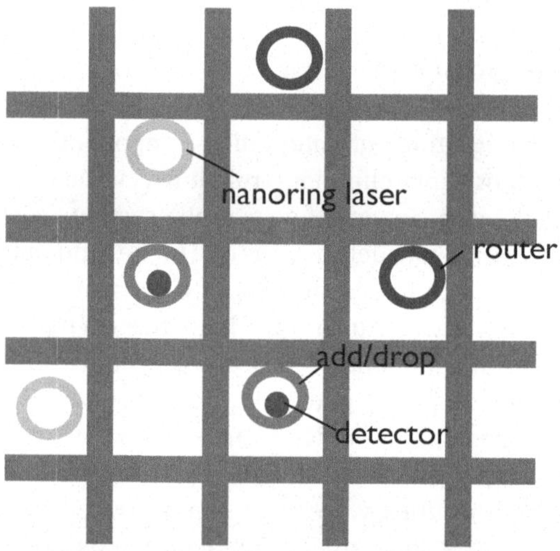

Figure 13.1 The illustration of photonics integration based on ring resonators that can function as light sources (lasers), routers, add/drop filters, and etc.

A laser comprises of a gain media and an optical cavity. To achieve a smaller laser, one must reduce the sizes of both the gain media and the optical cavity. In a typical semiconductor ridge waveguide, as one shrinks the width of the waveguide, a larger fraction of the optical mode will propagate outside the semiconductor region, leading to a reduced modal gain. A greater fraction of optical mode outside the waveguide core also results in a greater radiation loss when the mode undergoes a turn. Hence it is difficult to simultaneously maintain a low optical loss and a high optical gain when downsizing a semiconductor ring laser. The optical loss becomes especially severe as the dimension of the cavity approaches the optical diffraction limit that sets the lower bound of the optical modal volume to be at least $\lambda/2_n$ in every direction, where λ and n are the optical wavelength and the refractive index, respectively. To alleviate these challenges, one must improve the optical confinement in the semiconductor gain region. Because light cannot penetrate deep into metals, the use of metals, despite the inherent optical losses, can confine the light into a smaller space. It is perceivable that when the improved optical confinement outweighs the increased optical loss, the scaling of a ring laser can benefit from metals. Indeed, many state-of-the-art sub-wavelength nanolasers have been enabled by the use of metals. Examples include bowtie cavity lasers [8,9], metal-clad nanopillar and Fabry–Pérot lasers [10,11], plasmonic nanowire lasers [12], nanopatch laser [13], spacer-based nanolasers [14], and nanoring lasers [4]. In this chapter, we will first present the scaling properties of a nanoring laser and compare those properties to two similar nanolaser structures: nanopillar lasers and surface plasmon-enabled sub-wavelength injection lasers (SPESILs) [15]. In Section 13.3, we will discuss the design of a nanoring laser. We will discuss several design parameters that can affect the laser output including the lasing wavelength control and laser threshold. Last, we will present the fabrication of the first nanoring laser and discuss their experimental results.

13.2 Scaling Properties of Nanoring Lasers and the Role of Metals

The sub-wavelength nanoring laser can be formed by conformally depositing a metal layer over a semiconductor ring structure. The metal enables tight optical confinement and improves heat conduction of the device, resulting in a device that can be scaled to sub-wavelengths in all three directions. The metal-clad nanoring cavity supports a hybrid traveling-wave dielectric–plasmonic mode along the ring waveguide. The hybrid nature of the dielectric and surface plasmonic confinement at the metal–semiconductor interface is evident as manifested by the exponentially decaying electric field magnitudes on both sides of the interface. The field profile was calculated using MEEP, a finite-difference time-domain (FDTD) simulator [16]. The calculations were performed in three dimension (3D) with a passive gain region, that is, assuming the gain region neither attenuates nor amplifies light. The mode profile is expected to be slightly changed with the presence of optical gain. However, the hybrid dielectric–plasmonic nature is expected to be preserved.

We studied the scaling properties of the metal-clad nanoring laser by examining the laser cavity loss as a function of the laser dimension. Since the laser threshold is determined by the balance between the volume of gain and the total cavity loss, the more slowly the cavity loss increases with a decreasing laser dimension, the better the scaling behaviour will be [6]. The resonant frequencies of the passive laser cavities were calculated using the filter diagonalization method [17]. To reach the lasing threshold, the optical gain must overcome all the losses, including metal absorption loss, optical diffraction loss, and optical scattering loss. In the calculations, the scattering loss was simulated by the grid size in the FDTD model. The grid size was chosen to be 12 nm by observing the edge roughness of the fabricated ring cavity using scanning electron microscope. The epitaxial structure is shown in Fig. 13.2 with all materials lattice matched to InP. Complex refractive index for the metal gold (Au) was taken from Johnson and Christy [18].

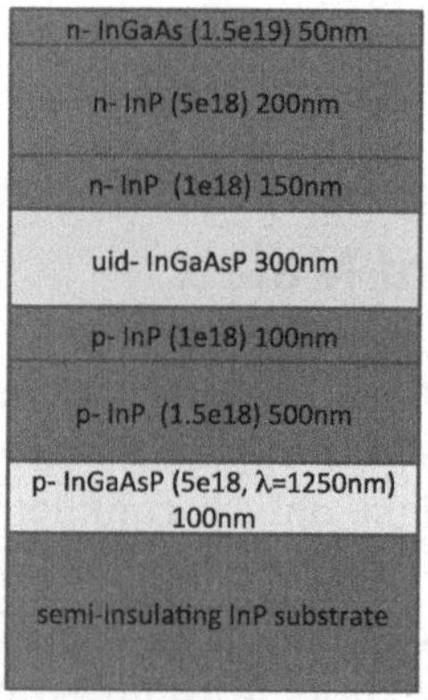

Figure 13.2 The epitaxial structure of the metal-clad nanoring lasers used in this study.

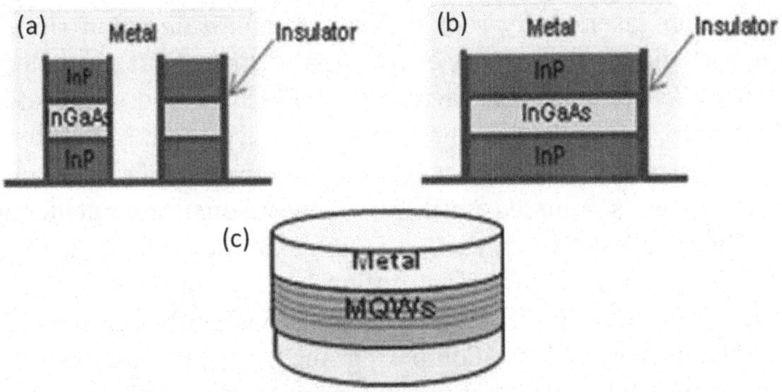

Figure 13.3 Three metal-clad nanolasers structures compared in the analysis: (a) Side view of the metal-clad nanoring laser. (b) Side view of the metal-clad nanopillar laser. (c) Surface plasmon-enhanced sub-wavelength injection laser [15]. The top layer is metal sitting in close proximity to quantum well gain region. From Kim and Ku [6].

We compared the scaling properties of the metal-clad nanoring laser with two other similar nanolaser structures: the metal-clad nanopillar lasers [10,11] and the SPESILs [15]. The cross-sectional views of all three laser structures are shown in Figs. 13.3 (a–c). We chose these three laser structures as they support three different types of optical modes: pure surface plasmonic mode in SPESILs, standing-wave hybrid dielectric-plasmonic mode in nanopillars, and traveling-wave hybrid dielectric-plasmonic in nanorings. The extents of optical modes into the metal regions in all three laser structures are similar as they depend on the skin depth of the metal. Hence, we expect the pure surface plasmonic mode in SPESILs to experience the largest modal loss since the fraction of optical energy in the metal region is the largest. However, because of the surface plasmonic nature, the optical mode in SPESILs also exhibits a higher effective index that lowers the bending loss and moves the optical diffraction limit to a smaller dimension. The combination of these properties allow the SPESILs to excel in scaling when $D/\lambda_0 < 0.5$, where D is the device diameter and λ_0 is the free-space lasing wavelength.

In a larger cavity ($D/\lambda_0 > 0.5$), shifting the peak of the electric field away from the metal region using the hybrid dielectric–plasmonic mode in both nanopillar, and nanoring lasers is able to lessen the burden of metal loss. The downside of the hybrid mode is a larger laser cavity required in the lateral direction since the electric field does not decay exponentially throughout the dielectric. Figure 13.4 compares the cavity quality (Q) factors versus the device diameters, normalized to the free-space wavelength λ_0 for all three laser structures. In the calculations, λ_0 is fixed while the diameter of the laser is adjusted until the cavity resonance coincides with λ_0. The result shows that when $D/\lambda_0 > 0.5$, the cavity Q factor in a SPESIL structure shows a saturation behavior as the propagation loss is no longer limited by the diffraction loss but rather by the metal absorption loss. Comparing the nanoring and nanopillar structures, it is seen that the nanoring exhibits a better scaling property. We attribute this superior scaling property of a nanoring laser to its extra design parameter. In the ring geometry, both the ring width and the ring diameter can be independently adjusted. With two freely adjustable dimensional parameters, we were able to optimize the Q factor without moving the resonance wavelength.

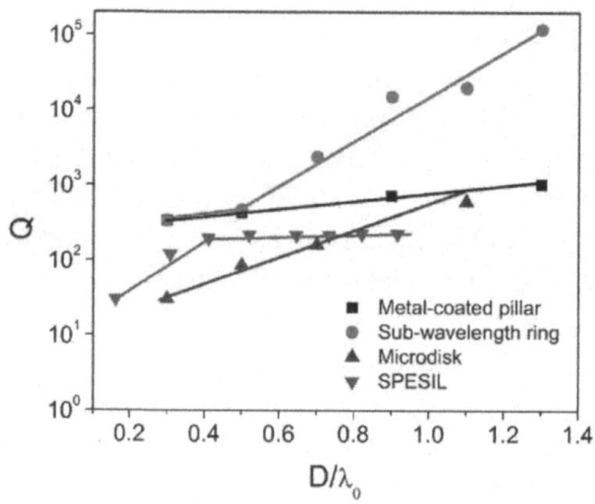

Figure 13.4 Calculated passive cavity Q factor versus D/lo for nanoring laser, nanopillar laser, and SPESIL. Lines are for visual guidance only. From Kim and Ku [6].

In addition to the three metal-clad laser structures as discussed above, we have also studied the scaling property of a metal-free dielectric microdisk laser with the result shown in Fig. 13.3. It is seen that the scaling property of a metal-free microdisk laser is worse than that of a similar metal-clad nanopillar laser. Although the metal introduces additional optical losses, a more tightly confined optical mode increases the modal gain while reducing the diffraction loss. The result is a higher cavity *Q* factor, which lowers the threshold required to reach the lasing condition. Experimentally, we demonstrated the role of the metal using the microring laser structure. Two 4-μm diameter microring lasers were fabricated side by side on the same sample with one coated with gold and the other uncoated. The epitaxial structure is the same as in Fig 13.2. Both laser structures were optically pumped. The output was collected using a micro-photoluminescence (μ-PL) setup. Figure 13.5 shows the light-in light-out (L–L) curve using the peak μ-PL intensity. The 4-μm diameter ring laser that is covered with Au (in black) shows the transition from spontaneous emission to stimulated emission at a threshold. In contrast, the device that is not covered with Au (in red) shows no such transition even at a much higher pumping intensity. Figure 13.5 inset shows the output spectra that shows clear coherent laser peak from the metal-clad structure.

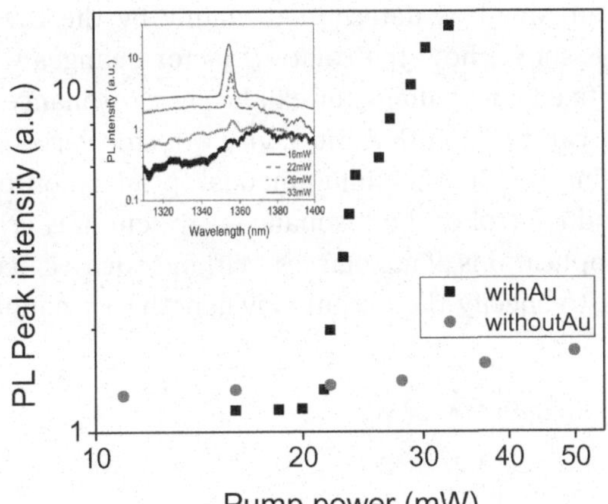

Figure 13.5 Comparison of peak intensity versus pump power for 4 μm ring laser devices covered with and without Au. Inset: The output spectra from the Au-covered ring laser showing the development of a coherent lasing peak. Reprinted with permission from Kim and Ku [5]. Copyright 2011, American Institute of Physics.

13.3 Design of Nanoring Lasers [6]

In this section, we will discuss the design and characteristics of metal-clad semiconductor nanoring lasers. Three design parameters will be described: the ring width, ring diameter, and metal thickness. Their impacts on the output wavelength, fabrication tolerance, and out-coupling will be discussed. Importantly, we will show that the laser output wavelength can be precisely controlled in a nanoring cavity but not easily attainable in a nanopillar structure. In addition, the small laser dimension can greatly increase the free-spectral range, allowing single longitudinal operation to be attained without additional feedback structures.

13.3.1 Control of Output Wavelength

In contrast to the microdisk or nanopillar cavity, the nanoring configuration allows one additional dimensional parameter, that is, the width of the ring. As will be shown, this extra parameter makes it possible to independently control the laser wavelength and optimize the laser threshold as determined mainly by the cavity Q factor. Figure 13.6 shows how the cavity Q factor changes with the ring width at a fixed a ring diameter exhibiting a resonance wavelength of 1 μm. It can be seen that the cavity Q factor varies significantly with the ring width. The simultaneous optimization of the cavity Q factor and control of the resonance wavelength is critical for the practical applications of nanolasers. Without such a capability, it will be difficult to specify the output wavelength of a nanolaser due to

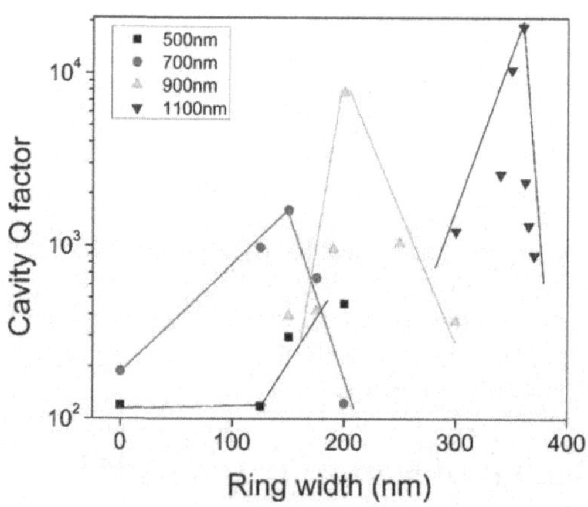

Figure 13.6 Cavity Q factor versus ring width for four different ring diameters. The cavity resonances are fixed at 1 μm. The lines are for visual guides only. After Oskooi et al. [6].

the large free spectral range that is typical for a "small" laser cavity. According to Fig. 13.6, as the laser dimension becomes larger, the optimal ring width that results in the largest Q factor also becomes larger. For example, the optimal ring width for a 700-nm diameter laser is 125 nm whereas that for a 900-nm diameter laser is 200 nm. For smaller devices, particularly those with diameters less than 500 nm, the optimal design of a nanoring laser becomes a nanopillar laser. This is not surprising as when the diameter becomes small enough, the optical mode can no longer be properly confined near the perimeter of the ring structure. The overlap of optical modes from different parts of the ring perimeter makes the optimal design the same as a nanopillar laser.

Figure 13.7 shows how tolerable a nanoring laser of a given diameter is to non-optimal ring widths. As the ring diameter increases, the ring width tolerance for full width at half maximum (FWHM) of the cavity Q factor decreases significantly. For a ring diameter of 700 nm, the Q decreases to half of the maximum value within 50 nm on either side of the optical ring width. However, as the diameter increases to 1100 nm, the tolerance is only 10 nm on either side. Therefore, the design requirement for a larger diameter nanoring laser is in fact more stringent.

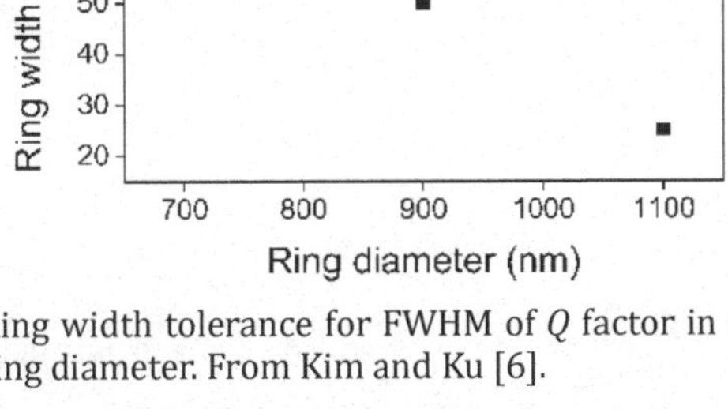

Figure 13.7 Ring width tolerance for FWHM of Q factor in nm for a given ring diameter. From Kim and Ku [6].

13.3.2 Metal Thickness

A metal-clad nanoring laser exhibits a better photon confinement than a metal-free microdisk laser. A central question in the design of any metal-incorporated nanolaser is how "much" metal is enough to create benefits, that is, the enhanced photon confinement, while outweighing the added optical losses due to metal absorption. In the metal-clad nanoring design, we studied the role of metal thickness. As shown in Fig. 13.8, when the metal layer is too thin, the optical mode cannot be tightly confined inside the semiconductor region, causing more light to leak into the metal region. The result is a larger metal absorption as well as a larger diffraction loss that lead to a lower cavity Q factor. As the metal layer thickness increases, more and more field is confined inside the semiconductor, therefore the cavity Q increases. Figure 13.9 shows the electric field profiles with three different metal layer thicknesses (50 nm, 90 nm, and 200 nm). In Fig. 13.9, $x = 0$ denotes the center of the ring and the semiconductor region is located between $x = -200$ nm and –400 nm. Due to the boundary condition and the exponentially decaying electric field profile inside the metal, more electric field lies within the semiconductor for devices with a thicker metal coverage. Furthermore, with a thinner metal thickness such as 50 nm, theoptical mode is concentrated more toward the inner perimeter of the ring. Because the grid size used in FDTD calculations, 12 nm in Fig. 13.9, simulates the surface scattering loss, this causes a higher loss for a ring with a thinner metal thickness.

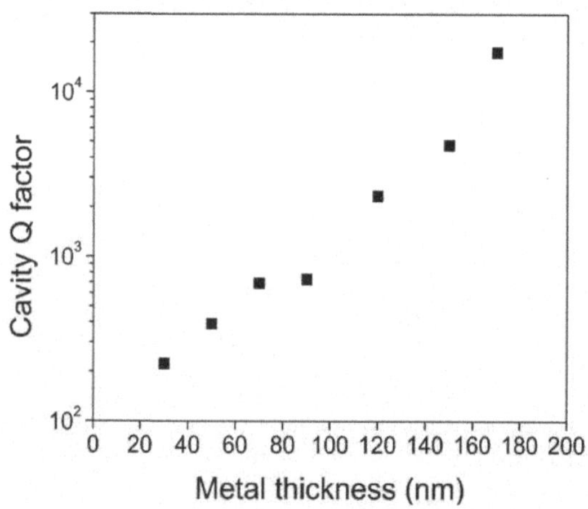

Figure 13.8 Cavity Q factor vs. metal thickness for a fixed diameter of 800 nm in the semiconductor region and ring width of 200 nm. The cavity resonance is chosen to be 1 µm. From Kim and Ku [6].

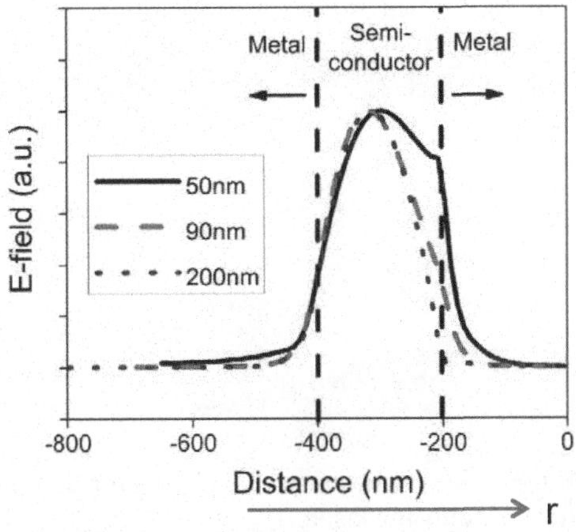

Figure 13.9 Lateral electric-field profile in metal-clad nanoring lasers with varying metal thicknesses. After Ref. [6].

13.3.3 Output Coupling

Conventional semiconductor lasers use bulky lenses for output coupling into an optical system such as an optical waveguide. In certain applications, an ideal nanolaser should have a nanoscale output coupling scheme so that it can be easily integrated with other nanoscale optoelectronic components. In microring lasers, the output coupling can be achieved by evanescent coupling via the placement of a waveguide in tangential to the ring cavity, separated by a narrow gap [19]. In metal-clad nanoring lasers, this scheme will be difficult to implement due to the presence of the metal layer. Figure 13.10 illustrates a possible out-coupling scheme for the metal-clad nanoring laser. An opening in the metal layer at the outer edge of the nanoring cavity is introduced with an output waveguide inserted into it. To keep the opening as small as possible and therefore reduce the optical loss in the laser cavity, the beginning segment of the output waveguide can also be metal-clad to improve confinement. The same output coupling scheme can be used in the nanopillar laser structure too. Inevitably, the output coupling of a nanolaser is going to decrease the cavity Q factor. Figures 13.11(a) and (b) show the cavity Q factors of both the nanoring and nanopillar lasers versus the width of the output waveguide. It can be seen that the nanoring laser exhibits a higher cavity Q factor than the nanopillar laser does with output waveguides attached to them for

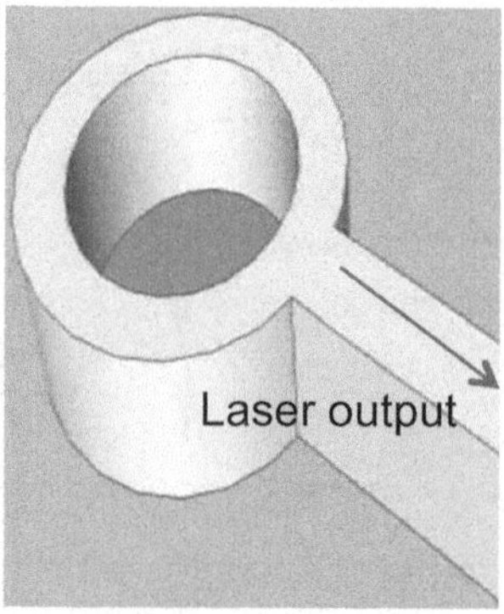

Figure 13.10 A possible output-coupling scheme for a metal-clad semiconductor nanoring laser.

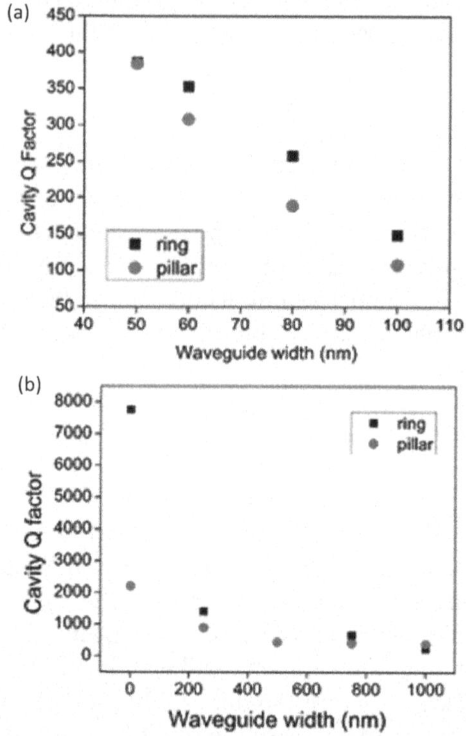

Figure 13.11 (a) Cavity Q factor versus the output waveguide width for $D/\lambda_0 = 0.6$ and ring width = 150 nm. (b) Cavity Q factor versus the output waveguide width for $D/\lambda_0 = 0.9$ and ring width = 200 nm. From Kim and Ku [6].

all sizes. In addition, a smaller nanoring laser has a better tolerance to the perturbation created by the presence of a waveguide. This is attributed to the results shown in Fig. 13.7 that a larger nanoring laser has a more stringent design requirement.

13.4 Fabrication and Characterization of Nanoring Lasers

In this Section, we describe the fabrication and characteristics of a metal-clad nanoring laser. The sample preparation is as follows. The epitaxial structure is the same as in Fig. 13.2. The sample was grown by the metal-organic chemical vapor deposition (MOCVD) technique. The optical gain is provide by the InGaAs layer that is lattice matched to InP. The bulk InGaAs gain region is chosen to maximize the optical confinement factor in the vertical direction. The vertical optical confinement is provided by the index contrast between InGaAs and the two InP claddings. To pattern the ring, 200 nm of silicon nitride followed by 50 nm of chromium were deposited as the hard mask via electron beam evaporation. Electron beam lithography (Raith 150) with a negative resist (Hydrogen Silsesquioxane, HSQ) was used for pattern definition. HSQ was spin-coated at 3000 rpm for 50 seconds and baked at 180°C for 2 minutes on a hot plate. The lithography conditions were as follows: 700 μC/cm^2 area dose, 20 kV acceleration voltage, and 10 μm aperture. The resist pattern was developed in AZ 400K:DI H_2O (1:4) for 40 seconds. Subsequently, the InGaAsP/InP epiwafer was dry etched in a $CH_4/H_2/Ar$ environment (4 sccm/28 sccm/14 sccm) using Plasma Therm 790. The pressure and power were maintained at 30 mTorr and 200 W, respectively. During the InP etching, polymer was deposited as a by-product and was removed using an O_2 ashing cycle (70 sccm, 200 mTorr, 100 W, 3 minutes). Once the semiconductor ring was formed, 10 nm of Al_2O_3 was deposited via atomic layer deposition as a spacer to separate the semiconductor from the subsequent metal coverage. This spacer is critical to avoid luminescence quenching [20]. As shown in Fig. 13.12, the PLs from two 2-μm diameter ring structures were compared. One ring was coated with Au and the other was not. There was no spacer layer in both structures. It was observed that the PL was completely quenched in the Au-clad sample without a spacer.

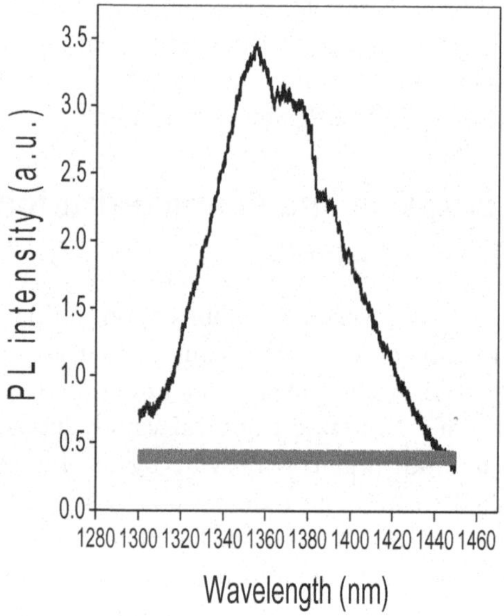

Figure 13.12 Photoluminescence of a 2-μm diameter ring with (red) and without (black) Au.

For conformal metal coverage, a combination of sputtering and electrodeposition processes was used. A thin (20 nm) layer of Au was first sputtered as a conductive layer. Then, electrodeposition was used to deposit 300 nm of Au. Figure 13.13 shows a 1.2-μm-diameter sub-wavelength nanoring laser. The cross sectional structure was obtained via focused ion beam (FIB). Due to the etching chemistry and limitation on small feature size, etch depth of the center hole is only about one third that of outer diameter. According to the FDTD simulation, the Q factor of an 800-nm-diameter ring with 200-nm ring width changes only from 7750 to 7130 with such a change in the center hole etch depth. This means that a nanoring laser does not need to have the center hole etched through the active region and filled with metal. Just as the electric field in a ridge waveguide is laterally confined via refractive index contrast due to a semiconductor ridge on top, a similar ridge of the ring shape on the top of the nanoring provides the necessary index contrast for lateral confinement in the nanoring laser without sacrificing the quality factor, as confirmed by the FDTD simulations shown in Fig. 13.14.

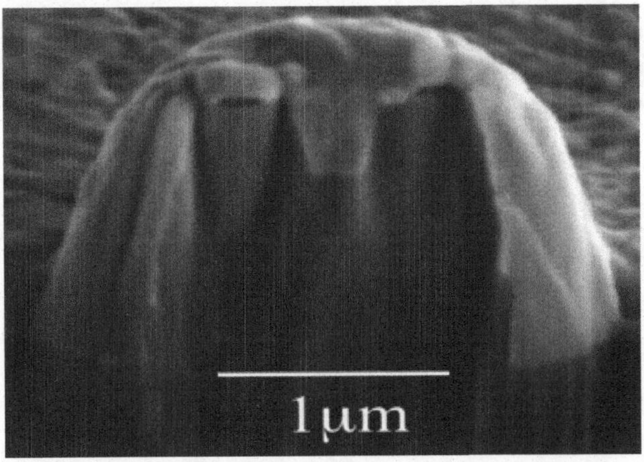

Figure 13.13 Scanning electron micrograph of the cross section of a 1.2-μm nanoring laser device, illustrating uniform metal coverage of a center trench via electrodeposition. From Kim and Ku [6].

To characterize the laser, we used optical pumping via a micro-PL setup. The measurements were carried out at 10 K to minimize non-radiative recombination. The sample was optically pumped and the PL signal was collected, both through the substrate. The PL results are shown in Fig. 13.15. Transition from spontaneous emission to stimulated emission as well as progression of a lasing peak is clearly seen. With the increasing pump intensity, the spontaneous emission background becomes both blue shifted and broader, both of which are attributed to the band filling in the bulk active region. The PL data were analyzed further, and the pump power versus peak PL intensity is plotted in Fig. 13.16. The laser threshold is estimated to be 50 μW according to the total incident power 10 mW, the laser beam size, and the nanoring dimension. Compared to a 4-μm metal-clad microring laser, which had the transition at an incident power of 21.5 mW (actual power absorbed was 1.2 mW), the smaller metal-clad nanoring cavity exhibits a much smaller laser threshold, which is expected as a result of a smaller optical modal volume. The modal volume of the metal-clad nanoring laser of 1.2 μm diameter is only 0.004 μm^3 or 0.56 times of the optical diffraction limit, that is, $(\lambda_0/2n)^3$ where n is the effective index of optical mode in the laser cavity. The physical size of the laser is also sub-wavelength in all three directions.

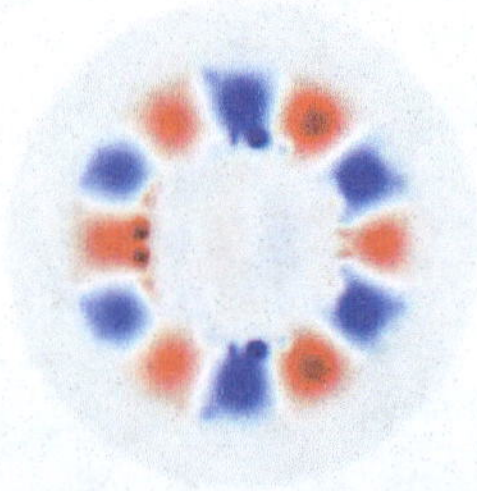

Figure 13.14 3D-FDTD simulation of a metal-clad nanoring laser with an incomplete center hole etch depth. The center hole is etched down to, but exclusive of, the active region.

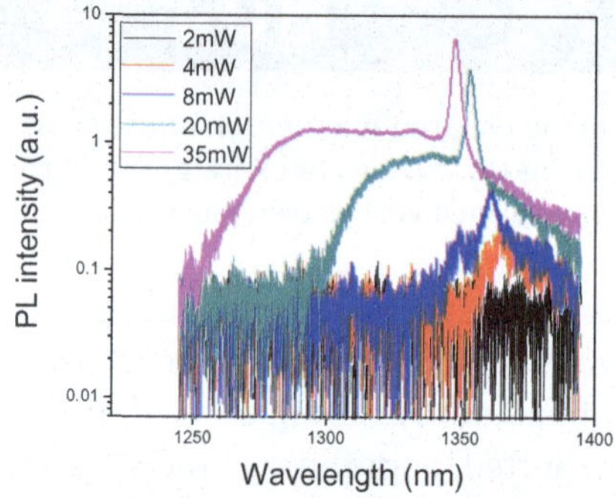

Figure 13.15 Intensity-dependent μ-PL of a 1.2-μm diameter metal-clad nanoring laser with 250-nm ring width. Reprinted with permission from Kim and Ku [4]. Copyright 2011, American Institute of Physics.

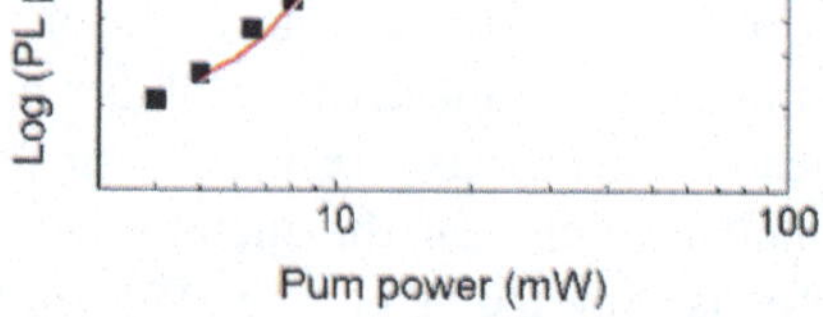

Figure 13.16 Pump power versus peak PL intensity of 1.2-μm diameter metal-clad nanoring. Fitting was done using laser rate equations. Reprinted with permission from Kim and Ku [4]. Copyright 2011, American Institute of Physics.

The L–L curve in Fig. 13.16 was fitted with the following laser rate equations [13]:

$$\frac{dN}{dt} = P - gS - \frac{1-\beta}{\tau_{sp}}N - \frac{F\beta}{\tau_{sp}}N - \frac{v_s S_a}{V_a}N \tag{13.1}$$

$$\frac{dS}{dt} = \Gamma gS - \frac{S}{\tau_{ph}} + \frac{\Gamma F\beta}{\tau_{sp}}N \tag{13.2}$$

where N, S are carrier and photon densities, respectively, P is the pumping rate, g is the optical gain, τ_{sp} is the spontaneous emission lifetime (τ_{sp} = 1.5 ns was used), F is the Purcell factor, β is the spontaneous emission coupling factor, S_a and V_a are surface area and volume of gain region, respectively, v_s is surface recombination velocity ($v_s = 2 \times 10^4$ cm/s was used), Γ is confinement factor (Γ = 0.65 was used), and τ_{ph} is the photon lifetime ($\tau_{ph} = Q/2\pi f$, where f is the resonant frequency). g was assumed to follow a linear model with $g = cG(N - N_o)/n_g = 1.09 \times 10^{-5}$ (N – 4 × 10^{17} cm^{-3}) s^{-1}, where N, N_o, c, n_g, and G are the carrier density in the active region, transparent carrier density, speed of light in vacuum, the group refractive index of the cavity, and the linear differential gain coefficient, respectively. N can be determined by the amount of absorbed laser power in the active region. The Purcell factor F was calculated to be 1.32 with a Q factor of 160 determined experimentally from far below the lasing threshold, and the calculated optical modal volume. Using the above parameters, the 1.2 μm nanoring laser has a β of 0.01. In comparison, the 4-μm-diameter metal-clad microring laser had a β of 5×10^{-4}. The higher β factor also contributes to the smaller laser threshold of the smaller laser. Additionally, the 1.2-μm diameter nanoring cavity shows a saturation behaviour as the pump power exceeds 40 mW possibly due to sample heating. The focused laser spot was about 10 μm in diameter and any laser power that hits outside the laser area can contribute to the sample heating.

Figure 13.17 shows both the linewidth and peak wavelength as a function of the pump power [4]. The PL data from Fig. 13.15 were fitted with a Lorenzian function to determine the lasing linewidth. A clear linewidth collapse was observed at the laser threshold and the linewidth was 2.3 nm above the laser threshold. The lasing wavelength continuously decreased with the increasing pump power, possibly due to band filling, although the lasing wavelength seemed to be settling at a high pump power. This is explained by the thermally induced red-shift at the high pump power that compensates the blue shift.

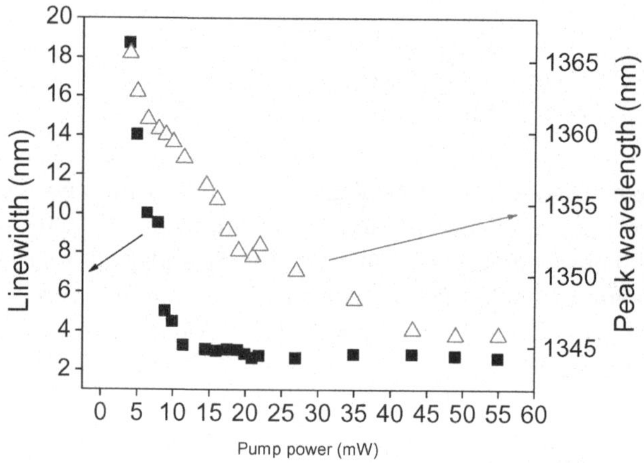

Figure 13.17 Linewidth and peak wavelength versus pump power of 1.2-μm diameter metal-clad nanoring. Reprinted with permission from Kim and Ku [4]. Copyright 2011, American Institute of Physics.

As discussed in the previous section, the nanoring laser has two design parameters, ring diameter and width that control the lasing wavelength. For example, semiconductor nanoring laser's outer diameter can stay constant while controlling the lasing wavelength using the ring width. The two parameters can also be used together to optimize the cavity Q factor. These enable lasers of the same sizes to exhibit different lasing wavelengths within a small range, making the semiconductor nanoring lasers attractive for on-chip applications: constant physical dimensions makes the placement of nanoring lasers exhibiting different output wavelengths in an ultradense layout easier; the possibility of multiple wavelengths within a small range makes on-chip wavelength-division multiplexing possible, enabling a more efficient on-chip optical interconnect network. While it is true that gratings have been a preferred choice for resonance wavelength control for larger laser cavities, it can be very difficult to integrate these wavelength-controlling feedback structures such as gratings into a nanoscale laser cavity. Therefore, as laser dimensions become smaller, multiple design parameters give an added advantage of wavelength control.

To demonstrate wavelength tuning, the metal-clad nanoring lasers of 1.2-μm diameter with various ring widths were fabricated and measured using the identical optical measurement scheme described above. The results are shown in Fig. 13.18. It can be clearly seen that, by altering the ring width between 240 nm and 340 nm, the lasing wavelength can be fine-tuned to within a 10 nm range. The oscillatory behavior of the lasing wavelength is due to the shift of nearby resonance peaks in and out of the spontaneous emission range of the gain spectrum. According to the FDTD simulations, the resonance peaks undergo red-shift behavior with increasing ring width in general. Resonance peaks from 3D-FDTD simulations are plotted along with the experimental data. The trend agrees well with the experiment. The discrepancy of the lasing wavelengths may be due to the passive active region assumed in the calculations.

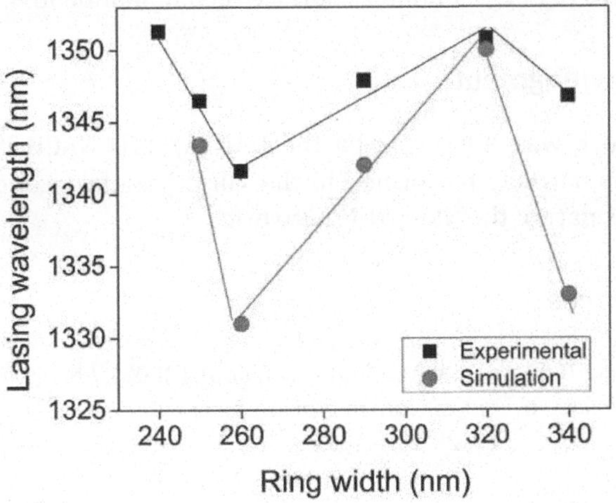

Figure 13.18 Lasing wavelength shift due to ring width variation. Lines are for visual guidance only. Reprinted with permission from Kim and Ku [4]. Copyright 2011, American Institute of Physics.

13.5 Summary

In summary, we explored the possibility of using metal cladding on semiconductor ring lasers to enable scaling of the ring diameter to a sub-wavelength dimension. Theoretically, we showed that the addition of metal cladding forms a hybrid dielectric–plasmonic mode that helps confine the electric field inside the semiconductor waveguide. This enhanced confinement outweighs the extra metal absorption loss and gives rise to a better laser scaling capability. Furthermore, we saw that the metal-clad nanoring lasers have two design parameters for independent control of cavity quality factor and resonance wavelength. Lastly, out-coupling capabilities of metal-clad nanoring lasers were investigated. Experimentally, lasing in a 1.2-μm diameter sub-wavelength metal-clad nanoring resonator has been demonstrated. We showed the advantageous role of metal-cladding in enabling the ring laser scaling. The output wavelength of the sub-wavelength metal-clad nanoring laser can be fined tuned for a 10 nm wavelength range by varying the ring width. It is possible to further lower the laser threshold and improve the operating temperature by adopting an electrical pumping scheme, improving etch profile and sidewall roughness, and using a quantum-dot active region to minimize surface recombination loss.

Acknowledgment

This work was supported by the DARPA/MTO NACHOS program and was partially performed in the Lurie Nanofabrication Facility, which is part of the NSF NNIN network.

References

1. Liang D, Fiorentino M, Okumura T, et al. (2009). Electrically-pumped compact hybrid silicon microring lasers for optical interconnects, *Opt. Express*, **17**, 20355–20364.
2. Matsumoto N, Kumabe K (1977). AlGaAs-GaAs semiconductor ring laser, *Jpn. J. Appl. Phys.*, **16**, 1395–1398.
3. Scifres DR, Burnham RD, Streifer W (1976). Grating-coupled GaAs single heterostructure ring laser, *Appl. Phys. Lett.*, **28**, 681–683.
4. Kim MW, Ku PC (2011). Semiconductor nanoring lasers, *Appl. Phys. Lett.*, **98**, 201105 (1–3).
5. Kim MW, Ku PC (2011). Lasing in a metal-clad microring resonator, *Appl. Phys. Lett.*, **98**, 131107 (1–3).
6. Kim MW, Ku PC (2011). The metal-clad semiconductor nanoring laser and its scaling properties, *Opt. Express*, **19**, 3218–3225.

7. Wang YG, Chang SW, Chen CC, et al. (2011). Room temperature lasing with high group index in metal-coated GaN nanoring, *Appl. Phys. Lett.*, **99**, 251111 (1–3).

8. Yu NF, Cubukcu E, Diehl L, et al. (2007). Bowtie plasmonic quantum cascade laser antenna, *Opt. Express*, **15**, 13272–13281.

9. Chang SW, Chuang SL (2008). Plasmonic nanolaser based on metallic bowtie cavity, *Proc. Conf. Lasers Electro-Optics*, Optical Society of America, San Jose, CA.

10. Hill MT, Oei YS, Smalbrugge B, et al. (2007). Lasing in metallic-coated nanocavities, *Nat. Photonics*, **1**, 589–594.

11. Hill MT, Marell M, Leong ESP, et al. (2009). Lasing in metal-insulator-metal sub-wavelength plasmonic waveguides, *Opt. Express*, **17**, 11107–11112.

12. Oulton RF, Sorger VJ, Zentgraf T, et al. (2009). Plasmon lasers at deep subwavelength scale, *Nature*, **461**, 629–632.

13. Yu K, Lakhani A, Wu MC (2010). Subwavelength metal-optic semiconductor nanopatch lasers, *Opt. Express*, **18**, 8790–8799.

14. Noginov MA, Zhu G, Belgrave AM, et al. (2009). Demonstration of a spaser-based nanolaser, *Nature*, **460**, 1110–1112.

15. Kim MW, Chen YH, Moore J, et al. (2009). Subwavelength surface plasmon optical cavity-scaling, amplification, and coherence, *IEEE J. Sel. Top Quantum Electron*, **15**, 1521–1528.

16. Oskooi AF, Roundy D, Ibanescu M, Bermel P, Joannopoulos JD, Johnson SG (2010). MEEP: A flexible free-software package for electromagnetic simulations by the FDTD method, *Comput. Phys. Commun.*, **181**, 687–702.

17. Mandelshtam VA, Taylor HS (1997). Harmonic inversion of time signals and its applications, *J. Chem. Phys.*, **107**, 6756–6769.

18. Johnson PB, Christy RW (1972). Optical constants of noble metals, *Phys. Rev. B*, **6**, 4370–4379.

19. Little BE, Chu ST, Pan W, Kokubun Y (2000). Microring resonator arrays for VLSI photonics, *IEEE Photon Technol. Lett.*, **12**, 323–325.

20. Ambati M, Genov DA, Oulton RF, Zhang X (2008). Active plasmonics: Surface plasmon interaction with optical emitters, *IEEE J. Sel. Top Quantum Electron*, **14**, 1395–1403.

Index